Introduction to
DIFFERENTIAL EQUATIONS
with Boundary Value Problems

Student:

To help you make the most of your study time and improve your grades, we have developed the following supplement designed to accompany Andrews/Introduction to Differential Equations with Boundary Value Problems:

Student's Solutions Manual by Andrews/DeVoe,
ISBN 0-06-500002-1

You can order a copy at your local bookstore or call HarperCollins Publishers directly at 1-800-782-2665.

Introduction to
DIFFERENTIAL EQUATIONS
with Boundary Value Problems

Larry C. Andrews

University of Central Florida

HarperCollins*Publishers*

Sponsoring Editor: Peter Coveney/George Duda
Project Editor: Ellen MacElree/Kristin Syverson
Art Direction: Heather A. Ziegler/Julie Anderson
Text Adaptation: Heather A. Ziegler
Cover Coordinator: Julie Anderson
Cover Design: Matthew J. Doherty
Cover Photo: © Pete Turner
Text Art: Syntax International
Photo Research: Nina Page (cover)
Director of Production: Jeanie A. Berke
Production Assistant: Linda Murray
Compositor: Syntax International
Printer/Binder: R. R. Donnelley & Sons Company
Cover Printer: Lehigh Press Lithographers

Introduction to Differential Equations with Boundary
Value Problems

Library of Congress Cataloging-in-Publication Data

Andrews, Larry C.
    Introduction to differential equations with boundary value problems
    Larry C. Andrews
        p.   cm.
    Includes index.
    ISBN 0-06-040293-8
        1. Differential equations.   2. Boundary value problems.
    I. Title.
    QA371.A58   1990
    515'.35—dc20
                                                                90-37829
                                                                    CIP

90 91 92 93   9 8 7 6 5 4 3 2 1

Contents

---

† [0] indicates an optional section.

CHAPTER 10   **Numerical Methods**   *405*

CHAPTER 11   **Boundary Value Problems and Fourier Series**   *429*

CHAPTER 12 Applications Involving Partial Differential Equations   *479*

Preface

This book is intended as a one-semester or two-semester introductory treatment of differential equations and their applications. It is designed for students in mathematics, engineering, or science who have successfully completed the basic sequence of courses in calculus.

Because of the importance of differential equations in a variety of engineering and science areas, there are a number of applications to problems in the physical sciences, as well as some in the social and life sciences, that are prominent throughout the text. However, I have tried to avoid the temptation of introducing a multitude of applications in many diverse areas of application. My experience is that students are often distracted by too many different types of applications. A few basic applications readily make the point about the important role of differential equations in the real world. Moreover, I have kept the discussions involving applications at an elementary level so that a minimal background in the various sciences is required of the student or instructor to follow or understand them. In addition, I have tried to maintain a close relationship between mathematical theories and applications, whenever possible, by providing physical interpretations of some of the mathematical results.

The text contains the standard material that is found in the majority of introductory texts on differential equations, but contains some distinctive features that we list below.

Distinctive Features

- **Linear first-order equations:** A separate discussion of the notions of homogeneous and particular solutions in addition to the standard integrating factor technique for solving first-order linear equations (Chapter 2). I believe this provides a unifying point of view that is useful in later chapters dealing with linear equations of higher order.
- **Green's function:** Another novel feature of the text is the introduction of the causal Green's function (Chapter 5) for handling nonhomogeneous initial value problems in a systematic and physically meaningful fashion. This function is linked in Chapter 6 to the impulse response function which is basic in engineering applications involving the analysis of linear systems.
- **Qualitative methods:** Also included is a brief discussion of the qualitative methods used in oscillation theory (Chapter 5) and the stability of solutions of nonlinear systems of equations (Chapter 9). These brief exposures to qualitative methods permit the student to see how the general behavior of solutions to certain differential equations can still be determined in the absence of an explicit solution function.

- **Worked examples:** There are over 220 numbered worked examples, each of which is generally indicative of typical problems to be found in the exercise sets.
- **Exercises:** Nearly 1800 problems are included in the exercise sets, containing a blend of drill-like problems, some more difficult and some that extend the theory and applications beyond that discussed in the exposition. Problems that are considered more challenging are marked by a star (∗).

Each chapter in the text begins with an overview of the topics to be covered in that chapter. In addition, each chapter contains a chapter summary in which the most important chapter topics are highlighted. The first eight chapters, which make up the bulk of material covered in most one-semester courses, also contain a review exercise set at the end of the chapter summary. Many of the chapters after the first four chapters are independent of each other so that various arrangements of topics can be made to suit individual course needs. Also, sections that can easily be omitted for a shorter course are marked [0].

Answers to all odd-numbered problems are provided at the end of the text to aid the students, while an instructor's Answer Key is available which contains answers to all problems (both odd and even-numbered problems). In addition, there is an accompanying Solutions Manual which contains detailed solutions to every other odd problem plus those even-numbered problems marked by a rectangular black bullet ■. Not included in the Solutions Manual are the review exercises at the end of the first eight chapters.

I am grateful to:

Harvey Greenwald	California Polytechnic State University San Luis Obispo
Gilbert Lewis	Michigan Technological University
Joaquin Loustaunau	New Mexico State University
Bernard Marshall	
Allan Krall	Pennsylvania State University
Seymour Goldberg	University of Maryland
Maurino Bautista	Rochester Institute of Technology
Hendrik Kuiper	Arizona State University

who served as reviewers. Also, I wish to acknowledge the hard work of Jody DeVoe who worked out the answers to all the exercises. Finally, I wish to thank my editor Peter Coveney and the entire production staff of HarperCollins for the fine job they did in getting this text published in a timely manner.

L. C. Andrews

CHAPTER 1
Basic Concepts

Differential equations play a fundamental role in engineering, mathematical and physical sciences, and the life sciences because they can be used in the formulation of many physical laws and relations. The development of the theory of differential equations is closely interlaced with the development of mathematics in general, and it is indeed difficult to separate the two. In fact, most of the famous mathematicians from the time of Newton and Leibniz had some part in the cultivation of this fascinating subject.

In a systematic development of the general theory of differential equations it is helpful to organize the various types of equations into **classes.** The reason is that all differential equations belonging to a particular class can often be solved by the same method. Therefore, in Section 1.2 we start with a discussion of some of the various classification schemes, emphasizing **order** and **linearity.**

In Section 1.3 we define what we mean by a **solution** of a differential equation. Here we discover that differential equations are peculiar in that they generally possess many different solutions, and for this reason we usually seek a specific function, called a **general solution,** with the property that all solutions can be obtained from it.

We introduce the notions of **initial value problem** and **boundary value problem** in Section 1.4. These are the names attached to those problems occurring in applications wherein the solution of a differential equation must satisfy certain additional **auxiliary conditions.** The study of these problems is so important in applications that we have devoted a significant portion of the text to developing the theory associated with them. In the last section of the chapter are some **historical comments** concerning differenfial equations.

1.1 Introduction

The theory of differential equations (DEs) has played an important role in science and engineering since the invention of the calculus by Newton[†] and Leibniz.[‡] Problems in the physical sciences have subsequently been investigated primarily by formulating them as DEs. The first problems studied from this point of view came mostly from the field of mechanics.

The role of DEs in solving problems in mechanics is nicely illustrated by considering **Newton's second law** of motion. In beginning physics courses this law is commonly introduced by the simple algebraic formula

$$F = ma$$

For a single "particle" in motion, F denotes the force acting on the particle (which may be simply its weight), m is the mass of the particle (generally assumed constant), and a is its *acceleration*. In solving problems in mechanics, however, we are usually concerned also with the *velocity* and *position* of the particle as a function of time, not just with its acceleration. We may recall from calculus that acceleration is the time derivative of velocity $v = v(t)$, that is, $a = dv/dt$. Hence, Newton's second law can also be expressed by the equation

$$F = m \frac{dv}{dt}$$

Moreover, if $y = y(t)$ is the position of the particle at time t, then its velocity is related by $v = dy/dt$. Using this relation, we see that $dv/dt = d^2y/dt^2$, and Newton's second law now assumes the additional form

$$F = m \frac{d^2y}{dt^2}$$

Because they involve derivatives of unknown quantities, these last two equations are examples of DEs. Clearly, DEs can evolve quite naturally in the formulation of even rather simple problems in physics. Of course, DEs are now used extensively in all areas of physics and engineering, and more recently have also found their way into the social and life sciences.

[†] Sir Isaac Newton (1642–1727) was born on Christmas Day in the countryside of England. Along with Leibniz, he is credited with the invention of the calculus (see also Section 1.6).

[‡] Gottfried Wilhelm von Leibniz (1646–1716) was born in Leipzig. Known as both a philosopher and mathematician, he is credited with building a remarkable system of modern philosophy and being co-developer of the calculus (see also Section 1.6).

1.2 Classification of DEs

By a **differential equation** we mean simply an equation that is composed of a single unknown function and a finite number of its derivatives. One of the simplest examples that occurs early in the calculus is to find all functions y for which

$$y' = f(x) \tag{1}$$

where $f(x)$ is a given function. For instance, if $f(x) = x^2$, the unknown function y is obtained through a simple integration to yield

$$y = \frac{x^3}{3} + C \tag{2}$$

where C is a constant of the integration which can assume any value.

Most of the DEs that concern us in this text are not of the simple variety as described by Equation (1). Typical examples, some of which we discuss in later chapters, include the following:

$$y' = x^2 y^3 \tag{3}$$

$$y'' + k^2 y = \sin x \tag{4}$$

$$y'' + b \sin y = 0 \tag{5}$$

$$y''' + xy'' + 5y^2 = 2e^x \tag{6}$$

$$(y')^2 + 3xy = 1 \tag{7}$$

$$a^2 u_{xx} = u_{tt} - ku_t \tag{8}$$

$$u_{xy} = 0 \tag{9}$$

$$u_{xxxx} + 2u_{xxyy} + u_{yyyy} = 0 \tag{10}$$

> ***Remark.*** Various notations for derivatives are commonly employed, depending upon which is convenient at the time. For instance, we recognize the equivalences $y' = dy/dx$, $y'' = d^2y/dx^2, \ldots, y^{(n)} = d^n y/dx^n$. For partial derivatives, comparable notation is $u_x = \partial u/\partial x$, $u_{xy} = \partial^2 u/\partial y \, \partial x$, and so forth.

In order to provide a framework in which to discuss various solution techniques for DEs, it is helpful to first introduce **classification schemes** for the equations. For example, if the unknown function y appearing in a DE depends on only a single independent variable, say x, the equation is said to be an **ordinary differential equation** (ODE). Most of the DEs in this text are ODEs. When the unknown function depends upon more than one independent variable, the derivatives will be partial derivatives and the equation is then called a **partial differential equation** (PDE). Examples of ODEs are given by (3) through (7) above, while (8) through (10) are PDEs.

The *order* of a DE is another type of classification that is useful to recognize for solution purposes.

Definition 1.1 The **order** of a DE is the order of the highest derivative appearing in the equation.

According to Definition 1.1, Equation (3) is *first-order*, Equations (4), (5), (8), and (9) are *second-order*, (6) is *third-order*, (10) is *fourth-order*, and (7) is *first-order* even though the highest derivative y' is squared.

DEs are further divided into two large classes—*linear* and *nonlinear*—for which we have the following definition.

Definition 1.2 A **linear** (ordinary) DE of order n is any equation that can be expressed in the form

$$A_n(x)y^{(n)} + A_{n-1}(x)y^{(n-1)} + \cdots + A_1(x)y' + A_0(x)y = F(x)$$

where $A_0(x)$, $A_1(x)$, ..., $A_n(x)$, and $F(x)$ are specified functions. Except for $F(x)$, these functions are called the coefficients of the DE. Any equation that cannot be put in this form is a **nonlinear** DE.

The essential features of a linear DE are (1) that the unknown function y and all its derivatives are of the first degree (algebraically) and (2) that each coefficient depends only upon the independent variable x. First-order and second-order linear DEs assume the specific forms, respectively,

$$A_1(x)y' + A_0(x)y = F(x) \tag{11}$$

and

$$A_2(x)y'' + A_1(x)y' + A_0(x)y = F(x) \tag{12}$$

Only Equation (4) above (of the ODEs) is *linear;* Equations (3), (5), (6), and (7) are *nonlinear.*

A single equation is sufficient if only one unknown function appears in the formulation of a problem. Most of the problems that we treat in this text fall into this category. However, if there is more than one unknown function, then a **system of equations** is necessary. For example, in discussing the motion of a particle of mass m in the xy plane under the influence of a force vector $\mathbf{F} = F_1\mathbf{i} + F_2\mathbf{j}$, we need to determine both x and y, where

$$m\frac{d^2x}{dt^2} = F_1$$

$$\tag{13}$$

$$m\frac{d^2y}{dt^2} = F_2$$

and t denotes time. We discuss systems of equations in Chapters 8 and 9.

1.2.1 Origin and Applications of DEs

As mentioned in the introduction, DEs originated out of a study of certain kinds of problems in mechanics. Today, however, their use is far more widespread. They occur in various branches of engineering and the sciences, and are used in

1. The study of particle motion.
2. The analysis of electric circuits and servomechanisms.
3. Continuum and quantum mechanics.
4. The theory of diffusion processes and heat flow.
5. Electromagnetic theory.
6. The theory of vibrations and sounds.

Disciplines such as economics and the biological sciences are also now making use of DEs to investigate problems in

7. Interest rates.
8. Population growth.
9. The ecological balance of systems.
10. The spread of epidemics.

among other types of problems.

The mathematical formulation of problems like those listed above gives rise to a DE. This occurs because the various scientific laws employed in the formulation of these problems involve certain *rates of change* of one or more quantities with respect to other quantities, and such rates of change are expressed mathematically by *derivatives*. Hence, any resulting mathematical equation will involve derivatives in the unknown quantity, and this is what we mean by a differential equation.

In formulating the DE, one must normally make certain simplifying assumptions so that the resulting DE is tractable. Determining just what assumptions are reasonable is often the most critical part of a problem. Sometimes certain aspects of a problem seem relatively unimportant and can be modified by assuming an approximate situation; sometimes an aspect of a problem may even be entirely eliminated. The DE resulting from any such assumptions will actually be that of an idealized situation.

Even after a number of simplifying assumptions are made, the mathematical formulation of a problem can still lead to a DE that is troublesome to solve. This is particularly true when the resulting DE is *nonlinear,* since they are usually difficult or impossible to solve exactly. Thus, when confronted with a nonlinear DE we often resort to a **numerical method** for approximating the actual solution, or rely on **qualitative methods** for obtaining information about the behavior of the actual solution. Some qualitative techniques are briefly discussed in Chapters 5 and 9 while numerical techniques are introduced in Chapter 10. *Linear* DEs are much easier to handle in many ways, mostly because various properties of their solutions can be characterized in a general sort of way and several standard solution techniques are available. For this reason, linear equations occupy a more prominent place in the theory and applications of DEs. However, even linear equations often defy solution

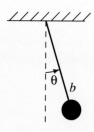

Figure 1.1
Oscillating pendulum.

in terms of elementary functions. In such cases we may again resort to a numerical procedure or, in some instances, use a **power series method** to develop the solution in the form of an infinite series. Power series methods applied to second-order linear DEs are discussed in Chapter 7.

[0] 1.2.2 Linear Approximation

Because of the difficulties encountered in solving most *nonlinear* DEs, they are often *approximated by linear* DEs when the resulting linear equations can describe certain fundamental characteristics of the nonlinear systems. For instance, the angle θ that an oscillating pendulum of length b makes with the vertical direction (see Figure 1.1) is governed by the nonlinear DE

$$\frac{d^2\theta}{dt^2} + \frac{g}{b}\sin\theta = 0 \tag{14}$$

where g is the gravitational constant. This DE is nonlinear because of the presence of the term $\sin\theta$. If we write $\sin\theta$ in terms of its Maclaurin series, that is,

$$\sin\theta = \theta - \frac{\theta^3}{3!} + \frac{\theta^5}{5!} - \frac{\theta^7}{7!} + \cdots \tag{15}$$

then for small angular displacements θ, it may be reasonable to use the approximation $\sin\theta \cong \theta$. In this case, (14) can be replaced with the linear equation

$$\frac{d^2\theta}{dt^2} + \frac{g}{b}\theta = 0 \tag{16}$$

The linear equation (16) in this problem gives fairly accurate information about the behavior of the pendulum when θ is small. There are a number of applications, however, for which this approach is fruitless, since the phenomenon being studied does not lend itself to any reasonable linear approximation. In such cases we try to solve the nonlinear DE itself.

Exercises 1.2

For each of the following DEs, state whether the equation is *linear* or *nonlinear,* and give its *order*.

1. $\dfrac{d^2y}{dt^2} + 3\dfrac{dy}{dt} + y = 0$

2. $y' + x^2y = 0$

3. $y' + a_0(x)y = f(x)$

4. $2xyy' = x^2 + y^2$

5. $\dfrac{dP}{dx} = aP - bP^2$

6. $y''' - 2y'' + 6y' = x$

7. $y' = \dfrac{x}{y}$

8. $\dfrac{d^4y}{dx^4} = q(x)$

9. $L\dfrac{di}{dt} + Ri = E(t)$

10. $xy''' - 2(y')^4 + y = 0$

11. $2x^2y'' + 3xy' - y = 0$

12. $x^2 \dfrac{dy}{dx} + y - xy - xe^x = 0$

13. $y^2 \dfrac{dx}{dy} + x = 0$

14. $\left| \dfrac{dy}{dx} \right| + y = 0$

15. $yy''' + y'' - y' + 15xy = 0$

16. $\sqrt{1 + x^2}\, y'' + (\sin x)y' = e^x$

17. $y' + 1 = (y')^2 - \sin x$

18. $\sqrt{1 + y''} + y' = x^2 + 3$

19. $(x^2 - y^2)y' + 3xy = 0$

20. $x^2 y'' + xy' - y = 3 \cos^2 x$

1.3 Solutions of DEs

One obvious requirement of a solution function of a DE is that it be continuous and differentiable. More precisely, we define *solution* in the following way.

Definition 1.3 A **solution** of a DE on a given interval I is a continuous function possessing all derivatives occurring in the equation that, when substituted into the DE, reduces it to an identity for all x in the interval I.

EXAMPLE 1

Verify that $y = e^{-x}$ is a solution of

$$y' + y = 0$$

and state the interval of validity.

Solution

We first note that the function $y = e^{-x}$ is continuous and has derivatives of all orders for all x. Furthermore,

$$y' + y = -e^{-x} + e^{-x}$$
$$= 0$$

for all values of x. ∎

It should be pointed out that a given DE will usually possess many solutions. For example, it is easily verified that $y = 3e^{-x}$ satisfies the DE in Example 1 as does $y = Ce^{-x}$ for any value of the constant C. Thus we obtain the *family of solutions* shown in Figure 1.2. Even the **trivial solution** ($y = 0$), obtained by setting $C = 0$, is a solution of $y' + y = 0$.

EXAMPLE 2

Verify that

$$y_1 = C_1 \cos 2x$$
$$y_2 = C_2 \sin 2x$$

where C_1 and C_2 are any constants, are both solutions of

$$y'' + 4y = 0$$

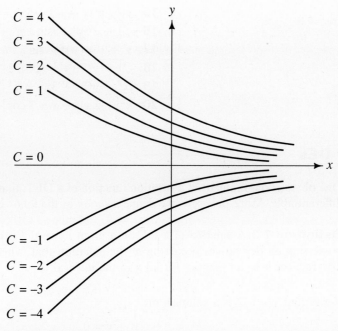

$C = 4$

$C = 3$

$C = 2$

$C = 1$

$C = 0$

$C = -1$

$C = -2$

$C = -3$

$C = -4$

Figure 1.2 Solutions of $y' + y = 0$.

Solution

For $y_1 = C_1 \cos 2x$, we find

$$y_1' = -2C_1 \sin 2x, \qquad y_1'' = -4C_1 \cos 2x$$

so that

$$y_1'' + 4y_1 = -4C_1 \cos 2x + 4C_1 \cos 2x$$
$$= 0$$

Similarly, it follows that

$$y_2'' + 4y_2 = -4C_2 \sin 2x + 4C_2 \sin 2x$$
$$= 0$$

We might also observe here that the functions

$$y = C_1 \cos 2x + C_2 \sin 2x$$

and

$$y = C_3 \sin x \cos x$$

are also solutions of the same DE, the verification of which is left to the reader. ■

Solutions of a DE that can be expressed in the form $y = \phi(x)$ are called **explicit solutions.** This is generally the most convenient form in which to represent the solution since values for y can be readily calculated for each value of x. *Linear* DEs always lead to explicit solution formulas, but this is rarely the case for *nonlinear* DEs. Thus, in solving nonlinear DEs we usually have to express the solution by a relation of the form $g(x, y) = 0$, called an **implicit solution.**

EXAMPLE 3

Verify that $g(x, y) = 0$ is a solution of

$$y' = \frac{x}{y}$$

where $g(x, y) = x^2 - y^2 - 1$.

Solution

By implicit differentiation of $x^2 - y^2 - 1 = 0$, we obtain

$$\frac{d}{dx}(x^2) - \frac{d}{dx}(y^2) - 0 = 0$$

or

$$2x - 2yy' = 0$$

Solving for y' yields the original DE

$$y' = \frac{x}{y}$$

which verifies that $g(x, y) = 0$ is a solution.   ■

The implicit relation $g(x, y) = 0$ need not define a single-solution function. For instance, in Example 3 we find that $g(x, y) = x^2 - y^2 - 1$ provides us with two solutions, namely,

$$y = \begin{cases} +\sqrt{x^2 - 1} \\ -\sqrt{x^2 - 1} \end{cases}$$

Once we are provided with a particular solution function for a given DE, verifying that it is indeed a solution is rather routine. That is, we simply substitute it into the DE. However, finding the solution function itself may be difficult except for a few special types of equations. Because of this, it may be useful to first ask the question—Does a solution **exist?** Not all DEs have solutions! For example, the simple-looking DE

$$(y')^2 = -1$$

clearly does not have a (real) solution. Knowing that a given DE has a solution is very important in applications. That is, the DE is a *mathematical model* of some physical situation, so if we can prove that the model does not have a solution, it is probably a poor model. Meaningful physical problems should always have solutions.

Once a solution has been found we have answered the question concerning existence. We might then ask—Is the solution **unique?** In the examples we have discussed, the solutions are not unique (see Example 2). Nonetheless, certain additional requirements can be imposed upon the solution function in most instances so that a unique solution can be found. More will be said about this situation later.

The existence and uniqueness questions are difficult to answer for DEs in general. Usually we discuss these questions with respect to rather narrow classes of DEs. Most of our efforts in the following chapters, however, are devoted to the more practical question—How do we find solutions? It turns out there are a variety of solution techniques that have been developed, but each is generally restricted to a particular class of DEs. We develop some of the more elementary and useful of these techniques throughout the text.

EXAMPLE 4

Determine the values of m such that $y = e^{mx}$ is a solution of

$$y'' - y' - 12y = 0$$

Solution

The direct substitution of $y = e^{mx}$ into the left-hand side of the DE yields

$$y'' - y' - 12y = m^2 e^{mx} - m e^{mx} - 12 e^{mx}$$

$$= (m^2 - m - 12)e^{mx}$$

Since $e^{mx} \neq 0$, it follows that the right-hand side of this last expression is zero if and only if

$$m^2 - m - 12 = (m - 4)(m + 3) = 0$$

Thus, $m = 4$ or $m = -3$, and we find two solutions

$$y_1 = e^{4x}$$

$$y_2 = e^{-3x}$$

The reader should verify that $y = C_1 e^{4x} + C_2 e^{-3x}$ is also a solution of the DE, where C_1 and C_2 are any constants.   ∎

EXAMPLE 5

Find a first-order DE involving both y and y' for which $y = x^2 - 1$ is a solution.

Solution

With $y = x^2 - 1$, we see that $y' = 2x$. Thus, one possible DE is

$$xy' - 2y = 2$$

However, we also see that another possible DE is

$$\tfrac{1}{4}(y')^2 - y = 1$$

In addition to these two DEs, we could construct many others, all of which have the same solution $y = x^2 - 1$. Therefore, we must conclude that knowing the solution of a DE does *not* uniquely determine the DE.   ∎

1.3.1 Particular and General Solutions

By choosing specific values of the parameter(s) in a family of solutions we can produce a variety of solutions known as **particular solutions.** For example, by setting $C = 1$, 3, and -5 in the family of solutions $y = Ce^{-x}$ belonging to $y' + y = 0$, we obtain particular solutions $y = e^{-x}$, $y = 3e^{-x}$, and $y = -5e^{-x}$, respectively. In practice we usually seek a family of solutions such that *all* particular solutions can be obtained from it. Such a family of solutions is called a **general solution,** and the number of constants appearing in it is always equal to the *order* of the DE. For instance, solving $y' = f(x)$ leads to *one* constant (of integration) and solving $y'' = f(x)$ leads to *two* constants, and so on. Thus, the *general solution* of an nth-order DE is an *n-parameter family of solutions* containing all solutions of the equation.

In solving certain *nonlinear* DEs there may exist solutions, known as **singular solutions** (see Problem 31 in Exercises 1.3), that are not part of the *n*-parameter family of solutions. Because of such possibilities, it is usually best to use the term "general solution" only when discussing *linear* DEs. For nonlinear DEs we simply refer to the "*n*-parameter family of solutions."

Another distinction between linear and nonlinear DEs is that the sum of two or more particular solutions is always a solution of certain kinds of linear DEs, called *homogeneous* equations. This property is referred to as the **linearity property** or **superposition principle,** but it does not apply in general to nonlinear equations (see the following example).

■————
EXAMPLE 6

Show that although $y_1 = e^{-x}$ and $y_2 = e^{2x}$ are solutions of both the *linear* equation

$$y'' - y' - 2y = 0$$

and the *nonlinear* equation

$$yy'' - (y')^2 = 0$$

the expression $y = C_1e^{-x} + C_2e^{2x}$ is a solution of only the linear DE for arbitrary *nonzero* values of C_1 and C_2.

Solution

By direct substitution of $y = C_1e^{-x} + C_2e^{2x}$ into the left-hand side of the linear DE, we see that

$$y'' - y' - 2y = (C_1e^{-x} + 4C_2e^{2x}) - (-C_1e^{-x} + 2C_2e^{2x}) - 2(C_1e^{-x} + C_2e^{2x})$$

$$= (C_1 + C_1 - 2C_1)e^{-x} + (4C_2 - 2C_2 - 2C_2)e^{2x}$$

$$= 0 \cdot e^{-x} + 0 \cdot e^{2x}$$

$$= 0$$

On the other hand, the nonlinear DE yields the result

$$yy'' - (y')^2 = (C_1 e^{-x} + C_2 e^{2x})(C_1 e^{-x} + 4C_2 e^{2x}) - (-C_1 e^{-x} + 2C_2 e^{2x})^2$$
$$= (C_1^2 - C_1^2)e^{-2x} + (5C_1 C_2 + 4C_1 C_2)e^x + (4C_2^2 - 4C_2^2)e^{4x}$$
$$= 9C_1 C_2 e^x$$

which is clearly not identically zero. In fact, it is zero only when either C_1 or C_2 is itself zero. ∎

Throughout the calculus we have become accustomed to the idea that the "solution" of a problem is an exact, closed-form, analytical expression in terms of elementary functions from which numerical values can be generated when needed. Unfortunately, this is seldom the case in practice, especially when seeking the solution of a DE. That is, comparatively few DEs occur in applications for which simple exact solutions can be found. The "solution" in such cases often consists of some type of approximation function, or a set of numerical values, approximating the true solution over some interval. Hence, because many of the DEs arising in modern science and engineering are becoming more complex and because of the greater capabilities of today's computers, approximation and numerical methods are playing an ever-increasing role in solving DEs.

While many practical problems do not lead to easily solved DEs, it is still instructive in learning the general theory of differential equations to practice with relatively simple DEs for which exact solutions can be found. Therefore, most of the problems in this text will be rather routine, permitting solution by well-established analytical means. Some numerical techniques are briefly discussed, but in general these numerical methods are better suited for courses whose primary emphasis is on numerical procedures.

Exercises 1.3

In Problems 1 to 14, verify that the given function is a solution of the specified DE. Assume that C_1 and C_2 are constants.

1. $y' + 2xy = 0$; $y = C_1 e^{-x^2}$

2. $xy' + y = 0$; $y = C_1/x$, $x \neq 0$

3. $xy' - y = x$; $y = x \log x$, $x > 0$†

*4. $(1 + x^2)y' + 1 + y^2 = 0$; $xy + x + y = 1$

5. $2xy' = 3y$; $x^3 - C_1 y^2 = 0$

6. $(y')^2 = \dfrac{y}{x}$; $\sqrt{y} = \sqrt{x} + 3$, $x > 0$

7. $\dfrac{dP}{dt} = aP - bP^2$;

$P = \dfrac{ae^{at}}{(1 + be^{at})}$ (a, b constants)

■8. $xy' + (1 - x)y = xe^x$;

$y = \frac{1}{2}xe^x + C_1 x^{-1}e^x$, $x \neq 0$

9. $y'' + 2y' - 3y = 0$; $y = C_1 e^x + C_2 e^{-3x}$

10. $y'' - 6y' + 9y = 0$; $y = (C_1 + C_2 x)e^{3x}$

† By log x, we mean the **natural logarithm** (also written as ln x).

11. $y'' - 2y' + 5y = 0$;
$y = C_1 e^x \cos 2x + C_2 e^x \sin 2x$

12. $y'' + y' - 2y = 2x - 40 \cos 2x$;
$y = C_1 e^x + C_2 e^{-2x} - \frac{1}{2} - x$
$+ 6 \cos 2x - 2 \sin 2x$

13. $x^2 y'' + 5xy' + 4y = 0$;
$y = C_1 x^{-2} + C_2 x^{-2} \log x, \ x > 0$

*14. $y' - 2xy = 1; \ y = C_1 e^{x^2} + e^{x^2} \int_0^x e^{-t^2} \, dt$

In Problems 15 to 20, determine for which value(s) of the constant m is $y = e^{mx}$ a solution of the given DE.

15. $y' - 2y = 0$

16. $y'' - y = 0$

17. $y'' + y' - 6y = 0$

■18. $y'' + 2y' = 0$

19. $y''' + 3y'' - 4y' = 0$

*20. $y''' + 6y'' + 11y' + 6y = 0$

21. Show that $x^3 + 3xy^2 - 1 = 0$ is an implicit solution of

$$2xyy' + x^2 + y^2 = 0, \quad 0 < x < 1$$

22. Show that $5x^2 y^2 - 2x^3 y^2 - 1 = 0$ is an implicit solution of

$$xy' + y = x^3 y^3, \quad 0 < x < \tfrac{5}{2}$$

In Problems 23 to 30, find a first-order DE involving both y and y' for which the given function is a solution.

23. $y = e^{-2x}$

24. $y = x^3 - 4$

25. $y = \frac{1}{2} x e^x$

26. $y = 2e^x - 2x - 1$

27. $y = \cos x$

28. $y = \sinh 3x$

29. $(x - 1)^2 + y^2 = 1$

■30. $y^2 = x^2 + y$

*31. Given the DE

$$(y')^2 - xy' + y = 0$$

(a) Verify that $y = Cx - C^2$ is a solution for any constant C.

(b) Determine a value of $K \ (\neq 0)$ such that $y = Kx^2$ is another solution of the DE. [Since $y = Kx^2$ cannot be obtained from the solution in (a) by specifying the constant C, it is called a **singular solution.**]

*32. Verify that $y = (\frac{1}{4} x^2 + C)^2$, where C is constant, is a solution of $y' = x\sqrt{y}$. Also, find a singular solution by inspection.

33. Verify that $y_1 = 1$ and $y_2 = x^2$ satisfy each of the DEs

$$xy'' - y' = 0$$

$$2yy'' - (y')^2 = 0$$

but that $y = C_1 + C_2 x^2$ satisfies only the first DE for arbitrary values of C_1 and C_2. Explain.

34. Verify that $y_1 = 1$ and $y_2 = \sqrt{x}$ satisfy each of the DEs

$$2xy'' + y' = 0$$

$$8x^3(y'')^2 - yy' = 0$$

but that $y = C_1 + C_2 \sqrt{x}$ satisfies only the first DE for arbitrary values of C_1 and C_2. Explain.

1.4 Initial and Boundary Conditions

In most applications the unknown function y must satisfy some **auxiliary conditions** in addition to satisfying the DE. These auxiliary conditions are usually based on *physical contraints* in the problem, the number of which ordinarily equals the order of the DE or number of arbitrary constants in its general solution. For example, first-order DEs require only one auxiliary condition while second-order DEs require two auxiliary conditions.

EXAMPLE 7

Given that $y = Ce^x$ is a general solution, solve

$$y' - y = 0, \qquad y(0) = 2$$

Solution

The general solution $y = Ce^x$ is a family of curves in the xy plane (see Figure 1.3). By specifying the auxiliary condition $y(0) = 2$, we are seeking that one particular curve that passes through the point $(0, 2)$. Thus, by substituting $x = 0$ into the general solution, we obtain

$$y(0) = Ce^0 = 2$$

We deduce, therefore, that $C = 2$, and the solution we seek is

$$y = 2e^x$$

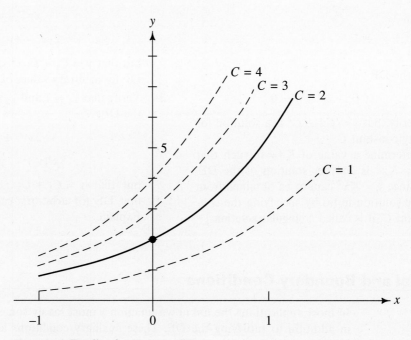

Figure 1.3 Family of curves $y = Ce^x$.

EXAMPLE 8

Given that $y = C_1 \cos x + C_2 \sin x + 3$ is a general solution, solve

$$y'' + y = 3 \qquad y(0) = 2 \qquad y'(0) = 0$$

Solution

This time the auxiliary conditions require that the solution we seek pass through the point $(0, 2)$ of the xy plane with zero slope at that point. Imposing these conditions on the general solution leads to (notice that $y' = -C_1 \sin x + C_2 \cos x$)

$$y(0) = C_1 \cdot 1 + C_2 \cdot 0 + 3 = 2$$
$$y'(0) = -C_1 \cdot 0 + C_2 \cdot 1 = 0$$

The first of these conditions yields $C_1 = -1$ while the second condition requires $C_2 = 0$. Hence, our solution becomes

$$y = -\cos x + 3 \qquad \blacksquare$$

EXAMPLE 9

Given that $y = C_1 \cos x + C_2 \sin x$ is a general solution, solve

$$y'' + y = 0, \qquad y(0) = 0, \qquad y(\pi/2) = 1$$

Solution

These auxiliary conditions require the solution function to pass through the two points $(0, 0)$ and $(\pi/2, 1)$. Applying the first condition to the general solution, we find

$$y(0) = C_1 \cdot 1 + C_2 \cdot 0 = 0$$

from which we deduce $C_1 = 0$. The second condition yields

$$y(\pi/2) = C_2 \sin(\pi/2) = 1$$

or $C_2 = 1$. Hence, the solution is

$$y = \sin x \qquad \blacksquare$$

When the auxiliary conditions are all specified at a single value of x, as in Examples 7 and 8, we call them **initial conditions** and refer to the problem as an **initial value problem.** Although we could choose the single value of x as any value, it is customary in practice to use $x = 0$. If the auxiliary conditions are specified at more than one point on the interval of interest, they are called **boundary conditions** and the resulting problem is called a **boundary value problem** (see Example 9). The boundary conditions are usually specified at only two points of the interval, and we refer to these problems as **two-point boundary value problems.** Such problems involve DEs that are at least second-order, since first-order DEs have but a single auxiliary condition and as such are classified as initial value problems.

In general, initial value problems are well behaved in that they almost always lead to unique solutions. Most of the chapters in this text are devoted to this class of problems. On the other hand, the general theory of boundary value problems is more complicated in that the problem may have (1) a *unique solution,* (2) *no solution,* or (3) *more than one solution.* Such matters are too involved for discussion here but will be discussed in detail in Chapter 11 (see also Problems 11 to 15 in Exercises 1.4).

Exercises 1.4

In Problems 1 to 10, use the given general solution to solve the initial value problem.

1. $y' - 2y = 0$, $y(0) = 1$; $y = Ce^{2x}$

2. $xy' + 2y = 0$, $y(1) = 4$; $y = C/x^2$

3. $y' + y = e^x$, $y(0) = 0$; $y = \frac{1}{2}e^x + Ce^{-x}$

4. $y'' + y = 0$, $y(0) = 1$, $y'(0) = -1$;
$y = C_1 \cos x + C_2 \sin x$

5. $y'' = 1$, $y(0) = 0$, $y'(0) = 0$;
$y = \frac{1}{2}x^2 + C_1 + C_2 x$

6. $y'' = e^x$, $y(0) = 2$, $y'(0) = -1$;
$y = e^x + C_1 + C_2 x$

7. $x^2 y'' - 2xy' + 2y = 0$, $y(1) = 0$, $y'(1) = 1$;
$y = C_1 x + C_2 x^2$

■8. $y'' - y = 0$, $y(0) = 0$, $y'(0) = 0$;
$y = C_1 e^x + C_2 e^{-x}$

9. $y'' - 2y' + 2y = 0$, $y(0) = 1$, $y'(0) = 0$;
$y = e^x(C_1 \cos x + C_2 \sin x)$

10. $x^2 y'' - xy' + y = 0$, $y(1) = 3$, $y'(1) = -1$;
$y = C_1 x + C_2 x \log x$

In Problems 11 to 15, use the given general solution to solve (if possible) the boundary value problem.

11. $y'' = 0$, $y(0) = 0$, $y(1) = 1$; $y = C_1 + C_2 x$

12. $y'' = 0$, $y(0) = 0$, $y(1) = 0$; $y = C_1 + C_2 x$

13. $y'' + y = 0$, $y'(0) = 0$, $y'(\pi) = 0$;
$y = C_1 \cos x + C_2 \sin x$

■14. $y'' + \pi^2 y = 0$, $y(0) = 1$, $y(1) = 0$;
$y = C_1 \cos \pi x + C_2 \sin \pi x$

15. $y'' - y = 0$, $y(-1) = 0$, $y'(1) = 0$;
$y = C_1 e^x + C_2 e^{-x}$

*16. Given the general solution $y = C_1 + C_2 x^2$, does the initial value problem

$$xy'' - y' = 0, \qquad y(0) = 0, \qquad y'(0) = 1$$

have a solution?

*17. For which nonzero values of k (if any) does the boundary value problem

$$y'' + k^2 y = 0, \qquad y(0) = 0, \qquad y(\pi) = 0$$

have nonzero solutions? The general solution is

$$y = C_1 \cos kx + C_2 \sin kx$$

1.5 Chapter Summary

In this chapter we have introduced the basic terminology for DEs. If the unknown function y depends upon only one independent variable, we say the DE is **ordinary;** otherwise, it is a **partial** DE. The **order** of a DE is the order of the highest derivative occurring in the DE. Equations are further classified as **linear** or **nonlinear.**

A **solution** is a continuous and differentiable function that, when substituted into the DE, reduces it to an identity. We call the solution an **explicit solution** if it can be written in the form $y = \phi(x)$; otherwise, it is called an **implicit solution.** Every solution of a DE not involving arbitrary constants is called a **particular solution.** The **general solution** of an nth-order DE is an n-parameter family of solutions containing all solutions of the DE.

If the solution of a DE is subject to **auxiliary conditions** all specified at a single point, the problem is called an **initial value problem.** If the auxiliary conditions are specified at more than one point, the problem is called a **boundary value problem.**

Review Exercises

In Problems 1 to 4, classify the DE as *ordinary* or *partial*. If ordinary, determine its *order* and whether it is *linear* or *nonlinear*.

1. $\dfrac{d^2y}{dx^2} + 5x^3\dfrac{dy}{dx} + 2y = 10e^x$

2. $\dfrac{\partial z}{\partial x} = x + 2y\dfrac{\partial z}{\partial y}$

3. $e^y y' + 3x^2 y = 1$

4. $(y'')^3 + 2xy' = 1$

In Problems 5 to 8, determine which of the given functions is a solution of the DE.

5. $y' + y = 0$; $y_1 = 3e^x$, $y_2 = 3e^{-x}$

6. $y' + y^2 = 0$; $y_1 = 1/x$, $y_2 = 1$

7. $y'' + 4y = 0$; $y_1 = 3\cos 2x - \sin 2x$, $y_2 = 5\sin 2x$

8. $y' - 2y = x$; $y_1 = -\frac{1}{4}(2x + 1)$, $y_2 = e^{2x}$

In Problems 9 and 10, show that the given relation is an implicit solution of the DE.

9. $4x(y')^2 + 2xy' - y = 0$; $(y - 2)^2 - 2x = 0$

10. $(x^2 - y^2)y' = 2xy$; $x^2 + y^2 - 6y = 0$

In Problems 11 to 16, determine whether the problem is an *initial value problem* or a *boundary value problem,* and use the given general solution to solve it.

11. $y' + y = 0$, $y(3) = 2$; $y = C_1 e^{-x}$

12. $y'' + 4y = 0$, $y(0) = 0$, $y'(0) = 1$; $y = C_1 \cos 2x + C_2 \sin 2x$

13. $y'' + y = 0$, $y(0) = 1$, $y'(0) = 2$; $y = C_1 \cos x + C_2 \sin x$

14. $y'' - y' - 2y = e^{3x}$, $y(0) = 1$, $y'(0) = 2$; $y = C_1 e^{-x} + C_2 e^{2x} + \frac{1}{4}e^{3x}$

15. $y'' + y = 0$, $y(0) = 1$, $y(\pi/2) = 1$; $y = C_1 \cos x + C_2 \sin x$

16. $y'' + y = 0$, $y(0) = 1$, $y'(\pi) = 1$; $y = C_1 \cos x + C_2 \sin x$

1.6 Historical Comments

The study of differential equations originated with the introduction of the calculus by Isaac Newton and Gottfried Wilhelm von Leibniz. Although mathematics began as a recreation for Newton, he became known as a great mathematician by the age of 24 after his invention of the calculus, discovery of the law of universal gravitation, and experimental proof that white light is composed of all colors. At the age of 23 he became professor of mathematics at Trinity College, Cambridge, the same place where he was educated. Newton's involvement with differential equations was mostly indirect, stemming primarily from his development of the calculus and laws of mechanics which provided a basis for their applications by others. His active research career basically ended in the early 1690s, although he did continue to work on problems formulated by other mathematicians. He spent the last two years of his life in constant pain, and after his death on March 20, 1727, was buried in Westminster Abbey.

Leibniz completed his doctorate in philosophy by the age of 20 at the University of Altdorf. Afterward he studied mathematics under the supervision of Christian Huygens (1629–1695) and, independently of Newton, helped develop the calculus. It is Leibniz who is responsible for the derivative notation (dy/dx) and the integral sign. Leibniz's involvement with the theory of differential equations was more direct than that of Newton. For example, he corresponded regularly with other prominent mathematicians concerning differential equations, and he developed several methods for solving first-order equations—both linear and nonlinear. The last years of his life were saddened by ill health, controversy, and neglect. A bitter controversy had been started by friends of Newton over the legitimacy of Leibniz's discovery of the calculus. Eventually both Newton and Leibniz themselves became occupied with the controversy, even though both were convinced that they had reached their results independently. Newton had actually discovered the calculus a few years earlier than Leibniz, but failed to immediately publish his results. Leibniz died on November 14, 1716, although many members of the scientific community were not aware of it until almost a year later.

Among other early mathematicians who contributed to the development of differential equations and their applications were members of the famous Bernoulli family of Switzerland, the most famous of which are James Bernoulli (1654–1705) and his younger brother John Bernoulli (1667–1748). The older Bernoulli brother was the first to be appointed professor of mathematics at Basel but, upon his death in 1705, he was subsequently replaced by John. The Bernoulli brothers were very competitive, especially with each other, and both made significant contributions to several areas of mathematics and mechanics.

James Bernoulli discovered in 1690 that a cycloid is the curve along which a pendulum moves to make a complete oscillation (in the same time), regardless of the length of arc through which it swings. His research interests also included probability theory and the calculus of variations, leading to many results that now bear his name. In 1691, John Bernoulli solved the problem of the catenary, the curve assumed by a cable hanging between two pegs, and in 1696 he proposed his now famous *brachistochrone problem,* that is, finding the path in a vertical plane down which a particle will fall from one point to another in the shortest time.

Throughout the years there have been many mathematicians who contributed to the general development of differential equations. In some of the chapters to follow we mention a few of the most prominent of these mathematicians and their specific contributions to this fascinating subject.

CHAPTER 2
First-Order Equations

In this chapter we consider certain types of **first-order** DEs for which exact solutions may be obtained by clearly designated techniques. Proficiency in solving such equations rests heavily on the ability of the practitioner to recognize various types of DEs and then apply the corresponding method of solution.

In Section 2.2 we discuss equations that can be solved by the method of **separation of variables.** This is one of the easiest techniques (theoretically) to apply and comes up often in applications. Also discussed here are **homogeneous equations** which, through a change of variable, can likewise be solved by separation of variables. The solution of **exact equations,** based on the concept of a **total** or **exact differential,** is featured in Section 2.3. The use of **integrating factors** for problems where the DE is not exact is also discussed.

In Sections 2.4 and 2.5 we give a detailed treatment of **first-order linear DEs.** The standard **integrating factor technique** is presented first, followed by a development of the **general theory** of linear DEs that introduces the notions of **homogeneous** and **particular solutions. Physical interpretations** of the homogeneous and particular solutions are also discussed.

Section 2.6 contains some **miscellaneous topics** and in Section 2.7 a **chapter summary** is provided. Some **historical comments** are given in the last section of the chapter.

2.1 Introduction

First-order DEs arise naturally in problems involving the determination of the *velocity of free-falling bodies subject to a resistive force,* finding the *current* or *charge in an electric circuit,* and finding *curves of population growth* or *radioactive decay,* among other applications. The type of DE that may occur in these applications usually falls into one or more categories, each of which demands a different method of solution. This is similar to the calculus problem of solving $y' = f(x)$, for which a variety of integration techniques may be used depending on the function $f(x)$. In this chapter we are concerned only with *solution techniques* for first-order DEs, of which there are several varieties. *Applications* are taken up in Chapter 3.

The DEs to be studied in this chapter may be expressed in either the **derivative form**

$$y' = F(x, y) \qquad\qquad (1)$$

or, through formal[†] algebraic manipulations, in the **differential form**

$$M(x, y)\, dx + N(x, y)\, dy = 0 \qquad\qquad (2)$$

depending on the type of DE and the solution technique. The equivalence of these two forms can be seen by writing (1) as

$$F(x, y)\, dx - dy = 0$$

or, provided $N(x, y) \neq 0$, by writing (2) as

$$\frac{dy}{dx} = -\frac{M(x, y)}{N(x, y)} = F(x, y)$$

In applications the solution of (1) or (2) is usually required to also satisfy an auxiliary condition of the form

$$y(x_0) = y_0 \qquad\qquad (3)$$

which geometrically specifies that the graph of the solution function pass through the point (x_0, y_0) of the xy plane. Solving (1) or (2) subject to the auxiliary condition (3) is called an **initial value problem** (IVP).

Many first-order DEs that arise in applications are *nonlinear*. Sometimes they can be solved by the special methods introduced in this chapter, or other known techniques, but at other times they cannot. Even when a known technique can be used, it often requires the evaluation of certain integrals which in practice cannot be evaluated by elementary means. Therefore, in trying to solve nonlinear DEs, either by an analytical method or by a numerical procedure, it can be important to know at the onset that the given problem indeed has a solution and moreover, if so, that it is the *only* solution (remember, not all DEs have solutions). To aid us in this regard, we have the following *existence-uniqueness theorem.*[‡]

[†] By "formal" we mean a mathematical process that is not rigorous.

[‡] For a proof of Theorem 2.1, see E. I. Ince, *Ordinary Differential Equations,* Dover, New York, 1956.

> **Theorem 2.1 Existence-uniqueness.** *If F is a continuous function in some rectangular domain $a \leq x \leq b$, $c \leq y \leq d$ containing the point (x_0, y_0), then the IVP*
>
> $$y' = F(x, y), \qquad y(x_0) = y_0$$
>
> *has at least one solution in some interval $|x - x_0| < h$ embedded in $a < x < b$. If, in addition, the partial derivative $\partial F / \partial y$ is continuous in that rectangle, then the IVP has a unique solution.*

Theorem 2.1 tells us that if the stated conditions on F and $\partial F / \partial y$ are satisfied, the given IVP has a *unique solution*. These conditions, however, are only **sufficient conditions**—not **necessary conditions**. That is, if these conditions are not satisfied, the problem may have (1) *no solution*, (2) *more than one solution*, or (3) a *unique solution*. Consider the following examples.

■─── EXAMPLE 1

Does Theorem 2.1 imply the existence of a unique solution for the IVP

$$y' = x^2 + y^3, \qquad x > 0, \qquad y(0) = 5?$$

Solution

The function $F(x, y) = x^2 + y^3$ is a continuous function for *all* values of x and y. The same is true for the partial derivative $\partial F / \partial y = 3y^2$. Therefore, based on Theorem 2.1, we can say the given problem has a *unique solution* on the specified interval. ■

■─── EXAMPLE 2

Does Theorem 2.1 imply the existence of a unique solution for the IVP

$$y' = y^{1/3}, \qquad y(0) = 0?$$

Solution

Although $F(x, y) = y^{1/3}$ is a continuous function for all x and y, the partial derivative

$$\frac{\partial F}{\partial y} = \frac{1}{3} y^{-2/3}$$

is not continuous in any rectangle containing the point $(0, 0)$ or any part of the x axis. Therefore, based on Theorem 2.1, we can say a solution *exists,* but cannot conclude anything about the *uniqueness* of the solution of the given problem.

By techniques soon to be discussed, it can be shown that

$$y_1 = \left(\frac{2x}{3}\right)^{3/2}, \qquad x \geq 0$$

is a solution of the IVP and, by inspection, we see that

$$y_2 = 0$$

is also a solution. Still other solutions can be found—in fact, it can be shown that this problem has *infinitely many* solutions!

Finally, it is worth noting that if the initial condition were prescribed at any point (x_0, y_0) not on the x axis, the resulting IVP would have a unique solution. ▪

In this chapter we discuss only analytical methods of solution for first-order DEs. Numerical techniques for these and other equations are presented in Chapter 10. However, readers who wish to study numerical techniques involving first-order DEs, before studying DEs of higher order, may skip over to Chapter 10 at the end of this chapter without loss of continuity.

2.2 Separation of Variables

The first-order DE

$$M(x, y)\, dx + N(x, y)\, dy = 0 \tag{4}$$

is said to be "separable" if $M(x, y) = p(x)q(y)$ and $N(x, y) = r(x)s(y)$. In this case it can be put in the form

$$f(x)\, dx = g(y)\, dy \tag{5}$$

where $f(x) = p(x)/r(x)$ and $g(y) = -s(y)/q(y)$. Observe that the left-hand side of (5) is a function of x alone while the right-hand side depends only on y.[†] A family of solutions of (5) can be obtained by integration to yield

$$\int f(x)\, dx = \int g(y)\, dy + C_1 \tag{6}$$

where C_1 is an arbitrary constant (only one constant is necessary since the constants of integration are additive). By (6), we mean a relation of the form

$$F(x) = G(y) + C_1$$

where

$$\frac{d}{dx}\, F(x) = f(x), \qquad \frac{d}{dx}\, G(y) = g(y)\, \frac{dy}{dx}$$

This general technique is called the method of **separation of variables.**

▪──────────────
EXAMPLE 3

Solve

$$(1 - x)\, dy + y\, dx = 0$$

Solution

We see that division by y and $(1 - x)$ leads to the "separated form"

$$\frac{dy}{y} = -\frac{dx}{1 - x}, \qquad x \neq 1 \qquad y \neq 0$$

──────────
[†] Either x or y may act as the independent variable in this setting.

Thus, integrating the left-hand side with respect to y and the right-hand side with respect to x, we get

$$\log|y| = \log|1 - x| + C_1$$

which, through properties of logarithms, can also be expressed in the form

$$y = \pm e^{C_1}(1 - x)$$

The $\pm$ signs are due to eliminating the absolute values that appeared in the logarithms. However, since C_1 is an arbitrary constant, it follows that $\pm e^{C_1}$ is also arbitrary, and we can redefine it as a new constant C. Introducing new constants in this way is very common in the solution of DEs. Our family of solutions now becomes

$$y = C(1 - x)^{\dagger}$$ ■

In Example 1 we are using the symbol $\log x$ to denote the **natural logarithm,** also commonly denoted by $\ln x$. Since integrals leading to logarithmic terms are quite prevalent in the solution of DEs, some care should be exercised in their evaluation. Recall from the calculus that

$$\int \frac{du}{u} = \log|u| + C \qquad u \neq 0$$

where the absolute value is usually retained unless we know in advance that $u > 0$.

<table>
<tr><td>■
EXAMPLE 4</td><td>Solve

$$y' = \frac{3y + 1}{x^2}, \qquad x > 0, \qquad y > -1/3$$</td></tr>
</table>

Solution

Upon separating the variables, we obtain

$$\frac{dy}{3y + 1} = \frac{dx}{x^2}, \qquad y \neq -1/3, \qquad x \neq 0$$

the integration of which yields

$$\frac{1}{3} \log|3y + 1| = -\frac{1}{x} + C_1$$

Exponentiation of this expression leads to

$$3y + 1 = \pm e^{3(C_1 - 1/x)}$$

or, writing $C = \pm \frac{1}{3} e^{3C_1}$, we obtain the explicit solution

$$y = Ce^{-3/x} - \frac{1}{3}$$ ■

---

† Notice that $C = 0$ yields the solution $y = 0$, even though $\pm e^{C_1} \neq 0$. The solution $y = 0$ was initially "lost" when we separated variables through division by y.

Although we technically have the (implicit) solution of the DE once the integration is complete, it is usually advantageous to simplify the algebra (when possible) in this solution function. If numerical values of the solution are desired, it is best to obtain an *explicit* solution function when this can be done. Our next example illustrates the gain in simplicity of the final solution function due to a little algebraic maneuvering.

■────── EXAMPLE 5

Solve the IVP

$$(x^2 + 1)y' + y^2 + 1 = 0, \qquad y(0) = 1$$

Solution

By rearranging terms, we have

$$\frac{dy}{y^2 + 1} = -\frac{dx}{x^2 + 1}$$

which, upon integration, leads to

$$\tan^{-1} y = -\tan^{-1} x + C_1$$

If we apply the prescribed initial condition ($x = 0$, $y = 1$), we see that $C_1 = \pi/4$, and thus

$$\tan^{-1} y + \tan^{-1} x = \frac{\pi}{4}$$

However, we can obtain a more convenient form of the solution by first using the trigonometric identity

$$\tan(A + B) = \frac{\tan A + \tan B}{1 - \tan A \tan B}$$

Thus, we find

$$\tan(\tan^{-1} y + \tan^{-1} x) = \frac{y + x}{1 - xy}$$

and consequently our above solution becomes

$$\frac{y + x}{1 - xy} = \tan(\pi/4) = 1$$

Finally, solving explicitly for y, we get

$$y = \frac{1 - x}{1 + x} \qquad\qquad ■$$

Notice that the solution $y = (1 - x)/(1 + x)$ in Example 5 becomes unbounded as $x \to -1$. This point, called a **singularity** of the solution, is not as easily identified from the implicit solution form in terms of inverse tangents. Again, this is an advantage of the explicit solution form. Also observe that the DE did not appear to give us any hint that $x = -1$ is a special point. This is because singularities in the solution of a nonlinear DE often depend upon the prescribed initial condition.

EXAMPLE 6

Solve

$$(x - 4)y^4 \, dx - x^3(y^2 - 3) \, dy = 0$$

Solution

We can separate variables through division by the factor $x^3 y^4$, finding

$$\left(\frac{x - 4}{x^3}\right) dx - \left(\frac{y^2 - 3}{y^4}\right) dy = 0, \qquad x \neq 0, \qquad y \neq 0$$

or

$$(x^{-2} - 4x^{-3}) \, dx - (y^{-2} - 3y^{-4}) \, dy = 0$$

Upon integrating, we get

$$-\frac{1}{x} + \frac{2}{x^2} + \frac{1}{y} - \frac{1}{y^3} = C_1$$

In separating the variables it was necessary to assume that $x \neq 0$ and $y \neq 0$. Note, however, that $y = 0$ is also a solution of the DE but cannot be obtained from the above family of solutions by selecting an appropriate value of the constant C_1. Thus, the solution $y = 0$ is called a **singular solution** of the DE. ∎

This last example illustrates one more troublesome feature associated with solving certain nonlinear DEs, namely, the possibility of singular solutions. Such solutions are frequently lost in the formal solution process, as in Example 6, and yet in some instances may be just the solution we seek in a particular IVP.

2.2.1 Homogeneous Equations

Let us now consider a class of DEs that are not of the type that can be solved directly by separating the variables, but through a change of variable can then be solved by this method.

If the DE

$$M(x, y) \, dx + N(x, y) \, dy = 0 \tag{7}$$

has the property that

$$M(tx, ty) = t^n M(x, y)$$
$$N(tx, ty) = t^n N(x, y) \tag{8}$$

for some real n, we say the functions $M(x, y)$ and $N(x, y)$ are **homogeneous functions of degree** n, and in this case the DE (7) is called a **homogeneous DE.**[†]

EXAMPLE 7

Determine which of the following functions is *homogeneous* and, if so, state its degree:

(a) $f(x, y) = 3x - y + 5x^2/y$

(b) $f(x, y) = \sqrt{x^5 - 2y^5}$

(c) $f(x, y) = x^2 + y^2 + 1$

Solution

(a) $f(tx, ty) = 3(tx) - (ty) + 5(tx)^2/(ty)$

$$= 3tx - ty + 5tx^2/y$$

$$= t(3x - y + 5x^2/y)$$

$$= tf(x, y)$$

Hence, the function is homogeneous of degree 1.

(b) $f(tx, ty) = \sqrt{t^5 x^5 - 2t^5 y^5}$

$$= t^{5/2}\sqrt{x^5 - 2y^5}$$

$$= t^{5/2}f(x, y)$$

Hence, the function is homogeneous of degree 5/2.

(c) $f(tx, ty) = t^2 x^2 + t^2 y^2 + 1$

$$\neq t^n f(x, y)$$

In this case the function is not homogeneous.   ■

If $M(x, y)$ and $N(x, y)$ are homogeneous, we can always express them as

$$M(x, y) = x^n M(1, y/x)$$

$$N(x, y) = x^n N(1, y/x)$$

and hence (7) can be written in the form

$$\frac{dy}{dx} = -\frac{M(x, y)}{N(x, y)} = -\frac{x^n M(1, y/x)}{x^n N(1, y/x)}$$

or, equivalently as

$$\frac{dy}{dx} = F(y/x) \tag{9}$$

[†] It is unfortunate that such equations are called "homogeneous" since this term is used in more than one context in the study of DEs; for example, see Section 2.4.

This last form of the DE suggests the substitution $y = vx$ (or $x = vy$), where v is a new variable. By setting $y = vx$, we see that

$$\frac{dy}{dx} = x\frac{dv}{dx} + v \tag{10}$$

through application of the product rule. Therefore, (9) becomes

$$x\frac{dv}{dx} + v = F(v)$$

or, in differential form,

$$x\,dv = \left[F(v) - v\right]dx$$

Finally, by "separating the variables" in this last expression, we obtain

$$\frac{dv}{F(v) - v} = \frac{dx}{x} \tag{11}$$

Hence, regardless of the form of the function F in (9), we can always reduce a homogeneous DE to one in which the variables can be separated by the substitution $y = vx$ (or $x = vy$).

EXAMPLE 8

Solve

$$(x^4 + y^4)\,dx + 2x^3y\,dy = 0$$

Solution

We first note that $M(x, y) = x^4 + y^4$ and $N(x, y) = 2x^3y$ are both homogeneous functions of degree 4. That is, $M(tx, ty) = t^4(x^4 + y^4)$ and $N(tx, ty) = t^4(2x^3y)$. If we make the substitution $y = vx$, it follows from (10) that $dy = v\,dx + x\,dv$, and thus the DE becomes

$$(x^4 + x^4v^4)\,dx + 2x^4v(v\,dx + x\,dv) = 0$$

If $x \neq 0$, we can divide by x^4 to obtain

$$(1 + 2v^2 + v^4)\,dx + 2xv\,dv = 0$$

where we have simplified the algebra. Separating variables now leads to

$$\frac{dx}{x} + \frac{2v}{(v^2 + 1)^2}\,dv = 0$$

which upon integration yields

$$\log|x| - (v^2 + 1)^{-1} = C_1$$

Finally, making the replacement $v = y/x$, we obtain

$$\log|x| - \frac{x^2}{x^2 + y^2} = C_1, \qquad x \neq 0$$

■ ────────────
EXAMPLE 9

Solve

$$y^2\, dx + (y^2 - xy + x^2)\, dy = 0$$

Solution

We first observe that $M(x, y) = y^2$ and $N(x, y) = y^2 - xy + x^2$ are both homogeneous functions of degree 2. Since the coefficient of dx is algebraically simpler than that of dy, it may be easier in this case to use the substitution $x = vy$ rather than $y = vx$. Doing so, and performing some routine algebraic manipulations, we get

$$\frac{dv}{1 + v^2} + \frac{dy}{y} = 0, \qquad y \neq 0$$

The integration of this expression leads to

$$\tan^{-1} v + \log|y| = C_1$$

or, writing $v = x/y$,

$$\tan^{-1} \frac{x}{y} + \log|y| = C_1, \qquad y \neq 0 \qquad\qquad ■$$

Exercises 2.2

In Problems 1 to 30, obtain a family of solutions by separating the variables. If an initial condition is given, find a particular solution that satisfies it.

1. $2y\, dx = 3x\, dy$

2. $y' = xy^2$

3. $y' = -\dfrac{x}{y}$, $y(1) = 3$

4. $y' = \dfrac{x - 4}{x - 3}$, $y(4) = 0$

5. $y' = \dfrac{y^2 - 1}{x^2 - 1}$

6. $\sin x \sin y\, dx + \cos x \cos y\, dy = 0$

7. $\sec^2 x\, dy + \csc y\, dx = 0$

8. $x^2(1 + y^3)\, dx + y^2(1 + x^3)\, dy = 0$

9. $\dfrac{dP}{dt} = aP - bP^2$

10. $\dfrac{dV}{dP} = -\dfrac{V}{P}$, $V(2) = \dfrac{1}{2}$

11. $\dfrac{dN}{dt} + N = Nte^{t+2}$, $N(1) = 1$

■**12.** $e^x(y - 1)\, dx + 2(e^x + 4)\, dy = 0$

***13.** $(xy + x)\, dx - (x^2y^2 + x^2 + y^2 + 1)\, dy = 0$

14. $x \cos^2 y\, dx + \tan y\, dy = 0$

15. $(1 + \log x)\, dx + (1 + \log y)\, dy = 0$

16. $x^2 yy' = e^y$

17. $2yy' - 1 = 2x$, $y(0) = 3$

18. $y' = e^{2x + 3y}$

19. $(y + 1)\, dy - y \log x\, dx = 0$

■**20.** $dy = y\sqrt{1 - y^2}\, dx$, $y(5) = 1$

21. $v\dfrac{dv}{dt} = g$, $v(t_0) = v_0$ (g constant)

22. $y' = -2xy$, $y(0) = y_0$

23. $xyy' = 1 + y^2$, $y(2) = 3$

24. $x \sin x \, e^{-y} \, dx - y \, dy = 0$, $y(0) = 1$

25. $y' = \dfrac{2x}{1 + 2y}$, $y(2) = 0$

26. $(x^2 - 1)y' = 2xy \log y$

*27. $(y')^2 = y^3 - y^2$

*28. $x(y')^2 = y$

*29. $y(y')^2 = 1$

*30. $y(y')^2 = e^{2x}$

*31. Solve $xy' - y^2 + 1 = 0$, subject to

    (a) $y(0) = 1$
    (b) $y(0) = -1$
    (c) $y(0) = 2$

*32. Solve $y' - x^k(y^2 + 1) = 0$, when

    (a) $k \neq -1$
    (b) $k = -1$

In Problems 33 to 40, determine if the given function is homogeneous and, if so, state its degree.

33. $f(x, y) = x^2 + y^4$

34. $f(x, y) = x^2 - x^3/y^2$

35. $f(x, y) = \dfrac{2x^3y + 5x^2y^2}{x^4 + y^4}$

36. $f(x, y) = \sqrt{x^2 - y^2} \, (3x - 4)$

37. $f(x, y) = xe^{y/x}$

38. $f(x, y) = \tan\left(\dfrac{x^2}{x^2 + y^2}\right)$

39. $f(x, y) = x - y \log y + y \log x$

■40. $f(x, y) = \log x^3 - 3 \log y$

In Problems 41 to 50, verify that the equation has homogeneous coefficients and find a family of solutions. If an initial condition is given, find a particular solution that satisfies it.

41. $(x - y) \, dx + x \, dy = 0$

42. $y' = \dfrac{2x + y}{x + 2y}$

43. $xy' = y + xe^{y/x}$, $y(1) = 0$

44. $(x^2 + y^2) \, dx + (x^2 - xy) \, dy = 0$, $y(2) = 0$

45. $(u + v) \, du + (v - u) \, dv = 0$

■46. $x \, dx + (y - 2x) \, dy = 0$, $y(1) = 3$

47. $xy \, dx + (x^2 + y^2) \, dy = 0$

*48. $(s - t)(4s + t) \, ds + s(5s - t) \, dt = 0$

*49. $(x - y \log y + y \log x) \, dx$
             $+ x(\log y - \log x) \, dy = 0$

*50. $[x - y \tan^{-1}(y/x)] \, dx + x \tan^{-1}(y/x) \, dy = 0$

2.3 Exact Equations

Closely associated with the notion of a total or exact differential is a class of DEs known as *exact equations*.

Recall from the calculus that if $f(x, y)$ and its first partial derivatives are continuous functions of x and y in some domain of the xy plane, its **total differential** is defined by

$$df = \frac{\partial f}{\partial x} \, dx + \frac{\partial f}{\partial y} \, dy \qquad (12)$$

■ EXAMPLE 10

Find the total differential of $f(x, y) = x^2y^3$.

Solution

Computing partial derivatives, we have

$$\frac{\partial f}{\partial x} = 2xy^3, \qquad \frac{\partial f}{\partial y} = 3x^2y^2$$

and therefore the total differential is given by

$$df = 2xy^3\, dx + 3x^2y^2\, dy$$ ∎

Now, if it should happen that the left-hand side of the DE

$$M(x, y)\, dx + N(x, y)\, dy = 0 \tag{13}$$

is the total differential of some function $f(x, y)$, that is, if

$$df = M(x, y)\, dx + N(x, y)\, dy$$

we say that $M(x, y)\, dx + N(x, y)\, dy$ is an **exact differential** and write (13) as simply

$$df = 0 \tag{14}$$

If the total differential of a function is zero, that function is a *constant*. It follows from (14), therefore, that the solution of (13) in such cases is given by the implicit family of functions

$$f(x, y) = C_1 \tag{15}$$

where C_1 is any constant.

For example, if we consider the DE

$$2xy^3\, dx + 3x^2y^2\, dy = 0 \tag{16}$$

then, based on Example 10, the left-hand side of this equation is the exact differential of the function $f(x, y) = x^2y^3$. Hence, we deduce that (16) has the family of solutions

$$x^2y^3 = C_1 \tag{17}$$

where C_1 is an arbitrary constant.

DEs like (13) for which the left-hand side is the exact differential of some function $f(x, y)$ are called **exact equations.** Of course, it is not always easy to determine whether a given DE is exact or not. We need to develop some kind of test. The following theorem provides us with such a test, and the proof actually provides us with a scheme for finding the function $f(x, y)$ in those cases for which we have an exact equation.

Theorem 2.2 *If $M(x, y)$ and $N(x, y)$ are continuous functions and have continuous first partial derivatives in some rectangular domain of the xy plane, then*

$$\frac{\partial M}{\partial y} = \frac{\partial N}{\partial x}$$

is both a necessary and sufficient condition that

$$M(x, y)\, dx + N(x, y)\, dy = 0$$

be an exact equation.

Proof. *To prove the necessity part of the theorem, we assume that*

$$M(x, y)\, dx + N(x, y)\, dy = 0$$

is an exact equation and show that this leads to $\partial M/\partial y = \partial N/\partial x$. *That is, if the left-hand side of the above DE is an exact differential, there exists a function* $f(x, y)$ *such that*

$$M(x, y)\, dx + N(x, y)\, dy = \frac{\partial f}{\partial x}\, dx + \frac{\partial f}{\partial y}\, dy$$

Hence, it follows that

$$\frac{\partial f}{\partial x} = M(x, y), \qquad \frac{\partial f}{\partial y} = N(x, y)$$

Next, forming mixed partial derivatives of $f(x, y)$ *from these expressions, we see that*

$$\frac{\partial^2 f}{\partial y\, \partial x} = \frac{\partial M}{\partial y}, \qquad \frac{\partial^2 f}{\partial x\, \partial y} = \frac{\partial N}{\partial x}$$

but owing to the assumed continuity of these partial derivatives, we deduce that

$$\frac{\partial M}{\partial y} = \frac{\partial N}{\partial x}$$

    In the proof of the sufficiency part, we assume that $\partial M/\partial y = \partial N/\partial x$, *and show that this implies*

$$M(x, y)\, dx + N(x, y)\, dy = 0$$

is an exact equation. To do this, we must construct a function $f(x, y)$ *such that*

$$\frac{\partial f}{\partial x} = M(x, y), \qquad \frac{\partial f}{\partial y} = N(x, y)$$

If we formally integrate the first of these relations with respect to x, *we obtain*

$$f(x, y) = \int M(x, y)\, dx + g(y)$$

where the function g *is the "constant" of integration (constant with respect to* x). *The derivative of this last expression with respect to* y *yields*

$$\frac{\partial f}{\partial y} = \frac{\partial}{\partial y} \int M(x, y)\, dx + g'(y)$$

which we must now show equals $N(x, y)$ *in order to complete the proof. If we assume that* $\partial f/\partial y = P(x, y)$, *then* g *is a solution of the first-order DE*

$$g'(y) = P(x, y) - \frac{\partial}{\partial y} \int M(x, y)\, dx$$

To determine g *through a single integration with respect to* y, *the right-hand side of this expression must be independent of* x. *We can prove this by showing that the derivative of it with respect to x is zero if* $P(x, y) = N(x, y)$. *That is,*

$$\frac{\partial}{\partial x}\left[P(x, y) - \frac{\partial}{\partial y}\int M(x, y)\,dx\right] = \frac{\partial P}{\partial x} - \frac{\partial^2}{\partial x\,\partial y}\int M(x, y)\,dx$$

$$= \frac{\partial P}{\partial x} - \frac{\partial^2}{\partial y\,\partial x}\int M(x, y)\,dx$$

where in the last step we have interchanged the order of the mixed partial derivative. However, from calculus we recall

$$\frac{\partial}{\partial x}\int M(x, y)\,dx = M(x, y)$$

so that the above expression becomes

$$\frac{\partial}{\partial x}\left[P(x, y) - \frac{\partial}{\partial y}\int M(x, y)\,dx\right] = \frac{\partial P}{\partial x} - \frac{\partial M}{\partial y}$$

which vanishes if $P(x, y) = N(x, y)$. *Hence, the theorem is proved.*

Remark.   In the sufficiency proof of Theorem 2.2 we could just as easily have started with $\partial f/\partial y = N(x, y)$ and integrated with respect to y to find

$$f(x, y) = \int N(x, y)\,dy + h(x)$$

In this case, $h(x)$ is determined by solving

$$h'(x) = M(x, y) - \frac{\partial}{\partial x}\int N(x, y)\,dy$$

The significant conclusion from Theorem 2.2 is that if $\partial M/\partial y = \partial N/\partial x$ in some rectangular domain, the DE

$$M(x, y)\,dx + N(x, y)\,dy = 0$$

is an exact equation, but if $\partial M/\partial y \neq \partial N/\partial x$, the DE is not exact. Hence, the condition

$$\frac{\partial M}{\partial y} \overset{?}{=} \frac{\partial N}{\partial x} \tag{18}$$

is a **conclusive test for exactness.**

■————————
EXAMPLE 11

Show that the following DE is exact and find its family of solutions

$$(4x - 2y + 5)\,dx - (2x - 2y)\,dy = 0$$

Solution

We identify

$$M(x, y) = 4x - 2y + 5$$

$$N(x, y) = -(2x - 2y) = 2y - 2x$$

from which we calculate

$$\frac{\partial M}{\partial y} = -2 = \frac{\partial N}{\partial x}$$

Hence, the equation is exact. By Theorem 2.2, a function $f(x, y)$ exists such that

$$\frac{\partial f}{\partial x} = 4x - 2y + 5$$

$$\frac{\partial f}{\partial y} = 2y - 2x$$

Integrating the first of these expressions with respect to x gives us

$$f(x, y) = 2x^2 - 2xy + 5x + g(y)$$

where $g(y)$ is to be determined by use of the second relation. Thus, by differentiating this last expression with respect to y and setting the result equal to $N(x, y)$, we obtain

$$\frac{\partial f}{\partial y} = -2x + g'(y) = 2y - 2x$$

From this relation we see that $g'(y) = 2y$, or

$$g(y) = y^2$$

where a constant of integration is not required since one appears in the final solution function. Therefore, we have shown that

$$f(x, y) = 2x^2 - 2xy + 5x + y^2$$

and the solution we seek is (rearranging terms)

$$2x^2 + y^2 - 2xy + 5x = C_1 \qquad \blacksquare$$

> **Remark.** Observe that the solution in Example 11 is *not* given by $f(x, y) = 2x^2 + y^2 - 2xy + 5x$, but by $f(x, y) = C_1$, where C_1 is any constant.

EXAMPLE 12

Solve

$$(y^2 - 1) \, dx + (2xy - \sin y) \, dy = 0$$

Solution

We first test for exactness and find that

$$\frac{\partial M}{\partial y} = 2y = \frac{\partial N}{\partial x}$$

proving that the DE is exact. Thus, there is a function $f(x, y)$ such that

$$\frac{\partial f}{\partial x} = y^2 - 1$$

$$\frac{\partial f}{\partial y} = 2xy - \sin y$$

For the sake of variety, let us integrate the second of these equations this time, which yields

$$f(x, y) = xy^2 + \cos y + h(x)$$

Taking the derivative of this expression with respect to x and equating it to $M(x, y)$ leads to

$$\frac{\partial f}{\partial x} = y^2 + h'(x) = y^2 - 1$$

We see that $h'(x) = -1$, or $h(x) = -x$. Thus, given that $f(x, y) = xy^2 - x + \cos y$, our family of solutions is

$$xy^2 - x + \cos y = C_1 \qquad \blacksquare$$

2.3.1 Integrating Factors

Sometimes we find that while the DE of interest

$$M(x, y)\, dx + N(x, y)\, dy = 0 \qquad\qquad (19)$$

is not exact, it can be made so by multiplying it by a suitable function $\mu(x, y)$, called an **integrating factor.** The resulting exact DE is then of the form

$$\mu(x, y)M(x, y)\, dx + \mu(x, y)N(x, y)\, dy = 0 \qquad\qquad (20)$$

which we consider to be "essentially equivalent" to (19) in that it has the same family of solutions. Integrating factors are often found by "inspection," but in some instances may be systematically determined for a given DE. Regardless of the method, once the integrating factor is found, the problem can then be solved as an *exact equation.*

> *Remark.*   Integrating factors for a given DE are not uniquely determined. There may exist many integrating factors for the same DE. In addition, any integrating factor multiplied by a nonzero constant is also an integrating factor.

EXAMPLE 13

Solve

$$(3x + 2y)\, dx + x\, dy = 0$$

Solution

Testing the DE for exactness reveals that

$$\frac{\partial M}{\partial y} = 2, \qquad \frac{\partial N}{\partial x} = 1$$

and thus the equation is not exact. Although at this point we have no way of systematically determining an integrating factor, notice that if we multiply the DE by $\mu(x, y) = x$, the new equation

$$(3x^2 + 2xy)\, dx + x^2\, dy = 0$$

is exact, that is, $\partial M/\partial y = 2x = \partial N/\partial x$. Solving as an exact equation, it can easily be shown that $f(x, y) = x^3 + x^2 y$, and hence we have found the family of solutions

$$x^3 + x^2 y = C_1 \qquad \blacksquare$$

In some cases we may be able to actually calculate the integrating factor for a particular DE, but mostly we determine them by "recognizing certain groups as differentials of known expressions." The following differential relations may be helpful in this regard:

$$\frac{x\,dy - y\,dx}{x^2} = d(y/x) \qquad (21)$$

$$\frac{y\,dx - x\,dy}{y^2} = d(x/y) \qquad (22)$$

$$y\,dx + x\,dy = d(xy) \qquad (23)$$

$$\frac{y\,dx - x\,dy}{xy} = d\left(\log \frac{x}{y}\right) \qquad (24)$$

$$\frac{y\,dx - x\,dy}{x^2 + y^2} = d\left(\tan^{-1} \frac{x}{y}\right) \qquad (25)$$

$$\frac{2x\,dx + 2y\,dy}{x^2 + y^2} = d[\log(x^2 + y^2)] \qquad (26)$$

For instance, based on (21) it appears that $\mu(x, y) = 1/x^2$ is an integrating factor for any DE of the form

$$x\,dy - y\,dx + a(x)\,dx = 0$$

since the related DE

$$\frac{x\,dy - y\,dx}{x^2} + \frac{a(x)}{x^2}\,dx = d(y/x) + \frac{a(x)}{x^2}\,dx = 0$$

can be immediately integrated. Unfortunately, there are no general rules for finding integrating factors—the basic procedure is often one of trial and error, and educated guesses.[†]

EXAMPLE 14

Solve

$$y^2\,dx - x(x\,dy - y\,dx) = 0$$

Solution

According to Equations (21) and (22) above, the combination $x\,dy - y\,dx$ suggests division by either y^2 or x^2. Here division by y^2, accompanied by a division by x,

---

[†] Systematic procedures are available for certain types of DEs (e.g., see Problems 37, 38, and 41 in Exercises 2.3).

leads to

$$\frac{dx}{x} + \frac{y\,dx - x\,dy}{y^2} = 0 \qquad x \neq 0, \qquad y \neq 0$$

In this case the integrating factor is actually $\mu(x, y) = 1/xy^2$. Expressing this last equation in the differential form

$$d(\log|x|) + d(x/y) = 0$$

direct integration leads to the family of solutions

$$\log|x| + \frac{x}{y} = C_1 \qquad \blacksquare$$

Once $\mu(x, y)$ has been determined, it may be easier in some cases to solve the resulting DE by a method other than the method of exact equations in Section 2.3. Notice, for instance, that once we had grouped terms as exact differentials in Example 14 we were able to directly integrate the resulting exact equation.

Exercises 2.3

In Problems 1 to 20, test for exactness and, if exact, solve the equation. If an initial condition is prescribed, find a particular solution that satisfies it.

1. $(3x^2 - 6xy)\,dx - (3x^2 + 2y)\,dy = 0$

2. $(2xy - \cos x)\,dx + (x^2 - 1)\,dy = 0$

3. $2xt\,dx + (x^2 - 1)\,dt = 0$

4. $(2y^2 x - 3)\,dx + (2yx^2 + 4)\,dy = 0$

5. $(y - x^3)\,dx - x\,dy = 0$

6. $(x^4 - y)\,dx + (x^2 y^2 + x)\,dy = 0$

7. $\cos y\,dx - (x \sin y - y^2)\,dy = 0$

■8. $(x - 2xy + e^y)\,dx + (y - x^2 + xe^y)\,dy = 0$

9. $(\cos \theta \sin \theta - \theta r^2)\,d\theta + r(1 - \theta^2)\,dr = 0$

10. $(1 + \log x + y/x)\,dx - (1 - \log x)\,dy = 0$

11. $(2u - e^{3v})\,du - 3(ue^{3v} - \cos 3v)\,dv = 0$

12. $3x(xy - 2)\,dx + (x^3 + 2y)\,dy = 0$

13. $(\tan x - \sin x \sin y)\,dx + \cos x \cos y\,dy = 0$

*14. $[x^{-1} + x^{-2} - y(x^2 + y^2)^{-1}]\,dx$
$$+ [ye^{-y} + x(x^2 + y^2)^{-1}]\,dy = 0$$

15. $(x + y)\,dx + (y - x)\,dy = 0, \; y(1) = 1$

■16. $x\,dx + [y - y^2(x^2 + y^2)]\,dy = 0, \; y(1) = 1$

17. $(3x^2 y + 8xy^2)\,dx + (x^3 + 8x^2 y$
$$+ 12y^2)\,dy = 0, \; y(0) = 1$$

18. $2xy\,dx + (x^2 + y^2)\,dy = 0, \; y(1) = 2$

19. $\sin x \cos y\,dx + \cos x \sin y\,dy = 0,$
$y(\pi/4) = \pi/4$

*20. $(y^2 e^{xy^2} + 4x^3)\,dx + (2xye^{xy^2} - 3y^2)\,dy = 0,$
$y(1) = 0$

In Problems 21 to 24, determine the constant B such that the equation is exact, and then solve the resulting exact equation.

21. $3y(x^2 + y)\,dx + x(x^2 + By)\,dy = 0$

22. $2(x^2 + y)\,dy + (3x^2 + Bxy)\,dx = 0$

23. $(Bx \cos y + 3x^2 y)\,dx$
$$- (y + x^2 \sin y - x^3)\,dy = 0$$

■24. $(2Bxy - y^3)\,dx + (4y + 3x^2 - Bxy^2)\,dy = 0$

In Problems 25 to 29, use the suggested integrating factor to solve the DE.

25. $2y(y - 1) dx + x(2y - 1) dy = 0$; $\mu(x, y) = x$

26. $(x + y) dx + x \log x \, dy = 0$; $\mu(x, y) = 1/x$

27. $y(x + y + 1) dx + (x + 2y) dy = 0$;
$\mu(x, y) = e^x$

*28. $(x^2 + 2xy - y^2) dx + (y^2 + 2xy - x^2) dy = 0$;
$\mu(x, y) = 1/(x^2 + y^2)$

29. $(x^2 + y^2 - x) dx - y \, dy = 0$;
$\mu(x, y) = 1/(x^2 + y^2)$

In Problems 30 to 36, find an integrating factor by inspection and solve the DE.

30. $x \, dx + y \, dy = y^2(x^2 + y^2) \, dy$

31. $y \, dx - x \, dy = x^3 dx$

■32. $(x + 3y) dx - (y - 3x) dy = 0$

33. $(x^4 - y) dx + (x^2 y^2 + x) dy = 0$

*34. $(x^3 y + 3x^2 + y) dx + (2xy^4 + 6x^3 + x) dy = 0$

*35. $(3x^5 y^4 + 4y) dx + (2x^6 y^3 + 3x) dy = 0$
Hint. Multiply the DE by $x^3 y^2$.

36. $(x + y) dx + (y - x) dy = 0$

In Problems 37 to 42, assume that the DE is $M(x, y) dx + N(x, y) dy = 0$.

*37. Prove that $\mu(x) = \exp[\int f(x) \, dx]$ is an integrating factor provided that

$$\frac{1}{N(x, y)} \left(\frac{\partial M}{\partial y} - \frac{\partial N}{\partial x} \right) = f(x)$$

*38. Prove that $\mu(y) = \exp[\int g(y) \, dy]$ is an integrating factor provided that

$$\frac{1}{M(x, y)} \left(\frac{\partial N}{\partial x} - \frac{\partial M}{\partial y} \right) = g(y)$$

*39. Use the results of Problems 37 and 38 to solve the given DE:

(a) $y \, dx + 4x \, dy = 0$
(b) $2xy \, dx + (y^2 - x^2) \, dy = 0$
(c) $y^2 \cos x \, dx + (3 + 4y \sin x) \, dy = 0$

*40. Prove that $\mu(x, y)$ is an integrating factor provided that

$$M(x, y) \frac{\partial \mu}{\partial y} - N(x, y) \frac{\partial \mu}{\partial x}$$
$$+ \left(\frac{\partial M}{\partial y} - \frac{\partial N}{\partial x} \right) \mu(x, y) = 0$$

*41. Using the result of Problem 40, show that $\mu(x, y) = x^p y^q$ is an integrating factor if p and q satisfy the relation

$$pyN(x, y) - qxM(x, y) = xy \left(\frac{\partial M}{\partial y} - \frac{\partial N}{\partial x} \right)$$

*42. Use the result of Problem 41 to obtain an integrating factor of the form $\mu(x, y) = x^p y^q$ and solve the given DE:

(a) $y(xy + 1) dx - x \, dy = 0$
(b) $(x + 2y) dx - x \, dy = 0$
(c) $2y \, dx + 3x \, dy = 3x^{-1} \, dy$

2.4 Linear Equations—Part I

In Chapter 1 we defined a **linear first-order DE** as one having the form

$$A_1(x)y' + A_0(x)y = F(x) \tag{27}$$

The functions $A_0(x)$ and $A_1(x)$ are called the **coefficients** of the DE, and $F(x)$ is the **input function.** In applications, the solution of (27) is usually required to satisfy the auxiliary condition

$$y(x_0) = y_0 \tag{28}$$

where x_0 is a point on I and y_0 is a specified constant. The point x_0 is usually the initial point in the interval I, and thus (28) is called an **initial condition.**

For developing the general theory of linear DEs, it is customary to put (27) in the **normal form**

$$y' + a_0(x)y = f(x) \tag{29}$$

obtained by dividing each term of (27) by $A_1(x)$. Hence, it follows that $a_0(x) = A_0(x)/A_1(x)$ and $f(x) = F(x)/A_1(x)$. Rearranging (29) in the form $y' = F(x, y)$, we obtain

$$y' = f(x) - a_0(x)y$$

Clearly, $F(x, y) = f(x) - a_0(x)y$ and $\partial F/\partial y = -a_0(x)$ are continuous functions in every rectangular domain containing an interval I along the x axis for which $a_0(x)$ and $f(x)$ are continuous. If x_0 lies in the interval I, it follows from Theorem 2.1 that the IVP consisting of (28) and (29) has a *unique solution* on some subinterval of I containing x_0. However, we will soon state a more comprehensive *existence-uniqueness* theorem (Theorem 2.3) concerning the interval of uniqueness for linear DEs. Moreover, unlike nonlinear DEs, we can derive a *solution formula* for the general solution of all first-order linear DEs.

There are essentially two methods of solution of first-order linear DEs, one of which involves the use of *integrating factors* (see Section 2.4.1). Since the integrating factor technique is restricted primarily to *first-order* DEs, a more encompassing technique is also discussed (in Section 2.5) which can be extended to *higher-order* linear DEs (see Chapter 4).

2.4.1 Method of Integrating Factors

To solve (29) by means of an integrating factor, we first express it in differential form

$$[a_0(x)y - f(x)] \, dx + dy = 0 \tag{30}$$

Testing (30) for exactness, we find that $\partial M/\partial y = \partial N/\partial x$ only when $a_0(x) \equiv 0$. Nonetheless, we can make it exact in all cases if we multiply it by a suitable integrating factor $\mu(x)$, which depends only on x. Doing so gives us

$$[\mu(x)a_0(x)y - \mu(x)f(x)] \, dx + \mu(x) \, dy = 0 \tag{31}$$

where we now find

$$\frac{\partial M}{\partial y} = \mu(x)a_0(x), \qquad \frac{\partial N}{\partial x} = \mu'(x)$$

For exactness, therefore, we select $\mu(x)$ such that

$$\mu'(x) = a_0(x)\mu(x) \tag{32}$$

By assuming $\mu(x) > 0$ and separating variables in (32), it follows that one possible solution is (ignoring the constant of integration)

$$\log \mu(x) = \int a_0(x) \, dx$$

which, through exponentiation, leads to the explicit form

$$\mu(x) = \exp\left[\int a_0(x)\,dx\right] \tag{33}$$

where $\exp(x) = e^x$.

Notice that the integrating factor $\mu(x)$ does not depend upon the function $f(x)$ in (30) but merely upon the coefficient $a_0(x)$. If $a_0(x)$ is a continuous function on I, it follows that $\mu(x) \neq 0$ on the interval I and is a continuous and differentiable function. Of course, any constant multiple of $\mu(x)$ given by (33) also serves as an integrating factor.

At this point we find there is a more direct method of solution than that illustrated in Section 2.3. That is, returning to the normal form (29) multiplied by $\mu(x)$, we have

$$\mu(x)y' + \mu(x)a_0(x)y = \mu(x)f(x)$$

which, by the use of (32), we can write as

$$\mu(x)y' + \mu'(x)y = \mu(x)f(x)$$

The left-hand side of this last equation is simply the *product rule* for differentiating $\mu(x)y$. Therefore, we can write it in the more suggestive form

$$\frac{d}{dx}\left[\mu(x)y\right] = \mu(x)f(x)$$

the direct integration of which leads to

$$\mu(x)y = \int \mu(x)f(x)\,dx + C_1 \tag{34}$$

where C_1 is a constant of integration. Finally, solving for y, we obtain the *family of solutions* of Equation (29) given by

$$y = \frac{1}{\mu(x)}\left[\int \mu(x)f(x)\,dx + C_1\right] \tag{35}$$

where $\mu(x)$ is defined by (33).

If $a_0(x)$ and $f(x)$ are continuous functions on some open interval I, then the integrals in (33) and (35) must exist. Thus, if Equation (29) has a solution, it must be given by (35). Moreover, given a point x_0 on I, there is a unique value of C_1 such that $y(x_0) = x_0$. Hence, *all* solutions are necessarily contained in (35); that is, (35) is a **general solution** of (29). In summary, we have the following existence-uniqueness theorem.

Theorem 2.3 Existence-uniqueness. *If $a_0(x)$ and $f(x)$ are continuous functions on the open interval I containing the point x_0, then the IVP*

$$y' + a_0(x)y = f(x), \qquad y(x_0) = y_0$$

has a unique solution on I given by Equation (35) with an appropriate choice of constant C_1.

■────────── Solve
EXAMPLE 15

$$xy' - 5y = 3x^6 e^x$$

Solution

First putting the DE in normal form [see Equation (29)], we get

$$y' - \frac{5}{x} y = 3x^5 e^x, \qquad x \neq 0$$

Next, by using (33) we determine the integrating factor

$$\mu(x) = \exp\left(-5 \int \frac{dx}{x}\right) = e^{-5 \log x} = x^{-5}$$

Finally, recognizing that $f(x) = 3x^5 e^x$, we can use (34) to find

$$x^{-5} y = \int x^{-5} (3x^5 e^x) \, dx + C_1$$

$$= 3 \int e^x \, dx + C_1$$

or, upon simplification,

$$y = x^5 (3e^x + C_1) \qquad\qquad ■$$

In Example 15 the functions $f(x)$ and $a_0(x)$ are continuous everywhere except at $x = 0$. Therefore, if we impose an initial condition $y(x_0) = y_0$, $x_0 \neq 0$, a *unique solution* will exist.

> **Remark.** A solution may or may not exist if we impose an initial condition at a point where $a_0(x)$ [and/or $f(x)$] is not defined. For instance, the problem in Example 15 has infinitely many solutions if we prescribe the condition $y(0) = y_0 = 0$, but no solution if $y_0 \neq 0$.

■────────── Solve the IVP
EXAMPLE 16

$$y' + 2xy = x, \qquad y(0) = 3$$

Solution

Because the DE is already in normal form, we can immediately calculate

$$\mu(x) = \exp\left(2 \int x \, dx\right) = e^{x^2}$$

Thus,

$$e^{x^2}y = \int xe^{x^2}\,dx + C_1$$

$$= \frac{1}{2}e^{x^2} + C_1$$

which simplifies to

$$y = \tfrac{1}{2} + C_1 e^{-x^2}$$

Because $a_0(x) = 2x$ and $f(x) = x$ are continuous functions everywhere, we conclude that the above family of solutions is the general solution on every interval.

Finally, setting $x = 0$ in the above general solution leads to

$$y(0) = \tfrac{1}{2} + C_1 = 3$$

from which we deduce $C_1 = \tfrac{5}{2}$. Hence, the unique solution we seek is

$$y = \tfrac{1}{2}(1 + 5e^{-x^2}) \qquad \blacksquare$$

When checking a given first-order DE for linearity, we should consider both variables since either variable may be considered the independent variable. That is, if the DE is not linear in one variable, it may be so in the other variable. Consider the following example.

EXAMPLE 17

Find the general solution of

$$y^2\,dx + (4xy + 1)\,dy = 0$$

Solution

Because of the term y^2, the DE is clearly *nonlinear in y*. However, if we rewrite the DE in the form

$$\frac{dx}{dy} + \frac{4}{y}x = -\frac{1}{y^2}, \qquad y \neq 0$$

we see that it is *linear in x*. Thus, we find

$$\mu(y) = \exp\!\left(4\int \frac{dy}{y}\right) = y^4$$

and the general solution assumes the form

$$y^4 x = C_1 + \int y^4\left(-\frac{1}{y^2}\right)dy$$

$$= C_1 - \int y^2\,dy$$

which reduces to

$$x = C_1 y^{-4} - \tfrac{1}{3}y^{-1}, \qquad y \neq 0 \qquad \blacksquare$$

Exercises 2.4

1. Show that

(a) $e^{-\log x} = \dfrac{1}{x}, x > 0$

(b) $e^{(1/2)\log x} = \sqrt{x}, x > 0$

(c) $e^{-\log|\sec x|} = |\cos x|$

(d) $e^{-(x^2 + 2\log x)} = \dfrac{e^{-x^2}}{x^2}, x > 0$

2. Find an integrating factor for the particular DE

$$y' + ay = f(x)$$

where a is a constant.

In Problems 3 to 30, find the general solution of the given DE by the method of integrating factors.

3. $y' + 2y = 0$

4. $y' + (\cos x)y = 0$

5. $4xy^3\, dy + dx = 0$

■**6.** $(x^2 + 9)\, dy + xy\, dx = 0$

7. $y' + 2xy = xe^{-x^2}$

8. $y' + (\cos x)y = e^{-\sin x}$

9. $x\, dy + y\, dx = (x \sin x)\, dx$

10. $y^2\, dx + x\, dy = 5\, dy$

11. $(y^2 + 1)\, dx + xy\, dy = dy$

*****12.** $(x^2 - 1)y' + 2xy = \cos x$

13. $y' = 1 - 5y$

14. $y' - 2y = 8x$

15. $xy' + y = x \sin x$

16. $xy' - y = x^2 \sin x$

17. $xy' = 4y + x^6 e^x$

■**18.** $(x^3 + 3y)\, dx - x\, dy = 0$

19. $z' = x - 4xz$

■**20.** $dy/dx = \csc x + y \cot x$

21. $dy/dx = \csc x - y \cot x$

22. $u\, dt + (3t - tu + 2)\, du = 0$

23. $(x^2 + 9)y' + xy = x$

24. $(1 + e^y)(dx/dy) + e^y x = y$

25. $y' - my = e^{mx}$

26. $y' - my = e^{kx}\ (k \neq m)$

27. $x^2 y' + x(x + 2)y = e^x$

*****28.** $(1 + x)y' - xy = x + x^2$

29. $xy' - ay = bx^k\ (x > 0, a \neq k)$

30. $xy' - ky = bx^k\ (x > 0)$

In Problems 31 to 40, find a solution of the given IVP and state the interval for which the solution is valid.

31. $y' + (\cos x)y = 0,\ y(0) = 2$

32. $y' + (\sin x)y = 0,\ y(\pi/2) = 1$

33. $L(di/dt) + Ri = E,\ i(0) = 0$
(L, R, and E constant)

34. $L(di/dt) + Ri = A \sin \omega t,\ i(0) = 0$
(L, R, and A constant)

35. $(2x + 3)y' - y = \sqrt{2x + 3},\ y(-1) = 0$

■**36.** $xy' + 2y = \sin x,\ y(\pi/2) = 1$

37. $y' - y = 2xe^{2x},\ y(0) = 1$

*****38.** $y' = x^3 - 2xy,\ y(1) = 1$

39. $x\, dy - y\, dx = x^2 e^x\, dx,\ y(1) = 0$

40. $y' = y/(y - x),\ y(2) = 3$

2.5 Linear Equations—Part II

The integrating factor method presented in Section 2.4 is restricted to first-order DEs and, as such, does not provide much insight into the general theory of linear DEs of order greater than 1. In this section we introduce terminology and techniques of solving first-order linear DEs that carry over to linear DEs of all orders, and thus can be used for motivation in developing the corresponding theory of higher-order linear DEs in Chapter 4.

A first-order linear DE in *normal form* is given by

$$y' + a_0(x)y = f(x) \tag{36}$$

If $f(x) \equiv 0$ on an interval I, we say the DE is **homogeneous;**[†] otherwise the equation is said to be **nonhomogeneous.** In solving (36) without the aid of an integrating factor, we first develop a general solution of the associated *homogeneous* equation that results by setting $f(x) \equiv 0$. Having done so, we then use this solution to help develop a general solution of the *nonhomogeneous* equation (36).

2.5.1 Homogeneous Equations

The **homogeneous equation** associated with (36) is

$$y' + a_0(x)y = 0 \tag{37}$$

One feature of *homogeneous* DEs in general is that $y = 0$ is a solution, called the **trivial solution.** However, we are mostly interested in **nontrivial solutions** (i.e., nonzero solutions).

To begin our search for nontrivial solutions, let us consider the special case of (37) for which $a_0(x)$ is constant, that is,

$$y' + ay = 0 \tag{38}$$

where a is a constant. By inspection, we see that $y = e^{-ax}$ is a solution of (38), as is

$$y = C_1 e^{-ax} \tag{39}$$

for any value of C_1. When $a_0(x)$ is not constant, we can solve (37) by separation of variables (see Section 2.2), which leads to

$$\frac{dy}{y} = -a_0(x)\, dx, \qquad y \neq 0$$

The direct integration of this expression then yields

$$\log|y| = -\int a_0(x)\, dx + C$$

where C is a constant of integration. Finally, solving for y, we obtain the explicit *family of solutions*

$$y = C_1 \exp\left[-\int a_0(x)\, dx \right] \tag{40}$$

where $C_1 = e^C$. It is customary to define the solution function

$$y_1(x) = \exp\left[-\int a_0(x)\, dx \right] \tag{41}$$

[†] As used here, the term "homogeneous" does not have the same meaning as in Section 2.2.1.

called the **fundamental solution,** and then (40) becomes simply

$$y = C_1 y_1(x) \tag{42}$$

> **Remark.** Notice that the fundamental solution $y_1(x)$ given by (41) is simply the reciprocal of the integrating factor $\mu(x)$ defined by (33), that is, $y_1(x) = 1/\mu(x)$.

If $a_0(x)$ is continuous on an open interval I, then the integral defining the fundamental solution y_1 *exists* on I. Moreover, y_1 represents the only nontrivial solution (to within a multiplicative constant) of the homogeneous equation (37). Thus, it is proper to call (42) a **general solution** of (37) since, by specializing the constant C_1, it contains all solutions, including the trivial solution ($y = 0$) arising when $C_1 = 0$.

■───────
EXAMPLE 18

Find a general solution of

$$y' + 2xy = 0$$

Solution

Separating the variables, we find

$$\frac{dy}{y} = -2x \, dx$$

integration of which yields

$$\log|y| = -x^2 + C$$

Solving for y, the solution can now be put in the simpler form

$$y = C_1 e^{-x^2}$$

where $C_1 = e^C$.   ■

■───────
EXAMPLE 19

Solve the IVP

$$xy' + y = 0, \qquad y(1) = 3$$

Solution

Separating variables once again, we obtain

$$\frac{dy}{y} + \frac{dx}{x} = 0, \qquad x \neq 0, \qquad y \neq 0$$

Thus, through routine integration it follows that

$$\log|y| + \log|x| = \log|C_1|$$

where we have expressed the constant as $\log|C_1|$ for convenience. Co. rithms and exponentiating the result leads to

$$yx = C_1$$

or

$$y = \frac{C_1}{x}, \qquad x \neq 0$$

By imposing the initial condition $y(1) = 3$, we obtain

$$y(1) = C_1 = 3$$

and hence the solution we seek is

$$y = \frac{3}{x}, \qquad x \neq 0$$

which is valid on any interval not containing $x = 0$. ∎

2.5.2 Nonhomogeneous Equations

We now consider the **nonhomogeneous equation**

$$y' + a_0(x)y = f(x) \tag{43}$$

Every solution of (43) is called a **particular solution.** Let us suppose that $y = y_P(x)$ is a particular solution of (43). Then, if $y = y_H(x) = C_1 y_1(x)$ denotes the general solution of the associated homogeneous equation

$$y' + a_0(x)y = 0$$

the linear combination of solutions

$$y = y_H(x) + y_P(x) = C_1 y_1(x) + y_P(x) \tag{44}$$

is also a solution of (43). Moreover, (44) is a **general solution** of (43). To see this, let us assume that Y is any solution of (43) and define $y = Y - y_P$. We then show that Y is contained in (44). The direct substitution of $y = Y - y_P$ into (43) yields

$$
\begin{aligned}
y' + a_0(x)y &= (Y - y_P)' + a_0(x)(Y - y_P) \\
&= [Y' + a_0(x)Y] - [y_P' + a_0(x)y_P] \\
&= f(x) - f(x) \\
&= 0
\end{aligned}
$$

Hence, the function $y = Y - y_P$ is a solution of the associated *homogeneous* equation and, as such, is contained in y_H, that is,

$$y = Y - y_P = y_H$$

It now follows that $Y = y_H + y_P$, and therefore belongs to the general family of solutions described by (44). Since Y could be any solution of (43), we conclude that (44) is a general solution of (43).

EXAMPLE 20   Verify that

$$y_P = -2x - 1$$

$$z_P = e^{2x} - 2x - 1$$

are both particular solutions of

$$y' - 2y = 4x$$

and find a general solution in each case.

Solution

For the function y_P, we have

$$y'_P - 2y_P = -2 + 4x + 2$$

$$= 4x$$

and thus conclude that it is a particular solution of the DE. Similarly, we see that

$$z'_P - 2z_P = 2e^{2x} - 2 - 2e^{2x} + 4x + 2$$

$$= 4x$$

so it is also a particular solution.

The associated homogeneous equation has the general solution $y_H = C_1 e^{2x}$. Therefore, the two general solutions of the nonhomogeneous DE are given by

$$y = y_H + y_P$$

$$= C_1 e^{2x} - 2x - 1$$

and

$$y = y_H + z_P$$

$$= C_1 e^{2x} + e^{2x} - 2x - 1$$

However, in the second case we note that

$$y = (C_1 + 1)e^{2x} - 2x - 1$$

$$= C_2 e^{2x} - 2x - 1$$

and so the two general solutions are actually equivalent.   ■

> **Remark.** As in the case of Example 20, the difference of two particular solutions of a linear DE is always a solution of the associated homogeneous DE.

To calculate a particular solution, we employ a method called **variation of parameters.** That is, we assume a particular solution of (43) exists that has the form

$$y_P = u(x)y_1(x) \tag{45}$$

where $u(x)$ is a function to be determined. The technique derives its name from the fact that the arbitrary constant in the homogeneous solution $y_H = C_1 y_1(x)$ is replaced by the unknown function $u(x)$.

The substitution of (45) into the left-hand side of the nonhomogeneous DE

$$y' + a_0(x)y = f(x)$$

leads to

$$y_P' + a_0(x)y_P = \frac{d}{dx}\left[u(x)y_1(x)\right] + a_0(x)u(x)y_1(x)$$

$$= u'(x)y_1(x) + u(x)\left[y_1'(x) + a_0(x)y_1(x)\right]$$

$$= u'(x)y_1(x) + 0$$

where we are using the fact that y_1 satisfies the associated homogeneous DE, that is, $y_1' + a_0(x)y_1 = 0$. Now, if y_P is indeed a particular solution of (43), then clearly the function $u(x)$ must be chosen such that

$$u'(x)y_1(x) = f(x) \tag{46}$$

or, upon integration,

$$u(x) = \int \frac{f(x)}{y_1(x)}\,dx + C \tag{47}$$

Technically, any constant C can appear in (47) and the resulting function $u(x)$ will lead to a particular solution of (43) (remember that particular solutions are not unique). It is customary, however, to set $C = 0$ and write simply

$$y_P = y_1(x)\int \frac{f(x)}{y_1(x)}\,dx \tag{48}$$

Finally, combining the particular solution (48) with the homogeneous solution $y_H = C_1 y_1(x)$ according to (44), we obtain the *general solution* of the nonhomogeneous DE (43) given by

$$y = C_1 y_1(x) + y_1(x)\int \frac{f(x)}{y_1(x)}\,dx \tag{49}$$

Remark.   Observe that, because $y_1(x) = 1/\mu(x)$, the solution formula (49) is the same as (35) found by the integrating factor method.

EXAMPLE 21

Find a general solution of

$$xy' + (1 - x)y = xe^x$$

Solution

We first rewrite the DE in normal form, which yields

$$y' + \left(\frac{1}{x} - 1\right)y = e^x, \qquad x \neq 0$$

Thus, using (41) with $a_0(x) = 1/x - 1$, we find that

$$y_1(x) = \exp\left[-\int\left(\frac{1}{x} - 1\right)dx\right] = \frac{1}{x}e^x$$

and, consequently, the homogeneous solution is

$$y_H = \frac{C_1 e^x}{x}$$

From the normal form, we identify $f(x) = e^x$. Hence, the particular solution is $y_P = u(x)e^x/x$, where

$$u(x) = \int \frac{e^x}{x^{-1}e^x}\,dx = \int x\,dx = \frac{x^2}{2}$$

This leads to

$$y_P = \tfrac{1}{2}xe^x$$

and our general solution is

$$y = y_H + y_P = \left(C_1\frac{1}{x} + \frac{1}{2}x\right)e^x, \qquad x \neq 0 \qquad\blacksquare$$

■──────
EXAMPLE 22

Find the solution of

$$\frac{dy}{dx} = \frac{1}{x + y^2}$$

which passes through the point $(-3, 0)$ of the xy plane.

Solution

The DE is not linear in y because of the presence of the term y^2. However, if we invert the DE, we get

$$\frac{dx}{dy} = x + y^2$$

or

$$\frac{dx}{dy} - x = y^2$$

which is linear in x. Hence, interchanging the roles of x and y in our previous results, we have

$$x_1(y) = \exp\left(\int dy\right) = e^y$$

which leads to $x_H = C_1 e^y$. Also,

$$x_P = e^y \int y^2 e^{-y}\,dy$$

$$= -y^2 - 2y - 2$$

which we obtain through two integrations by parts. Thus, the general solution is

$$x(y) = C_1 e^y - y^2 - 2y - 2$$

The prescribed initial condition $x(0) = -3$ is satisfied by setting $C_1 = -1$, and hence we finally obtain

$$x(y) = -(e^y + y^2 + 2y + 2) \qquad \blacksquare$$

Given that $a_0(x)$ and $f(x)$ are continuous functions on an interval I containing x_0, we know from Theorem 2.3 that the IVP

$$y' + a_0(x)y = f(x), \qquad y(x_0) = y_0 \tag{50}$$

has a unique solution on I. However, in practice these functions are not always continuous. For example, the function $f(x)$, which represents an external input function in applications, may exhibit a finite discontinuity at some point owing to, say, a switch being (instantaneously) turned on or off. Theorem 2.3 does not apply in problems such as these. Nonetheless, we may still be able to find a unique solution to the problem. Consider the following example.

EXAMPLE 23

Solve the IVP

$$y' + y = f(x), \qquad y(0) = 1$$

where

$$f(x) = \begin{cases} 0, & 0 \le x \le 1 \\ 1, & x > 1 \end{cases}$$

Solution

Here we see that the input function $f(x)$ has a finite jump discontinuity at $x = 1$. In this case we solve the problem in two parts, that is,

$$y' + y = 0, \qquad y(0) = 1, \qquad 0 \le x \le 1$$

and

$$y' + y = 1, \qquad x > 1$$

The solution of the first problem is readily found to be

$$y = e^{-x}, \qquad 0 \le x \le 1$$

while for the second problem we have

$$y = 1 + C_1 e^{-x}, \qquad x > 1$$

where C_1 is a constant to be determined. In order that the solution y may be a continuous function, we must select C_1 so that these two solutions agree at $x = 1$. That is, we require

$$\lim_{x \to 1^+} (1 + C_1 e^{-x}) = \lim_{x \to 1^-} e^{-x}$$

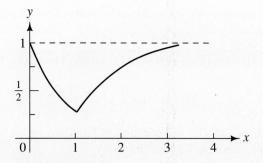

Figure 2.1

which leads to $C_1 = 1 - e$. Thus, our solution takes the form

$$y = \begin{cases} e^{-x}, & 0 \le x \le 1 \\ 1 + (1 - e)e^{-x}, & x > 1 \end{cases}$$

Notice that, while this solution is continuous for all $x \ge 0$, it is *not differentiable* at $x = 1$ (see Figure 2.1). Nevertheless, this is the best that we can do in producing a solution to this problem. This puzzling result is due to the "instantaneous" jump in the function $f(x)$ at $x = 1$. Realistically, such functions cannot exist, but it is mathematically convenient to model such jumps as occurring at a single point. The consequence of this model is a solution that is not everywhere differentiable.   ∎

[0] 2.5.3 Physical Interpretations

In most applications an initial condition is prescribed along with the DE, leading to an IVP of the form

$$y' + a_0(x)y = f(x), \qquad y(x_0) = y_0 \tag{51}$$

For discussion purposes we assume that $a_0(x)$ and $f(x)$ are continuous functions for all $x \ge x_0$. In order to derive a solution formula for (51) not involving an arbitrary constant, and also to anticipate physical interpretations, it is advantageous to select a "special" particular solution function y_P (recall that this function is not unique). For example, if we choose to write y_P in the form

$$y_P = y_1(x) \int_{x_0}^x \frac{f(t)}{y_1(t)} \, dt^\dagger \tag{52}$$

---

† It is customary to introduce a dummy variable of integration when one or both limits of integration are variable.

then at $x = x_0$ we have

$$y_P(x_0) = y_1(x_0) \int_{x_0}^{x_0} \frac{f(t)}{y_1(t)} \, dt = 0$$

Thus, y_P can be thought of as the solution of the IVP

$$y' + a_0(x)y = f(x), \qquad y(x_0) = 0 \tag{53}$$

wherein the initial condition is *homogeneous*. The physical interpretation of y_P, therefore, is that it *represents the response of a system in equilibrium* $(y = 0)$ *for* $x < x_0$, *but for* $x \geq x_0$ *is then subject to the external stimulus* $f(x)$. We previously referred to $f(x)$ as an **input function,** but because it often represents an external force in applications, it is also called a **forcing function.**

By selecting y_P as described by (52), the general solution of the DE in (51) takes the form

$$y = C_1 y_1(x) + y_1(x) \int_{x_0}^{x} \frac{f(t)}{y_1(t)} \, dt$$

If we now impose the initial condition $y(x_0) = y_0$, we see that

$$y(x_0) = C_1 y_1(x_0) + 0 = y_0$$

Hence, we deduce that

$$C_1 = \frac{y_0}{y_1(x_0)}, \qquad y_1(x_0) \neq 0 \tag{54}$$

and the solution formula for the IVP (51) that we seek is given by

$$y = \frac{y_0 y_1(x)}{y_1(x_0)} + y_1(x) \int_{x_0}^{x} \frac{f(t)}{y_1(t)} \, dt \tag{55}$$

Remark. It may be interesting to note that the solution formula (55) for the IVP (51) depends only upon the solution function $y_1(x)$ of the associated homogeneous DE and the input parameters y_0 and $f(x)$. This is a fundamental characteristic of all linear nonhomogeneous DEs and is the primary reason why so much effort in succeeding chapters on linear DEs is directed at finding solutions of homogeneous DEs.

Our choice of y_P, given by (52), has led to an important physical interpretation of this solution function, while at the same time requiring us to select the homogeneous solution function y_H to be a solution of the IVP

$$y' + a_0(x)y = 0, \qquad y(x_0) = y_0 \tag{56}$$

Physically, this suggests that $y_H = y_0 y_1(x)/y_1(x_0)$ *describes the behavior of* y *due entirely to the initial condition in the absence of an input function* $f(x)$.

■────────────
EXAMPLE 24

Solve the IVP

$$y' - 2xy = 1, \qquad y(0) = 3$$

by finding y_H and y_P as described above.

Solution

This time we first split the problem into two problems, namely,

$$y' - 2xy = 0, \qquad y(0) = 3$$

and

$$y' - 2xy = 1, \qquad y(0) = 0$$

and solve respectively for y_H and y_P.

The general solution of the homogeneous DE in the first problem is

$$y_H = C_1 e^{x^2}$$

and by imposing the initial condition $y(0) = 3$, we see that $C_1 = 3$; thus,

$$y_H = 3e^{x^2}$$

On the other hand, the nonhomogeneous DE has the particular solution [now using (52)]

$$y_P = e^{x^2} \int_0^x e^{-t^2} \, dt$$

but this integral is not an elementary integral. When this happens, we generally leave the solution in integral form and write the solution of the IVP as

$$y = y_H + y_P$$
$$= 3e^{x^2} + e^{x^2} \int_0^x e^{-t^2} \, dt$$

If numerical values of y are required, they can be obtained by numerical integration techniques. In this instance, however, the integral in the above solution is related to a *special function* known as the **error function,** which has been extensively tabulated. This function is defined by[†]

$$\mathrm{erf}(x) = \frac{2}{\sqrt{\pi}} \int_0^x e^{-t^2} \, dt$$

so that we may express the above solution of our IVP as

$$y = \left[3 + \frac{\sqrt{\pi}}{2} \, \mathrm{erf}(x) \right] e^{x^2} \qquad\qquad ■$$

───────────

[†] For a further discussion of the error function, see Chapter 3 in L. C. Andrews, *Special Functions for Engineers and Applied Mathematicians,* Macmillan, New York, 1985.

Example 24 illustrates that even simple-looking DEs may lead to nonelementary functions such as the error function. There are numerous other special functions of this nature that may occur in the solution of linear DEs, many of which have been tabulated like the error function. As long as these special functions are tabulated, or can be readily evaluated on a computer, they are only a little more difficult to work with than elementary functions such as exponential functions, logarithmic functions, and trigonometric functions.

Exercises 2.5

In Problems 1 to 12, use separation of variables to find the general solution of the homogeneous linear equation.

1. $y' + 2y = 0$

2. $ay' + by = 0$

3. $y' + (\tan x)y = 0$

4. $y' = xy$

5. $\dfrac{dy}{dx} = \dfrac{y}{x}$

6. $x^2 y' + y = 0$

7. $4x^3 y\, dx + dy = 0$

*8. $(\log x)y\, dx + dy = 0$

9. $t\, ds + s(2t + 1)\, dt = 0$

■10. $(1 + t)\dfrac{dw}{dt} + w = 0$

11. $4xy^3\, dy + dx = 0$

12. $(r^2 + 1)\, du + ru\, dr = 0$

In Problems 13 to 30, use the method of this section to find a general solution.

13. $y' = 1 - 5y$

14. $y' - 2y = 8x$

15. $xy' + y = x \sin x$

16. $xy' - y = x^2 \sin x$

17. $xy' = 4y + x^6 e^x$

18. $(x^3 + 3y)\, dx - x\, dy = 0$

19. $z' = x - 4xz$

■20. $\dfrac{dy}{dx} = \csc x + y \cot x$

21. $\dfrac{dy}{dx} = \csc x - y \cot x$

22. $u\, dt + (3t - tu + 2)\, du = 0$

23. $(x^2 + 9)y' + xy = x$

24. $(1 + e^y)\dfrac{dx}{dy} + e^y x = y$

25. $y' - my = e^{mx}$

26. $y' - my = e^{kx}$ $(k \neq m)$

27. $x^2 y' + x(x + 2)y = e^x$

*28. $(1 + x)y' - xy = x + x^2$

29. $xy' - ay = bx^k$ $(x > 0, a \neq k)$

30. $xy' - ky = bx^k$ $(x > 0)$

In Problems 31 to 40, find a solution of the given IVP and state the interval for which the solution is valid.

31. $y' + (\cos x)y = 0$, $y(0) = 2$

32. $y' + (\sin x)y = 0$, $y(\pi/2) = 1$

33. $L\dfrac{di}{dt} + Ri = E$, $i(0) = 0$ ($L, R,$ and E constant)

34. $L\dfrac{di}{dt} + Ri = A \sin \omega t$, $i(0) = 0$

($L, R,$ and A constant)

35. $(2x + 3)y' - y = \sqrt{2x + 3}$, $y(-1) = 0$

■36. $xy' + 2y = \sin x$, $y(\pi/2) = 1$

37. $y' - y = 2xe^{2x}$, $y(0) = 1$

*38. $y' = x^3 - 2xy$, $y(1) = 1$

39. $x\, dy - y\, dx = x^2 e^x\, dx$, $y(1) = 0$

40. $y' = \dfrac{y}{y - x}$, $y(2) = 3$

In Problems 41 to 43, solve the given IVP with *discontinuous* forcing function.

41. $y' + y = f(x)$, $y(0) = 0$, where

$$f(x) = \begin{cases} 1, & 0 \le x \le 1 \\ 0, & x > 1 \end{cases}$$

■ 42. $y' + 2xy = f(x)$, $y(0) = 2$, where

$$f(x) = \begin{cases} x, & 0 \le x \le 1 \\ 0, & x > 1 \end{cases}$$

43. $(1 + x^2)y' + 2xy = f(x)$, $y(0) = 0$, where

$$f(x) = \begin{cases} x, & 0 \le x < 1 \\ -x, & x \ge 1 \end{cases}$$

44. If $y = \phi(x)$ is a solution of

$$y' + a_0(x)y = f(x)$$

and $y = \psi(x)$ is a solution of

$$y' + a_0(x)y = g(x)$$

show that $y = \phi(x) + \psi(x)$ is a solution of

$$y' + a_0(x)y = f(x) + g(x)$$

*45. Show that the IVP

$$y' + 3y = \frac{1}{\sqrt{2\pi x}}e^{-3x}\cos x, \qquad y(0) = 2$$

has the solution

$$y = [2 + C_2(x)]e^{-3x}$$

where $C_2(x)$ is a **Fresnel integral** defined by

$$C_2(x) = \frac{1}{\sqrt{2\pi}}\int_0^x \frac{\cos t}{\sqrt{t}}\,dt$$

*46. Show that the IVP

$$x^2y' + xy = \sin x, \qquad y(1) = 5$$

has the solution

$$y = \frac{5}{x} + \frac{1}{x}[\mathrm{Si}(x) - \mathrm{Si}(1)]$$

where $\mathrm{Si}(x)$ is the **sine integral** defined by

$$\mathrm{Si}(x) = \int_0^x \frac{\sin t}{t}\,dt, \qquad x > 0$$

[0] 2.6 Miscellaneous Topics

In this section we consider a few additional techniques for solving nonlinear DEs, some of which appear in the exercises.

2.6.1 Bernoulli's Equation

The *nonlinear* DE

$$y' + a_0(x)y = f(x)y^n, \qquad n \ne 0, 1 \tag{57}$$

is called **Bernoulli's equation** after the Swiss mathematician J. Bernoulli.[†] It is assumed that n is any real number except as noted. The values $n = 0$ and $n = 1$ are excluded from the discussion since each case leads to a *linear* equation.

Bernoulli's equation is an example of a nonlinear equation that can be reduced to a linear equation by means of a suitable change of variable. For example, if we make the substitution

$$z = y^{1-n} \tag{58}$$

[†] See the historical comments in Sections 1.6 and 2.8 concerning J. Bernoulli.

then
$$z' = (1 - n)y^{-n}y'$$

and (57) becomes
$$\frac{y^n z'}{1 - n} + a_0(x)y^n z = f(x)y^n$$

or, upon simplifying,
$$z' + (1 - n)a_0(x)z = (1 - n)f(x) \qquad (59)$$

Because (59) is linear, we can solve it by previous methods.

EXAMPLE 25

Solve the Bernoulli equation
$$dy + 2xy\,dx = xe^{-x^2}y^3\,dx$$

Solution

Let us first divide by dx to put the DE in the standard form
$$y' + 2xy = xe^{-x^2}y^3$$

which we now recognize as a Bernoulli equation with $n = 3$. Therefore, we set
$$z = y^{1-n} = y^{-2}$$

from which we obtain the linear DE
$$z' - 4xz = -2xe^{-x^2}$$

Here we calculate
$$z = C_1 e^{2x^2} + e^{2x^2}\int(-2xe^{-3x^2})\,dx$$
$$= C_1 e^{2x^2} + \tfrac{1}{3}e^{-x^2}$$

and by changing back to the original variable y, we obtain the implicit solution
$$3y^{-2} = Ce^{2x^2} + e^{-x^2}$$

where C is an arbitrary constant.   ■

EXAMPLE 26

Solve the IVP
$$\frac{dP}{dt} = aP - bP^2, \qquad P(0) = P_0\ (P_0 \neq a/b)$$

Solution

This DE can be solved by separating variables (see Section 2.2), but this method leads to an implicit solution that is more cumbersome to simplify than that obtained by the Bernoulli method. To solve as a Bernoulli equation we first make the substitution $z = 1/P$, which reduces the given IVP to the related linear DE and initial condition

$$\frac{dz}{dt} + az = b, \qquad z(0) = 1/P_0$$

The general solution of this equation is readily found to be

$$z(t) = C_1 e^{-at} + e^{-at} \int b e^{at} \, dt$$

$$= C_1 e^{-at} + \frac{b}{a}$$

and by imposing the initial condition, we find that

$$z(0) = C_1 + \frac{b}{a} = \frac{1}{P_0}$$

or $C_1 = 1/P_0 - b/a = (a - bP_0)/aP_0$. Therefore,

$$z(t) = \left(\frac{a - bP_0}{aP_0} \right) e^{-at} + \frac{b}{a}$$

$$= \frac{(a - bP_0)e^{-at} + bP_0}{aP_0}$$

and by taking the reciprocal, we finally obtain

$$P(t) = \frac{aP_0}{(a - bP_0)e^{-at} + bP_0} \qquad \blacksquare$$

2.6.2 Picard's Method

A certain iterative technique, due to Picard,[†] provides a novel alternate approach to solving IVPs of the general form

$$y' = F(x, y), \qquad y(x_0) = y_0 \tag{60}$$

The primary use of the technique, however, is not for practical purposes but for theoretical purposes; that is, it forms the basis of the proof of Theorem 2.1 (also known as Picard's theorem).

To begin, let us integrate both sides of the DE in (60) from x_0 to x, obtaining

$$\int_{x_0}^{x} y'(x) \, dx = \int_{x_0}^{x} F[x, y(x)] \, dx$$

[†] Emile Picard (1856–1941) was a French mathematician who made several outstanding contributions to the field of mathematical analysis.

or

$$y(x) - y(x_0) = \int_{x_0}^{x} F[x, y(x)] \, dx \qquad (61)$$

However, since $y(x_0) = y_0$, we can rewrite (61) in the form

$$y(x) = y_0 + \int_{x_0}^{x} F[t, y(t)] \, dt \qquad (62)$$

where we have replaced the dummy variable of integration x with t on the right-hand side to avoid confusion. Equation (62), which is called an **integral equation,** is simply an equivalent formulation of the IVP (60). For example, by setting $x = x_0$ we obtain $y(x_0) = y_0$, and if we differentiate (62) we obtain the original DE.

We now seek a **sequence of approximations** $y_0(x), y_1(x), y_2(x), \ldots, y_n(x), \ldots$, hopefully such that

$$\lim_{n \to \infty} y_n(x) = y(x)$$

where $y(x)$ is the true solution. We start by choosing $y_0(x) = y_0$, and then set $y(t) = y_0$ on the right-hand side of (62) to obtain the next approximation

$$y_1(x) = y_0 + \int_{x_0}^{x} F(t, y_0) \, dt \qquad (63)$$

Repeating this procedure, we next set $y(t) = y_1(t)$ on the right-hand side of (62), which yields

$$y_2(x) = y_0 + \int_{x_0}^{x} F[t, y_1(t)] \, dt \qquad (64)$$

Continuing in this manner, it becomes clear that the approximation $y_n(x)$ is determined from $y_{n-1}(x)$ by

$$y_n(x) = y_0 + \int_{x_0}^{x} F[t, y_{n-1}(t)] \, dt, \qquad n = 1, 2, 3, \ldots \qquad (65)$$

The general iterative technique described by (65) is known as **Picard's method.**

EXAMPLE 27

Use Picard's method to solve

$$y' = 2xy, \qquad y(0) = 1$$

Solution

Here we see that $x_0 = 0$, $y_0 = 1$ and $F(x, y) = 2xy$, so that (65) becomes

$$y_n(x) = 1 + 2 \int_0^x t y_{n-1}(t) \, dt$$

For $n = 1$, we have $y_0(t) = 1$ and hence

$$y_1(x) = 1 + 2 \int_0^x t \, dt$$

$$= 1 + x^2$$

Continuing in this fashion, we obtain

$$y_2(x) = 1 + 2 \int_0^x t(1 + t^2) \, dt$$

$$= 1 + x^2 + \tfrac{1}{2}x^4$$

$$y_3(x) = 1 + 2 \int_0^x t(1 + t^2 + \tfrac{1}{2}t^4) \, dt$$

$$= 1 + x^2 + \frac{x^4}{2!} + \frac{x^6}{3!}$$

and

$$y_4(x) = 1 + 2 \int_0^x t\left(1 + t^2 + \frac{t^4}{2!} + \frac{t^6}{3!}\right) dt$$

$$= 1 + x^2 + \frac{x^4}{2!} + \frac{x^6}{3!} + \frac{x^8}{4!}$$

At this point the pattern is clear and we deduce that

$$\lim_{n \to \infty} y_n(x) = \lim_{n \to \infty} \sum_{k=0}^{n} \frac{x^{2k}}{k!} = e^{x^2}$$

That is, the solution of the IVP is

$$y = e^{x^2}$$

■

In general, of course, we don't expect to find the solution by Picard's method as easily as we did in Example 27. More often than not, if the DE is not easily solved by one of the earlier methods that we discussed, the integrals that evolve out of Picard's method quickly become formidable. It can be shown, however, by using concepts from advanced calculus that the sequence of successive approximations $\{y_n(x)\}$, developed out of Picard's method, converges uniformly to the true solution $y(x)$ provided $F(x, y)$ satisfies the conditions of Theorem 2.1. Unfortunately, the rate of convergence is usually so slow that the method of Picard is not very effective in practice.

Exercises 2.6

In Problems 1 to 10, use the method of Section 2.6.1 to solve the given Bernoulli equation.

1. $xy' + y = x^2y^{-1}$

2. $y' = 5y - 7y^3$

3. $x^2y' + 2xy = y^3$

■**4.** $xy' - (1 + x)y = xy^2$

*5. $6y^2 \, dx - x(2x^3 + y) \, dy = 0$

*6. $y' + y = (xy)^2$, $y(0) = 1/3$

7. $y' - y = xy^5$, $y(0) = 1$

8. $y' + 2xy + xy^4 = 0$

9. $y' + y = (\cos x - \sin x)y^2$, $y(\pi) = e^{-\pi}$

10. $x \, dy = [y + xy^3(1 + \log x)] \, dx$, $y(1) = 1$

In Problems 11 to 20, use Picard's method to find y_1, y_2, and y_3.

11. $y' = y$, $y(0) = 1$

12. $y' = y - 1$, $y(0) = 2$

13. $y' - y = x$, $y(0) = 1$

■**14.** $y' = x + y^2$, $y(0) = 0$

15. $y' = 1 + y^2$, $y(0) = 0$

*16. $y' + y = 2e^x$, $y(0) = 1$

17. $y' + 2xy = 1$, $y(0) = 1$

18. $y' = 1 + xy$, $y(1) = 2$

19. $y' = 2 + \sin x + y$, $y(0) = 0$

*20. $y' = x^2 + y^2$, $y(0) = 1$ (find only y_1 and y_2)

In Problems 21 to 26, solve by the indicated method.

*21. Any DE of the form

$$y = xy' + f(y')$$

is called an **equation of Clairaut.** Show that a general solution of this equation is the family of lines

$$y = mx + f(m)$$

where m is an arbitrary constant.

*22. Based on Problem 21, solve the following equations:

(a) $2y = 2xy' + (y')^2$

(b) $y = xy' + (y')^3$

(c) $xy' - y = e^{y'}$

*23. Show that the substitution $z = \log y$ transforms the nonlinear DE

$$y' + a_0(x)y = f(x)y \log y$$

into a linear DE in the variable z.

*24. The nonlinear DE

$$y' = P(x)y^2 + Q(x)y + R(x)$$

is called a **Riccati equation.** If $y = y_P$ is a particular solution of the equation, show that

$$y = y_P + \frac{1}{z}$$

is a general solution, where z is a general solution of the linear DE

$$z' + [2y_P P(x) + Q(x)]z = -P(x)$$

*25. Referring to Problem 24, find a general solution of

$$y' = y^2 + (1 - 2x)y + x^2 - x + 1$$

given that $y_P = x$ is a particular solution.

*26. When $P(x) = -1$, the Riccati equation in Problem 24 becomes

$$y' = -y^2 + Q(x)y + R(x)$$

Show that the substitution $y = u'/u$ reduces this equation to the *second-order linear* DE

$$u'' - Q(x)u' - R(x)u = 0$$

2.7 Chapter Summary

In this chapter we have introduced several methods of solution, each dependent upon the type of DE to be solved. In general, we can say that the IVP

$$y' = F(x, y), \qquad y(x_0) = y_0$$

has a unique solution on some interval about x_0 if $F(x, y)$ and $\partial F/\partial y$ are continuous in a rectangular region of the xy plane containing the point (x_0, y_0).

Listed below are the most important solution methods introduced in this chapter.

Separation of Variables

We say that the DE

$$M(x, y)\, dx + N(x, y)\, dy = 0 \tag{66}$$

is **separable** if it can be put in the form

$$f(x)\, dx = g(y)\, dy \tag{67}$$

The general solution is then obtained by integrating both sides of (67). If (66) is not separable but $M(tx, ty) = t^n M(x, y)$ and $N(tx, ty) = t^n N(x, y)$, we say that $M(x, y)$ and $N(x, y)$ are **homogeneous functions** of degree n. In this case the substitution $y = ux$ or $x = vy$ reduces (66) to a separable equation like (67).

Exact Equations

We say that (66) is an **exact equation** if and only if

$$\frac{\partial M}{\partial y} = \frac{\partial N}{\partial x} \tag{68}$$

When this is the case, there exists some function $f(x, y)$ such that

$$\frac{\partial f}{\partial x} = M(x, y), \qquad \frac{\partial f}{\partial y} = N(x, y) \tag{69}$$

the simultaneous solution of which yields the general solution $f(x, y) = C_1$, where C_1 is an arbitrary constant. When (66) is not exact, we may be able to find a function $\mu(x, y)$, called an **integrating factor,** such that

$$\mu(x, y)M(x, y)\, dx + \mu(x, y)N(x, y)\, dy = 0 \tag{70}$$

is an exact equation. We then solve (70) as an exact DE.

Linear Equations

Equations of the form

$$y' + a_0(x)y = f(x) \tag{71}$$

are called **linear equations.** They can be solved by either finding an **integrating factor** $\mu(x)$ or by solving separately for the **homogeneous solution** y_H and **particular solution** y_P. The integrating factor is always given by

$$\mu(x) = \exp\left[\int a_0(x)\, dx\right] \tag{72}$$

and leads to the general solution

$$y = \frac{1}{\mu(x)}\left[\int \mu(x)f(x)\, dx + C_1\right] \tag{73}$$

On the other hand, we find

$$y_H = C_1 y_1(x) = C_1 \exp\left[-\int a_0(x)\, dx\right] \qquad (74)$$

and

$$y_P = y_1(x) \int \frac{f(x)}{y_1(x)}\, dx \qquad (75)$$

leading to the general solution

$$y = y_H + y_P$$

$$= C_1 y_1(x) + y_1(x) \int \frac{f(x)}{y_1(x)}\, dx \qquad (76)$$

Review Exercises

In Problems 1 to 20, solve by any method.

1. $x^3\, dx + (y + 1)^2\, dy = 0$
2. $4x\, dy - y\, dx = x^2\, dy$
3. $(x^3 + y^3)\, dx - 3xy^2\, dy = 0$
4. $x\, dy - y\, dx = \sqrt{x^2 - y^2}\, dx$
5. $\sqrt{x}\, dy = y\, dx,\ y(0) = 5$
6. $2xy\, dx + (1 + x^2)\, dy = 0$
7. $(x^2 + y^2 + x)\, dx = -xy\, dy$
8. $(x + y)\, dx - x\, dy = 0$
9. $(4x^3 y^3 - 2xy)\, dx + (3x^4 y^2 - x^2)\, dy = 0$
10. $(2x^3 + 3y)\, dx + (3x + y - 1)\, dy = 0$

11. $x(y - 3)y' = 4y$
12. $(1 + x^3)\, dy - x^2 y\, dx = 0,\ y(1) = 2$
13. $(6x^5 y^3 + 4x^3 y^5)\, dx + (3x^6 y^2 + 5x^4 y^4)\, dy = 0$
14. $2ye^{x/y}\, dx + (y - 2xe^{x/y})\, dy = 0$
15. $(2x + 3y)\, dx + (y - x)\, dy = 0,\ y(1) = 0$
16. $(x + y)\, dy = y\, dx$
17. $y' = 1 + y \tan x$
18. $(x^2 - y)\, dx - x\, dy = 0$
19. $(x - y^2)\, dx + 2xy\, dy = 0$
*20. $(y \log y)\, dx + (x - \log y)\, dy = 0,\ y(1) = 1$

2.8 Historical Comments

The method of **separation of variables** was discovered by Leibniz. He was also responsible for reducing **homogeneous equations** to variables-separable type and for solving **first-order linear DEs.** Along with John Bernoulli, Leibniz further proposed using differential equations in finding curves intersecting a given family of curves at a given angle. The special case of **orthogonal trajectories** (see Section 3.2.1) was solved in 1715 by Newton in response to a challenge by Leibniz.

The nonlinear DE

$$y' + a_0(x)y = f(x)y^n, \qquad n \neq 0, 1$$

was studied by James Bernoulli and is now named **Bernoulli's equation** in his honor. However, it was Leibniz in 1696 who showed that it could be reduced to a linear equation by the substitution $z = y^{1-n}$, $n = 2, 3, 4, \ldots$.

Leonhard Euler (1707–1783), considered to be one of the five greatest mathematicians of all time, introduced the method for solving **exact equations** and he developed the theory of **integrating factors,** among other contributions to the general theory of differential equations. He was born in Basel, Switzerland, where his father, together with John Bernoulli, instructed a youthful Euler in mathematics as well as theology, astronomy, physics, medicine, and several languages. Euler first studied theology in college and received his master's degree at the age of seventeen, whereupon he changed from theology to the study of mathematics. In 1727, he accepted a faculty position at St. Petersburg Academy in Russia, first in medicine and later in the chair of natural philosophy from which he pursued his mathematical interests. Three years later he became professor of physics, and in 1732 he succeeded his close friend Daniel Bernoulli (1700–1782), one of John's sons, to the chair of mathematics where he remained until 1741. For the next twenty-five years he assumed the position of director of the Department of Mathematics at the Academy of Berlin, eventually returning to St. Petersburg in 1766.

In addition to differential equations, Euler made significant contributions in virtually every branch of pure and applied mathematics, including mechanics, number theory, geometry, fluid dynamics, astronomy, and optics. His collected works fill more than 70 volumes, making him the most prolific writer in the history of mathematics! Although he was blind during the last 17 years of his life, he still continued his work until his death on September 18, 1783, at the age of 76.

Around 1724, Jacopo Riccati (1676–1754) of Venice introduced the equation

$$y' = P(x)y^2 + Q(x)y + R(x)$$

which today is known as the **Riccati equation.** Riccati was also a physicist and philosopher and was responsible for bringing much of Newton's work to the attention of other Italian mathematicians. The **equation of Clairaut**

$$y = xy' + f(y')$$

was studied in 1734 by the French mathematician Alexis Claude Clairaut (1713–1765).

CHAPTER 3

Applications Involving First-Order Equations

First-order DEs can be applied in many diverse areas of the physical and life sciences. It often happens that the same DE arises in the mathematical formulation of two or more quite distinct problems, such as the motion of a free-falling body or the growth rate of a certain type of bacteria. However, once the problem has been mathematically formulated, the origin of the problem makes little difference until we need to interpret the solution.

In Section 3.2 we feature the geometric problem of finding the **orthogonal trajectories** of a given family of curves. Some simple mechanics problems involving the velocity of **free-falling bodies** and the **velocity of escape** from the earth (and other heavenly bodies) are discussed in Section 3.3; also included in this section are problems involving simple **electric circuits.** Problems of **growth and decay** are treated in Section 3.4, including the **half-life** of a radioactive substance and the **spread of an epidemic.**

Some miscellaneous applications are presented in the last section. Here we discuss problems involving **cooling bodies,** the **mixing of two solutions,** applying Torricelli's law to the **flow of water through an orifice,** and finding **curves of pursuit** that describe the path of a pursuer tracking its prey.

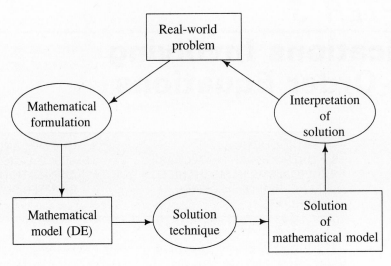

Figure 3.1

3.1 Introduction

DEs have always been applied to problems in mechanics, most of which involve quantities that change in a *continuous* manner such as distance, velocity, acceleration, and force. However, in problems in the life sciences where the quantity of interest may be, for example, the population size of a particular community, the quantity changes by *discrete amounts* rather than continuously. For that reason, we might not expect to describe such changes by derivatives, or DEs in general, since derivatives are meaningful only for continuous functions. This is indeed the case for small population sizes, but if a population size is "sufficiently large," it can often be *modeled* as a *continuous system*. That is, the continuous system may describe general characteristics of the discrete problem being studied and even predict certain results that compare well with experimental data. The justification for such approximations depends mostly on whether it works! Because continuous models have proved effective in a number of different problems in the life sciences, the use of DEs in this area has grown significantly in recent years.

In the application of DEs we are generally concerned with more than just solving a particular initial value or boundary value problem. Indeed, the starting point is usually some real-world problem that must be mathematically formulated before it can be solved. The complete *solution process* consists primarily of the following three steps (see also Figure 3.1):

1. *Construction of a mathematical model.* The variables involved must be carefully defined and the governing physical (or biological) laws identified. The mathematical model is then some equation(s) representing an idealization of the physical (or biological) laws, taking into account some simplifying assumptions in order to make the model tractable.

2. *Solution of the mathematical model.* When permitted, exact solutions are usually desired, but in many cases one must rely on approximate solutions; hence,

it is helpful, when possible, to establish the existence and uniqueness of the solution.

3. *Interpretation of the results.* The solutions obtained should be consistent with physical intuition and physical evidence. If a good model has been constructed, the solution should describe many of the essential characteristics of the system under study.

3.2 Orthogonal Trajectories

It is well known in analytic geometry that the slopes of *perpendicular* lines are negative reciprocals of one another, that is, $m_1 = -1/m_2$, where m_1, $m_2 \neq 0$. In a more general setting, we say that two intersecting curves are **orthogonal** if and only if their tangent lines are perpendicular at the point of intersection.

Suppose we have the one-parameter family of curves defined by

$$f(x, y, c) = 0 \tag{1}$$

where each member of the family corresponds to a particular value of the parameter c. In certain applications it is important to be able to obtain a second family of curves given by

$$g(x, y, k) = 0 \tag{2}$$

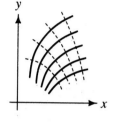

Figure 3.2 Curves and their orthogonal trajectories.

with the property that all intersections of the two families are orthogonal. The two families are then said to be **orthogonal trajectories** of each other. Except for the particular cases where the tangents are parallel to the x and y axes, this means that their slopes at the points of intersection are negative reciprocals (see Figure 3.2).

Orthogonal trajectories are important in the study of electric potentials, for example, where the lines of force between two bodies of opposite charge are perpendicular to the equipotential curves (i.e., curves for which the potential is constant). In Figure 3.3 we have shown the lines of force for such a problem by the dashed

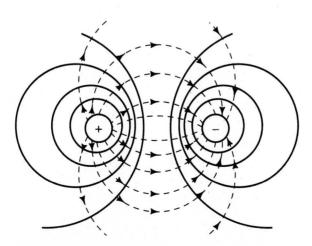

Figure 3.3 Equipotential curves.

curves while the solid curves are the equipotential curves. Similar orthogonal families of curves occur in the study of perfect fluids and in the construction of meteorological maps.

Suppose we imagine (1) to be the general solution of a DE having the form

$$y' = F(x, y) \tag{3}$$

which is independent of the arbitrary constant c. It then follows that the DE whose general solution is (2) must be

$$y' = -\frac{1}{F(x, y)} \tag{4}$$

The procedure, therefore, is to find a DE (3) for which family (1) is a general solution, and then obtain the orthogonal trajectories as solutions of (4).

EXAMPLE 1

Find the orthogonal trajectories of the family of circles

$$x^2 + y^2 = c^2$$

Solution

The constant c in the equation $x^2 + y^2 = c^2$ can be eliminated by implicit differentiation, leading to

$$\frac{d}{dx}(x^2) + \frac{d}{dy}(y^2) = 0$$

or

$$2x + 2yy' = 0$$

Solving for y', we find

$$y' = -\frac{x}{y}$$

and thus the DE for the family of orthogonal trajectories is

$$y' = \frac{y}{x}$$

We can solve this last DE by separating variables, that is,

$$\frac{dy}{y} = \frac{dx}{x}$$

from which we deduce

$$\log|y| = \log|x| + C_1$$

or

$$y = kx$$

where $k = \pm e^{C_1}$. Hence, the orthogonal trajectories are straight lines passing through the origin (see Figure 3.4).

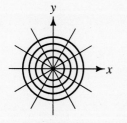

Figure 3.4

EXAMPLE 2

Find the orthogonal trajectories of the family of parabolas

$$y^2 = 4px \ (p \text{ constant})$$

Solution

If we rewrite the equation as

$$\frac{y^2}{x} = 4p$$

and then use implicit differentiation, we eliminate the constant $4p$. This action leads to

$$\frac{x(2yy') - y^2}{x^2} = 0$$

the simplification of which yields the governing DE

$$y' = \frac{y}{2x}$$

Thus, the orthogonal trajectories of the family of parabolas must satisfy the related DE

$$y' = -\frac{2x}{y}$$

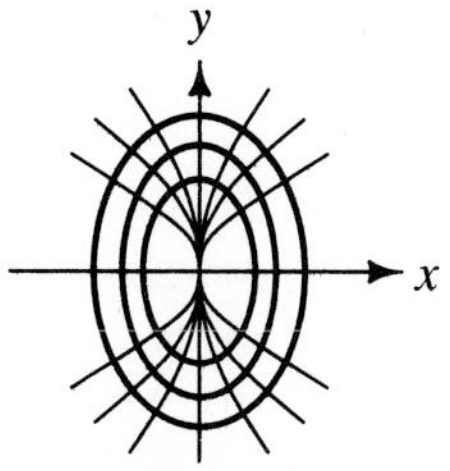

Separating the variables in this last equation, we obtain

$$y \, dy = -2x \, dx$$

and therefore deduce that the orthogonal trajectories are the family of ellipses described by (see Figure 3.5)

$$2x^2 + y^2 = k^2 \qquad \blacksquare$$

Figure 3.5

Exercises 3.2

In Problems 1 to 12, find the orthogonal trajectories of the given family of curves.

1. $xy = c$

2. $y^2 = cx^3$

3. $y^2 = x + c$

4. $x - 2y = c$

5. $x^2 - y^2 = c$

6. $e^y - e^{-x} = c$

7. $x^2 - y^2 = cx$

■**8.** Circles through the origin with centers on the y axis

9. $y^2 = 4x^2(1 - cx)$

10. $y = 3x - 1 + ce^{-3x}$

11. $x^{1/3} + y^{1/3} = c$

12. $y = \dfrac{1 + cx}{1 - cx}$

*13. In the calculus it is shown that the angle ψ measured in the counterclockwise direction from the radius vector to the tangent line at a point satisfies the relation [see Figure (a)]

$$\tan \psi = r \frac{d\theta}{dr}$$

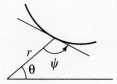

Problem 13(a)

where r and θ are polar coordinates. If two curves in polar coordinates are orthogonal, show that [see Figure (b)]

$$\tan \psi_1 = -\frac{1}{\tan \psi_2}$$

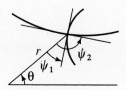

Problem 13(b)

In Problems 14 to 18, use the result of Problem 13 to find the orthogonal trajectories of the given family of curves.

14. $r = c(1 + \cos \theta)$

15. $r = c(1 - \sin \theta)$

■16. $r = 2c \cos \theta$

17. $r = c \cos^2 \theta$

18. $r^2 = c \sin 2\theta$

*19. A family of curves that intersects another family of curves at a constant angle α $(\alpha \neq \pi/2)$ are called **isogonal trajectories** of each other. If $y' = F(x, y)$ is the DE of the first family of curves, show that

$$y' = \frac{F(x, y) \pm \tan \alpha}{1 \mp F(x, y) \tan \alpha}$$

is the governing DE of the second isogonal family.

*20. Using the result of Problem 19, find the isogonal families of the following families of curves.

 (a) $y(x + c) = 1; \alpha = \pi/4$
 (b) $y = cx; \alpha = 45°$
 (c) $y = cx; \alpha = 30°$
 (d) $y^2 = x + c; \alpha = \tan^{-1}(4)$

21. A given family of curves is said to be **self-orthogonal** if it has the property that its family of orthogonal trajectories is the same as that of the given family. Verify that the family of parabolas $y^2 = 2cx + c^2$ is self-orthogonal.

3.3 Problems in Mechanics and Simple Electric Circuits

In this section we discuss some simple problems involving the *motion* of a *single particle* moving in a straight line and the *charge* and *current* in a *simple electric circuit*.

3.3.1 Free-Falling Bodies

The basic principle of mechanics used in studying particle motion is **Newton's second law of motion,** which states that the rate of change of momentum of a body with respect to time is equal to the resultant force acting on the body. The **momen-**

tum of a body is the product mv, where m is the *mass* of the body and v is its *velocity* (speed). Hence, in symbols, Newton's law becomes

$$\frac{d}{dt}(mv) = F$$

or, if the mass is *constant,*

$$m\frac{dv}{dt} = F \tag{5}$$

where F is the sum of forces acting on the body.

> **Remark.** Newton's second law is often formulated by $F = ma$, where $a = dv/dt$ is the **acceleration** of the body.

To begin, let us consider the problem of a free-falling body that is acted upon by only the force of gravity. That is, the only active force is the *weight W* of the body. If the body is "close" to the earth's surface, the weight of the body is essentially $W = mg$, where g is the **gravitational constant.** At sea level, experimental evidence indicates that $g \cong 32$ ft/s^2 (English system) or $g \cong 9.8$ m/s^2 (international system). Newton's law (5) in this case becomes

$$m\frac{dv}{dt} = W = mg$$

or simply

$$\frac{dv}{dt} = g$$

where we have chosen the *positive direction downward.*

In some problems it may be necessary to also take into account a **resistive force** F_R, such as the air resistance encountered by a parachutist as he falls or the resistive force of a viscous liquid into which a ball is dropped. The weight $W = mg$ in such cases is then offset somewhat by the resistive force F_R (see Figure 3.6), and the governing DE becomes

$$m\frac{dv}{dt} = mg - F_R \tag{6}$$

The amount of resistance encountered depends upon the velocity of the body, but a general law expressing this dependency is not known. We often make the assumption that

$$F_R = cv \quad \text{or} \quad F_R = cv^2$$

where c is some positive constant whose value is determined by the nature of the resistive medium. Thus, we obtain either the *linear* DE

$$m\frac{dv}{dt} = mg - cv, \qquad c > 0 \tag{7a}$$

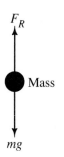

F_R

Mass

mg

Figure 3.6

or the *nonlinear* DE

$$m\frac{dv}{dt} = mg - cv^2, \qquad c > 0 \tag{7b}$$

Of course, in addition to solving (7a) or (7b), we need to prescribe the **initial velocity** of the body so that the problem has a unique solution, that is, $v(0) = v_0$.

EXAMPLE 3
Particle Motion

A skydiver weighing 150 lb opens his parachute when his downward velocity is 100 ft/s.

(a) If the force of air resistance is $5v$, find the velocity of the skydiver at any later time (prior to hitting the ground).

(b) What is the limiting velocity of the skydiver?

(c) If the skydiver opens his parachute at 2000 ft, how close to the ground will he be after 1 min?

Solution

(a) The governing DE and initial condition are

$$m\frac{dv}{dt} = mg - 5v, \qquad v(0) = 100$$

Since $g \cong 32$ ft/s^2 and the mass is $m = 150/32 = 75/16$ slugs,[†] the DE, in normal form, becomes

$$\frac{dv}{dt} + \frac{16}{15}v = 32, \qquad v(0) = 100$$

The general solution of this linear DE is readily found to be

$$v(t) = C_1 e^{-16t/15} + 30$$

and by imposing the prescribed initial condition, we find $C_1 = 70$. Hence, the velocity at any time t is

$$v(t) = 70e^{-16t/15} + 30, \qquad t \geq 0$$

(b) The **limiting** or **terminal velocity** can be found by letting $t \to \infty$ in (a), that is,

$$v_\infty = \lim_{t \to \infty} v(t) = 30 \text{ ft/s}$$

(c) The distance y of the skydiver above the ground is related to his velocity by

$$\frac{dy}{dt} = -v$$

$$= -(70e^{-16t/15} + 30)$$

where the negative sign indicates that y is decreasing. Direct integration of this result leads to

$$y(t) = \frac{525}{8}e^{-16t/15} - 30t + C_2$$

[†] A "slug" is a unit of mass equal to 1 lb/(ft/s^2).

and by applying the condition that $y(0) = 2000$, we see that $C_2 = 15{,}475/8 \cong$ 1934. Since 1 minute corresponds to 60 seconds, the position of the skydiver at this time is

$$y(60) \cong 0 - 1800 + 1934$$

$$\cong 134 \text{ ft}$$

above the ground.   ∎

> **Remark.** The limiting velocity can also be obtained directly from the DE by setting $dv/dt = 0$. For example, $v_\infty = mg/c$ in (7a) and $v_\infty = \sqrt{mg/c}$ in (7b).

EXAMPLE 4

Solve parts (a) and (b) of Example 3 under the condition that the air resistance is cv^2, where $c > 0$. Also, determine the value of c so that the limiting velocity here is the same as that in Example 3.

Solution

This time the IVP is described by

$$m \frac{dv}{dt} = mg - cv^2, \qquad v(0) = 100$$

or, upon dividing by c and setting $m = 75/16$ and $g = 32$,

$$\frac{75}{16c} \frac{dv}{dt} = \frac{150}{c} - v^2$$

Separating the variables, we obtain

$$\frac{dv}{150/c - v^2} = \frac{16c}{75} dt$$

the direct integration of which yields

$$\frac{1}{10} \sqrt{\frac{c}{6}} \log \left| \frac{v + 5\sqrt{6/c}}{v - 5\sqrt{6/c}} \right| = \frac{16c}{75} t + C_1$$

or, through exponentiation,

$$\frac{v + 5\sqrt{6/c}}{v - 5\sqrt{6/c}} = C_2 \exp\left(\frac{32}{5} \sqrt{\frac{2c}{3}} \, t \right)$$

where $C_2 = \pm \exp\left(10 \sqrt{\frac{6}{c}} \, C_1 \right)$.

Let us impose the initial condition $v(0) = 100$, which leads to

$$C_2 = \frac{100 + 5\sqrt{6/c}}{100 - 5\sqrt{6/c}}$$

Now solving explicitly for v, we get

$$v(t) = 5\sqrt{\frac{6}{c}}\,\frac{C_2 + \exp\left(-\frac{32}{5}\sqrt{\frac{2c}{3}}\,t\right)}{C_2 - \exp\left(-\frac{32}{5}\sqrt{\frac{2c}{3}}\,t\right)}, \qquad t \geq 0$$

In this case the limiting velocity is

$$v_\infty = \lim_{t \to \infty} v(t) = 5\sqrt{\frac{6}{c}}$$

and by equating the above result to 30, we deduce that $c = 1/6$. For this value of c, the above velocity is then given by

$$v(t) = 30\,\frac{13 + 7\exp(-32t/15)}{13 - 7\exp(-32t/15)}$$

Although theoretically the limiting velocity is approached only as $t \to \infty$, it always comes close to this value in a finite amount of time. For example, the above nonlinear model predicts $v \cong 30$ ft/s in approximately 3 seconds, while the linear model in Example 3 predicts $v \cong 30$ ft/s in approximately 7 seconds. ∎

3.3.2 Velocity of Escape

Let us now consider the problem of a body being projected **upward** from the earth's surface with a sufficient initial velocity v_0 such that it reaches a distance far above the earth's surface. At large radial distances r from the center of the earth (distances beyond sea level), the weight of a body differs from its weight at sea level since the acceleration a is not equal to the gravitational constant g at these distances. According to **Newton's law of gravitation,** the acceleration of a body is inversely proportional to the square of the distance from the center of the earth. In symbols we write

$$a = \frac{dv}{dt} = \frac{k}{r^2} \tag{8}$$

where k is a proportionality constant. In this problem we choose the *positive axis upward* so that velocity is positive in the upward direction. However, our sign convention is such that k is positive if the body is falling to earth and negative if the body is leaving the earth. (This implies that upward velocities will be decreasing so that acceleration is negative in the upward direction.) Neglecting resistive forces, we find that Newton's second law of motion becomes (after dividing by the mass m)

$$\frac{dv}{dt} = \frac{k}{r^2} \tag{9}$$

To determine the constant k in (9), we note that if the radius of the earth is defined by R, then $a = -g$ when $r = R$ (and the body is projected upward). Thus, substituting these results into (8), we find

$$-g = \frac{k}{R^2}$$

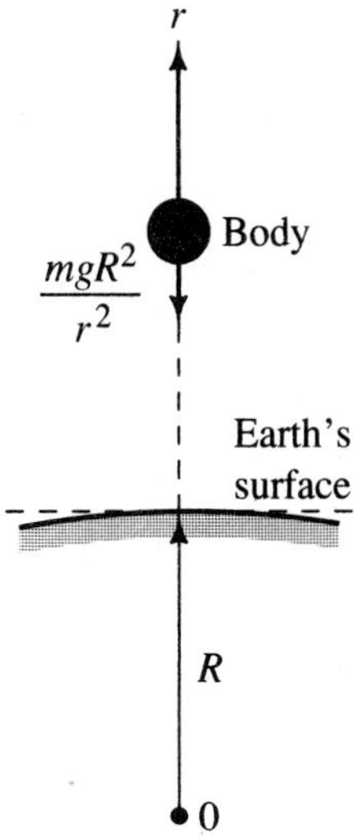

or

$$k = -gR^2 \qquad (10)$$

Equation (9) now assumes the form (see Figure 3.7)

$$\frac{dv}{dt} = -\frac{gR^2}{r^2}, \qquad v(0) = v_0 \qquad (11)$$

where we have also included the initial condition.

We wish to use (11) to determine the velocity of a body projected upward as a function of its radial distance r (over the range $r \geq R$). Hence, to solve the IVP (11), we must first express the velocity v in terms of r. Observe that

$$\frac{dv}{dt} = \frac{dv}{dr}\frac{dr}{dt} = v\frac{dv}{dr}$$

where we are using the fact that $v = dr/dt$. Substituting this result into (11), we get

$$v\frac{dv}{dr} = -\frac{gR^2}{r^2}, \qquad r > R, \qquad v(R) = v_0 \qquad (12)$$

where we now regard v as a function of distance r. Separating variables, we have

$$v\,dv = -\frac{gR^2}{r^2}\,dr$$

the integration of which yields

$$\frac{v^2}{2} = \frac{gR^2}{r} + C_1$$

Imposing the initial condition $v(R) = v_0$ leads to

$$\frac{v_0^2}{2} = gR + C_1$$

or $C_1 = v_0^2/2 - gR$, and thus our solution becomes

$$v^2 = \frac{2gR^2}{r} + v_0^2 - 2gR \qquad (13)$$

To calculate whether the initial velocity v_0 of the body is sufficient to escape the earth's gravity, we note that the velocity v must remain positive for this to happen; otherwise the body will stop and fall back to earth. Since $v^2 \geq 0$ and the first term on the right in Equation (13) is positive, the critical initial velocity v_0 must be such that $v_0^2 - 2gR \geq 0$, or $v_0 \geq \sqrt{2gR}$. The minimum value of v_0 is therefore

$$v_e = \sqrt{2gR} \qquad (14)$$

called the **escape velocity**.

> **Remark.** Recall that air resistance was neglected in deriving (14). The actual escape velocity, taking air resistance into account, will be greater than $\sqrt{2gR}$.

Figure 3.7

r

$\frac{mgR^2}{r^2}$

Body

Earth's surface

R

0

EXAMPLE 5
Escape Velocity of Earth

Neglecting air resistance and given that the radius of the earth is $R \cong 3960$ miles, determine the escape velocity for the earth.

Solution

Using English units, but converted to miles instead of feet, we see that $g = 32$ ft/s^2 corresponds roughly to $g = 0.006$ mi/s^2. Therefore, using (14), we find

$$v_e = \sqrt{2 \times 0.006 \times 3960}$$

$$\cong 6.9 \text{ mi/s}$$

Of course, the escape velocity for other heavenly bodies, such as the moon or Mars, will be different since both R and the gravitational constant are different for these bodies. ∎

3.3.3 Simple Electric Circuits

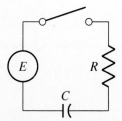

Figure 3.8 Simple RC circuit.

Consider a single-loop electric circuit containing a *resistor R* and *capacitor C* connected in series with a *voltage source* (battery) $E(t)$, as shown in Figure 3.8. **Kirchhoff's second law**[†] states that

> the sum of instantaneous voltage drops across each part of the circuit is equal to the voltage source.

From experimental observations, we have that

$$\text{Voltage drop across a resistor} = Ri$$

$$\text{Voltage drop across a capacitor} = \frac{q}{C}$$

where q denotes the electric charge on the capacitor and is related to the current i by $i = dq/dt$. Thus, the governing equation is

$$Ri + \frac{q}{C} = E(t)$$

or, in terms of the charge q, we find

$$R\frac{dq}{dt} + \frac{1}{C}q = E(t) \qquad (15)$$

In addition, we must prescribe either the initial current $i(0)$ or the initial charge $q(0)$.

EXAMPLE 6
Electric Circuit

If a resistance of 2000 ohms (Ω) and a capacitance of 5×10^{-6} farad (F) are connected in series with a voltage source of 100 volts (V) as shown in Figure 3.8, what is the current at any time, given that $i = 10^{-2}$ ampere at the time the switch is closed ($t = 0$). Also determine the initial charge on the capacitor.

[†] Named in honor of the German physicist Gustav R. Kirchhoff (1824–1887).

Solution

The governing DE in this case is

$$R\frac{dq}{dt} + \frac{1}{C}q = E_0$$

where $R = 2000\ \Omega$, $C = 5 \times 10^{-6}$ F, and $E_0 = 100$ V. The general solution of this first-order linear DE is readily found to be

$$q(t) = C_1 e^{-t/RC} + e^{-t/RC} \int \left(\frac{E_0}{R}\right) e^{t/RC}\, dt$$

$$= C_1 e^{-t/RC} + E_0 C$$

Since, $i = dq/dt$, it follows that

$$i(t) = -\frac{C_1}{RC} e^{-t/RC}$$

$$= -100\, C_1 e^{-100t}$$

where we have now substituted the actual values of the circuit parameters. Imposing the initial condition $i(0) = 10^{-2}$, we find that $-100\, C_1 = 10^{-2}$, and hence the current at any time is

$$i(t) = 10^{-2}\, e^{-100t} \qquad \text{ampere}$$

Using the value $C_1 = -10^{-4}$, the corresponding charge is given by

$$q(t) = -10^{-4}\, e^{-100t} + 5 \times 10^{-4}$$

$$= (5 - e^{-100t}) \times 10^{-4}$$

from which we deduce the initial value

$$q(0) = 4 \times 10^{-4} \qquad \text{V}$$

Exercises 3.3

1. A stone weighing $\frac{1}{4}$ lb falls from rest from a tall building. If the air resistance is known to be $v/160$, where v is the velocity of the stone, determine v at any time.

*2. If the building in Problem 1 is 200 ft high, what is the velocity of the stone when it hits the ground? (Approximate your answer.)

3. An object is thrown upward with an initial velocity v_0, and the air resistance is $F_R = cv$.

(a) Find the time required for the object to reach its maximum height.

(b) Find the maximum height if $c = 1/160$, $mg = 1/4$, and $v_0 = 8$.

■4. Solve Example 3 when the resistive force is $F_R = mv/4$. What is the limiting velocity in this case?

5. Solve Example 4 when the resistive force is $F_R = 5v^2$. What is the limiting velocity in this case?

6. Suppose a parachutist weighing 160 lb falls from rest toward the earth. When his speed is 30 ft/s (at the instant $t = 0$) the chute opens. Let the air resistance be $F_R = cv^2$, where c is a positive constant.
   (a) Determine the velocity v at any later time.
   (b) What is the chutist's limiting velocity?
   (c) Calculate the limiting velocity if the air resistance is cv instead of cv^2.

7. A skydiver weighing 160 lb falls from rest toward the earth. Before the parachute opens, the air resistance is $F_R = v/2$. The parachute opens 5 s later, and the air resistance changes to $F_R = 5v^2/8$. Find the velocity of the skydiver
   (a) before the parachute opens.
   (b) after the parachute opens.
   (c) What is the skydiver's limiting velocity?
   (d) If the parachute never opened, what would be the limiting velocity of the skydiver?

*8. A man and his parachute together weigh 192 lb. Assume a safe landing velocity is 16 ft/s and the air resistance is known to be proportional to the square of the velocity, equaling $\frac{1}{2}$ lb for each square foot of cross-sectional area of the parachute when it is moving 20 ft/s. What is the cross-sectional area of the parachute necessary for the chutist to make a safe landing?

9. If it takes time T for a ball thrown upward to reach its highest point, show that the return time is also T, given that $F_R = 0$. If the initial velocity of the ball is v_0, what is the return velocity?

10. Determine the escape velocity from the moon given that the moon's radius is roughly 1080 miles and the acceleration of gravity is $0.165g$, where g is the gravitational constant on earth.

11. Determine the escape velocity from Mars given that its radius is 2100 miles and the acceleration of gravity is $0.38g$, where g is the gravitational constant on earth.

12. Given that the force of gravity on Venus is about 85 percent of the earth's gravitational force and the radius of Venus is roughly 3800 miles, determine the escape velocity for Venus.

13. At 200 miles above the earth's surface, the atmosphere offers almost no resistance. What velocity should a rocket have at this altitude in order to reach a height of 4000 miles if all its fuel is exhausted at this point?

*14. If a body is shot straight up from the earth's surface with an initial velocity v_0 and no air resistance, show that the rising time t as a function of the distance r of the body from the center of the earth is given by

$$t = C - \frac{1}{B}\sqrt{r(A - Br)}$$

$$+ \frac{A}{2B\sqrt{B}}\sin^{-1}\left(\frac{2Br - A}{A}\right)$$

where $A = 2gR^2$, $B = 2gR - v_0^2 > 0$, and C is a constant to be determined from the initial condition $r = R$ when $t = 0$.
   Hint. Show that $dr/dt = 1/[r\sqrt{r(A - Br)}]$.

15. What is the charge on the capacitor in Example 6
   (a) after 1 s?
   (b) after a very long period of time?

■16. In the RC circuit shown in Figure 3.8, how long will it take the current $i(t)$ to decrease to one-half its original value if the voltage source is constant E_0?

17. Find the steady-state current in an RC circuit when the voltage source is $E(t) = E_0 \sin \omega t$.
   Hint. Let $t \to \infty$.

*18. A variable resistance $R = 1/(5 + t)\ \Omega$ and a capacitance $C = 5 \times 10^{-6}$ F are connected in series with a voltage source of 100 V. If the initial charge q is zero, what is the charge on the capacitor after 60 s?

19. The current $i(t)$ in a circuit containing a resistance of $R\ \Omega$ and an inductance of L henrys (H) in series with a voltage source $E(t)$ is

governed by (see accompanying figure)

$$L \frac{di}{dt} + Ri = E(t), \qquad i(0) = i_0$$

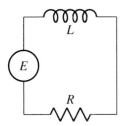

Problem 19

where L and R are known constants. Solve this IVP when

(a) $E(t) = E_0$ (constant).
(b) $E(t) = E_0 \sin \omega t$.

20. Solve Problem 19 under the conditions $R = 1\ \Omega$, $L = 10$ H, $i_0 = 6$ A, and

$$E(t) = \begin{cases} 6\ \text{V}, & 0 \le t \le 10 \\ 0\ \text{V}, & t > 10 \end{cases}$$

3.4 Growth and Decay

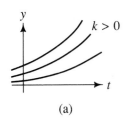

(a)

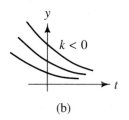

(b)

Figure 3.9

The simple *linear* DE

$$\frac{dy}{dt} = ky \qquad (16)$$

where k is a constant, arises in numerous physical theories concerning either growth or decay of some entity. For instance, Equation (16) might describe the rate at which a radioactive substance decomposes or the rate at which temperature changes in a cooling body. This same equation might be used to predict the population growth of certain small animals over short intervals of time or to describe the growth rate of certain bacteria.

The general solution of (16) has the exponential form $y = C_1 e^{kt}$, where C_1 is an arbitrary constant. Typical graphs of such solutions are shown in Figure 3.9 for cases when $k > 0$ and $k < 0$. Since dy/dt represents the slope of the function y, the sign of k tells whether y is *increasing* ($k > 0$) or *decreasing* ($k < 0$).

3.4.1 Radioactive Decay and Half-Life

Let us begin with a simple example of radioactive decay.

EXAMPLE 7
Radioactive Decay

Experimental evidence indicates that a radioactive substance decays at a rate directly proportional to the amount present. Starting at time $t = 0$ with y_0 grams of un-decayed matter, find the amount present at some later time.

Solution

If $y(t)$ denotes the amount of undecayed matter at any time t, then the experimental evidence is mathematically described by the IVP

$$\frac{dy}{dt} = -ky, \qquad y(0) = y_0, \qquad (k > 0)$$

where the negative sign indicates that y is a *decreasing function*. We recognize this DE as a homogeneous linear equation, with general solution

$$y(t) = C_1 e^{-kt}$$

Imposing the initial condition $y(0) = y_0$ requires $C_1 = y_0$, and thus

$$y(t) = y_0 e^{-kt}, \qquad t \geq 0$$

■

In physics, the stability of a radioactive substance is measured in terms of its **half-life,** that is, the time it takes for one-half of the atoms in an initial sample of the substance to decompose. Longer half-lifes correspond to more stable substances. For instance, radium decomposes quite rapidly with a half-life of approximately 1660 years (see Example 8), whereas the isotope uranium-238 has a half-life of about 4.5×10^9 years.

■EXAMPLE 8
Half-Life

If 0.5 percent of radium disappears in 12 years, what is the half-life of radium?

Solution

If we start with y_0 grams at time $t = 0$, then based on Example 7, the amount present at any later time is

$$y(t) = y_0 e^{-kt}$$

To establish the half-life of radium, we must first determine the constant k. To do so, we note that $0.005y_0$ grams of the radium will disappear in 12 years, leaving $0.995y_0$ grams. Hence, setting $t = 12$ in the above solution, we find

$$0.995y_0 = y_0 e^{-12k}$$

Canceling the common factor y_0 leads to

$$e^{-12k} = 0.995$$

or

$$e^{-k} = (0.995)^{1/12}$$

from which we deduce $k = -\frac{1}{12} \log(0.995) \cong 4.18 \times 10^{-4}$. The above solution now assumes the form

$$y(t) = y_0 e^{-(4.18 \times 10^{-4})t}$$

or, more conveniently, the alternate form

$$y(t) = y_0(0.995)^{t/12}$$

Half the radium will be gone when $y(t) = \frac{1}{2}y_0$, which leads to

$$\tfrac{1}{2}y_0 = y_0(0.995)^{t/12}$$

or

$$(0.995)^{t/12} = 0.5$$

Finally, taking logarithms of both sides, we see that the half-life is

$$t = 12 \frac{\log(0.5)}{\log(0.995)}$$

$$\cong 1659 \text{ years} \qquad \blacksquare$$

One of the interesting applications of radioactive decay involves approximating the age of fossils. It is known, for example, that cosmic radiation interacting with nitrogen produces the isotope carbon-14 in the atmosphere. Curiously, the ratio of carbon-14 to ordinary carbon in the atmosphere is roughly constant so that the proportion of carbon-14 found in the bodies of all breathing creatures at any time is the same as that found in the atmosphere. Thus, by comparing the amount of carbon-14 present in a fossil with the constant ratio found in the atmosphere, the age of the fossil (i.e., the time when the organism stopped breathing) can be reasonably predicted. The method, which is based upon knowledge of the half-life of carbon-14 (approximately 5600 years), was devised by Willard Libby and won him the Nobel Prize for chemistry in 1960 (see Problem 3 in Exercises 3.4).

3.4.2 Population Growth

Simple models of population growth are very similar to those governing radioactive decay. Again we illustrate with a simple example.

■ **EXAMPLE 9**
Population Growth

The population of a particular community is observed to increase at a rate proportional to the number of people present at any one time. In the last 5 years the population has doubled. How many years will it take for the population to triple?

Solution

The governing DE is

$$\frac{dP}{dt} = kP, \qquad k > 0$$

where $P(t)$ denotes the population at any time t and k is a positive constant (since P is increasing). Let us assume the population is P_0 at time $t = 0$, although we don't know its numerical value. The solution of this IVP is therefore

$$P(t) = P_0 e^{kt}$$

Now when $t = 5$ years, it is known that $P(5) = 2P_0$, which leads to

$$2P_0 = P_0 e^{5k}$$

or, upon canceling the common factor P_0,

$$e^k = 2^{1/5}$$

(The specific value of k is $k = \frac{1}{5} \log 2 \cong 0.1386$, although we only need e^k to solve the problem.) The above solution of the problem can now be expressed as

$$P(t) = P_0 2^{t/5}$$

which emphasizes the fact that the population doubles every 5 years. The population triples when $P(t) = 3P_0$, or

$$3P_0 = P_0 2^{t/5}$$

Taking logarithms of both sides of this expression, we see that

$$t = 5 \frac{\log 3}{\log 2}$$

$$\cong 7.9 \text{ years}$$ ∎

The simple DE

$$\frac{dP}{dt} = kP, \qquad k > 0 \tag{17}$$

used in Example 9 to describe population growth is not very realistic since prolonged exponential growth does not exist in most populations. It can describe the initial stages of growth for some populations, however, so it may be reasonable to use it for "short" time periods.

For example, the population growth of the United States from 1800 to 1860 can be closely approximated by the simple model

$$\frac{dP}{dt} = 0.03 \, P, \qquad P(0) = 5.3 \text{ million}$$

where the proportionality constant is $k = 0.03$. Starting in 1800 with a total population of approximately 5.3 million people, this model predicts that 32.1 million people were living in the United States in 1860, which compares well with the actual figure of 31.4 million. It also does remarkably well in predicting the population size for *every* year between 1800 and 1860. However, the same model predicts 106 million people were living in the United States in 1900, whereas the actual population was only 76 million people. The reason the model didn't continue to work, of course, is that it ignores such important factors as birth and death rates, wars (the Civil War started in 1860), diseases, immigration, and emigration, all of which have important effects on population growth.

To offset the rapid growth predicted by (17), an inhibitive factor, which is sometimes referred to as the "death rate," is usually introduced into the model. One of the earliest mathematical formulations of the problem employing such a term was suggested in 1844 by the Belgian mathematician P. F. Verhulst. He assumed that the inhibiting factor was proportional to $-P^2$, resulting in the IVP

$$\frac{dP}{dt} = kP(M - P), \qquad P(0) = P_0 \tag{18}$$

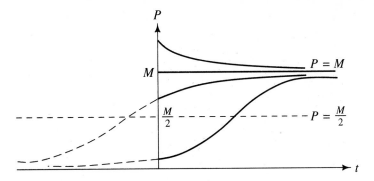

Figure 3.10 Solutions of the logistic equation.

where both k and M are positive constants and P_0 is the initial population. This equation is known as the **logistic equation.** Unlike those of Equation (17), all solutions of (18) are *bounded*.

The DE in (18) is a *Bernoulli equation* that can be solved by reducing it to a linear DE (see Example 26 in Section 2.5), and it can also be solved by *separation of variables.* By either method, the solution can be expressed in the form

$$P(t) = \frac{MP_0}{P_0 + (M - P_0)e^{-kMt}} \qquad (19)$$

The general shape of (19) is illustrated in Figure 3.10 for the cases when $0 < P_0 < M/2$, $M/2 < P_0 < M$, and $P_0 > M$. Regardless of the initial population P_0, however, we observe that all solutions approach the limiting value

$$\lim_{t \to \infty} P(t) = M \qquad \text{(saturation value)} \qquad (20)$$

When the initial population $P_0 < M$, the population $P(t)$ *increases* toward the saturation value M, whereas when $P_0 > M$, then $P(t)$ *decreases* toward M. Finally, if $P_0 = M$, it has this population size for all time t.

The logistic model (18) has proved quite effective in predicting growth patterns of fruit flies (in a limited space) and certain types of bacteria, the spread of an epidemic, and the population of the United States through the year 1930, based on 1790–1840 population data. Although the model is remarkably accurate up to 1930, the population of the United States did not follow the same trend after 1930.

EXAMPLE 10
Epidemic

A student carrying a flu virus returns after Christmas vacation to an isolated college campus of 2000 healthy students and begins to infect the other students. Suppose the rate at which the virus spreads among the students is proportional to the number of infected students P and is proportional as well to the number of students not infected, that is, $2000 - P$. The governing DE is therefore given by

$$\frac{dP}{dt} = kP(2000 - P)$$

where k is the proportionality constant. If 20 students are infected with the virus after 3 days, how many students will be infected at the end of a week?

Solution

Since only one student is infected at the beginning, the IVP we want to solve is described by

$$\frac{dP}{dt} = kP(2000 - P), \qquad P(0) = 1$$

which is the same as (18) with $M = 2000$ and $P_0 = 1$. Hence, the solution is [see (19)]

$$P(t) = \frac{2000}{1 + 1999\, e^{-2000kt}}$$

Using the information that $P(3) = 20$, we obtain

$$20 = \frac{2000}{1 + 1999\, e^{-6000k}}$$

or

$$e^{-2000k} = \left(\frac{99}{1999}\right)^{1/3}$$

Rather than solve this equation for k, we simply make a direct substitution into the above equation for $P(t)$ to obtain

$$P(t) = \frac{2000}{1 + 1999(99/1999)^{t/3}}$$

Finally, setting $t = 7$ (which corresponds to 1 week) leads to

$$P(7) = \frac{2000}{1 + 1999(99/1999)^{7/3}}$$

$$\cong 715 \text{ students}^\dagger$$

Exercises 3.4

1. The half-life of a certain radioactive substance is 1620 years. If 10 grams are initially present in a given sample, how much will be left after 162 years?

■2. A certain breeder reactor converts uranium-238 into the isotope plutonium-239. After 15 years, 0.043 percent of the initial amount of plutonium has decayed. What is the half-life of this isotope?

3. By comparing the amount of carbon-14 found in a fossil with the constant ratio found in the atmosphere, the age of a fossil can be esti-

---

† We have rounded the answer *up* to the nearest integer value.

mated. Assuming the half-life of carbon-14 is 5600 years, determine the approximate age of a fossil that is found to contain 0.1 percent of the original amount of carbon-14.

4. A certain chemical is converted into another substance at a rate proportional to the square of the amount of unconverted chemical. Starting at time $t = 0$ with an amount y_0 of unconverted chemical, determine the amount of unconverted chemical for all time.

5. The population of a certain country was 1 million in the year 1950, and the instantaneous growth rate since that time has been observed to be 3 percent of the current population. Assuming this trend continues, what population is expected by the year 2000?

6. Suppose that a certain population has grown to 10,000 after 3 years and that after 4 years the population will be double the original amount. What was the original population, and what will be the population after 10 years if it continues to grow at a rate proportional to the number of people present at any time?

*7. If we allow a population to change by either immigration into or emigration out of, the governing DE is modified to

$$\frac{dP}{dt} = kP + f(t)$$

where $f(t)$ is the rate that members of the population are being added to or subtracted from outside the system. Determine the population "growth" if $k = -3$ and the immigration is governed by the periodic function

$$f(t) = 1000(1 + b \sin t)$$

Assume the initial population is P_0, and discuss separately the cases $|b| < 1$ and $|b| > 1$; that is, which case represents immigration and which emigration? Finally, determine the **steady-state** population by considering the limit of $P(t)$ as $t \to \infty$.

8. The infusion of glucose into the bloodstream of an individual occurs at the constant rate of b grams per minute. At the same time, the glucose is converted and removed from

the bloodstream at a rate proportional to the amount of glucose present.
(a) Show that the amount of glucose $G(t)$ present at any time is governed by the DE

$$\frac{dG}{dt} = b - kG, \qquad (k \text{ constant})$$

(b) What concentration of glucose is attained after a sufficiently long period of time $(t \to \infty)$?

9. Suppose a sum of money M_0 is deposited into a bank that pays 6 percent interest.
(a) Show that the value $M(t)$ of the investment at the end of t years is given by the expression

$$M(t) = M_0 \left(1 + \frac{0.06}{k} \right)^{kt}$$

where k is the frequency (number of times each year) at which the interest is compounded.
(b) Determine the amount of money in the bank after 10 years if $k = 1$, $k = 4$, $k = 365$.
(c) If the interest is **compounded continuously,** write the governing DE for $M(t)$ illustrating this growth of investment, and show that its solution is the same as obtained by allowing $k \to \infty$ in the formula given in (a). Determine the amount of money in the bank at the end of 10 years as predicted by this model.

10. Referring to Problem 9, how long will it take to double the original investment if the interest is compounded continuously? If $k = 1$?

11. Referring to Example 10, how many days (approximately) will it take before 1500 students are infected with the virus?

■12. Assume that at time $t = 0$, half a population of 100,000 persons have a cold that is spreading among the population. If the number of people who have a cold is increased by 1000 persons at the end of the first day, how long will it take until 80 percent of the population have caught the cold?

3.5 Miscellaneous Applications

In this section we consider a number of additional applications involving first-order DEs, further illustrating the diversity of areas in which DEs play an important role.

3.5.1 Cooling and Mixing Problems

Newton's law of cooling states that the rate of change of temperature u in a cooling body is proportional to the difference between u and the temperature T_0 of the surrounding medium. In symbols, this law reads

$$\frac{du}{dt} = -k(u - T_0), \qquad k > 0 \tag{21}$$

where k is the proportionality constant.

**EXAMPLE 11
Cooling Body**

A metal ball is heated to a temperature of 100°C and then immersed in water at temperature 30° at time $t = 0$. Find its temperature at all later times if it is known that after 3 min the temperature of the ball is reduced to 70°C.

Solution

The governing DE is (21) with $T_0 = 30$°C. (It is assumed that the volume of water is sufficiently large that it can be maintained at 30°C even with the metal ball immersed.) Hence, we have the IVP

$$\frac{du}{dt} = -k(u - 30), \qquad u(0) = 100$$

This first-order DE can be solved as a linear equation or by using the method of separation of variables. Either way, we obtain the general solution

$$u(t) = C_1 e^{-kt} + 30$$

The prescribed initial condition $u(0) = 100$ requires $C_1 = 70$, and thus

$$u(t) = 70e^{-kt} + 30$$

What remains now is to determine the value of the proportionality constant k. Given that $u(3) = 70$, we find

$$70 = 70e^{-3k} + 30$$

or

$$e^{-k} = \left(\frac{4}{7}\right)^{1/3}$$

The exact value of k is therefore $k = \frac{1}{3}\log(7/4) \cong 0.1865$. Substituting this result into our solution yields

$$u(t) = 70\left(\frac{4}{7}\right)^{t/3} + 30$$

or, equivalently,

$$u(t) \cong 70e^{-0.1865t} + 30, \qquad t \geq 0$$

In Figure 3.11 we have plotted the temperature of the ball as a function of time. Observe that if we wait long enough ($t \to \infty$), the temperature of the metal ball will eventually approach 30°C, the temperature of the water. From a practical point of view, this will happen in approximately 15 min.

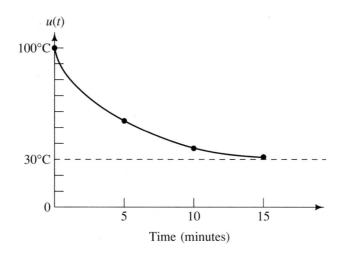

Figure 3.11

The mixing of two solutions can also give rise to a similar linear DE. In the next example we consider the mixing of pure water with a salt brine solution.

EXAMPLE 12
Mixing Solutions

4 gal/min

3 gal/min

Figure 3.12

A certain tank contains 100 gallons of a solution of dissolved salt and water, the mixture being kept uniform by stirring. If pure water is now allowed to flow into the tank at the rate of 4 gal/min, and the mixture flows out at the rate of 3 gal/min, how much salt will remain in the tank after t min if 15 lb of salt is initially in the mixture (see Figure 3.12)?

Solution

Let us denote the amount of salt present in the tank at any one time by $S(t)$. The net rate at which $S(t)$ changes is given by

$$\frac{dS}{dt} = (\text{rate of salt in}) - (\text{rate of salt out})$$

Since pure water is coming into the tank, the rate of salt entering the tank is zero. The rate at which salt is leaving the tank is the product of the amount of salt per gallon (called the **concentration**) and the number of gallons per minute leaving the tank, that is,

$$\text{Rate of salt out} = (3 \text{ gal/min})\left(\frac{S}{V} \text{ lb/gal}\right)$$

The governing DE is therefore given by

$$\frac{dS}{dt} = -\frac{3S}{V}, \qquad S(0) = 15$$

where we have also included the initial amount of salt. In this problem the volume V of the mixture is not constant since the rate of liquid in and the rate of liquid out are not equal. Thus, to solve our DE we must first express V as a function of t. This we do by recognizing that

$$\frac{dV}{dt} = (\text{rate of liquid in}) - (\text{rate of liquid out})$$

$$= 4 - 3$$

$$= 1$$

or

$$V(t) = t + C_1$$

At time $t = 0$ the volume is 100 gal, so that $C_1 = 100$. Substituting the resulting expression for $V(t)$ into the first DE yields

$$\frac{dS}{dt} + \frac{3}{t + 100} S = 0, \qquad S(0) = 15$$

This is a first-order, linear, homogeneous DE whose solution can be readily shown to be

$$S(t) = \frac{15 \times 10^6}{(t + 100)^3}, \qquad t \geq 0$$

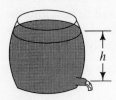

Figure 3.13

3.5.2 Flow of Water through an Orifice

Consider a tank filled with water that is pouring out near the bottom of the tank through an orifice (see Figure 3.13). **Torricelli's law**[†] states that the velocity (speed) v with which water issues from an orifice is

$$v = k\sqrt{2gh} \tag{22}$$

[†] Named in honor of the Italian physicist Evangelista Torricelli (1608–1647).

where k is a proportionality constant, g is the gravitational constant, and h is the height of water above the orifice at any time t. It has been determined experimentally that $k = 0.6$ for water. Also, if Q is the volume of water in the tank at any time t and A is the cross-sectional area of the orifice, it is known that the time rate change of volume is related to A by

$$\frac{dQ}{dt} = -Av = -0.6A\sqrt{2gh} \tag{23}$$

In solving Equation (23) it may be advantageous in some cases to express the volume Q in terms of the height h. Using the chain rule, we have

$$\frac{dQ}{dt} = \frac{dQ}{dh}\frac{dh}{dt} = B\frac{dh}{dt}$$

where $B(h)$ denotes the cross-sectional area of the tank at height h. Therefore, (23) can also be expressed in the form

$$B\frac{dh}{dt} = -0.6A\sqrt{2gh} \tag{24}$$

EXAMPLE 12   How long will it take to empty a cylindrical tank of radius 6 in. and vertical height of 2 ft if the tank is initially full of water and the orifice is a $\frac{1}{3}$-in.-diameter hole in the bottom of the tank?

Solution

The cross-sectional area of the orifice is

$$A = \pi(1/72)^2$$

where the $\frac{1}{6}$-in. radius (the diameter is $\frac{1}{3}$ in.) has been converted to feet. When the height of water in the tank is h ft, the volume of water is $Q = \pi(\frac{1}{2})^2 h$, where we have converted the 6-in. radius to feet. Therefore

$$B = \frac{dQ}{dh} = \frac{\pi}{4}$$

and (24) becomes

$$\frac{\pi}{4}\frac{dh}{dt} = -0.6\pi(1/72)^2\sqrt{2(32)h}$$

We can solve this DE by separation of variables, which leads to

$$2\sqrt{h} = C_1 - \frac{t}{270}$$

At time $t = 0$ the height of water is 2 ft, so the initial condition is $h(0) = 2$. Imposing this condition requires that $C_1 = 2\sqrt{2}$, and therefore, solving for t,

$$t = 540(\sqrt{2} - \sqrt{h})$$

The tank will be empty when $h = 0$, and this corresponds to

$$t = 540\sqrt{2} \cong 764 \text{ s}$$

Hence, it takes approximately 12 min and 44 s to empty the tank. ∎

3.5.3 Curves of Linear Pursuit

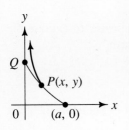

Figure 3.14 Pursuit curve.

A **curve of pursuit** is a path generated by a point P that is always moving in the direction of a second point Q constrained to move along a prescribed path. In other words, it is the path of a pursuer tracking its prey. The problem of finding such a curve seems to have originated with Leonardo da Vinci in the fifteenth century, but its curious difficulties still intrigue modern mathematicians.

The general problem of determining pursuit curves is very difficult, but certain special cases lend themselves to solution methods that we have discussed in Chapter 2. For simplicity, let us assume the "prey" is located at Q and constrained to move along the y axis (see Figure 3.14). The pursuer is assumed to be at the point $(a, 0)$ when the prey is at the origin. It is further assumed that the speeds of the pursuer and prey are constant.

To illustrate, let us imagine a large field in which a fox is located at the point $(a, 0)$. He spots a rabbit at $(0, 0)$ running along the y axis in the positive direction with constant speed v. The fox immediately runs toward the rabbit with speed w. After t seconds the rabbit is located at $Q(0, vt)$ and the fox is at $P(x, y)$. Because the fox is always running directly at the rabbit, the line $\overline{PQ}$ is always tangent to the path of the fox. The slope of this line is clearly

$$y' = \frac{y - vt}{x} \tag{25}$$

Our objective is to solve (25) for y as a function of x. To do so, we need to eliminate the parameter t by finding another relation. For example, observe that the length of the path traveled by the fox at speed w is given by the *arc length* formula

$$wt = \int_x^a \sqrt{1 + (dy/ds)^2}\, ds \tag{26}$$

where we use s as a dummy variable of integration. Now solving (25) and (26) for t and equating the resulting expressions, we find

$$\frac{y - xy'}{v} = \frac{1}{w} \int_x^a \sqrt{1 + (dy/ds)^2}\, ds \tag{27}$$

which is independent of t.

In its present form, Equation (27) seems quite formidable. However, by differentiating it with respect to x we can eliminate the integral on the right. This action leads to

$$\frac{y' - xy'' - y'}{v} = -\frac{1}{w}\sqrt{1 + (y')^2}$$

where we are using the relation

$$\frac{d}{dx} \int_x^a f(s) \, ds = -f(x)$$

Upon simplification, we get the *second-order* DE

$$xy'' = \frac{v}{w} \sqrt{1 + (y')^2} = k\sqrt{1 + (y')^2} \tag{28}$$

where k is the constant ratio of the two speeds. Although (28) is a second-order DE, we can reduce it to a first-order DE by making the substitution $p = y'$. Doing so, (28) becomes

$$xp' = k\sqrt{1 + p^2}$$

or, separating variables,

$$\frac{dp}{\sqrt{1 + p^2}} = \frac{k \, dx}{x} \tag{29}$$

To integrate the left-hand side of this last expression, we make the trigonometric substitution $p = \tan \theta$, which gives us

$$\int \frac{dp}{\sqrt{1 + p^2}} = \int \sec \theta \, d\theta$$

$$= \log|\sec \theta + \tan \theta|$$

$$= \log(p + \sqrt{1 + p^2})$$

Hence, the solution of (29) is given by

$$\log(p + \sqrt{1 + p^2}) = k \log x + C_1 \tag{30}$$

To determine the constant C_1, we note that at time $t = 0$ the slope of the pursuit curve is zero. Thus, $p = y' = 0$ when $x = a$ and (30) requires that $C_1 = -k \log a$. Making this substitution in (30) and letting $p = y'$, we have

$$\log[y' + \sqrt{1 + (y')^2}] = k \log \frac{x}{a} = \log\left(\frac{x}{a}\right)^k$$

or

$$y' + \sqrt{1 + (y')^2} = \left(\frac{x}{a}\right)^k \tag{31}$$

Isolating the square root on one side of the equation and squaring, we are led to

$$1 + (y')^2 = \left(\frac{x}{a}\right)^{2k} - 2y'\left(\frac{x}{a}\right)^k + (y')^2$$

which, solving for y', yields

$$y' = \frac{1}{2}\left[\left(\frac{x}{a}\right)^k - \left(\frac{x}{a}\right)^{-k}\right]$$

Finally, the integral of this last equation gives us the family of solutions

$$y = \frac{a}{2}\left[\frac{(x/a)^{k+1}}{1+k} - \frac{(x/a)^{1-k}}{1-k}\right] + C_2, \qquad k \neq 1 \qquad (32)$$

Recognizing that $y = 0$ when $x = a$, we deduce that $C_2 = ak/(1 - k^2)$, and thus

$$y = \frac{a}{2}\left[\frac{(x/a)^{k+1}}{1+k} - \frac{(x/a)^{1-k}}{1-k}\right] + \frac{ak}{1-k^2}, \qquad k \neq 1 \qquad (33)$$

The fox will catch the rabbit when $x = 0$, and this happens when

$$y = \frac{ak}{1-k^2} \qquad (34)$$

Of course, this expression makes sense only if the fox runs faster than the rabbit, that is, only if $k < 1$. The special case $k = 1$, which corresponds to the fox and rabbit running at the same speed, is taken up in the exercises (see Problems 17 and 18 in Exercises 3.5).

Exercises 3.5

1. A thermometer reading 20°F is brought into a room kept at 72°F. Two minutes later the thermometer reads 46°F.
   (a) Find the temperature reading for any time.
   (b) What is the temperature reading of the thermometer after 6 min?

2. A thermometer is taken from an inside room to the outside where the temperature is 7°F. After 1 min the thermometer reads 50°F, and after 4 min the reading is 32°F. What is the temperature of the inside room?

3. A 4-lb prime rib was initially warmed in a microwave oven and then immediately placed in a regular oven at 350°F. After 1 h the prime rib was at 120°F and after 2 h it was at 150°F. What was the temperature of the prime rib when it was removed from the microwave oven?

■**4.** A 5-lb roast is put into a 375°F oven at 6 P.M. After 75 min the roast is 125°F. If the roast was initially at 50°F, at what time will the roast be ready to eat medium rare (150°F)?

5. A pie is removed from a 350°F baking oven into a room at temperature 75°F.
   (a) How long will it take the pie to cool to 100°F if it cooled 150° in the first 4 min?
   (b) How long will it take the pie to reach 76°F?
   (c) How long will it take the pie to reach room temperature?

6. Fifty pounds of salt are initially dissolved in a tank holding 300 gal of water. A brine solution is pumped into the tank at the rate of 2 gal/min, and the (well-stirred) solution is allowed to flow out of the tank at the same rate. If the salt concentration entering the tank is 2 lb/gal, determine the amount of salt in the

tank
(a) at any time.
(b) after 60 min.
(c) How much salt will remain in the tank after a long period of time?

7. Solve Problem 6 if pure water instead of a brine mixture is pumped into the tank at the rate of 2 gal/min.

■8. Solve Problem 6 if the mixture is allowed to flow out of the tank at the slower rate of 1 gal/min.

9. Solve Problem 6 if the mixture is allowed to flow out of the tank at the faster rate of 3 gal/min.

10. How long will it take for 80 percent of the limiting amount of salt to accumulate in the tank of Problem 6?

11. A particular tank contains 200 liters of a dye solution with a concentration of 1 gal/liter. Fresh water is entering the tank at the rate of 2 liters/min, and the (well-stirred) solution is flowing out at the same rate. Find how much time will elapse before the dye concentration in the tank reaches 1 percent of its original value.

12. A 120-gal tank initially contains 90 lb of salt dissolved in 90 gal of water. A brine mixture containing 2 lb/gal of salt flows into the tank at the rate of 4 gal/min, and the mixture (kept well stirred) flows out at the rate of 3 gal/min. How long will it take for the tank to fill with the mixture and how much salt will be in the tank at the time?

13. How long does it take to empty a cylindrical tank of height 20 ft and radius 10 ft initially full of water if the cross-sectional area of the orifice in the bottom of the tank is 1 ft?

14. A conical tank of circular cross section standing on its apex whose angle is 60° has an outlet of cross-sectional area 0.5 cm². The tank is initially full of water and at time $t = 0$ the outlet is opened and the water flows out. Given that the height of the tank is 1 m, how long will it take for the tank to empty?

15. A water tank in the shape of a paraboloid of revolution measures 6 ft in diameter at the top and contains water 3 ft deep. When will the tank empty if a hole at the bottom has a 1-in. diameter?

*16. A large hemispherical cistern with a 25-ft radius is filled with water. A circular hole of radius 1 ft is cut into the bottom of the cistern. How long will it take for the cistern to empty?

17. In the fox-rabbit problem discussed in Section 3.5.3, find the path of the fox when $k = 1$, that is, when $v = w$.

18. In Problem 17, show that the fox will never catch the rabbit and, in fact, will never get even as close as $a/2$ to the rabbit.

19. Calculate the distance between the fox and the rabbit in terms of the variable x in the problem discussed in Section 3.5.3. Assume that $k > 1$, or $v > w$.

*20. A man standing at (0, 0) holds a rope of length L to which a weight is attached at the point $(L, 0)$. The man then walks along the positive y axis, dragging the weight after him (see figure).

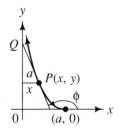

Problem 20

(a) Show that the slope of the path along which the weight moves is

$$y' = -\frac{\sqrt{L^2 - x^2}}{x}$$

Hint. Find $\tan \phi$ directly from the figure.

(b) Solve the DE in (a) to find the path of the weight. (This particular curve is called a **tractrix**.)

21. Two skaters are located on the x axis, skater Q at the origin and skater P at $(36, 0)$. Suppose that Q skates along the positive y axis and that P skates directly toward Q at all times. If P skates twice as fast as Q, how far will Q travel before being caught by P? Answer the question if P skates 3 times as fast as Q.

*22. A pilot always keeps the nose of his plane pointed toward a city C due west of his starting point at $(a, 0)$. If the plane's speed is v mi/h, and a wind is blowing from the south at the rate of w mi/h, show that the equation of the plane's path is

$$y = \frac{a}{2}\left[\left(\frac{x}{a}\right)^{1-k} - \left(\frac{x}{a}\right)^{1+k}\right]$$

where $k = w/v$.

23. In Problem 22, if the wind speed and plane speed are equal, show that the path of the plane is that of a parabola. Will the pilot ever reach city C in this case?

24. Find the equation of a curve that passes through the point $(4, 1)$ and has slope $-y/(x - 3)$ at the point (x, y) on the curve.

25. Find the shape of a curved mirror such that light from a distant source will be reflected to the origin.
Hint. The slope of such a curve in the xy plane is given by

$$y' = \frac{-x \pm \sqrt{x^2 + y^2}}{y}$$

*26. On a winter day it began snowing early in the morning, and the snow continued falling at a constant rate. The speed at which a snowplow can clear a road is inversely proportional to the height of the accumulated snow. Assume the snowplow started at 11 A.M. and had cleared 4 miles of road by 2 P.M. Another 2 miles was cleared by 5 P.M. At what time did it begin snowing?

*27. A cable of constant density hanging from two pegs (such as a telephone wire) assumes a shape determined by the DE

$$y + k\sqrt{1 + (y')^2} = 0$$

where k is a constant. Show that the cable assumes the shape of a hyperbolic cosine, called a **catenary.**

*28. One of the most famous problems in mechanics is called the **brachistochrone problem.** The problem is to determine the curve along which a particle will slide (without friction) from point 0 to point P (see figure) in the shortest time, where gravity is the only acting force. Point P is below 0, but not directly beneath it. The curve that solves the problem is a solution of the nonlinear DE

$$[1 + (y')^2]y = 2a$$

where a is a constant.

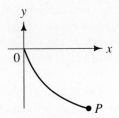

Problem 28

(a) Solve the DE for y', using the negative square root. (Why use the negative square root?)
(b) By letting $y = -2a \sin^2 t$, where t is a new parameter, solve the DE in (a).
(c) Show that the solution satisfying the auxiliary condition $y(0) = 0$ can be expressed in the parametric form

$$x = a(\theta - \sin \theta)$$

$$y = -a(1 - \cos \theta)$$

where $\theta = 2t$. The curve described by these equations is called a **cycloid.**

3.6 Chapter Summary

In this chapter we have illustrated how first-order DEs may be used in a variety of applications, including problems in geometry, mechanics, electric circuits, and population growth.

We say two families of curves are **orthogonal trajectories** of one another if every member of one family is orthogonal to every member of the second family. Given the one family $f(x, y, c) = 0$, we first seek its differential equation $y' = F(x, y)$. The governing DE of the second family is then $y' = -1/F(x, y)$, the solutions of which produce the orthogonal trajectories of the original family.

A **free-falling particle** of mass m, or one moving in a straight line, is governed by an equation of the form

$$m\frac{dv}{dt} = mg - F_R \tag{35}$$

where F_R represents a **resistive force,** often modeled by $F_R = cv$ or $F_R = cv^2$, where $c > 0$. The charge q in a **simple electric circuit** satisfies the equation

$$R\frac{dq}{dt} + \frac{1}{C}q = E(t) \tag{36}$$

where R is the resistance in the circuit, C is the capacitance, and $E(t)$ is a voltage source.

Radioactive decay or **population growth** may both be modeled by a simple linear DE of the form

$$\frac{dy}{dt} = ky \tag{37}$$

under suitable conditions. When an inhibiting factor is introduced into the population model, the population P may then satisfy an equation of the form

$$\frac{dP}{dt} = kP(M - P) \tag{38}$$

where k and M are positive constants. This same equation may also describe the flow of a virus among a certain population.

Review Exercises

1. Find the orthogonal trajectories of the family of curves $x^2 + y^2 = cx$.

2. Find the orthogonal trajectories of the family of curves $y = ce^x$.

3. A body of mass 2 slugs is dropped from a height of 800 ft with zero velocity. If the air resistance is $0.5v$ lb,

(a) What is the velocity of the body at all later times?

(b) What is the limiting velocity of the body?

4. An object weighing 192 lb falls from rest at time $t = 0$. If the resistance of the medium is $3v^2$,

(a) What is the velocity at all later times?

(b) What is the limiting velocity of the object?

5. An *RL* circuit has a voltage source of 5 V, a resistance of 50 Ω, an inductance of 1 H, and no initial current.

(a) What is the current at all later times?

(b) What is the steady-state current?

6. A resistor $R = 5\,\Omega$ and a capacitor $C = 0.02$ F are connected in series with a 100-V battery. At time $t = 0$ the charge on the capacitor is 5 C.

(a) What is the charge on the capacitor at all later times?

(b) What is the current in the circuit at all later times?

7. A certain culture of bacteria grows at a rate that is proportional to the number present. If B_0 bacteria are initially present and this number doubles in 4 h, how many bacteria will exist at the end of 12 h?

8. A certain radioactive substance is known to decay at a rate proportional to the amount present. If initially there is 100 mg of the substance present and after 2 years it is observed that 5 percent of the original amount has decayed, how long will it take for 10 percent of the original amount to decay?

9. If $\frac{1}{2}$ g of a radioactive substance is present at time $t = 0$ and 0.1 percent of the amount has decayed after 1 week, what is the half-life in years of the radioactive substance?

10. A 50-gal tank initially contains 10 gal of fresh water. At time $t = 0$, a brine solution containing 1 lb of salt per gallon enters the tank at the rate of 4 gal/min, while the well-stirred mixture leaves the tank at the rate of 2 gal/min.

(a) At what time will the tank overflow?

(b) How much salt will be in the tank at the time of overflow?

CHAPTER 4
Linear Equations of Higher Order

The study of **linear DEs** is of both theoretical and practical importance. In Chapter 2 we presented the general theory of first-order linear DEs and discussed some applications involving them in Chapter 3. Here we wish to build upon that theory and, in some instances, use it for motivational purposes in developing the corresponding theory for higher-order equations. Applications involving linear DEs will be taken up in subsequent chapters.

In Sections 4.2 to 4.5 we discuss only **second-order homogeneous DEs.** We first introduce the **superposition principle** and then discuss the notions of **linear independence** and **fundamental solution sets,** critical in the development of a **general solution.** In particular, we show how the linear independence of a set of solutions can be established by the evaluation of a determinant called the **Wronskian.** In Section 4.3 we introduce a method for producing a second linearly independent solution of an equation, given that one solution is known. In Section 4.4 we consider DEs with **constant coefficients.** Such DEs are special in that they can be solved by purely algebraic methods. The algebraic nature of this class of DEs is further emphasized in Section 4.5 by introducing **polynomial operators,** a special type of **differential operator.**

Constant-coefficient DEs of **higher order** (of order greater than 2) are solved in Section 4.6 by the same algebraic methods introduced in Section 4.4. Our treatment here mostly parallels that given in Sections 4.2 to 4.5.

We consider **nonhomogeneous DEs** in Sections 4.7 and 4.8, the **general solution** of which consists of the sum of a **homogeneous solution** and a **particular solution.** In Section 4.7 we discuss the **method of undetermined coefficients** for constructing a particular solution, while a more general method, called the **method of variation of parameters,** is presented in Section 4.8.

In Section 4.9 we extend the solution techniques for constant-coefficient DEs to a special **variable-coefficient DE** called the **Cauchy–Euler equation.** Some **historical comments** are given in the last section.

4.1 Introduction

The techniques introduced in Chapter 2 for solving first-order nonlinear DEs are generally not applicable to DEs of higher order. In fact, it is usually very difficult to find exact solutions of nonlinear DEs of order greater than 1. For this reason we restrict our attention here and in the remainder of the text almost exclusively to *linear* DEs. Also, most of our treatment will concern only *second-order* linear DEs since they occur more frequently in applications than equations of higher order.

Recall from Chapter 1 that the general **second-order linear DE** has the form

$$A_2(x)y'' + A_1(x)y' + A_0(x)y = F(x) \tag{1}$$

We call $A_0(x)$, $A_1(x)$, and $A_2(x)$ the **coefficients** of the equation, which are functions of x alone. Unless otherwise stated, we generally assume that the coefficients and $F(x)$ are continuous functions on some interval I, and that $A_2(x) \neq 0$ on I. If $A_0(x)$, $A_1(x)$, and $A_2(x)$ are all *constant*, we say the DE has **constant coefficients,** and **variable coefficients** otherwise.

The function $F(x)$ is called the **input function** or, in some cases, the **forcing function.** If $F(x)$ is not identically zero on I, then (1) is called a **nonhomogeneous** DE. On the other hand, if $F(x) \equiv 0$ on I, then (1) reduces to

$$A_2(x)y'' + A_1(x)y' + A_0(x)y = 0 \tag{2}$$

called the *associated* **homogeneous** equation. For example, the DE

$$xy'' + y' - 3y = 2 \sin x$$

is nonhomogeneous and its associated homogeneous equation is

$$xy'' + y' - 3y = 0$$

Second-order linear DEs are prominent in numerous areas of application. For instance, the forced and free oscillations of a spring-mass system are governed by the equation

$$m\frac{d^2y}{dt^2} + c\frac{dy}{dt} + ky = F(t)$$

where m, c, and k are system parameters (usually constant) and $F(t)$ is a prescribed external force. A similar equation arises in electric circuit problems where the circuit components may include resistors, coils, and capacitors. The static displacements of a string or wire supporting a distributed load (such as a telephone wire) can be described by a second-order linear DE, as can the steady-state temperatures in a rod or wire with insulated lateral surface. Problems involving circular or cylindrical-shaped regions often lead to the variable-coefficient DE

$$x^2y'' + xy' + (x^2 - v^2)y = 0$$

known as **Bessel's equation,**[†] whereas **Legendre's equation**[‡]

$$(1 - x^2)y'' - 2xy' + n(n + 1)y = 0$$

often occurs in problems featuring spherical-shaped regions.

For discussion purposes, it is sometimes convenient to divide each term of Equation (1) by the lead coefficient $A_2(x)$. Doing so, we obtain the **normal form**

$$y'' + a_1(x)y' + a_0(x)y = f(x) \qquad (3)$$

where $a_1(x) = A_1(x)/A_2(x)$, $a_0(x) = A_0(x)/A_2(x)$, and $f(x) = F(x)/A_2(x)$. The significant feature of Equation (3) is that the coefficient of the highest-order derivative is *unity*.

Because the solution family of a second-order DE contains two arbitrary constants, we must specify two auxiliary conditions in order to determine them. These are generally in the form of *initial conditions* where $y(x_0)$ and $y'(x_0)$ are specified. The following *existence-uniqueness* theorem, which is a natural generalization of Theorem 2.3 to the case of second-order DEs, is central in the theory of IVPs.

Theorem 4.1 Existence-Uniqueness. *If $a_0(x)$, $a_1(x)$, and $f(x)$ are continuous functions on an open interval I containing the point x_0, then the IVP*

$$y'' + a_1(x)y' + a_0(x)y = f(x)$$

$$y(x_0) = \alpha, \qquad y'(x_0) = \beta$$

has a unique solution on I for every α and β.[§]

As an illustration of Theorem 4.1, first observe that

$$y = C_1e^x + C_2xe^x$$

is a solution of the linear DE

$$y'' - 2y' + y = 0$$

for every choice of constants C_1 and C_2. If we prescribe the initial conditions

$$y(0) = 3, \qquad y'(0) = 1$$

then $y(0) = C_1 = 3$, $y'(0) = C_1 + C_2 = 1$, from which we deduce $C_1 = 3$ and $C_2 = -2$. Thus, we obtain the solution

$$y = (3 - 2x)e^x$$

[†] Named after Friedrich Wilhelm Bessel (1784–1846). See Section 7.7 and also the historical comments in Section 7.9.

[‡] Named after Adrien Marie Legendre (1752–1833). See also Section 7.6 and the historical comments in Section 7.9.

[§] For a proof of Theorem 4.1, see E. L. Ince, *Ordinary Differential Equations*, Dover, New York, 1956.

and based on Theorem 4.1, this is the only solution. On the other hand, had we prescribed the *homogeneous* initial conditions

$$y(0) = 0, \qquad y'(0) = 0$$

the only solution would be the **trivial solution** $y = 0$.

General solution formulas similar to those developed in Sections 2.4 and 2.5 for first-order linear DEs simply do not exist for second-order (and higher-order) linear DEs. There is, however, an extensive mathematical theory for these DEs that permits us to understand a great deal about their solutions and to readily solve many of them. Our initial development of this mathematical theory focuses on *homogeneous* equations, while the theory associated with *nonhomogeneous* equations begins in Section 4.7.

4.2 Second-Order Homogeneous Equations

Our development of the theory of second-order linear DEs starts with the associated *homogeneous* equation

$$y'' + a_1(x)y' + a_0(x)y = 0 \tag{4}$$

One of the most useful properties of this DE is the fact that the sum of two (or more) solutions is also a solution, as is any constant multiple of a solution. This remarkable property is known as the **superposition principle** (or **linearity property**), which we now state as a formal theorem.

Theorem 4.2 Superposition Principle. *If y_1 and y_2 are both solutions of the homogeneous DE*

$$y'' + a_1(x)y' + a_0(x)y = 0$$

on some interval I, then

$$y = C_1y_1(x) + C_2y_2(x)$$

is also a solution on I for any constants C_1 and C_2.

Proof. *The substitution of $y = C_1y_1(x) + C_2y_2(x)$ into the DE gives us*

$$y'' + a_1(x)y' + a_0(x)y = (C_1y_1'' + C_2y_2'') + a_1(x)(C_1y_1' + C_2y_2')$$
$$+ a_0(x)(C_1y_1 + C_2y_2)$$
$$= C_1[y_1'' + a_1(x)y_1' + a_0(x)y_1]$$
$$+ C_2[y_2'' + a_1(x)y_2' + a_0(x)y_2]$$
$$= C_1 \cdot 0 + C_2 \cdot 0$$

the last step a consequence of the fact that both y_1 and y_2 are solutions of the homogeneous DE. Hence, we have shown that $y = C_1y_1(x) + C_2y_2(x)$ is also a solution for any choice of constants C_1 and C_2.

To illustrate Theorem 4.2, we first observe that $y_1 = \cos x$ and $y_2 = \sin x$ are both solutions of $y'' + y = 0$. Hence, any linear combination of y_1 and y_2, such as

$$y = 5y_1 - 2y_2 = 5\cos x - 2\sin x$$

is also a solution.

4.2.1 Fundamental Solution Sets

The **general solution** of an nth-order linear DE is an n-parameter family of solutions. For example, if y_1 is any solution on an interval I of the homogeneous *first-order* linear DE

$$y' + a_0(x)y = 0 \tag{5}$$

then the general solution is given by (see Section 2.5)

$$y = C_1 y_1(x) \tag{6}$$

where C_1 is an arbitrary constant (parameter).[†] From Theorem 4.2, we also know that if y_1 and y_2 are both solutions on an interval I of the homogeneous *second-order* linear DE

$$y'' + a_1(x)y' + a_0(x)y = 0 \tag{7}$$

then, for any constants C_1 and C_2, the linear combination

$$y = C_1 y_1(x) + C_2 y_2(x) \tag{8}$$

is likewise a solution. Because it contains two constants, we might expect (8) to represent a general solution of (7). This is indeed the case provided y_1 and y_2 are "sufficiently different solutions" or, more precisely, *linearly independent solutions*.

Definition 4.1 We say two nonzero functions y_1 and y_2 are **linearly independent** functions on an interval I if neither one is a constant multiple of the other on I. Otherwise, we say that y_1 and y_2 are **linearly dependent** functions on I.

Based on Definition 4.1, we can say that y_1 and y_2 are *linearly independent* functions if

$$\frac{y_2(x)}{y_1(x)} \neq \text{constant} \tag{9}$$

while they are *linearly dependent* functions if

$$\frac{y_2(x)}{y_1(x)} = \text{constant} \tag{10}$$

[†] Also, $y = C_1/\mu(x)$, where $\mu(x)$ is an integrating factor.

EXAMPLE 1

Show that $y_1 = \sin x$ and $y_2 = 5 \sin x$ are linearly dependent on any interval I while $y_3 = 5 \sin 2x$ and y_1 are linearly independent on any interval I.

Solution

In the first case, we observe that (for $\sin x \neq 0$)

$$\frac{y_2(x)}{y_1(x)} = \frac{5 \sin x}{\sin x} = 5$$

hence, y_1 and y_2 are *linearly dependent*. In the second case,

$$\frac{y_3(x)}{y_1(x)} = \frac{5 \sin 2x}{\sin x} = 10 \cos x \neq \text{constant}$$

where we have used the identity $\sin 2x = 2 \sin x \cos x$; thus, we conclude that y_1 and y_3 are *linearly independent*. ■

EXAMPLE 2

Discuss the linear independence of $y_1 = x$ and $y_2 = |x|$ over the intervals $0 < x < \infty$ and $-\infty < x < \infty$ (see Figure 4.1).

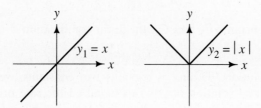

Figure 4.1

Solution

Here we see that $(x \neq 0)$

$$\frac{y_2(x)}{y_1(x)} = \frac{|x|}{x} = \begin{cases} 1, & 0 < x < \infty \\ -1, & -\infty < x < 0 \end{cases}$$

Thus, since the ratio $y_2(x)/y_1(x)$ remains constant over the first interval $0 < x < \infty$, the functions are *linearly dependent* on this interval. However, over the larger interval $-\infty < x < \infty$, the ratio $y_2(x)/y_1(x)$ changes values, and thus is not constant. [At $x = 0$ we see that $y_1(x) = k y_2(x)$ for any value k.] The functions are therefore *linearly independent* on $-\infty < x < \infty$. ■

Notice that although the trivial function $y_1 = 0$ is always a solution of any *homogeneous* DE, it is not linearly independent of any *nontrivial* function $y_2 = \phi(x)$ (on any interval) since $0 \cdot \phi(x) = 0$. Linearly independent solutions y_1 and y_2 of the same second-order homogeneous DE on an interval I are said to form a **fundamental set of solutions** on I, and in this case we say that (8) is a **general solution.** For a given DE, there are (infinitely) many possible fundamental solution sets. Nonetheless, if the conditions of Theorem 4.1 are met, every general solution will lead to the same particular solution of an IVP (see Example 4 below).

We now state the following important theorem.[†]

Theorem 4.3 *If y_1 and y_2 form a fundamental set of solutions of the homogeneous linear DE*

$$y'' + a_1(x)y' + a_0(x)y = 0$$

on some interval I where $a_1(x)$ and $a_0(x)$ are continuous functions, then every solution of this DE is given by

$$y = C_1 y_1(x) + C_2 y_2(x)$$

for appropriate choices of the constants C_1 and C_2. Furthermore, a fundamental set of solutions always exists.

■──────────
EXAMPLE 3

Show that both $y_1 = e^x$ and $y_2 = e^{-2x}$ are solutions of

$$y'' + y' - 2y = 0, \qquad -\infty < x < \infty$$

and find a general solution.

Solution

By substituting $y_1 = e^x$ into the left-hand side of the DE, we see that

$$y'' + y' - 2y = e^x + e^x - 2e^x$$
$$= 0$$

and thus $y_1 = e^x$ is a solution. Similarly, checking $y_2 = e^{-2x}$ leads to

$$y'' + y' - 2y = 4e^{-2x} - 2e^{-2x} - 2e^{-2x}$$
$$= 0$$

which shows that it also is a solution. Moreover,

$$\frac{y_2(x)}{y_1(x)} = e^{-3x} \neq \text{constant}, \qquad -\infty < x < \infty$$

───────

[†] For a proof of Theorem 4.2, see E. L. Ince, *Ordinary Differential Equations,* Dover, New York, 1956.

and thus $y_1 = e^x$ and $y_2 = e^{-2x}$ form a *fundamental solution set* on the interval $-\infty < x < \infty$. If follows therefore that a *general solution* of the given DE is

$$y = C_1 e^x + C_2 e^{-2x}$$

where C_1 and C_2 are arbitrary constants.   ■

EXAMPLE 4

Show, for arbitrary C_1 and C_2, that

$$y = C_1 e^{-x} + C_2 e^x \quad \text{and} \quad y = C_1 \cosh x + C_2 \sinh x$$

are both general solutions of $y'' - y = 0$, $-\infty < x < \infty$, and find a particular solution in both cases of the IVP

$$y'' - y = 0, \qquad y(0) = 1, \qquad y'(0) = -3$$

Solution

Checking, we find that $y_1 = e^{-x}$ and $y_2 = e^x$ are each solutions of $y'' - y = 0$. Also, we see that

$$\frac{y_2(x)}{y_1(x)} = e^{2x} \neq \text{constant}, \qquad -\infty < x < \infty$$

and deduce that $y_1 = e^{-x}$ and $y_2 = e^x$ form a fundamental solution set on the interval $-\infty < x < \infty$. Therefore,

$$y = C_1 e^{-x} + C_2 e^x$$

is a general solution of $y'' - y = 0$. Imposing the initial conditions, we find that

$$y(0) = C_1 + C_2 = 1$$
$$y'(0) = -C_1 + C_2 = -3$$

from which we deduce $C_1 = 2$ and $C_2 = -1$. Hence,

$$y = 2e^{-x} - e^x$$

Similarly, it can be shown that $y_1 = \cosh x$ and $y_2 = \sinh x$[†] are each solutions of $y'' - y = 0$, and further, that

$$\frac{y_2(x)}{y_1(x)} = \tanh x \neq \text{constant}, \qquad -\infty < x < \infty$$

Thus, $y_1 = \cosh x$ and $y_2 = \sinh x$ also form a fundamental solution set on the interval $-\infty < x < \infty$, so that

$$y = C_1 \cosh x + C_2 \sinh x$$

[†] Recall that $\cosh x = \frac{1}{2}(e^x + e^{-x})$ and $\sinh x = \frac{1}{2}(e^x - e^{-x})$.

is another (but equivalent) general solution. This time the initial conditions require that $C_1 = 1$ and $C_2 = -3$, leading to

$$y = \cosh x - 3 \sinh x$$
$$= \tfrac{1}{2}(e^x + e^{-x}) - \tfrac{3}{2}(e^x - e^{-x})$$
$$= 2e^{-x} - e^x$$

which is the same particular solution as above.   ∎

4.2.2 Wronskians

Observe that if constants C_1 and C_2 (not both zero) can be found such that $C_1 y_1(x) + C_2 y_2(x) = 0$ for all x in some interval I, then y_1 and y_2 are linearly dependent on I. That is, this condition is equivalent to ($C_1 \neq 0$)

$$y_1(x) = -\frac{C_2}{C_1} y_2(x) = k y_2(x)$$

showing that one function is a multiple of the other. Now let us consider the simultaneous equations

$$C_1 y_1(x) + C_2 y_2(x) = 0$$
$$C_1 y_1'(x) + C_2 y_2'(x) = 0 \tag{11}$$

the second of which is the derivative of the first. If we think of C_1 and C_2 as unknowns in (11), then *nonzero* values for C_1 and C_2 are possible (according to *Cramer's rule* from linear algebra) only when the coefficient determinant of (11) is zero for every x on some interval I, that is, only when

$$\begin{vmatrix} y_1(x) & y_2(x) \\ y_1'(x) & y_2'(x) \end{vmatrix} \equiv 0 \text{ on } I \tag{12}$$

This coefficient determinant is called the **Wronskian,** named after the Polish mathematician J. Wronski,[†] and denoted by

$$W(y_1, y_2)(x) = \begin{vmatrix} y_1(x) & y_2(x) \\ y_1'(x) & y_2'(x) \end{vmatrix}$$
$$= y_1(x)y_2'(x) - y_1'(x)y_2(x) \tag{13}$$

[†] Jozef M. H. Wronski (1778–1853) studied mathematics in Germany but lived most of his life in France. His most lasting contribution to mathematics appears to be the Wronskian determinant.

When confusion is not likely to occur, the Wronskian may also be written more compactly as simply $W(x)$.

Thus, if y_1 and y_2 are linearly *dependent* on an interval I, we have shown that $W(y_1, y_2)(x) \equiv 0$ on I. On the other hand, if $W(y_1, y_2)(x_0) \neq 0$ for some x_0 on I, does this mean that y_1 and y_2 are linearly independent? The answer is—not necessarily. That is, the condition $W(y_1, y_2)(x_0) \neq 0$ is only a sufficient condition, not a necessary condition, for y_1 and y_2 to be linearly independent. Nonetheless, if y_1 and y_2 are both *solutions* of the same DE, we can show that $W(y_1, y_2)(x_0) \neq 0$ for some x_0 on I is both a *necessary* and *sufficient condition* for their linear independence on I. To show this, we start with the following theorem.

Theorem 4.4 Wronskian of Solutions. *Suppose that y_1 and y_2 are solutions of*

$$y'' + a_1(x)y' + a_0(x)y = 0$$

on some open interval I for which $a_1(x)$ and $a_0(x)$ are continuous functions.

(a) If y_1 and y_2 are linearly dependent on I, then $W(y_1, y_2)(x) \equiv 0$ on I.

(b) If y_1 and y_2 are linearly independent on I, then $W(y_1, y_2)(x) \neq 0$ for every x on I.

Proof. *The proof of part (a) is already complete. For part (b), let us assume that $W(y_1, y_2)(x_0) = 0$ for some x_0 on I and show that this implies that y_1 and y_2 are linearly dependent.*

Based on Cramer's rule, it follows that the condition $W(y_1, y_2)(x_0) = 0$ implies that the system of equations

$$C_1 y_1(x_0) + C_2 y_2(x_0) = 0$$
$$C_1 y_1'(x_0) + C_2 y_2'(x_0) = 0$$

has a nontrivial solution; that is, C_1 and C_2 cannot both be zero. Now if we define the function

$$y = C_1 y_1(x) + C_2 y_2(x)$$

then the above conditions suggest that y satisfies the initial conditions

$$y(x_0) = 0, \qquad y'(x_0) = 0$$

Observe, however, that $y = 0$ satisfies these conditions and the homogeneous DE

$$y'' + a_1(x)y' + a_0(x)y = 0$$

Hence, by Theorem 4.1 it is the only solution. But the condition $y = 0$, or

$$C_1 y_1(x) + C_2 y_2(x) = 0$$

implies that y_1 and y_2 are linearly dependent solutions and the proof is complete.

The next theorem, due to Abel,[†] shows that, based on the DE alone, the Wronskian of two solutions can always be determined to within a multiplicative constant.

Theorem 4.5 Abel's Formula.   *If y_1 and y_2 are any two solutions of*

$$y'' + a_1(x)y' + a_0(x)y = 0$$

on an interval I where $a_1(x)$ and $a_0(x)$ are continuous functions, then the Wronskian of y_1 and y_2 is given by

$$W(y_1, y_2)(x) = C \exp\left(-\int a_1(x)\, dx\right)$$

where C is some constant.

Proof.   *Taking the derivative of the Wronskian $W(x) = W(y_1, y_2)(x)$ given by (13) leads to*

$$W'(x) = \frac{d}{dx}\left[y_1(x)y_2'(x) - y_1'(x)y_2(x)\right]$$

$$= y_1(x)y_2''(x) - y_1''(x)y_2(x)$$

However, since both y_1'' and y_2'' satisfy the relation

$$y'' = -a_1(x)y' - a_0(x)y$$

which is just the DE with all terms except for y'' shifted to the right-hand side, we can rewrite the derivative of the Wronskian as

$$W'(x) = y_1(x)\left[-a_1(x)y_2' - a_0(x)y_2\right]$$

$$- y_2(x)\left[-a_1(x)y_1' - a_0(x)y_1\right]$$

$$= -a_1(x)\left[y_1(x)y_2'(x) - y_1'(x)y_2(x)\right]$$

$$= -a_1(x)W(x)$$

Thus, we have shown that the Wronskian satisfies the first-order linear DE

$$W'(x) + a_1(x)W(x) = 0$$

the general solution of which is (see Section 2.4 or 2.5)

$$W(x) = C \exp\left(-\int a_1(x)\, dx\right)$$

where C is any constant.

[†] Niels H. Abel (1802–1829) was one of six children born into a poor Norwegian family. One of his early accomplishments was proving that the general fifth-degree algebraic equation has no radical solution. Stricken with tuberculosis in 1827, he died two years later at the age of 26.

EXAMPLE 5

To within a multiplicative constant, find the Wronskian associated with the DE

$$xy'' + y' + xy = 0$$

Solution

We first write the DE in normal form, that is,

$$y'' + \frac{1}{x}y' + y = 0, \qquad x \neq 0$$

Thus, we identify $a_1(x) = 1/x$, and using Abel's formula (Theorem 4.5), we find that

$$W(y_1, y_2)(x) = C \exp\left(-\int \frac{dx}{x}\right)$$

$$= C \exp(-\log|x|)$$

from which we deduce

$$W(y_1, y_2)(x) = \frac{C}{x}, \qquad x \neq 0 \qquad \blacksquare$$

Based on Abel's formula (Theorem 4.5), it is now easy to show that the Wronskian is either identically zero or else is never zero on I. That is, given the expression

$$W(y_1, y_2)(x) = C \exp\left(-\int a_1(x)\, dx\right)$$

observe that the exponential function cannot be zero for $a_1(x)$ continuous. Thus, if $C \neq 0$, we conclude that $W(y_1, y_2)(x) \neq 0$ for any x on I. On the other hand, if $C = 0$, then the Wronskian is identically zero on I.

Summarizing our results, we can draw the following conclusions:

> If y_1 and y_2 are solutions of the same second-order homogeneous DE on an interval I and, further,
>
> 1. If $W(y_1, y_2)(x_0) \neq 0$ at any point x_0 on I, then y_1 and y_2 are **linearly independent** functions and, as such, form a **fundamental set of solutions** on I.
> 2. If $W(y_1, y_2)(x_0) = 0$ at any point x_0 on I, then $W(y_1, y_2)(x) \equiv 0$ on I, and in this case y_1 and y_2 are **linearly dependent** functions.

EXAMPLE 6

Both $y_1 = \cos x$ and $y_2 = \sin x$ are solutions of

$$y'' + y = 0, \qquad -\infty < x < \infty$$

Use the Wronskian test to prove that y_1 and y_2 are linearly independent functions.

Solution

By definition, the Wronskian is

$$W(y_1, y_2)(x) = \begin{vmatrix} \cos x & \sin x \\ -\sin x & \cos x \end{vmatrix}$$

$$= \cos^2 x + \sin^2 x$$

$$= 1$$

for all x. Hence, the Wronskian cannot be zero and we deduce that y_1 and y_2 are linearly independent. ∎

EXAMPLE 7

Both $y_1 = \cos 2x$ and $y_2 = 2 \sin^2 x - 1$ are solutions of

$$y'' + 4y = 0, \qquad -\infty < x < \infty$$

Use the Wronskian test to determine whether they are linearly dependent or independent.

Solution

Here we find that

$$W(y_1, y_2)(x) = \begin{vmatrix} \cos 2x & 2\sin^2 x - 1 \\ -2\sin 2x & 4\sin x \cos x \end{vmatrix}$$

However, rather than attempting to simplify this expression, let us check to see whether it is zero or not by simply choosing a test value of x on the interval $-\infty < x < \infty$. For example, by setting $x = 0$, we see that

$$W(y_1, y_2)(0) = \begin{vmatrix} 1 & -1 \\ 0 & 0 \end{vmatrix} = 0$$

and thus deduce that the Wronskian is zero for all x. It follows, therefore, that y_1 and y_2 are linearly dependent in this case. ∎

Exercises 4.2

In Problems 1 to 10, use Definition 4.1 to determine whether the given functions are linearly dependent or independent.

1. $y_1 = 1$, $y_2 = x$

2. $y_1 = x^n$, $y_2 = x^{n+1}$, $n = 1, 2, 3, \ldots$

3. $y_1 = \log x$, $y_2 = 3 \log x^2$

4. $y_1 = e^x$, $y_2 = e^{2x}$

5. $y_1 = \cos x$, $y_2 = \sin x$

■**6.** $y_1 = \sin 2x$, $y_2 = \sin x \cos x$

7. $y_1 = e^x + e^{-x}$, $y_2 = e^x - e^{-x}$

8. $y_1 = e^x$, $y_2 = xe^x$

9. $y_1 = x$, $y_2 = e^{-\log x}$

10. $y_1 = 5(1 - x^2)$, $y_2 = (1 + x)(1 - x)$

*****11.** Show that the functions $y_1 = x^3$ and $y_2 = |x|^3$ are linearly independent on the interval

$-\infty < x < \infty$, but that the Wronskian of y_1 and y_2 is identically zero.

Hint. Show separately that $W(y_1, y_2)(x) \equiv 0$ for $x \geq 0$ and for $x < 0$.

12. Show that the functions $y_1 = x^2$ and $y_2 = x|x|$ are linearly dependent on any interval for which either $x > 0$ or $x < 0$, but are linearly independent on the larger interval $-\infty < x < \infty$.

In Problems 13 to 18, compute the Wronskian of the given pair of functions.

13. $y_1 = \sinh x$, $y_2 = \cosh x$

■14. $y_1 = e^x$, $y_2 = xe^x$

15. $y_1 = e^{ax}$, $y_2 = e^{bx}$

16. $y_1 = e^x \cos x$, $y_2 = e^x \sin x$

17. $y_1 = \sin^2 x$, $y_2 = \cos 2x - 1$

18. $y_1 = x \sin x$, $y_2 = \sin x$

In Problems 19 to 24, determine if the given functions form a fundamental set of solutions. If so, obtain a unique solution satisfying the IVP and state the interval of validity.

19. $2y'' - 5y' - 3y = 0$, $y(0) = 0$, $y'(0) = 1$;
$y_1 = e^{-x/2}$, $y_2 = e^{3x}$

20. $y'' - 4y' + 13y = 0$, $y(0) = -3$, $y'(0) = 0$;
$y_1 = e^{2x} \cos 3x$, $y_2 = e^{2x} \sin 3x$

21. $y'' - 6y' + 9y = 0$, $y(0) = 0$, $y'(0) = 0$;
$y_1 = e^{3x}$, $y_2 = xe^{3x}$

■22. $x^2 y'' + xy' - 4y = 0$, $y(1) = 1$, $y'(1) = -1$;
$y_1 = x^{-2}$, $y_2 = x^2$

23. $y'' + 4y = 0$, $y(0) = 5$, $y'(0) = 2$;
$y_1 = \sin x \cos x$, $y_2 = \sin 2x$

24. $y'' + 16y = 0$, $y(0) = 2$, $y'(0) = 1$;
$y_1 = \cos 4x$, $y_2 = \sin^2 2x - \cos^2 2x$

In Problems 25 to 28, use Abel's formula (Theorem 4.5) to determine the Wronskian (to within a multiplicative constant) of the solutions of the given DE.

25. $y'' - 4y' + 4y = 0$

26. $y'' - 3y' + 2y = 0$

27. $(1 - x^2)y'' - 2xy' + 6y = 0$

28. $x^2 y'' - xy' + (x^2 - 4)y = 0$

29. Given that y_1 and y_2 are linearly independent solutions of $x^2 y'' - 2xy' + 2y = 0$ and that $W(y_1, y_2)(3) = 9$, what is the value $W(y_1, y_2)(5)$?

■30. Given that y_1 and y_2 are linearly independent solutions of $y'' + y' - 2y = 0$ and that $W(y_1, y_2)(0) = -3$, what is the value $W(y_1, y_2)(2)$?

*31. Prove that if y_1 and y_2 are linearly independent functions, then so are the functions

$$y_3 = y_1 + y_2, \qquad y_4 = y_1 - y_2$$

*32. Prove that if y_1 and y_2 vanish at the same point in the interval $a \leq x \leq b$, they cannot form a set of linearly independent solutions of a second-order linear DE.

*33. The functions $y_1 = x$ and $y_2 = x^2$ are linearly independent solutions of $x^2 y'' - 2xy' + 2y = 0$ on the interval $-1 \leq x \leq 1$, but $W(y_1, y_2)(0) = 0$. Does this contradict Theorem 4.4? Explain.

*34. Show, by means of the Wronskian, that the homogeneous first-order DE

$$y' + a_0(x)y = 0$$

cannot have two linearly independent solutions y_1 and y_2.

4.3 Constructing a Second Solution From a Known Solution

It sometimes happens that a certain solution technique may lead to only one solution y_1 of a second-order linear DE. However, it is a curious fact that a second linearly independent solution y_2 can always be constructed (at least theoretically) from knowledge of only one solution. This being the case, we can always produce a fundamental solution set once we determine one solution. In this regard we have the following theorem.

Theorem 4.6 *If y_1 is a nontrivial solution of the homogeneous linear DE*

$$y'' + a_1(x)y' + a_0(x)y = 0$$

on some interval I where $a_0(x)$ and $a_1(x)$ are continuous functions, then a second linearly independent solution on I is given by

$$y_2 = y_1(x) \int \frac{\exp\left(-\int a_1(x)\,dx\right)}{y_1^2(x)}\,dx$$

Proof. *If y_1 and y_2 are linearly independent solutions of the given DE on the interval I, then, by definition*

$$y_1(x)y_2' - y_1'(x)y_2 = W(y_1, y_2)(x)$$

where $W(y_1, y_2)(x) \neq 0$ on I. We wish to interpret this equation as a linear first-order DE in y_2. Thus, after dividing the equation by $y_1(x)$ to put it in normal form, we recall that a solution is given by (see Section 2.4 or 2.5)

$$y_2 = y_1(x) \int \frac{W(y_1, y_2)(x)}{y_1^2(x)}\,dx$$

Based on Theorem 4.5, it now follows that

$$y_2 = Cy_1(x) \int \frac{\exp\left(-\int a_1(x)\,dx\right)}{y_1^2(x)}\,dx$$

Regardless of the value of C, this last expression is a solution of the given second-order linear DE, and so we simply set $C = 1$. (We leave it to the reader to verify directly that y_2 is a solution of the DE.)

EXAMPLE 8

Given that $y_1 = e^{-x}$ is a solution of

$$y'' - y' - 2y = 0, \qquad -\infty < x < \infty$$

find a general solution of the DE.

Solution

Using Theorem 4.6, we first obtain

$$y_2 = e^{-x} \int \frac{\exp\left(\int dx\right)}{e^{-2x}}\,dx$$

$$= e^{-x} \int e^{3x}\,dx$$

$$= \frac{1}{3} e^{2x}$$

Based on Theorem 4.6, we know that $y_2 = \frac{1}{3}e^{2x}$ is linearly independent of $y_1 = e^{-x}$. Moreover, since any multiple of y_2 is also a solution, we can ignore the multiplicative constant $\frac{1}{3}$ and write the general solution as

$$y = C_1 e^{-x} + C_2 e^{2x}$$ ■

■ **EXAMPLE 9**   Given that $y_1 = x^{-1}$ is a solution of

$$x^2 y'' + 3xy' + y = 0$$

find a general solution of the DE valid for $x > 0$.

Solution

Here we must first put the DE in normal form, that is,

$$y'' + \frac{3}{x} y' + \frac{1}{x^2} y = 0$$

Thus, we identify $a_1(x) = 3/x$, and by use of Theorem 4.6 we find that

$$y_2 = x^{-1} \int \frac{\exp\left[-3\int (dx/x)\right]}{x^{-2}} \, dx$$

$$= x^{-1} \int \frac{dx}{x}$$

$$= x^{-1} \log x$$

Therefore, we can construct the general solution

$$y = x^{-1}(C_1 + C_2 \log x), \qquad x > 0$$ ■

Exercises 4.3

In Problems 1 to 15, use the given solution and Theorem 4.6 to construct a general solution of the DE.

1. $y'' + 2y' = 0$; $y_1 = 1$
2. $y'' + 2y' - 3y = 0$; $y_1 = e^x$
3. $y'' - 6y' + 9y = 0$; $y_1 = e^{3x}$
■4. $y'' - 4y' + 4y = 0$; $y_1 = xe^{2x}$
5. $y'' + y = 0$; $y_1 = \sin x$
6. $y'' - y = 0$; $y_1 = \cosh x$
7. $y'' - 2y' + 5y = 0$; $y_1 = e^x \cos 2x$
8. $x^2 y'' + 2xy' - 6y = 0$; $y_1 = x^2$
9. $x^2 y'' - 6y = 0$; $y_1 = x^3$

10. $x^2 y'' - 20y = 0$; $y_1 = x^{-4}$
11. $4x^2 y'' + y = 0$; $y_1 = \sqrt{x} \log x$
■12. $x^2 y'' - xy' + 2y = 0$; $y_1 = x \sin(\log x)$
13. $(1 - x^2)y'' - 2xy' = 0$; $y_1 = 1$
*14. $(1 - x^2)y'' - 2xy' + 2y = 0$; $y_1 = x$
*15. $x^2 y'' + xy' + (x^2 - \frac{1}{4})y = 0$; $y_1 = x^{-1/2} \sin x$

16. Verify that $y_1 = e^x$ is a solution of

$$xy'' - (x + n)y' + ny, \qquad n = 0, 1, 2, \ldots$$

and find a second linearly independent solution for the case

(a) When $n = 1$.

(b) When $n = 2$.

(c) For $n = 1, 2, 3, \ldots$, verify that

$$y_2 = 1 + x + \frac{x^2}{2!} + \cdots + \frac{x^n}{n!}$$

which is simply the first $n + 1$ terms of the Maclaurin series for e^x.

*17. By assuming $y_2 = u(x)y_1(x)$ is a second solution of

$$y'' + a_1(x)y' + a_0(x)y = 0$$

given that y_1 is a known solution,

(a) Show that the function u satisfies

$$u'' + \left[a_1(x) + 2\frac{y_1'(x)}{y_1(x)} \right]u' = 0$$

(b) Let $v = u'$ and solve the resulting first-order DE in v to obtain the result

$$v = \frac{1}{y_1^2(x)} \exp\left(-\int a_1(x)\, dx \right)$$

(c) From (b), obtain an expression for u and verify that $y_2 = u(x)y_1(x)$ is the same solution as given in Theorem 4.6.

4.4 Homogeneous Equations with Constant Coefficients

Differential equations with *constant coefficients* are the easiest class of linear DEs to solve by any general method since the solution technique is primarily algebraic. We have previously found that the first-order linear DE

$$y' + ay = 0, \qquad -\infty < x < \infty \tag{14}$$

where a is constant, has the exponential solution $y = C_1 e^{-ax}$ valid on the interval $-\infty < x < \infty$. Because the derivative of an exponential function is simply a multiple of itself, it may seem natural to assume that higher-order linear DEs with constant coefficients also exhibit exponential solutions.

For example, suppose we consider the second-order DE

$$ay'' + by' + cy = 0, \qquad -\infty < x < \infty \tag{15}$$

where a, b, and c are constants. Let us assume that (15) has an exponential solution of the form

$$y = e^{mx} \tag{16}$$

for some value or values of the parameter m. Direct substitution of (16) into the left-hand side of (15) leads to

$$ay'' + by' + cy = am^2 e^{mx} + bm e^{mx} + c e^{mx}$$

$$= (am^2 + bm + c)e^{mx}$$

which, since $e^{mx} \neq 0$, can vanish if and only if

$$am^2 + bm + c = 0 \tag{17}$$

Observe that Equation (17), called the **auxiliary equation** or **characteristic equation** of (15), is dependent only upon knowledge of the constants a, b, and c in (15). By using the quadratic formula, we can obtain its two solutions

$$m_1 = \frac{-b + \sqrt{b^2 - 4ac}}{2a}, \qquad m_2 = \frac{-b - \sqrt{b^2 - 4ac}}{2a} \tag{18}$$

Clearly now, there are three separate cases to consider depending upon whether m_1 and m_2 are **real and distinct roots** ($b^2 - 4ac > 0$), **real but equal roots** ($b^2 - 4ac = 0$), or **complex conjugate roots** ($b^2 - 4ac < 0$).

Case I. Real and Distinct Roots: $b^2 - 4ac > 0$

When the roots m_1 and m_2 are real and $m_1 \neq m_2$, we find two solutions of the form (16), namely,

$$y_1 = e^{m_1 x}$$

$$y_2 = e^{m_2 x}$$

Applying the Wronskian test, we have

$$W(y_1, y_2)(x) = \begin{vmatrix} e^{m_1 x} & e^{m_2 x} \\ m_1 e^{m_1 x} & m_2 e^{m_2 x} \end{vmatrix} = (m_2 - m_1)e^{(m_1 + m_2)x} \neq 0$$

and thus conclude that these are linearly independent solutions of (15). In this case, we can express the general solution as

$$y = C_1 e^{m_1 x} + C_2 e^{m_2 x}, \qquad -\infty < x < \infty \tag{19}$$

EXAMPLE 10

Find a general solution of

$$2y'' + 5y' - 3y = 0$$

Solution

By inspection, we obtain the auxiliary equation

$$2m^2 + 5m - 3 = 0$$

which can also be written in the factored form

$$(2m - 1)(m + 3) = 0$$

Thus, the roots are $m_1 = 1/2$ and $m_2 = -3$, leading us to the solutions $y_1 = e^{x/2}$ and $y_2 = e^{-3x}$. The general solution in this case is

$$y = C_1 e^{x/2} + C_2 e^{-3x}$$

where C_1 and C_2 are arbitrary constants.

EXAMPLE 11

Solve the IVP

$$y'' + 3y' = 0, \qquad y(0) = 2, \qquad y'(0) = 1$$

Solution

The auxiliary equation is

$$m^2 + 3m = m(m + 3) = 0$$

with roots $m_1 = 0$ and $m_2 = -3$. Therefore, $y_1 = 1$ and $y_2 = e^{-3x}$, from which we deduce the general solution

$$y = C_1 + C_2 e^{-3x}$$

Computing the derivative, we find

$$y' = -3C_2 e^{-3x}$$

and hence imposing the initial conditions leads to

$$y(0) = C_1 + C_2 = 2$$
$$y'(0) = -3C_2 = 1$$

It now follows that $C_2 = -1/3$ and $C_1 = 7/3$, giving us the particular solution

$$y = \tfrac{1}{3}(7 - e^{-3x}) \qquad\qquad \blacksquare$$

Case II. Equal Roots: $b^2 - 4ac = 0$

When the two roots of the auxiliary equation are equal, that is, $m_1 = m_2 = m = -b/2a$, we initially obtain only one solution

$$y_1 = e^{mx} = e^{-bx/2a}$$

However, a second linearly independent solution can be found by application of Theorem 4.6. In this case we see that

$$y_2 = e^{-bx/2a} \int \frac{\exp\left[-(b/a) \int dx \right]}{e^{-bx/a}} \, dx$$

$$= e^{-bx/2a} \int dx = xe^{-bx/2a}$$

or

$$y_2 = xe^{mx}$$

The general solution then assumes the form

$$y = C_1 e^{mx} + C_2 xe^{mx}$$
$$= (C_1 + C_2 x)e^{mx}, \qquad -\infty < x < \infty \qquad\qquad (20)$$

EXAMPLE 12

Find a general solution of

$$y'' + 4y' + 4y = 0$$

Solution

This time the auxiliary equation is

$$m^2 + 4m + 4 = (m + 2)^2 = 0$$

with the double root $m = -2$. Hence, $y_1 = e^{-2x}$ and $y_2 = xe^{-2x}$ form a fundamental set of solutions and the general solution is

$$y = (C_1 + C_2x)e^{-2x}$$ ∎

EXAMPLE 13

Solve the IVP

$$y'' - 2y' + y = 0, \qquad y(0) = 3, \qquad y'(0) = 1$$

Solution

The auxiliary equation $m^2 - 2m + 1 = 0$ has the double root $m = 1$. Therefore, the general solution is

$$y = (C_1 + C_2x)e^x$$

In order to apply the prescribed initial conditions we also need the derivative

$$y' = (C_1 + C_2x)e^x + C_2e^x$$
$$= C_1e^x + C_2(1 + x)e^x$$

Now setting $x = 0$ in both y and y', we see that

$$y(0) = C_1 = 3$$
$$y'(0) = C_1 + C_2 = 1$$

Hence, $C_1 = 3$ and $C_2 = -2$, leading to the particular solution

$$y = (3 - 2x)e^x$$ ∎

4.4.1 Complex Exponential Functions

When the roots m_1 and m_2 given by (18) are complex (i.e., $b^2 - 4ac < 0$), they are complex conjugates that we can write as $m_1 = p + iq$ and $m_2 = p - iq$, where $i = \sqrt{-1}$, $p = -b/2a$, and $q = \sqrt{4ac - b^2}/2a$. The formal solutions are then given by the complex exponential functions

$$y_1 = e^{(p+iq)x}$$
$$y_2 = e^{(p-iq)x}$$ (21)

Our goal, however, is to produce *real* solutions of the DE. To see how this is done we must first investigate what we mean by exponential functions with complex arguments.

Recall from the calculus that the exponential function e^x has a Maclaurin series expansion about $x = 0$ given by

$$e^x = \sum_{n=0}^{\infty} \frac{x^n}{n!}, \qquad -\infty < x < \infty \tag{22}$$

The formal replacement of ix for x in (22) leads to

$$e^{ix} = \sum_{n=0}^{\infty} \frac{(ix)^n}{n!}$$

$$= \sum_{n=0}^{\infty} \frac{(-1)^n x^{2n}}{(2n)!} + i \sum_{n=0}^{\infty} \frac{(-1)^n x^{2n+1}}{(2n+1)!} \tag{23}$$

where we have separated the terms into even and odd powers, using the fact that $i^2 = -1$, $i^3 = -i$, $i^4 = 1$, and so on. The first series in (23) is the Maclaurin series for $\cos x$, while the second is the Maclaurin series for $\sin x$. Thus, we deduce that

$$e^{ix} = \cos x + i \sin x$$

$$e^{-ix} = \cos x - i \sin x \tag{24}$$

where the second expression arises from the first by replacing x with $-x$ and using the relations $\cos(-x) = \cos x$ and $\sin(-x) = -\sin x$. We henceforth interpret complex exponential functions $e^{\pm ix}$ according to Equations (24), which are collectively known as **Euler's formula.** It can easily be shown now that the usual rules of algebra and calculus applied to complex exponential functions follow that of real exponential functions, treating $\pm i$ as we would any real constant.

Case III. Complex Conjugate Roots: $b^2 - 4ac < 0$

When the roots of the auxiliary equation are given by $m = p \pm iq$, we can use Euler's formula given above to express the solutions (21) in terms of cosine and sine functions, and real exponential functions. That is, we simply write

$$y_1 = e^{(p+iq)x} = e^{px}e^{iqx}$$

$$y_2 = e^{(p-iq)x} = e^{px}e^{-iqx}$$

which by Euler's formula (24) leads to

$$y_1 = e^{px}(\cos qx + i \sin qx)$$

$$y_2 = e^{px}(\cos qx - i \sin qx) \tag{25}$$

To construct *real* solutions from (25), we can use the superposition principle (Theorem 4.2) and form the linear combination of solutions

$$y_3 = \frac{1}{2}(y_1 + y_2) = e^{px} \cos qx$$

$$y_4 = \frac{1}{2i}(y_1 + y_2) = e^{px} \sin qx \tag{26}$$

It follows that y_3 and y_4 are also solutions and, moreover, since they are not proportional, they are linearly independent solutions.[†] As such, they form a fundamental solution set from which we now construct the general solution

$$y = e^{px}(C_1 \cos qx + C_2 \sin qx), \qquad -\infty < x < \infty \qquad (27)$$

Remark. In practice it is not necessary to actually carry out all the algebraic steps above in obtaining the general solution (27). Once we have identified p and q, we simply write down a general solution in the form of (27).

EXAMPLE 14

Find a general solution of

$$y'' - 4y' + 13y = 0$$

Solution

The auxiliary equation is

$$m^2 - 4m + 13 = 0$$

which, by way of the quadratic formula, leads to the complex roots $m_1 = 2 + 3i$ and $m_2 = 2 - 3i$. Hence, $p = 2$ and $q = 3$ (by definition we choose q to be positive), from which we deduce $y_1 = e^{2x} \cos 3x$ and $y_2 = e^{2x} \sin 3x$. Based on (27), therefore, the general solution is

$$y = e^{2x}(C_1 \cos 3x + C_2 \sin 3x)$$

[0] 4.4.2 Hyperbolic Functions

In Case I where the roots of the auxiliary equation are real and distinct, we have shown that the general solution can always be expressed in terms of exponential functions. However, by using a technique similar to that in Case III above, it is also possible to express the general solution in terms of hyperbolic functions. Let us illustrate by way of some examples.

EXAMPLE 15

For $k \neq 0$, find general solutions of the DEs

$$y'' + k^2 y = 0$$
$$y'' - k^2 y = 0$$

[†] Observe also that

$$W(y_3, y_4)(x) = q e^{2px} \neq 0$$

the verification of which we leave to the reader (see Problem 20 in Exercises 4.4).

Solution

The first DE has the auxiliary equation $m^2 + k^2 = 0$ with pure imaginary roots $m_1 = ik$ and $m_2 = -ik$. Thus, the general solution is

$$y = C_1 \cos kx + C_2 \sin kx$$

For the second DE, the auxiliary equation $m^2 - k^2 = 0$ has real distinct roots $m_1 = k$ and $m_2 = -k$. We can therefore express the general solution in the form

$$y = C_1 e^{kx} + C_2 e^{-kx}$$

However, there is another representation of the general solution involving hyperbolic functions that is more convenient in applications featuring initial or boundary conditions. Recall that the hyperbolic cosine and hyperbolic sine functions are defined, respectively, by

$$\cosh kx = \tfrac{1}{2}(e^{kx} + e^{-kx})$$

$$\sinh kx = \tfrac{1}{2}(e^{kx} - e^{-kx})$$

which are simply linear combinations of the solutions e^{kx} and e^{-kx}. From the linearity property of homogeneous DEs (Theorem 4.2) it follows that $\cosh kx$ and $\sinh kx$ are also solutions of the same DE. Moreover, it is easily established that they are linearly independent solutions and thus we can express the general solution of the second DE in the alternate form

$$y = C_1 \cosh kx + C_2 \sinh kx \qquad \blacksquare$$

One of the reasons we want to express the general solution of a DE in terms of hyperbolic functions, as in Example 15, is that certain simplifications take place when applying initial conditions since $\cosh(0) = 1$ and $\sinh(0) = 0$. Other basic properties of these functions are reviewed in Problem 19 in Exercises 4.4. The following example illustrates the use of hyperbolic functions in a more general problem than above.

EXAMPLE 16

Find a general solution of

$$y'' + y' - y = 0$$

in terms of hyperbolic functions.

Solution

Through use of the quadratic formula, we find the roots of the auxiliary equation $m^2 + m - 1 = 0$ to be given by

$$m_1 = \tfrac{1}{2}(-1 + \sqrt{5}), \qquad m_2 = \tfrac{1}{2}(-1 - \sqrt{5})$$

Because these roots are real and distinct, we can express the general solution as

$$y = C_1 \exp[\tfrac{1}{2}(-1 + \sqrt{5})x] + C_2 \exp[\tfrac{1}{2}(-1 - \sqrt{5})x]$$

Also, by writing this solution in the form

$$y = e^{-x/2}(C_1 e^{\sqrt{5}x/2} + C_2 e^{-\sqrt{5}x/2})$$

and using a device similar to that used in Case III above for obtaining a general solution in terms of trigonometric functions, we find we can write the general solution of this problem in the alternate form

$$y = e^{-x/2}[C_1 \cosh(\tfrac{1}{2}\sqrt{5}x) + C_2 \sinh(\tfrac{1}{2}\sqrt{5}x)]$$

the verification of which is left to the exercises (see Problem 21 in Exercises 4.4). ∎

Exercises 4.4

In Problems 1 to 6, show that the auxiliary equation has distinct real roots and find the general solution.

1. $3y'' - y' = 0$

2. $y'' - 4y = 0$

3. $y'' + 2y' - 3y = 0$

4. $2y'' - 5y' - 3y = 0$

5. $y'' - 10y' + 17y = 0$

6. $3y'' - 10y' + 4y = 0$

In Problems 7 to 12, show that the auxiliary equation has a repeated root and find the general solution.

7. $y'' - 2y' + y = 0$

8. $y'' - 10y' + 25y = 0$

9. $9y'' + 6y' + y = 0$

10. $4y'' + 4y' + y = 0$

11. $9y'' - 12y' + 4y = 0$

∎12. $y'' - 2\sqrt{2}y' + 2y = 0$

In Problems 13 to 18, show that the auxiliary equation has complex roots and find the general solution.

13. $y'' + y' + y = 0$

14. $y'' + 25y = 0$

15. $y'' - 6y' + 25y = 0$

16. $y'' - 4y' + 13y = 0$

17. $2y'' - y' + y = 0$

∎18. $2y'' - 3y' + 10y = 0$

19. Show that the hyperbolic functions $\cosh x = \tfrac{1}{2}(e^x + e^{-x})$ and $\sinh x = \tfrac{1}{2}(e^x - e^{-x})$ satisfy the relations (see figure)

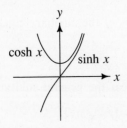

(a) $\cosh(0) = 1$

(b) $\sinh(0) = 0$

(c) $\cosh^2 x - \sinh^2 x = 1$

(d) $\dfrac{d}{dx} \cosh x = \sinh x$

(e) $\dfrac{d}{dx} \sinh x = \cosh x$

20. Given the functions $(q \neq 0)$

$$y_1 = e^{px} \cos qx, \qquad y_2 = e^{px} \sin qx$$

(a) Use the Wronskian to show that they are linearly independent on every interval.

(b) Verify that y_1 and y_2 both satisfy the DE

$$y'' - 2py' + (p^2 + q^2)y = 0$$

21. Given the functions $(q \neq 0)$

$$y_1 = e^{px} \cosh qx, \qquad y_2 = e^{px} \sinh qx$$

(a) Use the Wronskian to show that they are linearly independent on every interval.

(b) Verify that y_1 and y_2 both satisfy the DE

$$y'' - 2py' + (p^2 - q^2)y = 0$$

In Problems 22 to 27, write the general solution in terms of hyperbolic functions (see Problem 21).

22. $3y'' - y' = 0$

23. $y'' - 4y = 0$

24. $y'' + 2y' - 3y = 0$

25. $2y'' - 5y' - 3y = 0$

26. $y'' - 10y' + 17y = 0$

27. $3y'' - 10y' + 4y = 0$

In Problems 28 to 40, find a general solution of the given DE. If auxiliary conditions are prescribed, find a particular solution satisfying them.

28. $4y'' - 17y' - 15y = 0$

29. $y'' - 8y' + 16y = 0$

■**30.** $3y'' - 14y' - 5y = 0$

31. $4y'' - 12y' + 5y = 0$

32. $y'' - 4y' + 13y = 0$, $y(0) = -1$, $y'(0) = 2$

33. $y'' - 2y' - 3y = 0$, $y(0) = 0$, $y'(0) = -4$

34. $y'' + k^2y = 0$, $y(0) = y_0$, $y'(0) = v_0$

35. $y'' - 3y' + 2y = 0$, $y(1) = 0$, $y(2) = 1$

■**36.** $y'' - y' - 6y = 0$, $y(0) = 0$, $y(1) = e^3$

37. $y'' + y' - 6y = 0$, $y(0) = 3$, $y'(0) = 0$

38. $y'' + 6y' + 4y = 0$, $y(0) = 2$, $y'(0) = -3$

39. $y'' - 6y' + 9y = 0$, $y(0) = 0$, $y'(0) = 3$

40. $y'' - 2y' + 26y = 0$, $y(0) = 3/\sqrt{2}$, $y'(0) = 12/\sqrt{2}$

***41.** Show that the general solution of $y'' + y = 0$ can be expressed in the form

$$y = A \cos(x - \phi)$$

where A and ϕ are arbitrary constants.

***42.** Show that the solution of the IVP

$$y'' - 6y' + 25y = 0, \quad y(0) = -3, \quad y'(0) = -1$$

can be expressed in the form

$$y = \sqrt{13}e^{3x} \sin(4x - \theta)$$

where the angle θ is defined by the equations

$$\cos \theta = 2/\sqrt{13}, \qquad \sin \theta = 3/\sqrt{13}$$

4.5 Differential Operators

An **operator** is a function that transforms one function into another function, also called a **transformation.** There are many types of operators, familiar examples of which include the *integral* and *differential. Matrices* and *integral transforms* are further examples of operators, but our interest here concerns only differential operators.

The simplest example of a **differential operator** is $D = d/dx$, where $D[y] = y'$. Another example is xD for which $xD[y] = xy'$. A more general *first-order differential operator* is defined by

$$M_1 = A_1(x)D + A_0(x) \tag{28}$$

whereby, given any differentiable function y, we have

$$M_1[y] = A_1(x)y' + A_0(x)y \tag{29}$$

Similarly, the operator

$$M_2 = A_2(x)D^2 + A_1(x)D + A_0(x) \tag{30}$$

is a second-order differential operator for which

$$M_2[y] = A_2(x)y'' + A_1(x)y' + A_0(x)y \tag{31}$$

where y is any twice differentiable function. Among other reasons, the use of operators is notationally expedient in that DEs like

$$A_2(x)y'' + A_1(x)y' + A_0(x)y = F(x) \tag{32}$$

can be written more compactly as

$$M_2[y] = F(x) \tag{33}$$

where M_2 is defined by (30). This notation is particularly useful when dealing with DEs of order higher than 2 (see Section 4.6).

The differential operator M defined by either (28) or (30) is an example of a **linear operator**. That is, if y_1 and y_2 are any two twice differentiable functions, and C_1 and C_2 any arbitrary constants, then (see Problem 21 in Exercises 4.5)

$$M[C_1y_1 + C_2y_2] = C_1M[y_1] + C_2M[y_2] \tag{34}$$

All linear operators (differential and otherwise) satisfy this relation, which is called the **linearity property.**

4.5.1 Polynomial Operators

The general second-order, homogeneous, constant-coefficient DE is given by

$$ay'' + by' + cy = 0 \tag{35}$$

Associated with this particular DE is the *constant-coefficient operator* $M_2 = aD^2 + bD + c$. This type of operator is called a **polynomial operator** and is often designated by the special symbol

$$P_2(D) = aD^2 + bD + c \tag{36}$$

The interesting feature of the operator $P_2(D)$ is that it can be manipulated according to the basic rules of algebra applied to any polynomial. For instance, the product of two polynomial operators can be *commuted* (see the following example).

■──────────
EXAMPLE 17

Verify that

$$(D + 1)(2D - 3)y = (2D - 3)(D + 1)y$$

Solution

Applying the operators $D + 1$ and $2D - 3$ sequentially to y, we find that

$$(D + 1)(2D - 3)y = (D + 1)(2y' - 3y)$$
$$= 2y'' - y' - 3y$$
$$= (2D^2 - D - 3)y$$

Next, reversing the order of the operators, we find that

$$(2D - 3)(D + 1)y = (2D - 3)(y' + y)$$
$$= 2y'' - y' - 3y$$
$$= (2D^2 - D - 3)y$$

and thus conclude that

$$(D + 1)(2D - 3)y = (2D - 3)(D + 1)y$$ ∎

The commutativity property illustrated in Example 17 is peculiar to only the class of polynomial differential operators—it doesn't apply (in general) to **variable-coefficient** differential operators (see Section 4.5.2).

To better understand the algebraic nature of polynomial operators, let us apply the operator (36) to the exponential function $y = e^{mx}$, which leads to the relation

$$P_2(D)[e^{mx}] = (am^2 + bm + c)e^{mx}$$
$$= P_2(m)e^{mx} \tag{37}$$

Here we see that $P_2(D)$ and $P_2(m)$ have the same polynomial form and therefore *factor* exactly the same. This relation suggests that all constant-coefficient DEs, such as

$$(2D^2 - D - 3)y = 0 \tag{38}$$

can also be written in either of the equivalent forms

$$(D + 1)(2D - 3)y = 0$$

or

$$(2D - 3)(D + 1)y = 0$$

In other words, the polynomial operator $P_2(D) = 2D^2 - D - 3$ may be factored as if it were an ordinary polynomial function, that is,

$$2D^2 - D - 3 = (D + 1)(2D - 3) = (2D - 3)(D + 1)$$

Simple inspection of the factored polynomial operator, therefore, leads to the factored auxiliary equation. For example, from above we see that the auxiliary equation associated with (38) is

$$2m^2 - m - 3 = (m + 1)(2m - 3) = 0$$

Constant-coefficient differential operators of higher order can also be factored in the same manner. For instance, the third-order differential operator $D^3 - 1$ admits the factors

$$D^3 - 1 = (D - 1)(D^2 + D + 1) = (D^2 + D + 1)(D - 1)$$

[0] 4.5.2 Variable-Coefficient Operators

To more clearly see the difference between polynomial operators and general variable-coefficient operators, let us consider the following example.

■──────────   Show that
EXAMPLE 18

$$(xD + 1)(D - 2)y \neq (D - 2)(xD + 1)y$$

Solution

Here we see that

$$(xD + 1)(D - 2)y = (xD + 1)(y' - 2y)$$
$$= (xD)(y' - 2y) + y' - 2y$$
$$= xy'' + (1 - 2x)y' - 2y$$

or

$$(xD + 1)(D - 2)y = [xD^2 + (1 - 2x)D - 2]y$$

On the other hand,

$$(D - 2)(xD + 1)y = (D - 2)(xy' + y)$$
$$= D(xy') + Dy - 2xy' - 2y$$
$$= xy'' + 2(1 - x)y' - 2y$$

or

$$(D - 2)(xD + 1)y = [xD^2 + 2(1 - x)D - 2]y$$

Hence, we have shown that

$$(xD + 1)(D - 2)y \neq (D - 2)(xD + 1)y$$ ■

Based on Example 18, it is clear that variable-coefficient differential operators do not necessarily follow the rules of algebra. In particular, we have shown that

$$(D - 2)(xD + 1) = xD^2 + 2(1 - x)D - 2$$

which cannot be obtained from the usual rules of algebra. It is primarily for this reason that variable-coefficient DEs are more difficult to solve than constant-coefficient DEs (e.g., see Chapter 7).

Exercises 4.5

In Problems 1 to 10, factor the given differential operators.

1. $3D^2 - D$

2. $D^2 - 4$

3. $D^2 - 2D + 1$

4. $4D^2 + 4D + 1$

5. $D^2 - 6D + 25$

6. $2D^2 - 3D + 10$

7. $2D^3 + 3D^2 + D$

■8. $D^4 - D^2$

9. $D^3 + 1$

*10. $4D^3 - 3D + 1$

In Problems 11 to 20, perform the indicated multiplication.

11. $(D - 5)(4D + 3)$

12. $(2D + 7)(2D - 7)$

13. $(D - 1)(D^2 + D + 1)$

14. $(D + 2)(D - 1)^2$

15. $(D - x)(D + x)$

16. $(D + x)(D - x)$

17. $(xD + 3)(xD - 2)$

■**18.** $(xD - 2)(xD + 3)$

19. $(x^2D - 1)(D + x)$

20. $(D + x)(x^2D - 1)$

21. Verify that $M = A_2(x)D^2 + A_1(x)D + A_0(x)$ is a linear operator, that is, that

$$M[C_1y_1 + C_2y_2] = C_1M[y_1] + C_2M[y_2]$$

where y_1 and y_2 are any twice differentiable functions and C_1 and C_2 are arbitrary constants.

In Problems 22 to 25, find the general solution of the given DE, where $D = d/dx$.

22. $(D - 5)(4D + 3)y = 0$

23. $(D^2 - 8D + 16)y = 0$

■**24.** $(3D^2 - 14D - 5)y = 0$

25. $(4D^2 - 12D + 5)y = 0$

4.6 Higher-Order Homogeneous Equations

Although second-order linear DEs are far more prevalent in practice than equations of higher order, there are applications for which the mathematical model demands a DE of order greater than 2. For example, in studying the small deflections of a beam supporting a distributed load we find that the governing DE is fourth-order, as is the DE describing the buckling modes of a long slender column under an axial compressive load.

In this section we wish to extend the theory of second-order linear DEs to *linear DEs of higher order*. Much of the theory is, however, a natural generalization of that for second-order DEs; thus, our treatment will be somewhat briefer.

Any **linear DE of order n** can always be expressed in the form

$$A_n(x)y^{(n)} + A_{n-1}(x)y^{(n-1)} + \cdots + A_1(x)y' + A_0(x)y = F(x) \qquad (39)$$

We assume that the **coefficients** $A_0(x)$, $A_1(x)$, ..., $A_n(x)$ are continuous functions on some interval I, and that $A_n(x) \neq 0$ on I. Notational simplicity can be achieved by introducing the **nth-order linear differential operator**

$$M_n = A_n(x)D^n + A_{n-1}(x)D^{n-1} + \cdots + A_1(x)D + A_0(x) \qquad (40)$$

where $D = d/dx$. In terms of M_n, Equation (39) becomes simply

$$M_n[y] = F(x) \qquad (41)$$

Equation (41) is called **nonhomogeneous** if $F(x)$ is not identically zero on I, but if $F(x) \equiv 0$ on I, (41) reduces to the associated **homogeneous** DE

$$M_n[y] = 0 \qquad (42)$$

In developing the general theory of linear DEs, we sometimes put the DE in **normal form**

$$y^{(n)} + a_{n-1}(x)y^{(n-1)} + \cdots + a_1(x)y' + a_0(x)y = f(x) \qquad (43)$$

obtained by dividing each term of (39) by $A_n(x)$. If the coefficients in (43) are continuous on an interval I containing the point x_0 and we also prescribe the initial conditions

$$y(x_0) = k_0, \qquad y'(x_0) = k_1, \ldots, y^{(n-1)}(x_0) = k_{n-1} \qquad (44)$$

then the IVP consisting of (43) and (44) has a *unique solution* (recall Theorem 4.1). Other than this result, however, we consider only *homogeneous* DEs in the remainder of this section.

4.6.1 Fundamental Solution Sets

If $y_1, y_2, \ldots, y_n$ are all solutions on an interval I of a *homogeneous* linear DE of order n, the linear combination

$$y = C_1 y_1(x) + C_2 y_2(x) + \cdots + C_n y_n(x) \qquad (45)$$

is also a solution for any set of constants $C_1, C_2, \ldots, C_n$ (recall Theorem 4.2). If, in addition, the solutions $y_1, y_2, \ldots, y_n$ are *linearly independent* on I we then say they form a **fundamental set of solutions** on I, and in this case (45) is the **general solution.**

Definition 4.2 The set of functions $y_1, y_2, \ldots, y_n$ is said to be **linearly dependent** on an interval I if and only if there exists a set of constants $C_1, C_2, \ldots, C_n$, not all zero, such that

$$C_1 y_1(x) + C_2 y_2(x) + \cdots + C_n y_n(x) = 0$$

for all x in I. If the set of functions is not linearly dependent, it is said to be **linearly independent.**

If a set of functions $y_1, y_2, \ldots, y_n$ is linearly independent on an interval I, the only way that

$$C_1 y_1(x) + C_2 y_2(x) + \cdots + C_n y_n(x) = 0 \qquad (46)$$

can happen is for $C_1 = C_2 = \cdots = C_n = 0$. However, if the functions are linearly dependent, then at least one of them can be expressed as a linear combination of the others. For example, suppose that (46) is true where at least one of the C's, say C_1, is different from zero. This being the case, we can solve (46) for y_1, getting the result

$$y_1 = -\frac{C_2}{C_1} y_2 - \frac{C_3}{C_1} y_3 - \cdots - \frac{C_n}{C_1} y_n \qquad (47)$$

EXAMPLE 19

Show that $y_1 = 3x^2 - 8x$, $y_2 = x^2$, and $y_3 = 4x$ are linearly dependent on every interval I.

Solution

We see by inspection that

$$y_1 = 3y_2 - 2y_3$$

or $y_1 - 3y_2 + 2y_3 = 0$ for all x. Hence, it follows from Definition 4.2 that the functions are linearly dependent on every interval I. ∎

While it was fairly easy for us in Example 19 to prove that the given functions are linearly dependent by directly appealing to Definition 4.2, this is not always the case in practice, particularly when $n > 3$. In some cases it may be more convenient for us to determine the linear dependence or independence of a set of solutions of a given DE by using the *Wronskian*. In Section 4.2 we developed the Wronskian as a conclusive test of linear independence when the set of functions were solutions of the same second-order DE. A similar result applies to the case when more than two functions are involved. We define the **Wronskian** of the set of functions y_1, $y_2, \ldots, y_n$ by the nth-order determinant

$$W(y_1, \ldots, y_n)(x) = \begin{vmatrix} y_1(x) & y_2(x) & \cdots & y_n(x) \\ y_1'(x) & y_2'(x) & \cdots & y_n'(x) \\ \vdots & \vdots & & \vdots \\ y_1^{(n-1)}(x) & y_2^{(n-1)}(x) & \cdots & y_n^{(n-1)}(x) \end{vmatrix} \quad (48)$$

which, for $n = 2$, reduces to Equation (13). Hence, if $y_1, y_2, \ldots, y_n$ are all solutions of the same nth-order, homogeneous, linear DE, it can be shown that a necessary and sufficient condition for them to be linearly independent on an interval I is that $W(y_1, \ldots, y_n)(x_0) \neq 0$ for some x_0 on I.

Abel's formula (Theorem 4.5) for the Wronskian of a second-order linear DE can be extended to nth-order DEs as given in Theorem 4.7.

Theorem 4.7 *If $y_1, y_2, \ldots, y_n$ are solutions of the nth-order homogeneous DE*

$$y^{(n)} + a_{n-1}(x)y^{(n-1)} + \cdots + a_1(x)y' + a_0(x)y = 0$$

on some interval I for which $a_0(x)$, $a_1(x), \ldots, a_{n-1}(x)$ are continuous, the Wronskian of these solutions satisfies

$$W(y_1, \ldots, y_n)(x) = C \exp\left[-\int a_{n-1}(x)\, dx \right]$$

for an appropriate value of the constant C.

4.6.2 Constant-Coefficients DEs

We now wish to restrict our attention to linear DEs of the form

$$P_n(D)y \equiv (a_nD^n + a_{n-1}D^{n-1} + \cdots + a_1D + a_0)y = 0 \qquad (49)$$

in which the coefficients $a_0, a_1, \ldots, a_n$ are all *constants* and $a_n \neq 0$. By substituting $y = e^{mx}$ into (49) and simplifying, we obtain

$$P_n(m) \equiv a_nm^n + a_{n-1}m^{n-1} + \cdots + a_1m + a_0 = 0 \qquad (50)$$

which is the **auxiliary equation** associated with (49). Because of the many possible combinations of real and imaginary roots of (50), depending upon the value of n, it is difficult to individually list all such cases when $n > 2$. However, the following generalizations are fairly easy to establish:

1. If m_1 is a **real root** of the auxiliary equation $P_n(m) = 0$, there corresponds the single solution

$$y_1 = e^{m_1x}$$

2. If m_1 is a **real root of multiplicity** k of $P_n(m) = 0$, there correspond k linearly independent solutions given by

$$y_1 = e^{m_1x}, \qquad y_2 = xe^{m_1x}, \qquad \ldots, \qquad y_k = x^{k-1}e^{m_1x}$$

3. If $p \pm iq$ are **complex roots** of the auxiliary equation $P_n(m) = 0$, there correspond the linearly independent solutions

$$y_1 = e^{px}\cos qx, \qquad y_2 = e^{px}\sin qx$$

4. If $p \pm iq$ are **complex roots of multiplicity** k of $P_n(m) = 0$, there correspond the $2k$ linearly independent solutions

$$y_1 = e^{px}\cos qx, \qquad y_2 = xe^{px}\cos qx, \qquad \ldots, \qquad y_k = x^{k-1}e^{px}\cos qx$$

$$y_{k+1} = e^{px}\sin qx, \qquad y_{k+2} = xe^{px}\sin qx, \qquad \ldots, \qquad y_{2k} = x^{k-1}e^{px}\sin qx$$

We leave it to the reader to verify that in each case the given functions are indeed solutions of the DE and, moreover, are linearly independent solutions.

■ **EXAMPLE 20**

Find the general solution of the sixth-order DE

$$(D - 1)^3(D + 2)^2(3D - 2)y = 0$$

Solution

From inspection we obtain the auxiliary equation

$$(m - 1)^3(m + 2)^2(3m - 2) = 0$$

with roots $m = 1, 1, 1, -2, -2, 2/3$. Hence, the six fundamental solutions corresponding to these values of m are

$$y_1 = e^x, \qquad y_2 = xe^x, \qquad y_3 = x^2e^x, \qquad y_4 = e^{-2x}, \qquad y_5 = xe^{-2x}, \qquad y_6 = e^{2x/3}$$

providing us with the general solution

$$y = (C_1 + C_2x + C_3x^2)e^x + (C_4 + C_5x)e^{-2x} + C_6e^{2x/3}$$ ▪

EXAMPLE 21

Find the general solution of

$$D^4(D^2 - 2D + 5)^2 y = 0$$

Solution

Here we find the auxiliary equation

$$m^4(m^2 - 2m + 5)^2 = 0$$

with roots $m = 0, 0, 0, 0, 1 \pm 2i, 1 \pm 2i$. The fundamental set of solutions is therefore given by

$$y_1 = 1, \qquad y_2 = x, \qquad y_3 = x^2, \qquad y_4 = x^3,$$

$$y_5 = e^x \cos 2x, \qquad y_6 = xe^x \cos 2x, \qquad y_7 = e^x \sin 2x, \qquad y_8 = xe^x \sin 2x$$

from which we obtain the general solution

$$y = C_1 + C_2x + C_3x^2 + C_4x^3$$
$$+ e^x[(C_5 + C_6x) \cos 2x + (C_7 + C_8x) \sin 2x]$$ ▪

In practical applications the operator of the DE is not usually in factored form as in Examples 20 and 21. Hence, finding the roots of the auxiliary equation is often the most difficult part of solving the DE. According to the general theory of polynomials, if (50) has a real rational root of the form $m_1 = p/q$, where p and q are integers, then p must be a factor of a_0 and q a factor of a_n. For example, notice that $m_1 = 2/3$ is a rational root of

$$3m^2 + 10m - 8 = 0$$

where 2 is a factor of -8 and 3 a factor of 3. Of course, once we have found one root m_1, we can divide (50) (either directly or by using synthetic division) by the factor $(m - m_1)$ to obtain a polynomial of one less degree algebraically from which to determine the remaining roots. In general, however, it may be necessary to resort to a *numerical procedure* for finding the roots.

> ***Remark.*** The real roots of a polynomial do not have to be rational. For example, $m^2 + m - 1 = 0$ has irrational roots $m = \frac{1}{2}(-1 \pm \sqrt{5})$. Also recall that if the auxiliary polynomial is of even degree, it does not have to have any real roots. (All odd-degree polynomials must have at least one real root since complex roots occur in pairs.) Observe that the fourth-degree polynomial $m^4 + 2m^2 + 1 = 0$, for instance, has only the pure imaginary roots $m = \pm i$.

EXAMPLE 22

Find the general solution of

$$(4D^3 - 3D + 1)y = 0$$

Solution

The auxiliary equation is

$$4m^3 - 3m + 1 = 0$$

If there are any real rational roots, they are among the possibilities

$$m = \pm 1, \ \pm \tfrac{1}{2}, \ \pm \tfrac{1}{4}$$

Checking, we find that $m_1 = -1$ is one of the roots. Hence, by division, we obtain the result

$$4m^3 - 3m + 1 = (m + 1)(4m^2 - 4m + 1) = 0$$

from which we deduce that the remaining roots are $m_2 = m_3 = \tfrac{1}{2}$. The general solution in this case is

$$y = C_1 e^{-x} + (C_2 + C_3 x)e^{x/2} \qquad ■$$

EXAMPLE 23

Find the general solution of

$$(4D^4 - 15D^2 + 5D + 6)y = 0$$

Solution

For the auxiliary equation

$$4m^4 - 15m^2 + 5m + 6 = 0$$

the real rational root possibilities are

$$m = \pm 1, \ \pm 2, \ \pm 3, \ \pm 6, \ \pm \tfrac{1}{4}, \ \pm \tfrac{1}{2}, \ \pm \tfrac{3}{4}, \ \pm \tfrac{3}{2}$$

Systematically checking, we find that $m_1 = 1$ is a root, so we can write

$$4m^4 - 15m^2 + 5m + 6 = (m - 1)(4m^3 + 4m^2 - 11m - 6)$$

Hence, the remaining roots must satisfy the reduced equation

$$4m^3 + 4m^2 - 11m - 6 = 0$$

Checking the rational root possibilities this time, we see that $m_2 = \tfrac{3}{2}$, and thus

$$4m^4 - 15m^2 + 5m + 6 = (m - 1)(m - \tfrac{3}{2})(4m^2 + 10m + 4) = 0$$

By factoring, we obtain the remaining roots $m_3 = -\tfrac{1}{2}$ and $m_4 = -2$. Finally, based on the collection of distinct roots $m = 1, 3/2, -1/2, -2$, we are led to the general solution

$$y = C_1 e^x + C_2 e^{3x/2} + C_3 e^{-x/2} + C_4 e^{-2x} \qquad ■$$

Exercises 4.6

In Problems 1 to 5, determine whether the given functions are linearly dependent or independent.

1. $y_1 = x$, $y_2 = 3x^2$, $y_3 = x^2 - 7x$

2. $y_1 = x$, $y_2 = 3x^2$, $y_3 = x^2 - 7x + 1$

3. $y_1 = e^x$, $y_2 = xe^x$, $y_3 = x^2e^x$

■**4.** $y_1 = e^x$, $y_2 = e^{-x}$, $y_3 = \cosh x$

*****5.** $y_1 = \cos^2 x$, $y_2 = \sec^2 x$, $y_3 = \sin^2 x$, $y_4 = \tan^2 x$

In Problems 6 to 10, the given functions are solutions of the DE. Use the Wronskian test to determine whether the solutions are linearly dependent or independent and state the interval.

6. $y''' - 6'' + 5y' + 12y = 0$; $y_1 = e^{-x}$, $y_2 = e^{3x}$, $y_3 = e^{4x}$

7. $y''' + 4y' = 0$; $y_1 = 1$, $y_2 = \cos 2x$, $y_3 = \sin^2 x$

8. $y^{(4)} - y = 0$; $y_1 = e^x$, $y_2 = e^{-x}$, $y_3 = \sinh x$, $y_4 = \cos x$

9. $x^3y''' + x^2y'' - 2xy' + 2y = 0$; $y_1 = x$, $y_2 = x^2$, $y_3 = x^{-1}$

10. $y^{(4)} - 2y''' + y'' = 0$; $y_1 = 1$, $y_2 = x$, $y_3 = e^x$, $y_4 = xe^x$

In Problems 11 to 30, find the general solution. Assume that x is the independent variable.

11. $D(D - 1)^3y = 0$

12. $(D^2 + 1)^3y = 0$

13. $(D^2 + D - 1)^2y = 0$

14. $(D^2 - 6D + 25)^2y = 0$

15. $(D^3 - 1)y = 0$

■**16.** $(D^3 + 1)y = 0$

17. $(D^3 + 3D^2 + 3D + 1)y = 0$

18. $(D^3 + 3D^2 - 4D)y = 0$

19. $(D^4 + 3D^3 + D^2)y = 0$

20. $(D^4 - 1)y = 0$

21. $(4D^3 + 4D^2 + D)y = 0$

22. $(D^3 + D^2 - 2)y = 0$

23. $(3D^3 - 19D^2 + 36D - 20)y = 0$

24. $(D^4 - 5D^3 + 6D^2 + 4D - 8)y = 0$

25. $(4D^4 - 4D^3 - 23D^2 + 12D + 36)y = 0$

■**26.** $(D^4 - 4D^3 + D^2 + 6D)y = 0$

27. $(4D^4 - 8D^3 - 7D^2 + 11D + 6)y = 0$

28. $D^3(D^2 + 9)^2(D^2 - 2D + 1)^2y = 0$

*****29.** $(D^4 - 4D^3 + 10D^2 - 20D + 25)y = 0$

*****30.** $(D^5 + 2D^3 + D)y = 0$

In Problems 31 to 35, solve the given DE subject to the prescribed auxiliary conditions.

31. $(D^3 - 3D - 2)y = 0$, $y(0) = 0$, $y'(0) = 9$, $y''(0) = 0$

*****32.** $(D^3 + D^2 - D - 1)y = 0$, $\lim_{x \to \infty} y(x) = 0$, $y(0) = 1$, $y(2) = 0$

33. $(D^3 + 5D^2 + 17D + 13)y = 0$, $y(0) = 0$, $y'(0) = 1$, $y''(0) = 6$

■**34.** $D^4y = 0$, $y(0) = 2$, $y'(0) = 3$, $y''(0) = 4$, $y'''(0) = 5$

*****35.** $(D^4 + 6D^3 + 9D^2)y = 0$, $\lim_{x \to \infty} y'(x) = 1$, $y(0) = 0$, $y'(0) = 0$, $y''(0) = 6$

36. The roots of the auxiliary equation of some tenth-order linear DE are known to be 3, 3, 3, 3, $1 \pm 2i$, $1 \pm 2i$, 5, -1. Write the general solution of the DE.

37. The roots of the auxiliary equation of some third-order linear DE are known to be -4, $\frac{1}{2}$, $\frac{1}{2}$. What is the DE?

38. Repeat Problem 37 when the roots are 3, $2 \pm i$.

*****39.** Given that $y_1 = \sin x$ is one solution of

$$(D^4 + 2D^3 + 6D^2 + 2D + 5)y = 0$$

find the general solution.

*40. Given that $y_1 = e^{-x} \cos 2x$ is one solution of

$$(D^4 + 4D^3 + 14D^2 + 20D + 25)y = 0$$

find the general solution.

*41. Show, by means of the Wronskian, that the second-order DE

$$y'' + a_1(x)y' + a_0(x)y = 0$$

cannot have three linearly independent solutions y_1, y_2, and y_3.

4.7 Nonhomogeneous Equations—Part I

The general **nth-order nonhomogeneous DE** can be written as

$$M_n[y] = F(x) \tag{51}$$

where

$$M_n = A_n(x)D^n + A_{n-1}(x)D^{n-1} + \cdots + A_1(x)D + A_0(x) \tag{52}$$

For the special case when $n = 1$, we have shown in Section 2.5 that the **general solution** of (51) has the form

$$y = y_H + y_P \tag{53}$$

where y_P is any **particular solution** of (51) and y_H is a general solution of the associated homogeneous DE, $M_1[y] = 0$. Not only is this the case when $n = 1$, but it is the form of the general solution of (51) for *all n*.

Theorem 4.8 *If y_P is any particular solution of the nonhomogeneous DE, $M_n[y] = F(x)$, and y_H is a general solution of the associated homogeneous DE, $M_n[y] = 0$, then every solution of $M_n[y] = F(x)$ is contained in $y = y_H + y_P$.*

Proof. *Suppose that y_P is a particular solution of $M_n[y] = F(x)$ and that Y is any other particular solution of the same DE. Then, if we set $y = Y - y_P$ and apply the operator M_n, this action leads to*

$$M_n[y] = M_n[Y] - M_n[y_P]$$
$$= F(x) - F(x)$$
$$= 0$$

where we have used the linearity property of the operator M_n. Hence, we have shown that $y = Y - y_P$ is a solution of the homogeneous DE and therefore must be contained in y_H for suitable choices of the constants $C_1, C_2, \ldots, C_n$. That is,

$$Y - y_P = C_1 y_1(x) + C_2 y_2(x) + \cdots + C_n y_n(x)$$

or

$$Y = C_1 y_1(x) + C_2 y_2(x) + \cdots + C_n y_n(x) + y_P$$
$$= y_H + y_P$$

Since Y must necessarily be contained in the sum $y_H + y_P$, as we have shown above, and Y is any particular solution of $M_n[y] = F(x)$, it follows that every particular solution is contained in the sum $y_H + y_P$.

To find the general solution of a nonhomogeneous linear DE, we must find the *general solution* y_H of the associated homogeneous DE and any *particular solution* y_P of the nonhomogeneous DE. Because we know how to construct y_H when the coefficients of the DE are constant, we now concentrate on techniques for constructing y_P. There are essentially two methods available to us. The first one, called the *method of undetermined coefficients*, is discussed in Section 4.7.1, and a second more general technique, called *variation of parameters*, is taken up in Section 4.8.

4.7.1 Method of Undetermined Coefficients

In this section we restrict our attention to *constant-coefficient* DEs of the form

$$P_n(D)y = F(x)$$

where the nonhomogeneous term $F(x)$ is composed of functions whose nth derivative is zero or else reproduces the function $F(x)$. When this happens, we can construct the particular solution y_P by a relatively simple technique called the **method of undetermined coefficients.** To satisfy these restrictions, we find that $F(x)$ must be composed of functions that are either

1. A polynomial in x (including constants)
2. e^{px}
3. $\cos qx$ or $\sin qx$
4. A finite sum and/or product of these functions

Functions not suitable for this solution technique are $\log x$, $\tan x$, e^{x^2}, $\sec x$, and so on.

To illustrate the method, suppose we wish to find a particular solution of

$$y'' + y = 3e^x \tag{54}$$

Since differentiation of an exponential function merely reproduces the function again with at most a multiplicative constant, it seems natural to "guess" that a particular solution exists of the form

$$y_P = Ae^x \tag{55}$$

where A is a constant to be determined, that is, an "undetermined coefficient." We substitute (55) into (54), getting

$$Ae^x + Ae^x = 3e^x$$

or

$$2Ae^x = 3e^x$$

which reduces to an identity if $A = 3/2$. Thus, we deduce that

$$y_P = \tfrac{3}{2}e^x \tag{56}$$

is a particular solution of (54). Recognizing that

$$y_H = C_1 \cos x + C_2 \sin x \tag{57}$$

is the general solution of the homogeneous DE, $y'' + y = 0$, it now follows that the general solution of (54) is

$$y = y_H + y_P$$

$$= C_1 \cos x + C_2 \sin x + \tfrac{3}{2}e^x \qquad (58)$$

As a second example, let us consider the similar DE

$$y'' + y = 3x^2 \qquad (59)$$

Proceeding as before, we might guess that

$$y_P = Ax^2$$

However, this time the substitution of y_P into (59) yields

$$2A + Ax^2 = 3x^2$$

which cannot be satisfied for any choice of the constant A. The problem is that the derivatives of Ax^2 produce new functions that are linearly independent of it. Therefore, we should try a particular solution that also contains all derivatives of Ax^2, that is, assume

$$y_P = Ax^2 + Bx + C \qquad (60)$$

where we need to determine A, B, and C. The substitution of (60) into (59) gives us

$$2A + Ax^2 + Bx + C = 3x^2$$

or

$$Ax^2 + Bx + (2A + C) = 3x^2$$

By equating like coefficients in this last identity, we have

$$A = 3$$

$$B = 0$$

$$2A + C = 0$$

which yields the simultaneous solution

$$A = 3, \quad B = 0, \quad C = -6$$

Thus, a particular solution this time is

$$y_P = 3x^2 - 6 \qquad (61)$$

which, combined with (57), leads to the general solution

$$y = C_1 \cos x + C_2 \sin x + 3x^2 - 6 \qquad (62)$$

The general rule illustrated by these examples is the following.

> **Rule 1** *Assume y_P has the same functional form as the nonhomogeneous term $F(x)$, plus all linearly independent derivatives of $F(x)$.*

EXAMPLE 24

Find the general solution of

$$y'' + y = 3 \cos 2x$$

Solution

Since the derivative of $\cos 2x$ leads to $\sin 2x$ (but no other functional forms), the use of Rule 1 leads to the choice

$$y_P = A \cos 2x + B \sin 2x$$

Differentiating twice, we obtain $y_P'' = -4A \cos 2x - 4B \sin 2x$, and the substitution of y_P and y_P'' into the DE yields

$$-4A \cos 2x - 4B \sin 2x + A \cos 2x + B \sin 2x = 3 \cos 2x$$

or

$$-3A \cos 2x - 3B \sin 2x = 3 \cos 2x$$

In this case we see that $A = -1$ and $B = 0$, giving us

$$y_P = -\cos 2x$$

Since $y_H = C_1 \cos x + C_2 \sin x$, the general solution is

$$y = y_H + y_P$$
$$= C_1 \cos x + C_2 \sin x - \cos 2x \qquad \blacksquare$$

Suppose we now wish to solve

$$y'' + y = 3 \cos x \qquad (63)$$

which looks very much like Example 24. However, following Rule 1 and assuming the particular solution

$$y_P = A \cos x + B \sin x$$

we find that substituting this expression into (63) gives us

$$-A \cos x - B \sin x + A \cos x + B \sin x = 3 \cos x$$

or

$$0 = 3 \cos x$$

which cannot be satisfied for all x. To correct for this situation, we must assume y_P to be a function *linearly independent* of any function in the homogeneous solution. One way to produce such a function is to multiply y_P (obtained from Rule 1) by x, which is to now assume

$$y_P = x(A \cos x + B \sin x) \tag{64}$$

From this new form of y_P it follows that

$$y_P' = A \cos x + B \sin x + x(-A \sin x + B \cos x)$$

$$y_P'' = -2A \sin x + 2B \cos x - x(A \cos x + B \sin x)$$

When these expressions for y_P, y_P', and y_P'' are substituted into (63), we obtain

$$-2A \sin x + 2B \cos x = 3 \cos x$$

Equating like coefficients yields

$$-2A = 0$$

$$2B = 3$$

or $A = 0$ and $B = 3/2$. Hence,

$$y_P = \tfrac{3}{2}x \sin x \tag{65}$$

and a general solution of (63) is

$$y = C_1 \cos x + C_2 \sin x + \tfrac{3}{2}x \sin x \tag{66}$$

Although this example at first seems very much like Example 24, it clearly is different. Of course, we now recognize that the difference is that the input function $F(x) = 3 \cos x$ in (63) is contained in the homogeneous solution $y_H = C_1 \cos x + C_2 \sin x$ while the input function $F(x) = 3 \cos 2x$ in Example 24 is not. Thus, we are led to our second and last rule.

Rule 2 *If $F(x)$ or any of its derivatives satisfy the associated homogeneous equation $P_n(D)y = 0$, that is, correspond to a root of multiplicity k of the auxiliary equation, then the choice of y_P following Rule 1 must be multiplied by x^k.*

The method of solution outlined above is summarized in Table 4.1. If the DE has the general form

$$P_n(D)y = F_1(x) + F_2(x) + \cdots + F_r(x) \tag{67}$$

where each $F_i(x)$, $i = 1, 2, \ldots, r$ is a different type of function occurring in one of the six categories in Table 4.1, we first replace the DE with the equivalent system of

TABLE 4.1   METHOD OF UNDETERMINED COEFFICIENTS

DE: $P_n(D)y = F(x)$

Auxiliary equation: $P_n(m) = 0$

I. $F(x) = b_j x^j + \cdots + b_1 x + b_0, \qquad j = 0, 1, 2, \ldots$

(a) $m \neq 0$

$y_P = A_j x^j + \cdots + A_1 x + A_0$

(b) $m = 0, \quad k$ times

$y_P = x^k(A_j x^j + \cdots + A_1 x + A_0)$

II. $F(x) = be^{cx}$

(a) $m \neq c$

$y_P = Ae^{cx}$

(b) $m = c, \quad k$ times

$y_P = Ax^k e^{cx}$

III. $F(x) = (b_j x^j + \cdots + b_1 x + b_0)e^{cx}, \qquad j = 0, 1, 2, \ldots$

(a) $m \neq c$

$y^P = (A_j x^j + \cdots + A_1 x + A_0)e^{cx}$

(b) $m = c, \quad k$ times

$y_P = x^k(A_j x^j + \cdots + A_0)e^{cx}$

IV. $F(x) = a \cos qx + b \sin qx$

(a) $m \neq \pm iq$

$y_P = A \cos qx + B \sin qx$

(b) $m = \pm iq, \quad k$ times

$y_P = x^k(A \cos qx + B \sin qx)$

V. $F(x) = (b_i x^i + \cdots + b_1 x + b_0) \cos qx + (c_j x^j + \cdots + c_0) \sin qx$

(a) $m \neq \pm iq$

$y_P = (A_r x^r + \cdots + A_0) \cos qx$
$\quad + (B_r x^r + \cdots + B_0) \sin qx$
$r = \max(i, j)$

(b) $m = \pm iq, \quad k$ times

Multiply y_P in (a) by x^k

VI. $F(x) = ae^{px} \cos qx + be^{px} \sin qx$

(a) $m \neq p \pm iq$

$y_P = Ae^{px} \cos qx + Be^{px} \sin qx$

(b) $m = p \pm iq, \quad k$ times

Multiply y_P in (a) by x^k

DEs

$$P_n(D)y = F_1(x)$$

$$P_n(D)y = F_2(x)$$

$$\vdots$$

$$P_n(D)y = F_r(x) \tag{68}$$

We then determine a y_P for each of the above DEs, and sum them to obtain the particular solution for (67). That this can be done is based on the following theorem, which we state for general differential operators M_n (not simply polynomial operators).

Theorem 4.9 *If $y = \phi(x)$ is a solution of*

$$M_n[y] = F_1(x)$$

and $y = \psi(x)$ is a solution of

$$M_n[y] = F_2(x)$$

then $y = \phi(x) + \psi(x)$ is a solution of

$$M_n[y] = F_1(x) + F_2(x)$$

Proof.   *Let us apply the operator M_n directly to $y = \phi(x) + \psi(x)$, which yields*

$$M_n[y] = M_n[\phi(x)] + M_n[\psi(x)]$$
$$= F_1(x) + F_2(x)$$

and the theorem is proved.

EXAMPLE 25

Determine the proper form of y_P for

$$y'' + 2y' + y = 3x^2 e^{-x}$$

but do not solve for the undetermined coefficients.

Solution

The roots of the auxiliary equation $m^2 + 2m + 1 = 0$ are readily found to be $m = -1$, -1, leading to the homogeneous solution

$$y_H = (C_1 + C_2 x)e^{-x}$$

Following Rule 1 we would start by assuming

$$y_P = (Ax^2 + Bx + C)e^{-x}$$

But because the homogeneous solution e^{-x} corresponds to a *double* root of the auxiliary equation and appears among the derivatives of $F(x) = 3x^2 e^{-x}$, we must multiply y_P above by x^2, which leads to the proper choice (see also Category III in Table 4.1)

$$y_P = x^2(Ax^2 + Bx + C)e^{-x}$$
$$= (Ax^4 + Bx^3 + Cx^2)e^{-x}$$ ∎

EXAMPLE 26

Find a general solution of

$$D^2(D - 1)y = 2 \sin x - 5e^x$$

Solution

By inspection, we see that the auxiliary equation has roots $m = 0, 0, 1$. Therefore, the homogeneous solution is

$$y_H = C_1 + C_2 x + C_3 e^x$$

Next, we rewrite the DE as the system of DEs

$$D^2(D - 1)y = 2 \sin x$$
$$D^2(D - 1)y = -5e^x$$

The term $2 \sin x$ is not included in the homogeneous solution; so by Rule 1 we assume (see also Category IV in Table 4.1)

$$y_P = A \cos x + B \sin x$$

However, based on the term $5e^x$ in the second equation above, we need to write

$$y_P = x(Ce^x) = Cxe^x$$

since e^x appears in the homogeneous solution corresponding to a root (of multiplicity 1) of the auxiliary equation (see also Category II in Table 4.1). Thus, taking the sum of these particular solutions, we have the total particular solution

$$y_P = A \cos x + B \sin x + Cxe^x$$

where A, B, and C are yet to be determined.
Computing derivatives, we have

$$y_P' = -A \sin x + B \cos x + C(xe^x + e^x)$$
$$y_P'' = -A \cos x - B \sin x + C(xe^x + 2e^x)$$
$$y_P''' = A \sin x - B \cos x + C(xe^x + 3e^x)$$

and substituting these expressions into the original DE yields (after simplification)

$$(A + B) \sin x + (A - B) \cos x + Ce^x = 2 \sin x - 5e^x$$

By comparing like coefficients, we see that

$$A + B = 2$$
$$A - B = 0$$
$$C = -5$$

Therefore, $A = B = 1$, $C = -5$, and

$$y_P = \cos x + \sin x - 5xe^x$$

The general solution we seek is

$$y = y_H + y_P$$

$$= C_1 + C_2 x + C_3 e^x + \cos x + \sin x - 5xe^x \qquad \blacksquare$$

■ **EXAMPLE 27**   Solve the IVP

$$y'' + y = x \cos x - \cos x, \qquad y(0) = 1, \qquad y'(0) = \tfrac{1}{4}$$

Solution

The roots of the auxiliary equation are $m = \pm i$, and thus

$$y_H = C_1 \cos x + C_2 \sin x$$

Since $\cos x$ and $\sin x$ are among the derivatives of $F(x)$ and are also contained in the homogeneous solution (corresponding to a root of multiplicity 1), the proper form to assume for y_P is (see also Category IV in Table 4.1)

$$y_P = x[(Ax + B)\cos x + (Cx + D)\sin x]$$

$$= (Ax^2 + Bx)\cos x + (Cx^2 + Dx)\sin x$$

The substitution of y_P into the left-hand side of the given DE yields

$$y_P'' + y_P = [-(Ax^2 + Bx)\cos x - (4Ax + 2B)\sin x + 2A\cos x$$

$$- (Cx^2 + Dx)\sin x + (4Cx + 2D)\cos x + 2C\sin x]$$

$$+ (Ax^2 + Bx)\cos x + (Cx^2 + Dx)\sin x$$

Simplifying this expression and equating it to the right-hand side of the given DE leads to

$$4Cx\cos x + 2(A + D)\cos x - 4Ax\sin x + 2(C - B)\sin x = x\cos x - \cos x$$

Next, equating like coefficients, we see that

$$4C = 1$$

$$2(A + D) = -1$$

$$-4A = 0$$

$$2(C - B) = 0$$

and therefore deduce that $A = 0$, $B = 1/4$, $C = 1/4$, and $D = -1/2$. Combining y_P with y_H, we finally obtain the general solution

$$y = y_H + y_P$$

$$= C_1 \cos x + C_2 \sin x + \tfrac{1}{4}x^2 \sin x + \tfrac{1}{4}x \cos x - \tfrac{1}{2}x \sin x$$

Applying the prescribed initial conditions to this general solution, we obtain

$$y(0) = C_1 = 1$$
$$y'(0) = \tfrac{1}{4} + C_2 = \tfrac{1}{4}$$

and hence, $C_1 = 1$ and $C_2 = 0$. The solution of the IVP is therefore

$$y = \cos x + \tfrac{1}{4}x^2 \sin x + \tfrac{1}{4}x \cos x - \tfrac{1}{2}x \sin x$$
$$= (1 + \tfrac{1}{4}x) \cos x + \tfrac{1}{2}x(\tfrac{1}{2}x - 1) \sin x \qquad ■$$

Exercises 4.7

In Problems 1 to 10, set up the proper form for y_P but do not solve for the undetermined coefficients.

1. $(D^2 - 2)y = 7x^2$

2. $(D^2 - 3D - 4)y = 3xe^{2x}$

3. $y'' - 2y' + y = 5x^2 - 7x + 4x^2e^x$

4. $y'' + 2y' + y = 3xe^{-x} + 4x^2$

5. $y'' - 4y' = 5x^3e^{2x} - x + 10$

■**6.** $y'' + 4y = 3(x + 1)e^x \cos x$

7. $(D^3 + 2D^2 + D)y = 3xe^{-x} + 5x^2$

8. $D(D^2 - 4D + 4)y = 7x^3e^{2x} - x + 1$

9. $(D^4 - 2D^2 + 1)y = e^x + 2 \sin x$

10. $D^2(D^2 - 4)y = 3 \sin x + 5xe^{2x} + 2x^3$

In Problems 11 to 32, obtain a general solution of the given DE.

11. $y'' + y' = -\cos x$

■**12.** $y'' + 9y = 18$

13. $y'' - 6y' + 9y = e^x$

14. $y'' + 3y' + 2y = 6x^3$

15. $y'' + 8y = 2e^{-x} + 5x$

16. $(D^2 + 9)y = x \sin 3x$

17. $(D^2 + 4)y = 3 \sin x + 4 \cos x - 8$

18. $(D^2 + 4)y = 2 \sin x \cos x - 7$

19. $(D^2 - 3D - 4)y = 30e^{4x}$

20. $(D^2 - 1)y = 8xe^x$

21. $(D^2 - 2D + 5)y = e^x \sin x$

22. $D(D + 1) = (x + 1)^3$

23. $(D^2 + D + 1)y = 2 \sin^2 x$

■**24.** $D(D^2 + 4D + 4)y = xe^{-x}$

25. $(D^3 + D^2 - 4D - 4)y = 3e^{-x} - 4x - 6$

26. $D^2(D + 1)y = 1$

*****27.** $(2D^3 - 3D^2 - 3D + 2)y = 4 \cosh^2 x$

28. $(D^3 - 3D - 2)y = \cos 2x$

29. $(D^3 + D^2 - 2)y = x^2 + 10 \sin 2x$

30. $D^4y = 7$

31. $(16D^4 - 1)y = 6e^{x/2}$

32. $D^2(D^2 - 1)y = 8e^{-2x} + 3e^x - x + 8$

In Problems 33 to 40, solve the given IVP.

33. $y'' + y = \sin x,\ y(0) = -1,\ y'(0) = 1$

34. $y'' - 5y' = x - 2,\ y(0) = 0,\ y'(0) = 2$

35. $y'' + y = 8 \cos 2x - 4 \sin x,\ y(\pi/2) = 1,$
$y'(\pi/2) = 0$

36. $(D^2 - 3D - 4)y = e^{-x},\ y(2) = 3,\ y'(2) = 0$

37. $(D^3 - D^2 + 4D - 4)y = x,\ y(0) = 2,\ y'(0) = 1,$
$y''(0) = -1$

■**38.** $(2D^3 + 3D^2 - 3D - 2)y = e^{-x},\ y(0) = 0,$
$y'(0) = 0,\ y''(0) = 1$

39. $(D^4 - 1)y = x,\ y(0) = 0,\ y'(0) = 0,\ y''(0) = 0,$
$y'''(0) = 0$

40. $D(D^2 - 3D + 2)y = x + e^x,\ y(0) = 1,$
$y'(0) = -\tfrac{1}{4},\ y''(0) = -\tfrac{3}{2}$

41. Show that when the nonhomogeneous term is a *constant K*, the resulting DE

$$ay'' + by' + cy = K$$

has the particular solution $y_P = K/c$, $c \neq 0$. Does this result hold if the DE is of higher order?

In Problems 42 to 48, solve the given boundary value problem.

42. $y'' + 2y' + y = x$, $y(0) = -3$, $y(1) = -1$

43. $y'' + y = x + 1$, $y(0) = 1$, $y(1) = 1/2$

■**44.** $(D^2 + 1)y = 2 \cos x$, $y(0) = 0$, $y(\pi) = 0$

45. $(D^2 + 1)y = \sin x$, $y(0) = 1$, $y(\pi/2) = -1$

46. $y'' + 4y = e^{-x}$, $y(0) = -1$, $y'(2) = 3$

47. $y'' - y = x^2$, $y(0) = 0$, $y(1) = 0$

*__48.__ $(D^4 - 1)y = x^3/2$, $y(0) = 0$, $y''(0) = 0$, $y''(1) = 0$, $y'''(1) = 0$

4.8 Nonhomogeneous Equations—Part II

When the nonhomogeneous term $F(x)$ is not of the form assumed in Section 4.7.1, or when the DE has variable coefficients, a more general method of constructing y_P is needed. This more general method, called **variation of parameters,** is analogous to the method bearing that name that was used in Section 2.5.2 for solving nonhomogeneous first-order linear DEs. For simplicity, we restrict our development here to only *second-order* DEs

$$A_2(x)y'' + A_1(x)y' + A_0(x)y = F(x) \qquad (69)$$

The method in theory, however, is also applicable to higher-order DEs (e.g., see Problem 34 in Exercises 4.8).

4.8.1 Variation of Parameters

To begin, we need to put (69) in normal form. If we assume $A_2(x) \neq 0$ on the interval of interest I, then dividing each term of (69) by $A_2(x)$ gives us the **normal form**

$$y'' + a_1(x)y' + a_0(x)y = f(x) \qquad (70)$$

Next, let us assume that the associated *homogeneous* DE

$$y'' + a_1(x)y' + a_0(x)y = 0$$

has the general solution

$$y_H = C_1 y_1(x) + C_2 y_2(x) \qquad (71)$$

where y_1 and y_2 form a fundamental set of solutions on I. We then seek a particular solution of the form

$$y_P = u(x)y_1(x) + v(x)y_2(x) \qquad (72)$$

obtained from (71) by replacing the arbitrary constants with unknown functions u and v. The idea is to select u and v in such a way that (72) is a particular solution of (70).

To determine u and v, we need to produce two relations that they satisfy. Only one such relation is obtained by requiring (72) to satisfy (70), and so we must come up with a second relation that will lead to a simple determination of these functions. Specifically, we choose this second relation in such a way that u and v will be determined by solving two *first-order* DEs.

From (72), we first calculate

$$y_P' = u(x)y_1'(x) + v(x)y_2'(x) + u'(x)y_1(x) + v'(x)y_2(x)$$

which can be simplified (and hence, y_P' will be simplified) if we equate the above terms involving u' and v' to zero. That is, we set

$$u'(x)y_1(x) + v'(x)y_2(x) = 0 \tag{73}$$

which is the first of our conditions on u and v. Thus, we now have

$$y_P' = u(x)y_1'(x) + v(x)y_2'(x) \tag{74}$$

and by differentiating again, we get

$$y_P'' = u(x)y_1''(x) + v(x)y_2''(x) + u'(x)y_1'(x) + v'(x)y_2'(x) \tag{75}$$

Observe that, because of (73), this expression for y_P'' contains only first derivatives of u and v. The substitution of (72), (74), and (75) for y_P, y_P', and y_P'', respectively, into the left-hand side of (70) now leads to

$$y_P'' + a_1(x)y_P' + a_0(x)y_P = u(x)\left[y_1'' + a_1(x)y_1' + a_0(x)y_1\right]$$
$$+ v(x)\left[y_2'' + a_1(x)y_2' + a_0(x)y_2\right]$$
$$+ u'(x)y_1'(x) + v'(x)y_2'(x)$$

However, since y_1 and y_2 are solutions of the homogeneous DE, it follows that the terms above in brackets are zero and this expression reduces to

$$y_P'' + a_1(x)y_P' + a_0(x)y_P = u'(x)y_1'(x) + v'(x)y_2'(x) \tag{76}$$

In order that y_P be a solution of (70), the right-hand side of (76) must equal $f(x)$, that is,

$$u'(x)y_1'(x) + v'(x)y_2'(x) = f(x) \tag{77}$$

Hence, we have shown that (72) is a particular solution of (70) provided that u and v satisfy the relations (73) and (77). Rewriting these equations as a set of simultaneous equations, that is,

$$u'(x)y_1(x) + v'(x)y_2(x) = 0$$
$$u'(x)y_1'(x) + v'(x)y_2'(x) = f(x) \tag{78}$$

it is known (by Cramer's rule) that nontrivial solutions for u' and v' exist provided the coefficient determinant

$$\begin{vmatrix} y_1(x) & y_2(x) \\ y_1'(x) & y_2'(x) \end{vmatrix} \neq 0$$

But this determinant is precisely the Wronskian $W(y_1, y_2)(x)$ of the linearly independent solutions y_1 and y_2. Thus, it can never be zero on the interval I, and the simultaneous solution of (78) yields

$$u'(x) = -\frac{y_2(x)f(x)}{W(y_1, y_2)(x)}$$

$$v'(x) = \frac{y_1(x)f(x)}{W(y_1, y_2)(x)}$$

(79)

Any indefinite integral of the expressions in (79) provides suitable choices for u and v. We conclude, therefore, that the particular solution (72) can now be expressed in the form

$$y_P = u(x)y_1(x) + v(x)y_2(x)$$

$$= -y_1(x) \int \frac{y_2(x)f(x)}{W(y_1, y_2)(x)}\, dx$$

$$+ y_2(x) \int \frac{y_1(x)f(x)}{W(y_1, y_2)(x)}\, dx$$

(80)

In using the above method to construct a suitable y_P for a given DE, it is generally advisable to substitute the expression $y_P = u(x)y_1(x) + v(x)y_2(x)$ directly into the DE and proceed as above, rather than relying on Equation (80) as a formula. Finally, we remark that in practice the integrals in (80) frequently cannot be evaluated explicitly. In such cases the solution is either left in integral form or evaluated numerically.

EXAMPLE 28   Find the general solution of

$$2y'' + 18y = \csc 3x$$

Solution

We first rewrite the DE in normal form

$$y'' + 9y = \tfrac{1}{2} \csc 3x$$

which identifies the function $f(x) = \tfrac{1}{2} \csc 3x$. The solution of the associated homogeneous equation

$$y'' + 9y = 0$$

is

$$y_H = C_1 \cos 3x + C_2 \sin 3x$$

where we have now selected $y_1 = \cos 3x$ and $y_2 = \sin 3x$.

By assuming a particular solution of the form

$$y_P = u(x) \cos 3x + v(x) \sin 3x$$

and proceeding as above by substituting it into the DE, we find

$$u'(x) \cos 3x + v'(x) \sin 3x = 0$$

$$-3u'(x) \sin 3x + 3v'(x) \cos 3x = \tfrac{1}{2} \csc 3x$$

the simultaneous solution of which yields

$$u'(x) = -\tfrac{1}{6} \sin 3x \csc 3x = -\tfrac{1}{6}$$

and

$$v'(x) = \tfrac{1}{6} \cos 3x \csc 3x = \tfrac{1}{6} \cot 3x$$

Thus, integration of these relations gives us

$$u(x) = -\frac{1}{6} \int dx = -\frac{x}{6}$$

and

$$v(x) = \frac{1}{6} \int \cot 3x \, dx = \frac{1}{18} \log|\sin 3x|$$

leading to the particular solution

$$y_P = u(x) \cos 3x + v(x) \sin 3x$$

$$= -\frac{x}{6} \cos 3x + \frac{1}{18} \log|\sin 3x| \sin 3x$$

The general solution $y = y_H + y_P$ in this case can be written as

$$y = \left(C_1 - \frac{x}{6} \right) \cos 3x + \left(C_2 + \frac{1}{18} \log|\sin 3x| \right) \sin 3x \qquad \blacksquare$$

EXAMPLE 29

Find the general solution of

$$y'' - 3y' + 2y = \frac{1}{1 + e^{-x}}$$

Solution

The associated homogeneous DE has general solution

$$y_H = C_1 e^x + C_2 e^{2x}$$

with Wronskian $W(e^x, e^{2x}) = e^{3x}$. Hence, based on (79) we find

$$u'(x) = -\frac{e^{2x}}{e^{3x}(1 + e^{-x})}$$

$$= -\frac{e^{-x}}{1 + e^{-x}}$$

and

$$v'(x) = \frac{e^x}{e^{3x}(1 + e^{-x})}$$

$$= \frac{e^{-2x}}{1 + e^{-x}}$$

Integrating the above expressions, we arrive at

$$u(x) = -\int \frac{e^{-x}}{1 + e^{-x}} \, dx = \log(1 + e^{-x})$$

and

$$v(x) = \int \frac{e^{-2x}}{1 + e^{-x}} \, dx = \int \left(e^{-x} - \frac{e^{-x}}{1 + e^{-x}} \right) dx$$

where in the last expression we have carried out long division in the integrand. Completing the integration above, we deduce that

$$v(x) = -e^{-x} + \log(1 + e^{-x})$$

from which we obtain the particular solution

$$y_P = u(x)e^x + v(x)e^{2x}$$
$$= e^x \log(1 + e^{-x}) - e^x + e^{2x} \log(1 + e^{-x})$$
$$= e^x(1 + e^x) \log(1 + e^{-x}) - e^x$$

Finally, combining y_H and y_P, we see that the general solution is

$$y = C_1 e^x + C_2 e^{2x} + e^x(1 + e^x) \log(1 + e^{-x}) - e^x$$
$$= C_3 e^x + C_2 e^{2x} + e^x(1 + e^x) \log(1 + e^{-x})$$

where $C_3 = C_1 - 1$. ■

As a final example here, let us consider the case where the integrals are non-elementary but define some *special function*.

<hr>

EXAMPLE 30

Solve the boundary value problem

$$y'' + y = x^{-1}, \qquad y(\pi/2) = 0, \qquad y(\pi) = 0$$

Solution

Here we see that $y_1 = \cos x$ and $y_2 = \sin x$ form a fundamental set of solutions, so that we may assume

$$y_P = u(x) \cos x + v(x) \sin x$$

From (79), we have

$$u'(x) = -\frac{\sin x}{x}, \qquad v'(x) = \frac{\cos x}{x}$$

the integrals of which are nonelementary. In this case we take the specific anti-derivatives

$$u(x) = -\int_0^x \frac{\sin t}{t}\, dt = -\operatorname{Si}(x)$$

$$v(x) = \int_\infty^x \frac{\cos t}{t}\, dt = \operatorname{Ci}(x)$$

where $\operatorname{Si}(x)$ and $\operatorname{Ci}(x)$ are known as the **sine integral** and **cosine integral,** respectively. Although they are nonelementary functions, their numerical values are tabulated; for example, see Table 5.1 in M. Abramowitz and I. A. Stegun, *Handbook of Mathematical Functions,* Dover, New York, 1965.

The general solution of the DE now assumes the form

$$y = y_H + y_P$$

$$= C_1 \cos x + C_2 \sin x - \operatorname{Si}(x) \cos x + \operatorname{Ci}(x) \sin x$$

Imposing the first boundary condition leads to

$$y(\pi/2) = C_2 + \operatorname{Ci}(\pi/2) = 0$$

from which we deduce $C_2 = -\operatorname{Ci}(\pi/2)$. Likewise, from the second condition, we find that

$$y(\pi) = -C_1 + \operatorname{Si}(\pi) = 0$$

or $C_1 = \operatorname{Si}(\pi)$. Hence, the particular solution we seek satisfying the prescribed conditions is given by

$$y = [\operatorname{Si}(\pi) - \operatorname{Si}(x)] \cos x + [\operatorname{Ci}(x) - \operatorname{Ci}(\pi/2)] \sin x \qquad \blacksquare$$

Exercises 4.8

In Problems 1 to 20, use variation of parameters to construct y_P and use it to determine a general solution.

1. $y'' - y = e^x$

2. $y'' - y = x$

3. $y'' + 9y = \sin 3x$

4. $y'' + y = \csc x$

5. $y'' + y = \sec x$

■**6.** $y'' + y = \sec^2 x$

7. $y'' + y = \sec^4 x$

8. $y'' + y = \tan x$

9. $y'' + y = \cot x$

10. $y'' + y = \tan^2 x$

11. $y'' + y = \sec x \csc x$

12. $y'' + y = \csc x \cot x$

13. $(D^2 - 4D + 4)y = (x + 1)e^{2x}$

***14.** $(D^2 - 3D + 2)y = \dfrac{e^{3x}}{1 + e^x}$

15. $(D^2 + 2D + 1)y = e^{-x} \log x$

*16. $(D^2 - 2D + 1)y = \dfrac{e^{2x}}{(1 + e^x)^2}$

17. $(D^2 - 1)y = e^{-2x}\sin(e^{-x})$

■18. $(D^2 - 3D + 2)y = \sin(e^{-x})$

19. $(D^2 + 1)y = \csc^3 x \cot x$

*20. $(4D^2 - 4D + 1)y = \sqrt{1 - x^2}\, e^{x/2}$

In Problems 21 to 26, solve the given IVP using variation of parameters to construct the particular solution.

21. $y'' - y = xe^x$, $y(0) = 2$, $y'(0) = 0$

■22. $y'' + 2y' - 8y = 2e^{-2x} - e^{-x}$, $y(0) = 1$,
    $y'(0) = 0$

23. $y'' + y = 2\csc x \cot x$, $y(\pi/2) = 1$, $y'(\pi/2) = 1$

24. $(2D^2 + D - 1)y = x + 1$, $y(0) = 1$, $y'(0) = 0$

*25. $y'' + y = 3/x$, $y(\pi/2) = 0$, $y'(\pi/2) = 0$

*26. $y'' + y = 1/\sqrt{2\pi x}$, $y(\pi) = 0$, $y'(\pi) = 0$

Hint. Let

$$C_2(x) = \frac{1}{\sqrt{2\pi}}\int_0^x \frac{\cos t}{\sqrt{t}}\, dt,$$

$$S_2(x) = \frac{1}{\sqrt{2\pi}}\int_0^x \frac{\sin t}{\sqrt{t}}\, dt$$

known as **Fresnel integrals.**

In Problems 27 to 32, use the given homogeneous solution to obtain a general solution of the nonhomogeneous DE.

27. $(1 - x)y'' + xy' - y = 2(x - 1)^2 e^{-x}$;
    $y_H = C_1 x + C_2 e^x$

28. $x^2 y'' - x(x + 2)y' + (x + 2)y = x^3$;
    $y_H = C_1 x + C_2 x e^x$

29. $x(x - 2)y'' - (x^2 - 2)y' + 2(x - 1)y = 3x^2(x - 2)^2 e^x$; $y_H = C_1 x^2 + C_2 e^x$

■30. $xy'' - (1 + 2x^2)y' = x^5 e^{x^2}$; $y_H = C_1 + C_2 e^{x^2}$

31. $(1 - x^2)y'' - 2xy' = 2x$;
    $y_H = C_1 + C_2 \log\dfrac{1 + x}{1 - x}$

*32. $x^2 y'' + xy' + (x^2 - \frac{1}{4})y = 3x^{3/2}\sin x$, $x > 0$;
    $y_H = C_1 x^{-1/2}\cos x + C_2 x^{-1/2}\sin x$

*33. Find a general solution of the DE

$$(1 + x^2)y'' - 4xy' + 6y = 3(1 + x^2)^3$$

given that $y_1 = 1 - 3x^2$ is a solution of the associated homogeneous DE.

*34. Let y_1, y_2, y_3 form a fundamental solution set of the third-order homogeneous DE

$$y''' + a_2(x)y'' + a_1(x)y' + a_0(x)y = 0.$$

(a) By assuming a particular solution of the form

$$y_P = u(x)y_1(x) + v(x)y_2(x) + w(x)y_3(x)$$

where u, v, and w are unknown functions, show that substituting y_P into the related

nonhomogeneous DE

$$y''' + a_2(x)y'' + a_1(x)y' + a_0(x)y = f(x)$$

and following the scheme for second-order DEs leads to the system of simultaneous equations

$$y_1(x)u'(x) + y_2(x)v'(x) + y_3(x)w'(x) = 0$$

$$y_1'(x)u'(x) + y_2'(x)v'(x) + y_3'(x)w'(x) = 0$$

$$y_1''(x)u'(x) + y_2''(x)v'(x) + y_3''(x)w'(x) = f(x)$$

(b) Show that the above system of equations have the solution (by Cramer's rule)

$$u'(x) = \frac{\begin{vmatrix} 0 & y_2(x) & y_3(x) \\ 0 & y_2'(x) & y_3'(x) \\ f(x) & y_2''(x) & y_3''(x) \end{vmatrix}}{W(y_1, y_2, y_3)(x)}$$

$$v'(x) = \frac{\begin{vmatrix} y_1(x) & 0 & y_3(x) \\ y_1'(x) & 0 & y_3'(x) \\ y_1''(x) & f(x) & y_3''(x) \end{vmatrix}}{W(y_1, y_2, y_3)(x)}$$

$$w'(x) = \frac{\begin{vmatrix} y_1(x) & y_2(x) & 0 \\ y_1'(x) & y_2'(x) & 0 \\ y_1''(x) & y_2''(x) & f(x) \end{vmatrix}}{W(y_1, y_2, y_3)(x)}$$

where $W(y_1, y_2, y_3)(x)$ is the Wronskian.

In Problems 35 to 39, find a general solution of the given DE by using the result of Problem 34 to obtain a particular solution.

35. $D(D^2 - 1)y = 2x$

36. $3D(D^2 + 1)y = \tan x$, $0 < x < \pi/2$

37. $(D^3 - 2D^2 - D + 2)y = e^{3x}$

38. $D(D^2 - 2D + 1)y = x$

39. $(D^3 + 3D^2 - 4)y = e^{2x}$

*40. Find a general solution of

$$(x^3 D^3 + x^2 D^2 - 2xD + 2)y = x^3 \sin x,$$

$$x > 0$$

given that $y_1 = x$, $y_2 = x^{-1}$, and $y_3 = x^2$ form a fundamental set of solutions of the associated homogeneous DE.

*41. Find a general solution of

$$(x^3 D^3 - 3x^2 D^2 + 6xD - 6)y = x^{-1},$$

$$x > 0$$

given that $y_1 = x$, $y_2 = x^2$, and $y_3 = x^3$ form a fundamental set of solutions of the associated homogeneous DE.

*42. Find a general solution of

$$(x^3 D^3 - x^2 D^2 + 2xD - 2)y = x^3, \qquad x > 0$$

given that $y_1 = x$, $y_2 = x \log x$, and $y_3 = x^2$ form a fundamental set of solutions of the associated homogeneous DE.

4.9 Cauchy–Euler Equations

Thus far we have solved only *constant-coefficient* DEs, although most of the theory developed for higher-order linear DEs is applicable as well to *variable-coefficient* DEs. Unfortunately, we usually cannot solve variable-coefficient DEs as easily as those with constant coefficients. In most cases we must resort to some sort of power series method as discussed in Chapter 7. An exception to this is the class of **Cauchy–Euler equations** (or **equidimensional equations**).

The general form of a second-order Cauchy–Euler equation is given by

$$ax^2 y'' + bxy' + cy = F(x) \tag{81}$$

where a, b, and c are constants. The significant feature here is that the power of x in each coefficient corresponds to the order of the derivative of y. Equations of this type can always be transformed into constant-coefficient DEs by means of a change of independent variable.

To begin, we assume that $x > 0$ and make the change of variable $x = e^t$. (If $x < 0$, we set $x = -e^t$.) By the chain rule, we have

$$y' = \frac{dy}{dt}\frac{dt}{dx} = \frac{1}{x}\frac{dy}{dt}$$

and

$$y'' = \frac{d}{dx}\left(\frac{dy}{dx}\right) = \frac{d}{dt}\left(\frac{dy}{dx}\right)\frac{dt}{dx} = \frac{1}{x^2}\left(\frac{d^2 y}{dt^2} - \frac{dy}{dt}\right)$$

which leads to

$$xy' = Dy, \qquad x^2y'' = D(D-1)y \tag{82}$$

where $D = d/dt$. Under this transformation, Equation (81) becomes

$$[aD(D-1) + bD + c]y = F(e^t)$$

or

$$[aD^2 + (b-a)D + c]y = F(e^t) \tag{83}$$

which has constant coefficients. Generalizations to higher-order DEs are taken up in the exercises (see Problems 24 to 30 in Exercises 4.9).

EXAMPLE 31

Find the general solution of

$$4x^2y'' + y = 0, \qquad x > 0$$

Solution

Using the transformation $x = e^t$, we obtain the constant-coefficient DE

$$(4D^2 - 4D + 1)y = 0$$

The auxiliary equation $4m^2 - 4m + 1 = 0$ has the double root $m_1 = m_2 = 1/2$, and therefore the general solution of this transformed DE is

$$y(t) = (C_1 + C_2t)e^{t/2}$$

Transforming back to the original variable by setting $t = \log x$ yields

$$y(x) = (C_1 + C_2 \log x)\sqrt{x}, \qquad x > 0 \qquad \blacksquare$$

EXAMPLE 32

Find the general solution of

$$x^2y'' - xy' + 5y = 0, \qquad x > 0$$

Solution

The related constant-coefficient DE is given by

$$(D^2 - 2D + 5)y = 0$$

with $1 \pm 2i$ as the roots of the auxiliary equation. Thus,

$$y(t) = e^t(C_1 \cos 2t + C_2 \sin 2t)$$

and transforming back to the variable x, we obtain

$$y(x) = x[C_1 \cos(2 \log x) + C_2 \sin(2 \log x)], \qquad x > 0 \qquad \blacksquare$$

EXAMPLE 33

Find a general solution of

$$x^2 y'' - 3xy' + 3y = 2x^4 e^x, \qquad x > 0$$

Solution

This time the DE is nonhomogeneous and so we must find both y_H and y_P. Under the transformation $x = e^t$, the DE becomes

$$(D^2 - 4D + 3)y = 2e^{4t}e^{e^t}$$

The auxiliary equation has roots $m = 1, 3$, giving us

$$y_H(t) = C_1 e^t + C_2 e^{3t}$$

Because the forcing term in the transformed DE does not lend itself to the method of undetermined coefficients, we must solve for y_P by variation of parameters. Here now we have a choice. That is, we can either find $y_P(t)$ using the transformed DE or find $y_P(x)$ using the original DE. We choose the latter.

The homogeneous solution of the original DE is obtained by transforming $y_H(t)$ above back to the variable x, which leads to

$$y_H(x) = C_1 x + C_2 x^3$$

Hence, we have $y_1 = x$ and $y_2 = x^3$ with Wronskian

$$W(x, x^3) = \begin{vmatrix} x & x^3 \\ 1 & 3x^2 \end{vmatrix} = 2x^3$$

We now seek a particular solution of the form $y_P = u(x)x + v(x)x^3$. Putting the Cauchy–Euler equation in normal form,

$$y'' - \frac{3}{x} y' + \frac{3}{x^2} y = 2x^2 e^x$$

we identify $f(x) = 2x^2 e^x$. It follows that

$$u'(x) = -\frac{x^3(2x^2 e^x)}{2x^3} = -x^2 e^x$$

$$v'(x) = \frac{x(2x^2 e^x)}{2x^3} = e^x$$

integrations of which yield

$$u(x) = -x^2 e^x + 2xe^x - 2e^x$$

$$v(x) = e^x$$

The particular solution is therefore given by

$$y_P(x) = u(x)x + v(x)x^3$$

$$= 2x(x - 1)e^x$$

from which we deduce the general solution

$$y(x) = C_1 x + C_2 x^3 + 2x(x - 1)e^x, \qquad x > 0 \qquad \blacksquare$$

Exercises 4.9

In Problems 1 to 15, find the general solution valid for $x > 0$.

1. $x^2 y'' - 5xy' + 5y = 0$

2. $x^2 y'' - 4xy' + 6y = 0$

3. $x^2 y'' - xy' + y = 0$

4. $9x^2 y'' + 3xy' + y = 0$

5. $3xy'' + 2y' = 0$

6. $x^2 y'' + xy' + y = 0$

7. $x^2 y'' - 5xy' + 25y = 0$

■8. $x^2 y'' - 2xy' + 2y = x^2$

9. $x^2 y'' - 5xy' + 9y = 2x^3$

10. $x^2 y'' + xy' + 4y = 2x \log x$

11. $x^2 y'' - 4xy' + 6y = 4x - 6$

12. $x^2 y'' - xy' + y = 4x \log x$

13. $x^2 y'' + xy' - y = x$

14. $x^2 y'' - 2xy' + 2y = x^3 e^x$

15. $x^2 y'' - 2xy' + 2y = x^3 \log x^2$

In Problems 16 to 20, solve the given IVP.

16. $x^2 y'' - 2xy' - 10y = 0$, $y(1) = 5$, $y'(1) = 4$

17. $x^2 y'' - 4xy' + 6y = 0$, $y(2) = 0$, $y'(2) = 4$

18. $4x^2 y'' + 8xy' + y = 0$, $y(1) = 1$, $y'(1) = 0$

19. $x^2 y'' - 3xy' + 13y = x^3$, $y(1) = 1$, $y'(1) = 0$

■20. $x^2 y'' + xy' + 4y = \sin(\log x)$, $y(1) = 1$, $y'(1) = 0$

In Problems 21 to 23, solve the DE by assuming a solution of the form $y = x^m$, where m must be determined.

21. $2x^2 y'' + 3xy' - y = 0$

22. $x^2 y'' - 2xy' + 2y = 0$

23. $x^2 y'' + 7xy' + 5y = 0$

*24. Show that under the transformation $x = e^t$ $(x > 0)$,

$$x^3 y''' = D(D - 1)(D - 2)y$$

$$x^4 y^{(4)} = D(D - 1)(D - 2)(D - 3)y$$

where $D = d/dt$, and deduce that in general

$$x^n y^{(n)} = D(D - 1)(D - 2) \cdots (D - n + 1)y$$

In Problems 25 to 30, use the result of Problem 24 to find the general solution of each DE for $x > 0$.

25. $x^2 y''' - xy'' + y' = 0$

26. $x^3 y''' + x^2 y'' - 2xy' + 2y = 0$

27. $x^3 y''' + 2x^2 y'' + xy' - y = 0$

*28. $x^3 y''' - 3x^2 y'' + 6xy' - 6y = 3 + \log x^3$

*29. $x^3 y''' - x^2 y'' + 2xy' - 2y = x^3$

*30. $x^4 y^{(4)} + 6x^3 y''' + 15x^2 y'' + 9xy' + 16y = 0$

In Problems 31 and 32, solve the DE by first making an appropriate change of independent variable to reduce the given DE to a Cauchy–Euler DE.

*31. $(x + 5)^2 y'' - (x + 5)y' - 3y = 0$

*32. $(3x - 2)^2 y'' - 6(3x - 2)y' + 12y = 0$

4.10 Chapter Summary

In this chapter we have studied primarily **linear second-order DEs** of the form

$$A_2(x)y'' + A_1(x)y' + A_0(x)y = F(x) \tag{84}$$

When $F(x) \equiv 0$ on an interval I we obtain the associated **homogeneous** equation

$$A_2(x)y'' + A_1(x)y' + A_0(x)y = 0 \tag{85}$$

otherwise, we say that (84) is **nonhomogeneous.** If $A_0(x)$, $A_1(x)$, $A_2(x)$, and $F(x)$ are continuous functions on some interval I containing the point x_0, and $A_2(x) \neq 0$ on I, then for any α and β, (84) has a *unique solution* satisfying $y(x_0) = \alpha$, $y'(x_0) = \beta$.

If y_1 and y_2 are two solutions of the homogeneous equation (85), we define their **Wronskian** by the determinant

$$W(y_1, y_2)(x) = \begin{vmatrix} y_1(x) & y_2(x) \\ y_1'(x) & y_2'(x) \end{vmatrix} \tag{86}$$

If $W(y_1, y_2)(x) \neq 0$ for at least one point in an interval I, we say that y_1 and y_2 are **linearly independent** and as such form a **fundamental solution set** on I. If y_1 and y_2 are **linearly dependent** on I, then $W(y_1, y_2)(x) = 0$ for every x on I. Finally, if y_1 and y_2 form a fundamental solution set on an interval I, then the general solution of (85) on I is given by

$$y = C_1 y_1(x) + C_2 y_2(x) \tag{87}$$

When the DE has **constant coefficients,** that is, when

$$ay'' + by' + cy = 0 \tag{88}$$

where a, b, and c are constants, we first find the roots m_1 and m_2 of the auxiliary equation $am^2 + bm + c = 0$. Depending upon the nature of these roots, there are three forms of the general solution that can occur:

Case I. $m_1 \neq m_2$ (real)

$$y = C_1 e^{m_1 x} + C_2 e^{m_2 x}$$

Case II. $m_1 = m_2 = m$

$$y = (C_1 + C_2 x)e^{mx}$$

Case III. $m_1, m_2 = p \pm iq$

$$y = e^{px}(C_1 \cos qx + C_2 \sin qx)$$

The **general solution** of the nonhomogeneous equation (84) is given by $y = y_H + y_P$, where $y_H = C_1 y_1(x) + C_2 y_2(x)$ is the general solution of the homogeneous

DE (85) and y_P is any **particular solution** of (84). To find y_P, we use either the method of **undetermined coefficients** or **variation of parameters.** The former method is restricted primarily to constant-coefficient DEs and those for which $F(x)$ is a polynomial, e^{px}, $\cos qx$, $\sin qx$, or finite sums and products of these functions.

Review Exercises

In Problems 1 and 2, determine whether the given solutions are linearly independent or dependent using (a) Definition 4.1 and (b) the Wronskian.

1. $y'' = 0$; $y_1 = 3x$, $y_2 = 5x$
2. $y'' + y = 0$; $y_1 = \sin x$, $y_2 = 2 \sin x - \cos x$

In Problems 3 to 20, find a general solution to the given DE.

3. $y'' - y' - 2y = 0$
4. $y'' - 4y' + y = 0$
5. $y'' + 8y' + 25y = 0$
6. $y'' + 6y' + 9y = 0$
7. $(16D^2 + 8D + 1)y = 0$
8. $(2D^3 - D^2 - 5D - 2)y = 0$
9. $(D^3 + 6D^2 + 12D + 8)y = 0$
10. $(D^3 + 2D^2 + 5D - 26)y = 0$
11. $y''' + 4y' = 0$

12. $(D^4 + D^3 + D^2 + 2)y = 0$
13. $y'' - 7y' = 6e^{6x}$
14. $y'' - y' - 2y = e^{2x}$
15. $(D^4 + D^2)y = 16$
16. $(D^2 - 2D)y = e^x \sin x$
17. $y'' - 2y' + y = x^{-1}e^x$
18. $y'' + 64y = \sec 8x$
19. $y'' - y = (1 + e^{-x})^{-2}$
20. $y'' - y = e^{-x} \sin(e^{-x}) + \cos(e^{-x})$

In Problems 21 to 25, solve the given IVP.

21. $y'' - 3y' + 4y = 0$, $y(0) = 1$, $y'(0) = 5$
22. $y'' + y' - 2y = \sin x$, $y(0) = 0$, $y'(0) = 0$
23. $y'' + y = x$, $y(0) = 1$, $y'(0) = -2$

24. $y'' + 2y' + 5y = e^{-x} \sin x$, $y(0) = 0$, $y'(0) = 1$
25. $y'' - 3y' + 2y = 4e^{2x}$, $y(0) = -3$, $y'(0) = 5$

4.11 Historical Comments

In 1743, Leonhard Euler produced general solutions of **homogeneous linear equations with constant coefficients,** and later he extended the results to certain **nonhomogeneous equations.** However, it was Joseph Louis Lagrange (1736–1813) who showed that the general solution of an **nth-order linear homogeneous DE** is a linear combination of n independent solutions. Lagrange was also responsible for the development of the method of **variation of parameters** used to solve nonhomogeneous DEs.

Lagrange, of French and Italian heritage, was born in Turin, Italy. He was one of the two greatest mathematicians of the eighteenth century—the other being Euler. Lagrange was educated in Turin and became a professor of mathematics at the military academy there at the age of 19. Later he succeeded Euler in the chair of mathematics at the Berlin Academy. In addition to his work on ordinary linear differential equations, Lagrange was also well known for his contributions to partial differential equations, algebra, number theory, acoustics, and the calculus of variations. His most famous publications, however, were in the field of Newtonian mechanics.

The **Cauchy–Euler equation** was studied by both Leonhard Euler and the famous French mathematician Augustin Louis Cauchy (1789–1857). Among other contributions, Cauchy provided the first systematic study of the theory of limits and the first rigorous proof of the existence of solutions to first-order DEs. He also developed the concept of convergence of an infinite series and the theory of functions of a complex variable. For his demand for precise definitions and rigorous proofs, he is known as the father of modern mathematical analysis.

CHAPTER 5
Initial Value Problems

In the present chapter we consider **initial value problems** (IVPs) with applications involving mechanical vibrations and electric circuits. Because many vibration problems are mathematically similar, it is helpful to start with a simple mechanical system consisting of a coil spring (with mass attached) suspended from a rigid support.

Newton's law of motion combined with **Hooke's law** is used in Section 5.2 to derive the DE governing the small vertical movements of a **spring-mass system.** **Free motions** of the mass are discussed for cases of both **undamped** (simple harmonic motion) and **damped systems.** The latter type of motion results when frictional or resistive forces in the system cannot be ignored. In Section 5.3 we examine some **forced motions** of the spring-mass system that occur when an external force is applied to the mass. Analogous systems involving **simple electric circuits** are briefly discussed in Section 5.4.

The method of **Green's function,** which provides us with a systematic method for solving nonhomogeneous IVPs, is introduced in Section 5.5. Finally in Section 5.6 we discuss the **oscillatory behavior** (in the absence of explicit solutions) of equations with variable coefficients. Such DEs arise in practice when some of the system parameters vary over time.

5.1 Introduction

Linear second-order DEs arise in the study of *particle motion, population dynamics,* and *electric circuits,* among other areas of application. The general problem is to solve the linear DE

$$A_2(t)y'' + A_1(t)y' + A_0(t)y = F(t), \qquad t > t_0 \tag{1}$$

subject to the initial conditions

$$y(t_0) = k_0, \qquad y'(t_0) = k_1 \tag{2}$$

Because the initial conditions are both specified at the same point t_0, we call (1) and (2) an IVP. Although the value of t_0 is usually chosen as zero, that is, $t_0 = 0$, the general theory of such equations does not require this choice. The function $F(t)$ is called the **forcing function** or **input function,** and physically corresponds to an external stimulus (e.g., a force or voltage) to the system described by (1). If y represents the linear movement of some point mass, then k_0 and k_1 correspond, respectively, to the initial *position* and *velocity* of the mass. In an electric circuit, k_0 and k_1 correspond to the initial *charge* and *current,* respectively.

> **Remark.** Because IVPs generally involve time as the independent variable, we use t as the independent variable in this chapter rather than x.

For theoretical discussions, it is convenient to put Equation (1) in **normal form.** This we do by dividing the equation by $A_2(t)$, assuming that $A_2(t) \neq 0$ for $t > t_0$, which yields

$$y'' + a_1(t)y' + a_0(t)y = f(t), \qquad t > t_0 \tag{3}$$

where $a_1(t) = A_1(t)/A_2(t)$, $a_0(t) = A_0(t)/A_2(t)$, and $f(t) = F(t)/A_2(t)$. An interesting and important property of IVPs is that, under rather general conditions, they always possess *unique* solutions. We first stated this result in Chapter 4 (Theorem 4.1), but, for the sake of completeness in our discussion of IVPs, we restate it here as Theorem 5.1.

> **Theorem 5.1 Existence-uniqueness.** *If $f(t)$, $a_0(t)$, and $a_1(t)$ are continuous functions on the interval $t \geq t_0$, then for any choice of constants k_0 and k_1, there exists a unique solution of the IVP*
>
> $$y'' + a_1(t)y' + a_0(t)y = f(t), \qquad t > t_0$$
>
> $$y(t_0) = k_0, \qquad y'(t_0) = k_1$$

As an immediate consequence of Theorem 5.1, we have the following corollary for the case when the forcing function $f(t)$ is identically zero on $t \geq t_0$ and $k_0 = k_1 = 0$.

Corollary 5.1. *If $a_0(t)$ and $a_1(t)$ are continuous functions on the interval $t \geq t_0$, then the IVP*

$$y'' + a_1(t)y' + a_0(t)y = 0, \qquad y(t_0) = 0, \qquad y'(t_0) = 0$$

has only the trivial solution $y = 0$.

Proof. By inspection we see that $y = 0$ is a solution of the DE and initial conditions. Hence, based on Theorem 5.1 it is the only solution.

Remark. Corollary 5.1 can be interpreted physically as saying that a physical system at rest in its equilibrium configuration will remain so if not subject to an external stimulus.

5.2 Small Free Motions of a Spring-Mass System

When different weights are attached to an elastic spring suspended from a fixed support, the spring will stretch by an amount s that varies with the weight. **Hooke's law**[†] states that the spring will exert an upward restoring force proportional to the amount of stretch s (within reason); that is,

$$F = ks \tag{4}$$

The constant of proportionality, denoted by k, depends upon the "stiffness" of the spring and thus is different for each spring. For example, if a 10-lb weight stretches a spring 6 in. (1/2 ft), then

$$10 = k(1/2)$$

or $k = 20$ lb/ft, whereas if the weight only stretches the spring 2 in. we find $k = 60$ lb/ft.

The two most common systems of units are the *International* and *English* (or *British*), both of which (and their abbreviations) are listed in Table 5.1.

Suppose the natural length of a spring is b units and a body of weight $W = mg$ is attached to the spring, where m is the mass and g the gravitational constant. The body, also called simply the "weight" or the "mass," will then attain a position of equilibrium at $y = 0$, which is s units from the equilibrium position of the spring itself (see Figure 5.1). Based on Hooke's law (4), the upward restoring force is ks, which is offset by the weight mg, that is, $ks = mg$. If the system is now subjected to a (downward) external force of magnitude $F(t)$, the weight will move in the vertical direction. We assume that such motions are "small" so that Hooke's law remains valid.

In addition to an external force, there frequently exists a retarding force caused by resistance of the medium in which the motion takes place, or possibly by friction. For example, the weight could be suspended in a viscous medium (such as oil), connected to a dashpot damping device (i.e., a "shock absorber" or pistonlike device as shown in Figure 5.1c), and so on. In practice, many such retarding forces are approxi-

---

[†] Named in honor of the English physicist Robert Hooke (1635–1703).

TABLE 5.1   COMMON SYSTEMS OF UNITS AND THEIR ABBREVIATIONS

System of units	Force	Length	Mass	Time
International	newton (N)	meter (m)	kilogram (kg)	second (s)
English	pound (lb)	foot (ft)	slug	second (s)

Note: To convert from one system to another, we use the following relations:
$$1\ N = 1\ kg{\cdot}m/s^2 = 0.2248\ lb \qquad 1\ kg = 0.0685\ slug$$
$$1\ m = 3.2808\ ft \qquad\qquad 1\ lb = 1\ slug{\cdot}ft/s^2 = 4.4482\ N$$

mately proportional to the velocity y'. Hence, we assume that the resistive force is cy', where $c \geq 0$, and that this force always acts in a direction *opposing* the motion.

Now, as a consequence of **Newton's second law of motion,** that is, $\sum F = ma = my''$, we see that summing forces yields

$$mg - k(y + s) - cy' + F(t) = my''$$

or, since $mg = ks$, this reduces to (after rearranging terms)

$$my'' + cy' + ky = F(t), \qquad t > 0, \qquad c \geq 0 \tag{5}$$

> **Remark.**   Note that the terms on the left-hand side of (5) represent system forces, such as restoring and damping forces, while the function $F(t)$ on the right-hand side represents an external force to the system.

Equation (5) describes the general motions of a spring-mass system. Observe that it is a *nonhomogeneous, second-order, linear* DE with *constant coefficients*. As such, it can be solved by the methods of Chapter 4. The motion is said to be **undamped**

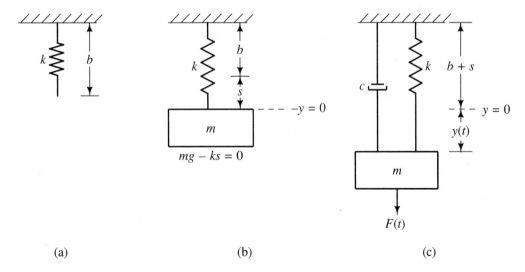

(a)   (b)   (c)

Figure 5.1 (a) Unstretched spring. (b) Equilibrium position with mass. (c) Spring-mass system with a forcing function and damping.

when $c = 0$ and **damped** when $c \neq 0$. The motion is further classified as **free** when $F(t) \equiv 0$ and **forced** when $F(t)$ is not identically zero.

EXAMPLE 1

Write the differential equation of motion for a 2-kg mass suspended by a spring having spring constant $k = 32$ N/m. A force of $0.5 \sin 3t$ N is applied to the mass, and a dashpot damping mechanism is such that $c = 5$ kg/s.

Solution

The relevant constants are $m = 2$, $c = 5$, and $k = 32$, leading to the equation of motion

$$2y'' + 5y' + 32y = 0.5 \sin 3t$$

■

EXAMPLE 2

A 10-lb weight stretches a given spring by 6 in., all of which is suspended in a highly viscous medium for which the damping force is 5 times the instantaneous velocity. If the weight (mass) is drawn 3 in. below the equilibrium position and then released, find the governing DE and initial conditions describing the subsequent motions of the mass.

Solution

To determine the spring constant we use Hooke's law, that is,

$$10 = k(1/2)$$

or $k = 20$ lb/ft. The mass m is determined by dividing the weight $W = mg$ by the gravitational constant g ($g \cong 32$ ft/s²); thus, $m = 10/g = 5/16$. The damping constant is given by $c = 5$, and hence the governing DE is

$$\frac{5}{16} y'' + 5y' + 20y = 0$$

The mass is initially pulled down in the positive direction, and so the initial position of the mass is $y(0) = 3$ in. $= \frac{1}{4}$ ft, and the initial velocity is zero. Therefore, the complete IVP governing the motion is (after simplification)

$$y'' + 16y' + 64y = 0, \qquad y(0) = \tfrac{1}{4}, \qquad y'(0) = 0$$

■

5.2.1 Undamped Motions

For the case when c is sufficiently small compared with mk and the time span is short, it may be acceptable to neglect the damping term cy'. (All systems have a certain amount of damping, no matter how small the motions or how short the period of time.) If this is done and if no external force acts on the mass, then the **free oscillations** of the mass are described by solutions of

$$my'' + ky = 0, \qquad t > 0, \qquad y(0) = y_0, \qquad y'(0) = v_0 \qquad (6a)$$

where y_0 is the initial displacement of the mass from equilibrium and v_0 is the initial velocity at that point. Dividing the DE by m and setting $\omega_0 = \sqrt{k/m}$, we have

$$y'' + \omega_0^2 y = 0, \qquad y(0) = y_0, \qquad y'(0) = v_0 \qquad (6b)$$

The general solution of (6b) is

$$y = C_1 \cos \omega_0 t + C_2 \sin \omega_0 t$$

By imposing the prescribed initial conditions, we see that

$$y(0) = C_1 = y_0$$
$$y'(0) = C_2 \omega_0 = v_0$$

from which we deduce $C_1 = y_0$ and $C_2 = v_0/\omega_0$. The motion of the mass is therefore described by

$$y = y_0 \cos \omega_0 t + \frac{v_0}{\omega_0} \sin \omega_0 t \qquad (7)$$

It is often useful to write (7) in the more compact form

$$y = A \cos(\omega_0 t - \phi) \qquad (8)$$

where A, which gives the maximum displacement of the motion from its equilibrium position, is called the **amplitude** of the motion and ϕ is the **phase angle.** To see the equivalence of (7) and (8), and also to identify A and ϕ, we use the identity

$$\cos(a - b) = \cos a \cos b + \sin a \sin b$$

Hence (8) becomes

$$y = A \cos \omega_0 t \cos \phi + A \sin \omega_0 t \sin \phi$$

from which it is clear that the *phase angle* ϕ satisfies

$$\cos \phi = \frac{y_0}{A}, \qquad \sin \phi = \frac{v_0}{\omega_0 A} \qquad (9)$$

or $\tan \phi = v_0/y_0\omega_0$, and that the *amplitude* A is given by

$$A = \sqrt{y_0^2 + \left(\frac{v_0}{\omega_0}\right)^2}^{\,†} \qquad (10)$$

Motion described by (8) is called **simple harmonic motion.** It is clearly *periodic* motion since the mass will oscillate continuously between $y = -A$ and $y = A$. The time between successive maxima, or the length of time required to complete one cycle of the motion, is called the **period** of the motion and is given by

$$T = \frac{2\pi}{\omega_0} = 2\pi \sqrt{\frac{m}{k}} \qquad (11)$$

The reciprocal of the period is called the **natural frequency** of the system and is measured in hertz (Hz). Thus,

$$f_0 = \frac{1}{T} = \frac{\omega_0}{2\pi} \qquad (12)$$

---

† If, in general, $y = C_1 \cos \omega_0 t + C_2 \sin \omega_0 t$, then $A = \sqrt{C_1^2 + C_2^2}$.

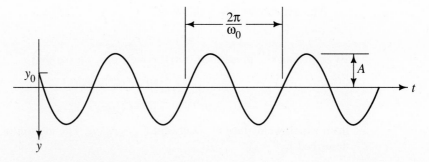

Figure 5.2 Simple harmonic motion.

which describes the number of complete cycles per second (cps). The value $\omega_0 = \sqrt{k/m}$ is called the **angular frequency;** it is measured in units of radians per second.

Regardless of the values of the input parameters y_0 and v_0, the graph of (8) is simply that of a cosine curve (see Figure 5.2). If either y_0 or v_0 is changed, the cosine curve will appear to shift a certain amount along the t axis and possibly change amplitude.

EXAMPLE 3 Find the period of oscillation of a spring-mass system for which the spring is stretched 4 in. by a 6-lb weight.

Solution

Since 4 in. corresponds to 1/3 ft, Hooke's law leads to

$$6 = k(1/3)$$

or $k = 18$ lb/ft. The mass $m = W/g = 6/32$ slug, and hence the period of oscillation is

$$T = 2\pi \sqrt{\frac{m}{k}} = \frac{\pi\sqrt{6}}{12} \quad \text{s} \qquad \blacksquare$$

EXAMPLE 4 Suppose a 16-lb weight is attached to the spring in Example 3 and released 3 in. above the equilibrium point of the spring-mass system with velocity 2 ft/s directed upward. Find the equation of motion of the mass. Also find the amplitude, phase, period, and frequency.

Solution

In Example 3 we found that $k = 18$ lb/ft. Also, the 16-lb weight corresponds to a mass of 0.5 slug. Thus, the governing IVP is

$$0.5y'' + 18y = 0, \qquad y(0) = -\tfrac{1}{4}, \qquad y'(0) = -2$$

or, putting the DE in normal form,

$$y'' + 36y = 0, \qquad y(0) = -\tfrac{1}{4}, \qquad y'(0) = -2$$

The negative values on the initial position and velocity reflect the fact that the positive y axis has been chosen *downward*. The general solution of the DE is

$$y = C_1 \cos 6t + C_2 \sin 6t$$

and by imposing the initial conditions we see that $C_1 = -1/4$ and $C_2 = -1/3$. Hence, the equation of motion is

$$y = -\tfrac{1}{4} \cos 6t - \tfrac{1}{3} \sin 6t$$

To put this equation in the form of Equation (8), we find the amplitude of the motion is

$$A = \sqrt{\left(-\frac{1}{4}\right)^2 + \left(-\frac{1}{3}\right)^2} = \frac{5}{12}\text{ ft}$$

whereas the phase angle satisfies $\cos \phi = -3/5$, $\sin \phi = -4/5$, or $\phi \cong 4.07$ rad (233°). In this case, we have (approximately)

$$y = \frac{5}{12} \cos(6t - 4.07)$$

for which the period of motion is $T = 2\pi \sqrt{m/k} = \pi/3$ s and the natural frequency is $f_0 = 1/T = 3/\pi$ Hz.　■

Sometimes it is useful to know the values of time for which the graph of $y(t)$ crosses the positive t axis. This corresponds physically to the mass passing through its equilibrium position. Writing the solution in the form of Equation (8) enables us to determine these times by setting

$$\cos(\omega_0 t - \phi) = 0 \tag{13}$$

For instance, using the solution in Example 4, we observe that $\cos(6t - 4.07) = 0$ when

$$6t - 4.07 = \frac{(2n - 1)\pi}{2} \qquad n = 0, \pm 1, \pm 2, \ldots$$

The first positive value of t that satisfies this relation is found to be $t_1 \cong 0.42$ s ($n = 0$), whereas the next value is $t_2 \cong 0.94$ s ($n = 1$), and so on (see Figure 5.3).

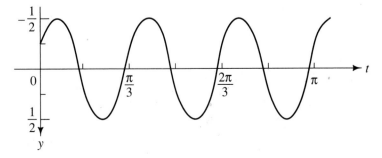

Figure 5.3 Location of the positive zeros of $y = \tfrac{5}{12} \cos(6t - 4.07)$.

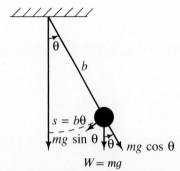

Figure 5.4 Oscillating pendulum.

[0] 5.2.2 The Pendulum Problem

A mass m is suspended from the end of a rod of constant length b whose weight is negligible (see Figure 5.4). We wish to determine the equation of motion of the mass in terms of the angle of displacement θ, given an initial displacement and velocity.

Summing forces makes it clear that the weight component $mg \cos \theta$, acting in the normal direction to the path, is offset by the force of constraint in the rod. Therefore, the only weight component contributing to the motion is $mg \sin \theta$, which acts in the direction of the tangent to the path. If we denote the arc length of the path by s, then Newton's second law of motion leads to

$$m \frac{d^2 s}{dt^2} = -mg \sin \theta$$

where the minus sign reflects the fact that the tangential force component opposes the motion for s increasing. The arc length s of a circle of radius b is related to the central angle θ by $s = b\theta$, and so $d^2s/dt^2 = bd^2\theta/dt^2$. Therefore, assuming an initial displacement θ_0 and initial velocity v_0, the above equation of motion (after simplification) and initial conditions become

$$b\theta'' + g \sin \theta = 0, \qquad \theta(0) = \theta_0, \qquad \theta'(0) = v_0 \tag{14}$$

The equation of motion in (14) is *nonlinear* owing to the term $\sin \theta$. To solve this equation exactly would necessitate introducing a special function called the **Jacobian elliptic function**,[†] since the equation has no solution that can be expressed in terms of elementary functions. However, recall that $\sin \theta \approx \theta$ when θ is "small," that is, less than $\pi/12$ (or $15°$). In this case we may replace $\sin \theta$ with θ in (14) to obtain the *linear* IVP

$$b\theta'' + g\theta = 0, \qquad \theta(0) = \theta_0, \qquad \theta'(0) = v_0 \tag{15}$$

We recognize (15) as being equivalent in form to the equation of motion of the undamped spring-mass system. Hence, like the undamped spring-mass system, the

---

[†] See, for example, pp. 110–113 in L. C. Andrews, *Special Functions for Engineers and Applied Mathematicians,* Macmillan, New York, 1985.

solutions of (15) will lead to *simple harmonic motion*. In Section 5.3 we find that this same DE also occurs in the study of certain electric circuit problems, once again illustrating the fact that the same DE can be used to describe several contrasting physical phenomena.

5.2.3 Damped Motions

When damping effects cannot be reasonably ignored, the **free motions** of the spring-mass system are described by solutions of

$$my'' + cy' + ky = 0, \qquad t > 0, \qquad c > 0 \tag{16}$$

The associated auxiliary equation is[†]

$$m\lambda^2 + c\lambda + k = 0$$

with roots

$$\lambda_1, \lambda_2 = \frac{-c \pm \sqrt{c^2 - 4mk}}{2m} \tag{17}$$

The solution of (16) obviously takes on three different forms, depending on the magnitude of the damping term. The three cases are: $c^2 > 4mk$, $c^2 = 4mk$, $c^2 < 4mk$. Let us discuss each case separately.

Case I. Overdamping: $c^2 > 4mk$

Damping is large compared with the spring constant. The roots λ_1, λ_2 of the auxiliary equation are *real, negative,* and *distinct,* leading to the general solution form

$$y = e^{-ct/2m}(C_1 e^{\alpha t} + C_2 e^{-\alpha t}) \tag{18}$$

where $\alpha = \sqrt{c^2 - 4mk}/2m$. This equation represents a smooth, **nonoscillatory** type of motion. Typical graphs of this motion are illustrated in Figure 5.5.

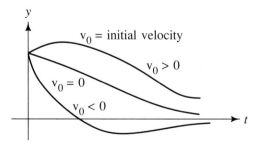

Figure 5.5 Overdamped motion.

[†] We use the parameter λ here since m denotes mass.

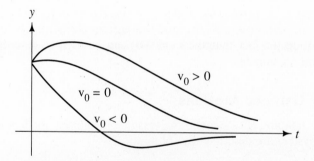

Figure 5.6 Critically damped motion.

Case II. Critical Damping: $c^2 = 4mk$

For this case the roots of the auxiliary equation are *equal*, that is, $\lambda_1 = \lambda_2 = -c/2m$, so the general solution takes the form

$$y = e^{-ct/2m}(C_1 + C_2 t) \tag{19}$$

The motions here are similar to those of the overdamped case, as shown in Figure 5.6.

Case III. Underdamping: $c^2 < 4mk$

This time the roots λ_1, λ_2 of the auxiliary equation are *complex conjugates,* and thus the general solution is

$$y = e^{-ct/2m}(C_1 \cos \mu t + C_2 \sin \mu t) \tag{20}$$

where $\mu = \sqrt{4mk - c^2}/2m$. Because of the sinusoidal functions in (20), the mass in this case will **oscillate** back and forth across the equilibrium position. However, the amplitude of the motion will steadily decrease in time (see Figure 5.7) because of the presence of the multiplicative factor $e^{-ct/2m}$ (and the fact that $\cos \mu t$ and $\sin \mu t$ are bounded).

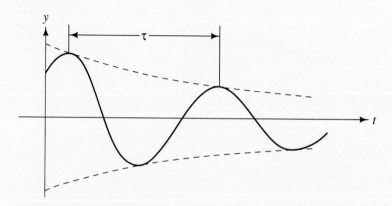

Figure 5.7 Underdamped motion.

In general, damped motion is characterized by the fact that the solution steadily tends to zero (i.e., approaches equilibrium) for increasing time. Hence, regardless of the prescribed initial conditions, all solutions of (16) satisfy the limit condition (see Problem 26 in Exercises 5.2)

$$\lim_{t \to \infty} y(t) = 0 \tag{21}$$

Actually, this situation confirms what we intuitively expect—without damping (friction), the motion of the system continues indefinitely, but with damping the motion eventually dies out.

Although damped motion in the case of underdamping is oscillatory, it is not truly periodic (since the amplitude of oscillation is not constant). Nonetheless, it is sometimes convenient in this case to introduce the notion of a **quasiperiod,** defined as the time τ between successive maxima of the displacement (see Figure 5.7). That is, we write

$$\tau = \frac{2\pi}{\mu} = \frac{4m\pi}{\sqrt{4mk - c^2}} \tag{22}$$

which, through some simple algebraic manipulations, can also be expressed in terms of T [given by (11)], that is,

$$\tau = 2\pi\sqrt{\frac{m}{k}}\left(1 - \frac{c^2}{4mk}\right)^{-1/2} = T\left(1 - \frac{c^2}{4mk}\right)^{-1/2} \tag{23}$$

Here we see that when damping is "small," that is, when $c^2/4mk \ll 1$, the quasiperiod τ is approximately equal to the period T of simple harmonic motion.

EXAMPLE 5

Consider the spring-mass system in Example 4 under the assumption that the mass is now subject to a resistive force equal to the instantaneous velocity (i.e., $c = 1$). Find the new position function, frequency, and quasiperiod.

Solution

This time the governing DE is $0.5y'' + y' + 18y = 0$, or, in normal form with the prescribed initial conditions,

$$y'' + 2y' + 36y = 0, \qquad y(0) = -\tfrac{1}{4}, \qquad y'(0) = -2$$

The roots of the auxiliary equation are $\lambda_1, \lambda_2 = -1 \pm i\sqrt{35}$ (underdamped case), leading to the general solution

$$y = e^{-t}(C_1 \cos \sqrt{35}t + C_2 \sin \sqrt{35}t)$$

As before, the first initial condition requires that $C_1 = -1/4$, but since $y'(0) = -C_1 + \sqrt{35}C_2 = -2$ in this case, it follows that $C_2 = -9/4\sqrt{35}$. Hence, the new position function is

$$y = -\frac{1}{4}e^{-t}\left(\cos \sqrt{35}t + \frac{9}{\sqrt{35}} \sin \sqrt{35}t\right)$$

Comparing this solution with that in Example 4, we first recognize by inspection that the new frequency of oscillation is $f_0 = \sqrt{35}/2\pi \cong 0.942$ Hz, and therefore the quasiperiod is $\tau = 1/f_0 \cong 1.062$ s. (Recall that $f_0 = 3/\pi \cong 0.955$ Hz and $T = 1/f_0 \cong 1.047$ s in Example 4.) Thus, owing to a fairly small damping coefficient (i.e., $c^2/4mk \cong 0.028 \ll 1$), the new frequency and quasiperiod in the damped system do not differ greatly from the undamped case. ∎

Exercises 5.2

In the following problems take $g = 32$ ft/s^2 in the English system or $g = 9.8$ m/s^2 in the International (MKS) system.

In Problems 1 to 4, solve the given IVP.

1. $y'' + 16y = 0$, $y(0) = 3$, $y'(0) = 9$

2. $25y'' + y = 0$, $y(0) = 0$, $y'(0) = 5$

3. $2y'' + 5y = 0$, $y(0) = -1$, $y'(0) = 5$

■**4.** $3y'' + 2y = 0$, $y(0) = 2$, $y'(0) = -1$

In Problems 5 to 8, use the given information to determine the (a) *period*, (b) *natural frequency*, (c) *amplitude*, and (d) *phase* of the undamped free motion.

5. A mass of 4 kg is attached to a spring with spring constant 100 N/m. The mass is released from rest 0.5 m below the equilibrium position.

6. A 24-lb weight stretches a spring 4 in. It is then released from a point 3 in. above the equilibrium position with a downward velocity of 2 ft/s.

7. A spring is stretched 6 in. by a 12-lb weight. The mass is subsequently pulled downward 4 in. below the equilibrium position and started upward with a velocity of 2 ft/s.

8. A mass of 5 kg stretches a spring 2 m. It is then pulled down 0.1 m below the equilibrium point and given an upward velocity of 0.1 m/s.

9. Find the first two positive values of time for which $y = 0$ in the spring-mass system in Problem 7.

■**10.** Calculate the time necessary for a 0.03-kg mass hanging from a spring with spring constant 0.5 N/m to undergo one full oscillation. What is the natural frequency of the system?

11. The period of free, undamped oscillations of a spring-mass system is observed to be $\pi/4$ s. If the spring constant is given by 16 lb/ft,

what is the numerical value of the weight in pounds?

12. Find the solution of $y'' + \omega_0^2 y = 0$ in the form $y = A \cos(\omega_0 t - \phi)$ when the initial conditions are

(a) $y(0) = y_0$, $y'(0) = 0$

(b) $y(0) = 0$, $y'(0) = v_0$

13. A mass of 1 slug is attached to a spring with $k = 9$ lb/ft. The mass initially starts moving from a point 1 ft above the equilibrium position with velocity $\sqrt{3}$ ft/s directed downward. Find the first positive value of time for which the mass is moving downward with a velocity of 3 ft/s.

14. Prove that the maximum value of the speed of a mass undergoing simple harmonic motion occurs when $y = 0$.

15. A clock has a pendulum 1 m long. The clock ticks once for each time the pendulum makes a complete swing, returning to its original position.

(a) How many ticks will the clock make in 1 min?

(b) What is the new natural period of the pendulum if its length is doubled?

16. Solve the pendulum problem [Equation (15)] when the weight is 8 lb and the rod is 1 ft long. Assume the weight is released from an angle of 0.5 rad with a positive velocity of 1.5 rad/s.

17. At what time does the pendulum in Problem 16 first pass through the angle $\theta = 0$? What is the velocity at this time?

■**18.** An 8-lb weight, attached to the end of a vertical spring, is pulled y_0 ft below its equilibrium position and released at time $t = 0$ with a downward velocity of 3 ft/s. Determine the spring constant k and the initial displacement y_0 if the amplitude of the resulting motion is known to be $\sqrt{5}$ and the period is $\pi/2$.

In Problems 19 to 22, solve the given IVP and identify the motion as *overdamped, critically damped,* or *underdamped.* If underdamped, find the *frequency* and *quasiperiod.*

19. $y'' + 8y' + 16y = 0$, $y(0) = 0$, $y'(0) = -5$

20. $y'' + \frac{1}{8}y' + y = 0$, $y(0) = 0$, $y'(0) = 1$

21. $y'' + y' + y = 0$, $y(0) = 1$, $y'(0) = -1$

22. $y'' + 10y' + 17y = 0$, $y(0) = 3$, $y'(0) = 0$

23. For what values of c are the motions governed by

$$2y'' + cy' + 12y = 0$$

(a) Overdamped?
(b) Critically damped?
(c) Underdamped?

24. A spring-mass system has $m = 3$ kg and $k = 12$ N/m. If the damping constant c and initial conditions vary according to the following, find the motion of the system in each case. If the system is underdamped, find the quasiperiod.

(a) $c = 13$; $y(0) = 1/4$, $y'(0) = 0$
(b) $c = 12$; $y(0) = 0$, $y'(0) = 1$
(c) $c = 15$; $y(0) = 0$, $y'(0) = -2$

25. A spring-mass system has $m = 1/4$ slug and $k = 4$ lb/ft. If the damping constant c and initial conditions vary according to the following, find the motion of the system in each case. If the system is underdamped, find the quasiperiod.

(a) $c = 3$; $y(0) = 1$, $y'(0) = -1$
(b) $c = 1$; $y(0) = 0$, $y'(0) = 1$
(c) $c = 2$; $y(0) = 2$, $y'(0) = -1$

26. Show that all roots of the auxiliary equation (17) are negative in the cases of overdamping and critical damping. Hence, deduce

that in both cases the solution satisfies

$$\lim_{t \to \infty} y(t) = 0$$

27. A 10-lb weight stretches a steel spring 2 in.
(a) Determine the natural period of the spring-mass system.
(b) If a damping force of magnitude $5y'$ exists, find the ratio of the quasiperiod of the damped motion to the natural period in (a).

28. Under what conditions on y_0 and v_0, where $y(0) = y_0$ and $v_0 = y'(0)$, will critically damped free motion have a maximum or a minimum for $t > 0$?

29. A certain straight-line motion is described by the IVP

$$y'' + 2by' + 169y = 0, \qquad y(0) = 0$$
$$y'(0) = 8, \qquad b > 0$$

(a) Find the value of b that leads to critical damping.
(b) Find the solution for that value of b.
(c) At what time does the motion momentarily stop, if at all?

■**30.** A spring is stretched 6 in. by a 3-lb weight, which is started from its equilibrium position with an upward velocity of 12 ft/s. If a retarding force equal in magnitude to $0.03v$ exists, find the resulting motion.

31. Consider a spring-mass system experiencing a viscous damping term. If the mass m is given an upward initial velocity of 50 m/s from the equilibrium position, find the motion given that $m = 4$ kg, $k = 64$ N/m, and $c = 40$ kg/s.

32. Determine the maximum displacement of the free motion of a spring-mass system governed by the IVP

$$y'' + 5y' + 4y = 0, \qquad y(0) = 1, \qquad y'(0) = 1$$

Does the graph of y ever cross the t axis?

33. Determine the time between consecutive maximum displacements of the mass of a damped free motion spring-mass system when $m = 30$ kg, $k = 2000$ N/m, and $c = 300$ kg/s.

34. A 4-lb weight is attached to a spring with spring constant 2 lb/ft. If the weight is released from 1 ft above the equilibrium position with a downward velocity of 8 ft/s, determine the time that the weight passes through the equilibrium position, assuming the retarding force is equal in magnitude to the instantaneous velocity. Find the time and position of the weight at its maximum displacement after passing through equilibrium.

*35. Show that *overdamped* free motion has the following characteristics:

   (a) The mass cannot pass through $y = 0$ more than once.

   (b) If the initial conditions are such that the constants C_1 and C_2 in the general solution have the same sign, the mass never passes through $y = 0$.

*36. Show that *underdamped* free motion has the following characteristics:

   (a) The characteristic angular frequency μ is independent of the initial conditions but decreases as c increases.

   (b) The natural logarithm of the ratio of two consecutive maximum amplitudes is the constant $\delta = \pi c/m\mu$, called the **logarithmic decrement** of the oscillation.

   (c) Find δ in the case $y = e^{-t} \cos t$, and determine which values of t correspond to maximum and minimum displacements.

5.3 Forced Motions of a Spring-Mass System

If an external force $F(t)$ is applied to the spring-mass system discussed in Section 5.2, the governing DE becomes *nonhomogeneous.* For discussion purposes we once again find it convenient to separately consider undamped motion and damped motion.

5.3.1 Undamped Motions

When the motion is *undamped,* we set $c = 0$ in the governing DE (5), which then reduces to

$$my'' + ky = F(t), \qquad t > 0 \tag{24}$$

Since we are primarily interested in undamped motions due to the input function $F(t)$, we assume the system is initially at rest in its equilibrium position before $F(t)$ is applied. Hence, we impose the *homogeneous* initial conditions

$$y(0) = 0, \qquad y'(0) = 0 \tag{25}$$

The general solution of (24) takes the form $y = y_H + y_P$, where y_P is any particular solution of (24) and y_H is the general solution of the associated homogeneous DE. In all cases, we have

$$y_H = C_1 \cos \omega_0 t + C_2 \sin \omega_0 t \tag{26}$$

where $\omega_0 = \sqrt{k/m}$, whereas the form of y_P will depend upon the nature of the forcing function $F(t)$.

Constant Force

To begin, let us suppose the external force is $F(t) = P$, where P is constant, giving us the IVP

$$my'' + ky = P, \qquad y(0) = 0, \qquad y'(0) = 0 \qquad (27)$$

Assuming a particular solution of the form $y_P = A$, for some constant A, we find that $A = P/k$. Combining $y_P = P/k$ with (26) yields the general solution

$$y = y_H + y_P$$

$$= C_1 \cos \omega_0 t + C_2 \sin \omega_0 t + P/k$$

The initial conditions in (27) require that $C_1 = -P/k$ and $C_2 = 0$, which leads to

$$y = \frac{P}{k}(1 - \cos \omega_0 t) \qquad (28)$$

In this case the mass oscillates at the natural frequency of the spring-mass system between the points $y = 0$ and $y = 2P/k$.

Sinusoidal Force ($\omega \neq \omega_0$)

Next, we assume that the forcing function is the simple periodic function $F(t) = P \cos \omega t$, where $\omega \neq \omega_0$. Hence, the IVP we want to solve takes the form

$$my'' + ky = P \cos \omega t, \qquad y(0) = 0, \qquad y'(0) = 0 \qquad (29)$$

Using the method of undetermined coefficients (see Section 4.7), we assume a particular solution exists of the form

$$y_P = A \cos \omega t + B \sin \omega t$$

where the constants A and B are to be determined. The substitution of y_P into the DE in (29) yields (upon simplification)

$$(k - m\omega^2)(A \cos \omega t + B \sin \omega t) = P \cos \omega t$$

Hence, equating like coefficients of $\cos \omega t$ and $\sin \omega t$, we obtain $A = P/(k - m\omega^2) = P/m(\omega_0^2 - \omega^2)$ and $B = 0$, so our general solution is

$$y = y_H + y_P$$

$$= C_1 \cos \omega_0 t + C_2 \sin \omega_0 t + \frac{P}{m(\omega_0^2 - \omega^2)} \cos \omega t$$

If we now impose the initial conditions in (29), we see that $C_1 = -P/m(\omega_0^2 - \omega^2)$ and $C_2 = 0$. Therefore, our solution becomes

$$y = \frac{P}{m(\omega_0^2 - \omega^2)} (\cos \omega t - \cos \omega_0 t), \qquad \omega \neq \omega_0 \qquad (30)$$

In this case the motion consists of the superposition of two modes of vibration—the **natural mode** at frequency ω_0 and the **forced mode** at frequency ω.

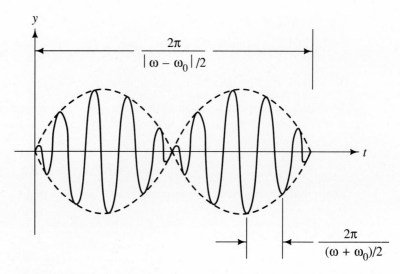

Figure 5.8 Phenomenon of beats.

An interesting phenomenon occurs when the forcing frequency ω in (30) is close to the natural frequency ω_0, that is, when $|\omega_0 - \omega|$ is "small." By setting $\omega_0 t = a + b$ and $\omega t = a - b$, we can use the trigonometric identities

$$\cos(a \pm b) = \cos a \cos b \mp \sin a \sin b$$

to rewrite (30) in the form

$$y = \frac{2P}{m(\omega_0^2 - \omega^2)} \sin\left[(\omega_0 - \omega)\frac{t}{2}\right] \sin\left[(\omega_0 + \omega)\frac{t}{2}\right] \tag{31}$$

Since we assume $0 < |\omega_0 - \omega| \ll 1$, the period of the sine wave $\sin[(\omega_0 - \omega)t/2]$ is large compared with the period of the sine wave $\sin[(\omega_0 + \omega)t/2]$. The motion described by (31) can then be visualized as a rapid oscillation with angular frequency $(\omega_0 + \omega)/2$, but with a slowly varying sinusoidal amplitude known as the *envelope* (see Figure 5.8). Motion of this type, possessing a periodic variation of amplitude, exhibits what is called a **beat.** The phenomenon of beats can most easily be demonstrated with acoustic waves—for example, when two tuning forks of nearly the same frequency are sounded at the same time.

Sinusoidal Force ($\omega = \omega_0$)

In the special case when the system is excited at its natural frequency, that is, when the input function is $F(t) = P \cos \omega_0 t$, we can obtain the response of the system directly from (30) by use of L'Hôpital's rule. To do so, we consider the limit

$$y = \lim_{\omega \to \omega_0} \left[\frac{P}{m(\omega_0^2 - \omega^2)} (\cos \omega t - \cos \omega_0 t)\right]$$

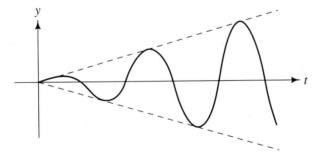

Figure 5.9 Phenomenon of resonance.

from which we deduce

$$y = \frac{P}{2m\omega_0} t \sin \omega_0 t \tag{32}$$

The interesting observation here is that the amplitude of motion described by (32) will become unbounded as $t \to \infty$. Thus, in this case we have the phenomenon of **resonance** (see Figure 5.9). Of course, we recognize that in a physical problem the amplitude cannot become unbounded. That is, a certain amount of damping, however small, is always present, and this has the effect of limiting the amplitude (see also the discussion in Section 5.3.2). At any rate, if the amplitude should become too large, the system is likely to fail (i.e., fall apart). This situation has actually caused certain bridges to collapse, such as the Tacoma Narrows bridge at Puget Sound in the state of Washington. On November 7, 1940, only 4 months after its grand opening, a hugh portion of the bridge collapsed into the water below. From the very beginning the bridge had experienced large undulations, which were later attributed to the wind blowing across the superstructure. In designing such structures, it is very important to make the natural frequency of the structure different (if possible) from the frequency of any probable forcing function.

> *Remark.* We might also point out that the reason soldiers do not march in step across bridges is to avoid the possibility of resonance occurring between the natural frequency of the bridge and the frequency of the uniformly stomping feet.

5.3.2 Damped Motions

When damping effects cannot be reasonably ignored, we must solve the general IVP described by

$$my'' + cy' + ky = F(t), \qquad t > 0, \qquad c > 0$$
$$y(0) = y_0, \qquad y'(0) = v_0 \tag{33}$$

The general solution of (33) is composed of $y = y_H + y_P$, where the homogeneous solution assumes one of the forms [see (18) to (20)]

$$y_H = \begin{cases} e^{-ct/2m}(C_1 e^{\alpha t} + C_2 e^{-\alpha t}), & \alpha = \sqrt{c^2 - 4mk}/2m \\ e^{-ct/2m}(C_1 + C_2 t) \\ e^{-ct/2m}(C_1 \cos \mu t + C_2 \sin \mu t), & \mu = \sqrt{4mk - c^2}/2m \end{cases}$$

depending upon the state of damping. Regardless of the prescribed initial conditions, it follows that $y_H \to 0$ as $t \to \infty$ and hence y_H contributes only initial effects to the motion of the system. For this reason, we often refer to y_H as a **transient solution** in physical applications. The particular solution y_P, for all forcing functions, dominates the response of the system after initial effects diminish and therefore contains what we call the **steady-state solution.**[†] Since the transient solution y_H does not have long-term effects, we ignore the initial conditions in the following discussion and concentrate on finding only the steady-state solution.

Let us start by assuming the forcing function is a simple sinusoidal function so that the governing DE becomes

$$my'' + cy' + ky = P \cos \omega t \tag{34}$$

where P is a constant. Using the method of undetermined coefficients, we assume a particular solution of the form

$$y_P = A \cos \omega t + B \sin \omega t$$

where A and B must be determined. When y_P is substituted in (34) and the resulting expression is simplified, we have

$$[(k - m\omega^2)A + \omega cB] \cos \omega t + [(k - m\omega^2)B - \omega cA] \sin \omega t = P \cos \omega t$$

Equating like coefficients of $\cos \omega t$ and $\sin \omega t$ gives the simultaneous solution

$$A = \frac{P(k - m\omega^2)}{(k - m\omega^2)^2 + \omega^2 c^2}, \qquad B = \frac{P\omega c}{(k - m\omega^2)^2 + \omega^2 c^2}$$

The particular solution, or steady-state solution, is then

$$y_P = \frac{P}{(k - m\omega^2)^2 + \omega^2 c^2} [(k - m\omega^2) \cos \omega t + \omega c \sin \omega t] \tag{35}$$

Since the steady-state solution is the sum of two sinusoids of the same frequency, we can put it in the amplitude-phase form of simple harmonic motion (see Section 5.2.1). By writing $\omega_0 = \sqrt{k/m}$, we find that (35) becomes

$$y_P = R \cos(\omega t - \phi) \tag{36}$$

where the amplitude R is (see Problem 14 in Exercises 5.3)

$$R = \frac{P}{\sqrt{(k - m\omega^2)^2 + \omega^2 c^2}} = \frac{P}{\sqrt{m^2(\omega_0^2 - \omega^2)^2 + \omega^2 c^2}} \tag{37}$$

[†] By "steady-state solution," we mean only that part of y_P that does not vanish as $t \to \infty$.

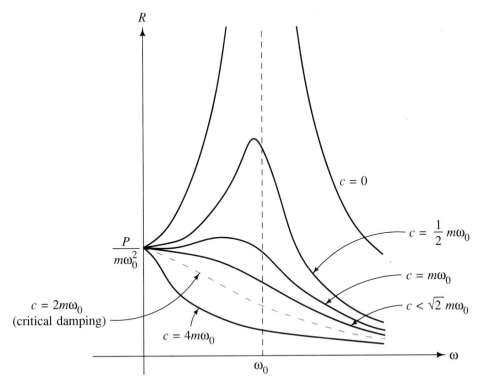

Figure 5.10 The amplitude R as a function of ω.

and the phase angle ϕ is such that

$$\tan \phi = \frac{\omega c}{k - m\omega^2} = \frac{\omega c}{m(\omega_0^2 - \omega^2)} \tag{38}$$

The amplitude R given by (37) is plotted in Figure 5.10 as a function of the input frequency ω. Here it is interesting to observe that, unlike the case of undamped motion, the maximum value of R does not occur when the forcing frequency is equal to the natural frequency of the system, that is, when $\omega = \omega_0$. To find the critical frequency for which R may assume a maximum, we can simply differentiate the expression under the square root in the denominator of (37) and set it to zero. This action leads to (see Problem 20 in Exercises 5.3)

$$\omega^2 = \omega_0^2 - \frac{c^2}{2m^2} = \frac{2m^2\omega_0^2 - c^2}{2m^2} \tag{39}$$

Hence, for damping coefficients satisfying $c < \sqrt{2}m\omega_0$, there will be a maximum amplitude given by (see Problem 21 in Exercises 5.3)

$$R_{\max} = \frac{2Pm}{c\sqrt{4m^2\omega_0^2 - c^2}}, \qquad c < \sqrt{2}m\omega_0 \tag{40}$$

However, for sufficiently large damping such that $c > \sqrt{2}m\omega_0$, there is no value of ω satisfying (39), and hence no maximum amplitude. This means that large amplitudes leading to resonance in the system can be avoided by a sufficient amount of damping. Hence, damping effects play a very important role in the design of mechanical structures (such as bridges) which might undergo oscillations due to external forces.

■
EXAMPLE 6

Interpret and solve the IVP
$$0.5y'' + 2y' + 10y = 5\cos 2t, \qquad y(0) = 0.5, \qquad y'(0) = 0$$

Solution

We can interpret the problem as representing a spring-mass system consisting of a mass of 0.5 unit, spring constant equal to 10 units, a damping term equal to twice the instantaneous velocity of the mass, and a periodic forcing term equal to $5\cos 2t$. The mass is released from rest 0.5 unit below the equilibrium position.

To solve the DE, let us first rewrite it in the form
$$y'' + 4y' + 20y = 10\cos 2t$$

The auxiliary equation of the associated homogeneous DE is $\lambda^2 + 4\lambda + 20 = 0$ with roots $\lambda = -2 \pm 4i$. Hence, the transient solution is
$$y_H = e^{-2t}(C_1 \cos 4t + C_2 \sin 4t)$$

To find the steady-state solution, we assume
$$y_P = A\cos 2t + B\sin 2t$$

Upon substituting this expression into the nonhomogeneous DE and comparing the coefficients of like terms, we find that
$$-8A + 16B = 0$$
$$16A + 8B = 10$$

Calculating, we have $A = 0.5$ and $B = 0.25$, so that
$$y_P = 0.5\cos 2t + 0.25\sin 2t$$

Combining y_H and y_P leads to the general solution
$$y = e^{-2t}(C_1 \cos 4t + C_2 \sin 4t) + 0.5\cos 2t + 0.25\sin 2t$$

Finally, by applying the initial conditions, we see that
$$y(0) = C_1 + 0.5 = 0.5$$
$$y'(0) = -2C_1 + 4C_2 + 0.5 = 0$$

from which we deduce $C_1 = 0$ and $C_2 = -0.125$. Therefore, we have
$$y = -0.125e^{-2t}\cos 4t + 0.5\cos 2t + 0.25\sin 2t$$

After a short period of time $(t > 3)$, the dominant part of the solution is the *steady-state* term
$$y \cong 0.5\cos 2t + 0.25\sin 2t, \qquad t > 3$$

which represents simple harmonic motion with amplitude

$$R = \sqrt{(0.5)^2 + (0.25)^2} \cong 0.56$$

and period $T = \pi$.

■

Exercises 5.3

In the following problems take $g = 32$ ft/s^2 in the English system or $g = 9.8$ m/s^2 in the International (MKS) system.

1. Solve the DE

$$y'' + 4y = 3$$

subject to the following sets of initial conditions:

(a) $y(0) = 0$, $y'(0) = 0$
(b) $y(0) = 0$, $y'(0) = 1$
(c) $y(0) = y_0$, $y'(0) = v_0$

2. Solve the DE

$$y'' + 25y = 10 \cos 7t$$

subject to the following sets of initial conditions:

(a) $y(0) = 0$, $y'(0) = 0$
(b) $y(0) = 0$, $y'(0) = -3$
(c) $y(0) = y_0$, $y'(0) = v_0$

3. Solve the DE

$$y'' + 25y = 2 \sin 5t$$

subject to the following sets of initial conditions:

(a) $y(0) = 0$, $y'(0) = 0$
(b) $y(0) = 1$, $y'(0) = 0$
(c) $y(0) = y_0$, $y'(0) = v_0$

■ **4.** Solve the DE

$$y'' + y = \begin{cases} 2, & 0 < t < 1 \\ 0, & t > 1 \end{cases}$$

subject to the following sets of initial conditions:

(a) $y(0) = 0$, $y'(0) = 0$
(b) $y(0) = 0$, $y'(0) = 1$
(c) $y(0) = y_0$, $y'(0) = v_0$

5. Verify the solution (32) by solving directly the IVP

$$my'' + ky = P \cos \omega_0 t,$$
$$y(0) = y'(0) = 0, \qquad \omega_0 = \sqrt{k/m}$$

6. A spring stretches 6 in. when a 4-lb weight is attached. If the weight is started from the equilibrium position with an upward velocity of 4 ft/s and has an impressed force of $0.5 \cos 8t$ lb acting on the weight, determine the position of the weight for all time. Neglect damping.

7. A 2-lb weight stretches a spring 6 in. An impressed force $16 \sin 8t$ lb is acting upon the spring, and the weight is pulled down 3 in. below the equilibrium position and released from rest. Determine the subsequent motion if damping is neglected.

8. A 0.03-kg mass is attached to a spring with $k = 0.5$ N/m. If a force $\frac{1}{8} \cos 4t$ N is imposed on the system at rest in equilibrium, determine the subsequent motion. Neglect damping.

*9. A spring with spring constant $k = 0.75$ lb/ft has a weight of 6 lb attached and the system comes to rest in the equilibrium position. A 1.5-lb force is applied to the weight in the downward direction for 4 s and then removed. Determine the subsequent motion if damping is neglected.

10. Show that the solution in Problem 2(a) can be expressed in the form $y = \frac{5}{6} \sin t \sin 6t$, and determine how many seconds there are between **beats.**

11. A 20-N weight is suspended by a frictionless spring for which $k = 98$ N/m. An external force of $2 \cos 7t$ N acts on the weight and damping is negligible.

(a) Find the frequency of the beat.

(b) Determine the amplitude of the motion, given that it starts from rest in equilibrium.

(c) How many seconds are there between the beats?

*12. A 2-kg mass is attached to a spring with $k = 32$ N/m, and a force of $0.1 \cos 4t$ N is applied to the mass while at rest in its equilibrium position. Neglect damping and calculate the time required for failure to occur if the spring breaks when the amplitude of oscillation exceeds 0.5 m.

13. Assuming $m = 1$, $c = 4$, $k = 13$, determine the solution of

$$my'' + cy' + ky = P \cos \sqrt{13}t$$

satisfying the following sets of initial conditions:

(a) $y(0) = y_0$, $y'(0) = 0$

(b) $y(0) = 0$, $y'(0) = v_0$

(c) $y(0) = y_0$, $y'(0) = v_0$

14. Verify that the amplitude of the solution (35) is that given by Equation (37).

15. The motion of a body rising with a drag force proportional to its velocity is described by

$$my'' + cy' + mg = 0$$

(a) If the initial velocity of the body is 100 m/s upward, $c = 0.4$ kg/s, and $m = 2$ kg, determine the time required for the body to reach its maximum height.

(b) How high will the body rise?

16. Determine both the transient and steady-state solutions of the IVP

$$y'' + 2y' + 17y = 2 \cos 3t,$$
$$y(0) = 0, \qquad y'(0) = 0$$

17. Determine both the transient and steady-state solutions of the IVP

$$y'' + 2y' + 2y = 4 \cos t + 2 \sin t,$$
$$y(0) = 0, \qquad y'(0) = 3$$

■18. A mass of 0.5 slug is attached to a spring with $k = 6$ lb/ft. A damping force numerically equal to twice the instantaneous velocity acts on the system.

(a) Find the steady-state response of the system to an external driving force $F(t) = 40 \sin t$ lb.

(b) What is the amplitude of the motion? Will R_{max} [defined by (40)] occur?

19. Find the steady-state response of the system in Problem 18 if the driving force is constant, that is, if $F(t) = P$.

20. Show that the maximum amplitude of R, given by (37), occurs when $\omega^2 = \omega_0^2 - c^2/2m$.

21. Derive Equation (40).

*22. The ratio of successive maximum amplitudes of a particular underdamped spring-mass system is found to be 1.25 when the system undergoes free motion. If $k = 100$ N/m, $m = 4$ kg, and a driving force of $F(t) = 10 \cos 4t$ N is imposed on the system, determine the amplitude of the steady-state motion.

5.4 Simple Electric Circuits

Let us consider the electric circuit in Figure 5.11, which is composed of a **resistor** with a resistance of R ohms (Ω), **capacitor** with a capacitance of C farads (F), and an **inductor** with an inductance of L henrys (H), connected in series with a **voltage source** of $E(t)$ volts (V). When the switch is closed at time $t = 0$, a **current** of $i(t)$ amperes (A) will flow in the loop and a **charge** of $q(t)$ coulombs (C) will accumulate on the capaci-

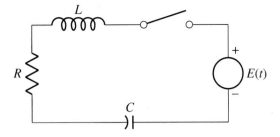

Figure 5.11 *RLC* circuit.

tor at time *t*. The charge and current are related by

$$i = \frac{dq}{dt} \tag{41}$$

The behavior in the network featured here, called an *RLC* **circuit,** is determined at each point in the network by solving appropriate DEs that result from applying **Kirchhoff's laws:**

1. The sum of the currents into (or away from) any point is zero.
2. The sum of the instantaneous voltage drops in a specified direction is zero around any closed path.

The first Kirchhoff law indicates that the current is the same throughout the circuit. To apply the second law, we must know the voltage drop across each of the idealized elements in the *RLC* circuit. From experimental observations, we have that

$$\text{Voltage drop across a resistor} = Ri$$

$$\text{Voltage drop across an inductor} = L\frac{di}{dt}$$

$$\text{Voltage drop across a capacitor} = \frac{q}{C}$$

while the impressed electromotive force *E(t)* contributes to a voltage gain. Thus, by applying Kirchhoff's second law to the circuit shown in Figure 5.11, we obtain the governing DE

$$L\frac{di}{dt} + Ri + \frac{q}{C} = E(t) \tag{42}$$

To express (42) in terms of only the charge *q*, we can use (41) to eliminate the current *i*. In this case we have

$$L\frac{d^2q}{dt^2} + R\frac{dq}{dt} + \frac{q}{C} = E(t)$$

or, using prime notation,

$$Lq'' + Rq' + C^{-1}q = E(t), \qquad t > 0 \tag{43}$$

We recognize that (43) is the same DE in form as that which governs the damped motions of a spring-mass system [(see Equation (5)]. Based on this comparison, we can make the following analogies between mechanical and electrical systems:

1. **Charge** q corresponds to **position** y.
2. **Current** i corresponds to **velocity** y'.
3. **Inductance** L corresponds to **mass** m.
4. **Resistance** R corresponds to **damping constant** c.
5. **Inverse capacitance** C^{-1} corresponds to **spring constant** k.
6. **Electromotive force** $E(t)$ corresponds to **input function** $F(t)$.

Such analogies between mechanical and electrical systems prove very useful in practice. For example, in the study of a certain mechanical system that is either too complicated or too expensive to build, the electrical counterpart may be constructed instead for the purpose of analysis. Interestingly, the phenomenon of *resonance* also occurs in electrical systems, but it does not have undesired side effects as in mechanical systems. Quite the contrary, it is primarily because of electrical resonance that we can tune a radio to the frequency of the transmitting radio station in order to obtain reception.

> **Remark.**   If we differentiate (42), we obtain
>
> $$Li'' + Ri' + C^{-1}i = E'(t)$$
>
> which can be solved directly in terms of the current i. In such cases, the initial value $i'(0)$ that is needed can be related to $i(0)$ and $q(0)$ (see Problems 1 and 4 in Exercises 5.4)

EXAMPLE 7

The circuit shown in Figure 5.11 contains the components $L = 1$ H, $R = 1000$ Ω, and $C = 4 \times 10^{-6}$ F. At time $t = 0$, both current and charge are zero, and a battery supplying a constant voltage of 24 V is instantaneously switched on. Find the charge q on the capacitor and the current i at any later time.

Solution

The governing IVP is

$$q'' + 1000q' + \tfrac{1}{4} \times 10^6 q = 24, \qquad q(0) = 0, \qquad q'(0) = i(0) = 0$$

the homogeneous solution of which is readily found to be

$$q_H = (C_1 + C_2 t)e^{-500t}$$

To obtain a particular solution q_P, we substitute $q_P = A$ into the nonhomogeneous DE, finding $A = 9.6 \times 10^{-5}$. Combining solutions $q = q_H + q_P$, we obtain the general solution

$$q = (C_1 + C_2 t)e^{-500t} + 9.6 \times 10^{-5}$$

The homogeneous initial conditions lead to

$$q(0) = C_1 + 9.6 \times 10^{-5} = 0$$

$$q'(0) = -500C_1 + C_2 = 0$$

from which we deduce $C_1 = -9.6 \times 10^{-5}$ and $C_2 = -4.8 \times 10^{-2}$. Thus, the charge on the capacitor is given by

$$q = 9.6 \times 10^{-5}(1 - e^{-500t}) - 4.8 \times 10^{-2}te^{-500t} \qquad \text{coulombs}$$

and the current by

$$i = q' = 24te^{-500t} \qquad \text{amperes} \qquad \blacksquare$$

Exercises 5.4

1. If the resistance R is not included in the RLC circuit of Figure 5.11, we have what is called an LC **circuit.** Show that the current $i(t)$ of the LC circuit satisfies the relation

$$i'(0) = \frac{E(0)}{L} - \frac{q(0)}{LC}, \qquad LC \neq 0$$

where $q(0)$ is the charge on the capacitor at time $t = 0$.

2. Using the result of Problem 1, find the current in the LC circuit when $i(0) = 0$, $q(0) = 0$, and
   (a) $L = 1$ H, $C = 0.25$ F, $E(t) = 30 \sin t$ V
   (b) $L = 10$ H, $C = 0.1$ F, $E(t) = 10t$ V

3. A **steady-state current** in the RLC circuit may result after a sufficient length of time $(t \to \infty)$. Find the steady-state current when
   (a) $R = 4\,\Omega$, $L = 1$ H, $C = 2 \times 10^{-4}$ F,
       $E(t) = 220$ V
   (b) $R = 10\,\Omega$, $L = 2$ H, $C = 0.5$ F,
       $E(t) = 10.8 \cos 2t$ V

4. Show that the current in the RLC circuit satisfies

$$i'(0) = \frac{E(0)}{L} - \frac{R}{L}i(0) - \frac{q(0)}{LC}, \qquad LC \neq 0$$

where $q(0)$ is the charge on the capacitor at time $t = 0$.

5. Find the current in the RLC circuit assuming $q(0) = i(0) = 0$, and
   (a) $R = 6\,\Omega$, $L = 1$ H, $C = 0.04$ F,
       $E(t) = 24 \cos 5t$ V
   (b) $R = 80\,\Omega$, $L = 20$ H, $C = 0.01$ F,
       $E(t) = 100$ V

6. What conditions on the parameters R, L, and C must be satisfied for the charge variation to be (see Section 5.2.3)
   (a) Underdamped?
   (b) Overdamped?
   (c) Critically damped?

7. A series RLC circuit has components $L = 0.5$ H, $R = 10\,\Omega$, $C = 0.01$ F, and $E(t) = 150$ V. Determine the instantaneous charge on the capacitor for $t > 0$ if initially $q(0) = 1$ and $i(0) = 0$. What charge persists after a long time?

■8. An electric circuit has components $L = 0.001$ H, $C = 2 \times 10^{-5}$ F, and a resistor R. Determine the critical resistance necessary to lead to an oscillatory current if the elements are connected in series.

*9. The amplitudes of two successive maximum currents in a series circuit with $L = 10^{-4}$ H and $C = 10^{-6}$ F are measured to be 0.2 and 0.1 A. Determine the resistance R.

10. A particular RLC circuit connected in series has an electromotive force given by $E(t) = E_0 \sin \omega t$.

(a) Show that the steady-state current (as $t \to \infty$) is

$$i = \frac{E_0}{Z}\left(\frac{R}{Z}\sin \omega t - \frac{X}{Z}\cos \omega t\right)$$

where $X = L\omega - 1/C\omega$ and $Z = \sqrt{X^2 + R^2}$. The quantity X is called the **reactance** of the circuit and Z is the **impedance** of the circuit.

(b) Show that when

$$\omega = \frac{1}{\sqrt{LC}}$$

the amplitude of the steady-state current is a maximum (i.e., **electrical resonance** occurs for this value of ω).

[0] 5.5 Method of Green's Function

We now wish to develop a solution technique for solving IVPs that leads to a very general solution formula. The method of attack is attributed to George Green[†] and is based upon the construction of a particular function known as the **Green's function.**

Since most of the DEs that commonly occur in practice are second-order, we develop the following theory for only this class of equations (we generalize the results to DEs of order three in the exercises). Thus, we confine our attention to IVPs of the form

$$N[y] = f(t), \qquad t > t_0, \qquad y(t_0) = k_0, \, y'(t_0) = k_1 \qquad (44)$$

where t_0 is not necessarily zero and where N is the **normal differential operator**

$$N = D^2 + a_1(t)D + a_0(t) \qquad (45)$$

Unless otherwise stated, we assume that $a_0(t)$, $a_1(t)$, and $f(t)$ are continuous functions on the interval $t \geq t_0$. Under these conditions, Theorem 5.1 assures us that (44) has a *unique* solution.

For solution purposes it is convenient to separate (44) into two simpler problems described by

$$N[y] = 0, \qquad y(t_0) = k_0, \qquad y'(t_0) = k_1 \qquad (46)$$

and

$$N[y] = f(t), \qquad y(t_0) = 0, \qquad y'(t_0) = 0 \qquad (47)$$

Notice that in (46) the DE is *homogeneous* while the initial conditions are *nonhomogeneous,* whereas in (47) it is the DE that is *nonhomogeneous* and the initial conditions *homogeneous.* Not only does splitting the original problem (44) into two simpler problems facilitate the solution process, but the solutions of each of the individual

---

[†] George Green (1793–1841) gained recognition for his important works concerning the reflection and refraction of sound and light waves. He also extended the work of S. D. Poisson (1781–1840) in the theory of electricity and magnetism.

problems (46) and (47) have important physical interpretations. For example, the solution of (46), which we denote by y_H, physically represents the *response of the system* described by (44) *due to the initial conditions* in the absence of other external disturbances [i.e., $f(t) \equiv 0$]. On the other hand, the solution of (47), which we denote by y_P, represents the *response of the same system* which is at rest in equilibrium until time $t = t_0$, after which it is *subjected to the external input* $f(t)$. The sum of solutions $y = y_H + y_P$ is the solution of the original problem (44) (see Problem 26 in Exercises 5.5).

The homogeneous DE in (46) has the general solution

$$y_H = C_1 y_1(t) + C_2 y_2(t) \tag{48}$$

where y_1 and y_2 form a fundamental solution set (recall Section 4.2). Hence, their Wronskian $W(y_1, y_2)(t) \neq 0$, $t \geq t_0$. Imposing the initial conditions in (46) upon this solution, we find

$$y_H(t_0) = C_1 y_1(t_0) + C_2 y_2(t_0) = k_0$$

$$y_H'(t_0) = C_1 y_1'(t_0) + C_2 y_2'(t_0) = k_1$$

the simultaneous solution of which leads to

$$C_1 = \frac{k_0 y_2'(t_0) - k_1 y_2(t_0)}{W(y_1, y_2)(t_0)}$$

$$C_2 = \frac{k_1 y_1(t_0) - k_0 y_1'(t_0)}{W(y_1, y_2)(t_0)} \tag{49}$$

Based on Theorem 5.1 we know that y_H is the unique solution of (46). Also, the special case where $k_0 = k_1 = 0$ leads to $C_1 = C_2 = 0$, or $y_H = 0$, in agreement with Corollary 5.1.

EXAMPLE 8

Solve the IVP

$$y'' + y = 0, \qquad y(0) = 1, \qquad y'(0) = -1$$

Solution

The general solution of the DE is

$$y_H = C_1 \cos t + C_2 \sin t$$

and imposing the initial conditions, we see that

$$y_H(0) = C_1 = 1$$

$$y_H'(0) = C_2 = -1$$

Hence, the unique solution is

$$y_H = \cos t - \sin t$$

5.5.1 One-Sided Green's Function

The solution of (46), given by (48) and (49), has been based on techniques developed in Chapter 4. To solve (47), that is,

$$N[y] = f(t), \qquad y(t_0) = 0, \qquad y'(t_0) = 0$$

which contains a *nonhomogeneous* DE, we also rely primarily on techniques developed in Chapter 4, but with suitable modifications.

Let us begin by assuming that the nonhomogeneous DE in (47) has a particular solution of the form

$$y_P = u(t)y_1(t) + v(t)y_2(t) \tag{50}$$

where $y_1(t)$ and $y_2(t)$ form the fundamental solution set in the solution of (46) and where $u(t)$ and $v(t)$ are functions to be determined. Using the method of **variation of parameters** (see Section 4.8), we previously showed that

$$u(t) = -\int \frac{y_2(t)f(t)}{W(y_1, y_2)(t)} \, dt, \qquad v(t) = \int \frac{y_1(t)f(t)}{W(y_1, y_2)(t)} \, dt$$

where $u(t)$ and $v(t)$ were the functions that resulted by setting the constants of integration to zero. Here now, however, we select $u(t)$ and $v(t)$ in a specific manner so that the prescribed homogeneous initial conditions in (47) are satisfied by (50). To do this, we find that the proper choice is

$$u(t) = -\int_{t_0}^{t} \frac{y_2(\tau)f(\tau)}{W(y_1, y_2)(\tau)} \, d\tau$$
$$v(t) = \int_{t_0}^{t} \frac{y_1(\tau)f(\tau)}{W(y_1, y_2)(\tau)} \, d\tau \tag{51}$$

which we verify below. By substituting these expressions back into (50) we obtain

$$y_P = -y_1(t) \int_{t_0}^{t} \frac{y_2(\tau)f(\tau)}{W(y_1, y_2)(\tau)} \, d\tau + y_2(t) \int_{t_0}^{t} \frac{y_1(\tau)f(\tau)}{W(y_1, y_2)(\tau)} \, d\tau$$
$$= \int_{t_0}^{t} \frac{y_1(\tau)y_2(t) - y_1(t)y_2(\tau)}{W(y_1, y_2)(\tau)} f(\tau) \, d\tau \tag{52}$$

However, by introducing the function

$$g_1(t, \tau) = \frac{y_1(\tau)y_2(t) - y_1(t)y_2(\tau)}{W(y_1, y_2)(\tau)} = \frac{\begin{vmatrix} y_1(\tau) & y_2(\tau) \\ y_1(t) & y_2(t) \end{vmatrix}}{W(y_1, y_2)(\tau)} \tag{53}$$

called the **one-sided Green's function** (or more simply, the **Green's function**), we can write our solution of (47) more compactly as

$$y_P = \int_{t_0}^{t} g_1(t, \tau)f(\tau) \, d\tau \tag{54}$$

Before discussing this function any further, let us first show that (54) indeed satisfies the homogeneous initial conditions prescribed in (47). Observe that when $t = t_0$, we get the immediate result

$$y_P(t_0) = \int_{t_0}^{t_0} g_1(t, \tau) f(\tau) \, d\tau = 0 \tag{55}$$

To show that the derivative also vanishes we need to use the Leibniz formula (from calculus)

$$\frac{d}{dt} \int_{t_0}^{t} F(t, \tau) \, d\tau = \int_{t_0}^{t} \frac{\partial F}{\partial t} (t, \tau) \, d\tau + F(t, t) \tag{56}$$

Hence, we see that

$$y_P'(t_0) = \int_{t_0}^{t_0} \frac{\partial g_1}{\partial t} (t_0, \tau) f(\tau) \, d\tau + g_1(t_0, t_0) f(t_0) = 0 \tag{57}$$

where we recognize that $g_1(t_0, t_0) = 0$ by definition. The initial conditions in (47) are therefore satisfied by (54).

It is interesting to notice that the construction of the one-sided Green's function, as defined by (53), depends only upon knowledge of the homogeneous solutions y_1 and y_2. That is, it is completely determined by the normal operator $N = D^2 + a_1(t)D + a_0(t)$. Since these homogeneous solutions always exist in the case that N has continuous coefficients, it follows that $g_1(t, \tau)$ **exists** under these conditions. Moreover, because of uniqueness of the solutions of IVPs (Theorem 5.1), the function $g_1(t, \tau)$ is also **unique.**

> **Remark.** For proper identification of the input function $f(t)$ occurring in the solution (54), it is important that the nonhomogeneous DE be put in normal form.

Finally, to summarize, we conclude that the solution of the IVP

$$N[y] = f(t), \qquad t > t_0, \qquad y(t_0) = k_0, \qquad y'(t_0) = k_1 \tag{58}$$

is given by

$$y = y_H + y_P$$

$$= C_1 y_1(t) + C_2 y_2(t) + \int_{t_0}^{t} g_1(t, \tau) f(\tau) \, d\tau \tag{59}$$

where the constants C_1 and C_2 are defined by (49).

> **Remark.** Although (59) represents a general solution formula for the given IVP, it does not always represent the simplest approach to finding the solution. The method of Green's function is important for developing and understanding some of the general theory of DEs and physical interpretations. For example, the one-sided Green's function is equivalent to the **impulse response function**[†] which plays a fundamental role in engineering applications. Also, it can be useful in those situations where the same DE must be solved a number of times with various input functions.

[†] See Section 6.7.1.

EXAMPLE 9

Determine the one-sided Green's function for the operator $N = D^2 + 1$.

Solution

Linearly independent solutions of $N[y] = 0$ are

$$y_1(t) = \cos t, \qquad y_2(t) = \sin t$$

with Wronskian

$$W(y_1, y_2)(t) = \begin{vmatrix} \cos t & \sin t \\ -\sin t & \cos t \end{vmatrix} = 1$$

Hence, based on (53), we obtain

$$g_1(t, \tau) = \begin{vmatrix} \cos \tau & \sin \tau \\ \cos t & \sin t \end{vmatrix} = \sin t \cos \tau - \cos t \sin \tau$$

or

$$g_1(t, \tau) = \sin(t - \tau)$$

With $g_1(t, \tau) = \sin(t - \tau)$ as given in Example 9, we are now in position to solve all DEs of the form $y'' + y = f(t)$, a fact that clearly illustrates the power and economy of using the Green's function. See Example 10 below.

EXAMPLE 10

Use the method of Green's function to solve

$$y'' + y = f(t), \qquad y(0) = 1, \qquad y'(0) = -1$$

Find a specific solution for the case $f(t) = \sin t$.

Solution

We first split the given problem into two problems:

$$y'' + y = 0, \qquad y(0) = 1, \qquad y'(0) = -1$$

and

$$y'' + y = f(t), \qquad y(0) = 0, \qquad y'(0) = 0$$

The first problem was solved in Example 8 and thus

$$y_H = \cos t - \sin t$$

To solve the second problem, we recognize that the one-sided Green's function is $g_1(t, \tau) = \sin(t - \tau)$ (see Example 9). Hence,

$$y_P = \int_0^t \sin(t - \tau) f(\tau) \, d\tau$$

whereas the complete solution is

$$y = y_H + y_P$$

$$= \cos t - \sin t + \int_0^t \sin(t - \tau) f(\tau) \, d\tau$$

For the special case where $f(t) = \sin t$, we find

$$y_P = \int_0^t \sin(t - \tau) \sin \tau \, d\tau$$

$$= \sin t \int_0^t \cos \tau \sin \tau \, d\tau - \cos t \int_0^t \sin^2 \tau \, d\tau$$

$$= \frac{1}{2} (\sin t - t \cos t)$$

and therefore in this case the complete solution is

$$y = \cos t - \sin t + \frac{1}{2} (\sin t - t \cos t)$$

$$= \left(1 - \frac{t}{2}\right) \cos t - \frac{1}{2} \sin t \qquad \blacksquare$$

EXAMPLE 11

Use the method of Green's function to solve

$$t^2 y'' - 3ty' + 3y = 2t^4 e^t, \qquad y(1) = 0, \qquad y'(1) = 2$$

Solution

The two problems we need to solve are

$$t^2 y'' - 3ty' + 3y = 0, \qquad y(1) = 0, \qquad y'(1) = 2$$

$$y'' - \frac{3}{t} y' + \frac{3}{t^2} y = 2t^2 e^t, \qquad y(1) = 0, \qquad y'(1) = 0$$

where we have divided the DE in the second problem by t^2 to put it in *normal form*. We recognize the DE as a Cauchy–Euler equation (see Section 4.9), whose homogeneous solution is

$$y_H = C_1 t + C_2 t^3$$

Imposing the initial conditions in the first problem above, we find that $C_2 = -C_1 = 1$, and thus

$$y_H = t^3 - t$$

The Wronskian of $y_1(t) = t$ and $y_2(t) = t^3$ is easily found to be $W(y_1, y_2)(t) = 2t^3$. Therefore, the one-sided Green's function is

$$g_1(t, \tau) = \frac{1}{2\tau^3} \begin{vmatrix} \tau & \tau^3 \\ t & t^3 \end{vmatrix} = \frac{\tau t^3 - t\tau^3}{2\tau^3}$$

From the normal form of the DE we recognize that $f(t) = 2t^2 e^t$, and hence the solution of the second problem above is

$$y_P = \int_1^t g_1(t, \tau)(2\tau^2 e^\tau) \, d\tau$$

$$= t^3 \int_1^t e^\tau \, d\tau - t \int_1^t \tau^2 e^\tau \, d\tau$$

or

$$y_P = (t - t^3)e + 2t(t - 1)e^t$$

By writing $y = y_H + y_P$, we obtain the desired solution

$$y = (t^3 - t)(1 - e) + 2t(t - 1)e^t \qquad \blacksquare$$

It should be observed that one of the distinct features of the Green's function technique is that the nonhomogeneous initial conditions are imposed only upon the solution y_H of the associated homogeneous DE. This is in sharp contrast with the methods employed in Chapter 4, wherein it was first necessary to construct a general solution of the nonhomogeneous DE before applying the prescribed initial conditions. The reason for this situation, of course, is that we are now seeking a particular solution y_P that always satisfies *homogeneous* initial conditions.

5.5.2 Table of One-Sided Green's Functions

Construction of the one-sided Green's function using (53) is mostly mechanical. Since the one-sided Green's function is not dependent upon the forcing function or initial conditions, all IVPs with the same operator N have the same one-sided Green's function. For this reason it is convenient to tabulate some of the most common differential operators and their corresponding one-sided Green's function. A few such entries are provided in Table 5.2.

TABLE 5.2 TABLE OF ONE-SIDED GREEN'S FUNCTIONS

	$M(D)$	$g_1(t, \tau)$
1.	D^2	$t - \tau$
2.	$D^n, n = 1, 2, 3, \ldots$	$\dfrac{(t - \tau)^{n-1}}{(n - 1)!}$
3.	$D^2 + b^2$	$\dfrac{1}{b} \sin b(t - \tau)$
4.	$D^2 - b^2$	$\dfrac{1}{b} \sinh b(t - \tau)$
5.	$(D - a)(D - b), a \neq b$	$\dfrac{1}{a - b} \left[e^{a(t-\tau)} - e^{b(t-\tau)} \right]$
6.	$(D - a)^2$	$(t - \tau)e^{a(t-\tau)}$
7.	$(D - a)^n, n = 1, 2, 3, \ldots$	$\dfrac{(t - \tau)^{n-1}}{(n - 1)!} e^{a(t-\tau)}$
8.	$D^2 - 2aD + a^2 + b^2$	$\dfrac{1}{b} e^{a(t-\tau)} \sin b(t - \tau)$
9.	$D^2 - 2aD + a^2 - b^2$	$\dfrac{1}{b} e^{a(t-\tau)} \sinh b(t - \tau)$
10.	$t^2 D^2 + tD - b^2$	$\dfrac{\tau}{2b} \left[\left(\dfrac{t}{\tau} \right)^b - \left(\dfrac{\tau}{t} \right)^b \right]$

Exercises 5.5

In Problems 1 to 10, determine the one-sided Green's function for the given operator $(D = d/dt)$.

1. D^2

2. $D^2 - 5$

3. $D^2 + 5$

4. $D^2 + 4D + 4$

5. $4D^2 - 8D + 5$

6. $D^2 - D - 2$

7. $t^2D^2 + tD - 16$

■**8.** $t^2D^2 - tD + 1$

*****9.** $D[(1 - t^2)D]$

*****10.** $tD^2 - (1 + 2t^2)D$

In Problems 11 to 25, use the one-sided Green's function to solve the given IVP.

11. $y'' - y = 1$, $y(0) = 0$, $y'(0) = 1$

■**12.** $y'' + y = e^{t-1}$, $y(1) = 0$, $y'(1) = 0$

13. $y'' - 3y' - 4y = e^{-t}$, $y(2) = 3$, $y'(2) = 0$

14. $y'' + y = 2 \csc t \cot t$, $y(\pi/2) = 1$, $y'(\pi/2) = 1$

15. $y'' + y' - 2y = -4$, $y(0) = 2$, $y'(0) = 3$

16. $2y'' + y' - y = t + 1$, $y(0) = 1$, $y'(0) = 0$

17. $y'' + 4y' + 6y = 1 + e^{-t}$, $y(0) = 1$, $y'(0) = -4$

18. $y'' + y' + 2y = \sin 2t - 2 \cos 2t$, $y(0) = 0$, $y'(0) = 0$

19. $y'' + 2y' + y = 3te^{-t}$, $y(0) = 4$, $y'(0) = 2$

■**20.** $y'' - 6y' + 9y = t^2 e^{3t}$, $y(0) = 2$, $y'(0) = 6$

21. $t^2 y'' + 7ty' + 5y = t$, $y(1) = 0$, $y'(1) = 0$

22. $t^2 y'' - 5ty' + 8y = 2t^3$, $y(1) = 1$, $y'(1) = 0$

23. $t^2 y'' - 6y = \log t$, $y(1) = 1/6$, $y'(1) = -1/6$

24. $t^2 y'' + ty' + 4y = \sin(\log t)$, $y(1) = 1$, $y'(1) = 0$

25. $4t^2 y'' + 8ty' + y = 3t^{-1/2} \log t^2$, $y(1) = 0$, $y'(1) = 1$

26. If y_H and y_P satisfy, respectively, the IVPs

$$N[y] = 0, \qquad y(t_0) = k_0, \qquad y'(t_0) = k_1$$

and

$$N[y] = f(t), \qquad y(t_0) = 0, \qquad y'(t_0) = 0$$

verify that the sum $y = y_H + y_P$ is a solution of

$$N[y] = f(t), \qquad y(t_0) = k_0, \qquad y'(t_0) = k_1$$

27. Show that the one-sided Green's function associated with the undamped spring-mass system described by

$$my'' + ky = F(t), \qquad y(0) = y_0, \qquad y'(0) = v_0$$

is

$$g_1(t, \tau) = \frac{1}{\omega_0} \sin[\omega_0(t - \tau)], \qquad \omega_0 = \sqrt{k/m}$$

28. Using the one-sided Green's function given in Problem 27, find the response of the spring-mass system described there when the system is initially at rest and subject to the forcing function

(a) $F(t) = P$ (constant)

(b) $F(t) = P \cos \omega t$, $\omega \neq \omega_0$

(c) $F(t) = P \cos \omega_0 t$

*****29.** Show that the one-sided Green's function associated with the damped spring-mass system described by

$$my'' + cy' + ky = F(t),$$

$$y(0) = y_0, \qquad y'(0) = v_0$$

for each of the following cases of damping is given by

(a) Underdamped $(c^2 < 4mk)$:

$$g_1(t, \tau) = \frac{1}{\mu} e^{-c(t-\tau)/2m} \sin[\mu(t - \tau)]$$

where $\mu = \sqrt{4mk - c^2}/2m$

(b) Critically damped $(c^2 = 4mk)$:

$$g_1(t, \tau) = (t - \tau)e^{-c(t-\tau)/2m}$$

(c) Overdamped $(c^2 > 4mk)$:

$$g_1(t, \tau) = \frac{1}{\alpha} e^{-c(t-\tau)/2m} \sinh[\alpha(t - \tau)]$$

where $\alpha = \sqrt{c^2 - 4mk}/2m$

*30. Using the one-sided Green's function given in Problem 29(a), find the **steady-state** response of the system described there when the forcing function is

(a) $F(t) = P(\text{constant})$

(b) $F(t) = P \cos \omega t$

*31. The one-sided Green's function associated with the third-order IVP

$$y''' + a_2(t)y'' + a_1(t)y' + a_0(t)y = f(t),$$

$$t > t_0, \qquad y(t_0) = k_0, \qquad y'(t_0) = k_1,$$

$$y''(t_0) = k_2,$$

is given by

$$g_1(t, \tau) = \frac{\begin{vmatrix} y_1(\tau) & y_2(\tau) & y_3(\tau) \\ y_1'(\tau) & y_2'(\tau) & y_3'(\tau) \\ y_1(t) & y_2(t) & y_3(t) \end{vmatrix}}{W(y_1, y_2, y_3)(\tau)}$$

where y_1, y_2, y_3 form a fundamental solution set of the associated homogeneous DE. Use this result to construct the one-sided Green's function for the operators

(a) D^3

(b) $D(D^2 + 4)$

(c) $D^2(D - 1)$

*32. Use the one-sided Green's function (see Problem 31) to solve

$$y''' - y'' + 4y' - 4y = 1,$$

$$y(0) = 0, \qquad y'(0) = 1, \qquad y''(0) = -1$$

Hint. The solution has the form

$$y = C_1 y_1(t) + C_2 y_2(t) + C_3 y_3(t)$$

$$+ \int_0^t g_1(t, \tau) f(\tau)\, d\tau$$

[0] 5.6 Elementary Oscillation Theory

We now turn our attention to equations more general than the constant-coefficient DEs arising in most spring-mass systems or electric circuits. In particular, we want to discuss the DE

$$A_2(t)y'' + A_1(t)y' + A_0(t)y = 0 \tag{60}$$

in which certain parameters of the system under study vary over time. Although in general we cannot obtain explicit elementary solutions of (60),[†] we can study the *qualitative behavior* of the solutions by directly analyzing the equation itself. For example, information concerning the existence and relative positions of the **zero crossings** can be obtained without formal solutions. In spring-mass systems the zero crossings correspond to the times when the mass is in the equilibrium position of the system. The zero crossings, therefore, describe the **oscillatory** characteristics of certain systems, and also they are related to the **eigenvalues** of certain types of boundary value problems (see Chapter 11).

In the theorems that follow, we treat the coefficients of (60) as continuous functions for $t > 0$ and assume $A_2(t) \neq 0$ on this interval.

[†] In Chapter 7 we introduce a *series method* for producing solutions of certain DEs with variable coefficients.

Theorem 5.2 Sturm Separation Theorem. *If y_1 and y_2 are linearly independent solutions of*

$$A_2(t)y'' + A_1(t)y' + A_0(t)y = 0$$

on some interval I, then y_1 has precisely one zero between any two consecutive zeros of y_2 on I.

Proof. *Let $t = a$ and $t = b$ be consecutive zeros of y_2 (see Figure 5.12). The linear independence of y_1 and y_2 implies that the Wronskian $W(y_1, y_2)(t) \neq 0$, $a < t < b$. And since $y_2(a) = y_2(b) = 0$, it follows that*

$$W(y_1, y_2)(a) = y_1(a)y_2'(a) \neq 0$$

$$W(y_1, y_2)(b) = y_1(b)y_2'(b) \neq 0$$

The Wronskian is continuous on $a \leq t \leq b$, and because it does not vanish on this interval, it must have the same sign at all points. In particular, it has the same sign at $t = a$ and $t = b$. On the other hand, since a and b are consecutive zeros of y_2, the function y_2' must have opposite signs at $t = a$ and $t = b$. In order to prevent the Wronskian from changing signs, it follows that $y_1(a)$ and $y_1(b)$ have opposite signs. Based on the continuity of the solution y_1, we now conclude that y_1 has at least one zero between the zeros of y_2 at a and b.

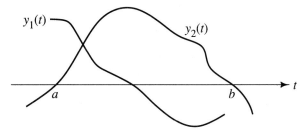

Figure 5.12

If we reverse the roles of y_1 and y_2 in the preceding argument, we deduce that y_2 must have at least one zero between consecutive zeros of y_1. Hence, y_1 cannot vanish more than once between $t = a$ and $t = b$.

Put another way, Theorem 5.2 states that the zeros of y_1 and y_2 occur alternately, and that on any finite interval the number of zeros of y_1 and y_2 can differ by at most one. For example, the solutions $\cos t$ and $\sin t$ of $y'' + y = 0$ have zeros that alternate on the entire axis. Of course, $\cos t$ and $\sin t$ are periodic functions whose consecutive zeros occur at regular intervals (separated by a distance π). In general, this is not the case.

The Sturm separation theorem does not assert the existence of any zeros of the solutions y_1 and y_2 of a given DE. Furthermore, it says nothing about the spacing between consecutive zeros (rate of oscillation) of those solutions. To pursue these matters it becomes convenient to put the differential equation in what we call the *Liouville normal form* or *standard form,* which does not contain a first derivative term.

If we divide Equation (60) by the leading coefficient $A_2(t)$, the resulting DE has the form

$$y'' + a(t)y' + b(t)y = 0 \tag{61}$$

which we previously called the normal form. Next, we put $y = u(t)v(t)$, so that

$$y' = uv' + u'v$$

$$y'' = uv'' + 2u'v' + u''v$$

In terms of u and v, therefore, Equation (61) becomes

$$vu'' + (2v' + av)u' + (v'' + av' + bv)u = 0 \tag{62}$$

The coefficient of u' can be made to vanish by selecting

$$2v' + a(t)v = 0$$

or

$$v(t) = \exp\left[-\frac{1}{2} \int a(t)\, dt \right] \tag{63}$$

Hence, (62) reduces to

$$u'' + Q(t)u = 0 \tag{64}$$

where

$$Q(t) = b(t) - \tfrac{1}{4}[a(t)]^2 - \tfrac{1}{2}a'(t) \tag{65}$$

We henceforth refer to (64) as the **standard form.**

Clearly, the solutions of (60) and (64) are not the same. However, because the function v [defined by (63)] cannot vanish, both u and $y = uv$ have the same zeros. Hence, we can now confine our attention to (64) in order to investigate the oscillation phenomena associated with solutions of (60).

■————————
EXAMPLE 12

Find the standard form of *Bessel's equation*[†]

$$t^2 y'' + ty' + (t^2 - v^2)y = 0$$

Solution

We first rewrite the equation as

$$y'' + \frac{1}{t}y' + \left(1 - \frac{v^2}{t^2}\right)y = 0$$

which identifies $a(t) = 1/t$ and $b(t) = 1 - v^2/t^2$. Hence,

$$Q(t) = 1 - \frac{v^2}{t^2} - \frac{1}{4t^2} + \frac{1}{2t^2} = 1 + \frac{1 - 4v^2}{4t^2}$$

———————
[†] The solutions of Bessel's equation are derived in Section 7.7.

and Bessel's equation in standard form becomes

$$u'' + \left(1 + \frac{1 - 4v^2}{4t^2}\right)u = 0$$

Although we don't need it, we see that $v(t) = t^{-1/2}$. ∎

We can now show that if $Q(t) < 0$ on a given interval, the solutions of $u'' + Q(t)u = 0$ do not oscillate on that interval. To better understand this situation, consider the solutions of $u'' - u = 0$.

Theorem 5.3 *If $Q(t) < 0$ on the interval I and u is a nontrivial solution of $u'' + Q(t)u = 0$, then u has at most one zero on I.*

Proof. *Let $t = a$ be a point on the interval I such that $u(a) = 0$. Since u is a nontrivial solution of the DE, it follows that $u'(a) \neq 0$ (Corollary 5.1). For definiteness, let us assume that $u'(a) > 0$ so that u is positive as t increases from $t = a$. Since $Q(t) < 0$, we see that $u'' = -Q(t)u$ must be positive for increasing t. And since u'' is the rate at which the slope of u changes, we conclude that the slope is increasing. Hence, u cannot have a zero for $t > a$. A similar argument can show that u has no zero for $t < a$, and the proof is the same in the case $u'(a) < 0$.*

Since our present interest concerns oscillation behavior, this last result suggests we confine our attention to only those cases for which $Q(t) > 0$. Even in this case, however, the solutions may not be oscillatory if $Q(t)$ decreases too rapidly for increasing t. The relevant theorem regarding this situation is stated below without proof (see Problem 10 in Exercises 5.6).

Theorem 5.4 *If u is a nontrivial solution of $u'' + Q(t)u = 0$, $Q(t) > 0$ for all $t > 0$, and*

$$\int_1^\infty Q(t)\, dt = \infty$$

then u has infinitely many zeros on the positive t axis.

We now state and prove the Sturm comparison theorem, which is fundamental in oscillation theory.

Theorem 5.5 Sturm Comparison Theorem. *Let u_1 and u_2 represent nontrivial solutions, respectively, of*

$$u'' + Q_1(t)u = 0$$
$$u'' + Q_2(t)u = 0$$

and suppose that $Q_1(t) > Q_2(t)$ everywhere on an interval I. Then there exists at least one zero of u_1 between every two consecutive zeros of u_2.

Proof. *Let $t = a$ and $t = b$ be consecutive zeros of u_2, and assume that u_1 does not vanish on the interval $a < t < b$. We further assume that both $u_1(t) > 0$ and $u_2(t) > 0$ on the interval $a < t < b$. Since the zeros of u_1 are the same as $-u_1$, this assumption is justified. Following the argument presented in the proof of Theorem 5.2, we have*

$$W(u_1, u_2)(a) = u_1(a)u_2'(a)$$

$$W(u_1, u_2)(b) = u_1(b)u_2'(b)$$

However,

$$\frac{d}{dt}\left[W(u_1, u_2)(t)\right] = \frac{d}{dt}(u_1 u_2' - u_1' u_2)$$

$$= u_1 u_2'' - u_1'' u_2$$

$$= -u_1 Q_2 u_2 + u_2 Q_1 u_1$$

$$= u_1 u_2 (Q_1 - Q_2)$$

and since $Q_1(t) > Q_2(t)$, we conclude that the Wronskian is an increasing function on $a < t < b$. Also, because we have assumed that $u_2(t) > 0$ on this interval, it follows that $u_2'(a) \geq 0$ and $u_2'(b) \leq 0$. Thus,

$$W(u_1, u_2)(a) \geq 0$$

$$W(u_1, u_2)(b) \leq 0$$

But this condition implies that the Wronskian cannot be an increasing function. Thus, contrary to our assumption, we must conclude that $u_1(t) = 0$ at least once on the interval $a < t < b$.

An immediate consequence of the Sturm comparison theorem is that the *larger* the coefficient of u in the DE $u'' + Q(t)u = 0$, the *more rapidly* the solution oscillates.

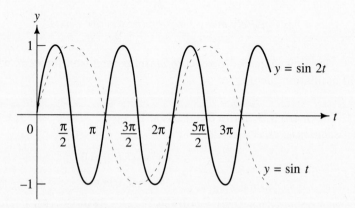

Figure 5.13 Graph of $y = \sin t$ and $y = \sin 2t$.

For example, let us examine the solutions $\sin t$ and $\sin 2t$, respectively, of the two equations

$$u'' + \quad u = 0$$

$$u'' + 4u = 0$$

The solution of the second equation oscillates more rapidly than that of the first equation, as can be seen from their graphs (see Figure 5.13). That is, the zeros of $\sin 2t$ are located at $t = 0, \pi/2, \pi, 3\pi/2, \ldots$, whereas the zeros of $\sin t$ are located at $t = 0, \pi, 2\pi, 3\pi, \ldots$.

EXAMPLE 13

Show that every solution of

$$y'' + t^2 y = 0$$

has infinitely many zeros on the interval $t > 1$.

Solution

The equation $y'' + y = 0$ has one solution $\sin t$ with infinitely many zeros at $t = n\pi$, $n = 0, 1, 2, \ldots$. Since $t^2 > 1$, it follows from the Sturm comparison theorem that any nontrivial solution of $y'' + t^2 y = 0$ has at least one zero between $t = n\pi$ and $t = (n + 1)\pi$, $n = 1, 2, 3, \ldots$, and hence has infinitely many zeros. This result is also a consequence of Theorem 5.4. ∎

5.6.1 Zeros of Bessel Functions

The solutions of *Bessel's equation* are called **Bessel Functions** (see Section 7.7). The zeros of these functions are of both practical and theoretical importance, and are found to satisfy the conditions of the following theorem by comparison with the zeros of solutions of $y'' + y = 0$. The standard form of Bessel's equation was derived in Example 12.

Theorem 5.6 Zeros of Bessel Functions. *Every nontrivial solution of Bessel's equation, which has the standard form*

$$u'' + \left(1 + \frac{1 - 4v^2}{4t^2}\right)u = 0$$

has infinitely many zeros on the positive t axis. Moreover, the distance between consecutive zeros is

(a) *Less than π for $0 < v < 1/2$*
(b) *Equal to π for $v = 1/2$*
(c) *Greater than π for $v > 1/2$*

Theorem 5.4 can be used to prove that the solutions of Bessel's equation have infinitely many zeros on the positive t axis and Theorem 5.5 is the basis for establishing the distance between consecutive zeros. We leave the details of the proof, however, to the reader (see Problem 22 in Exercises 5.6).

Interestingly, the zeros of the solutions of Bessel's equation coincide with the zeros of the sinusoidal functions when $v = 1/2$. The reason for this situation is that the solutions of Bessel's equation are directly related to the sinusoidal functions for this particular value of v (see Problem 9 in Exercises 7.7).

Exercises 5.6

1. Prove that the zeros of any solution of the second-order DE

$$A_2(t)y'' + A_1(t)y' + A_0(t)y = 0$$

are simple; that is, if $y(t_0) = 0$, then $y'(t_0) \neq 0$.
Hint. Use Corollary 5.1.

2. Prove that if y_1 and y_2 are linearly independent solutions of

$$A_2(t)y'' + A_1(t)y' + A_0(t)y = 0$$

they cannot vanish at the same point.
Hint. Assume $y_1(a) = y_2(a) = 0$ at some point $t = a$, and show that this leads to a contradiction that the Wronskian cannot vanish anywhere on the interval of interest.

3. Prove that between every pair of consecutive zeros of $\sin t$ there is one zero of $\sin t + \cos t$.

■**4.** Show that the zeros of $y_1 = \sin(\log t)$ and $y_2 = \cos(\log t)$ alternate.
Hint. Find a DE for which y_1 and y_2 are both solutions.

In Problems 11 to 14, put the DE in standard form.

11. $(1 - t^2)y'' - 2ty' + 6y = 0$
12. $t^2y'' - 2t^3y' - (4 - t^2)y = 0$

In Problems 15 to 18, show that every nontrivial solution has infinitely many zeros on the interval $t > 0$.

15. $y'' + ty = 0$
16. $y'' + (1/t)y = 0$
17. $y'' - 4y' + 13y = 0$
■**18.** $y'' - 2ty' + t^2y = 0$
19. Which of the equations

$$y'' + (1 + t^2)y = 0, \qquad y'' + 2ty = 0$$

has the most rapidly oscillating solution in the interval $0 \leq t \leq 10$?

5. Explain how the Sturm separation theorem applies to $y'' - y = 0$, if indeed it does.

6. Show directly that the substitution $y = t^{-1/2}u$ transforms Bessel's equation (see Example 12) to standard form.

7. Let y be a nontrivial solution of

$$y'' + (\sinh t)y = 0.$$

Show that
(a) y has at most one zero in the interval $t < 0$.
(b) y has infinitely many zeros in the interval $t > 0$.

*****8.** Show that every nontrivial solution of

$$y'' - 2ty' + ky = 0$$

has
(a) At most one zero when $k \leq 1$.
(b) Only finitely many zeros when $k > 0$.

9. Put the DE $my'' + cy' + ky = 0$ in standard form and show that it does not possess oscillatory solutions unless $c^2 < 4mk$.

*****10.** Prove Theorem 5.4.

13. $(t^3 - 2)y'' - t^2y' - 3y = 0$
14. $(t + 1)y'' - y' + 2ty = 0$

20. Show that the distance between consecutive zeros of a solution of $y'' + t^2y = 0$ approaches zero as $t \to \infty$.

21. Show that the distance between consecutive zeros of a solution of $y'' + ty = 0$ is less than π for at least $t > \pi$.

22. Prove Theorem 5.6.

23. Prove that every nontrivial solution of

$$y'' - 2ty' + 2ny = 0, \qquad n = 0, 1, 2, \ldots$$

has at most finitely many zeros on $-\infty < t < \infty$.

24. Prove that every nontrivial solution of

$$ty'' + (1 - t)y' + ny = 0, \qquad n = 0, 1, 2, \ldots$$

has at most finitely many zeros on $t > 0$.

5.7 Chapter Summary

In this chapter we have studied primarily the small movements of a spring-mass system subject to resistive and external forces. **Hooke's law** $F = ks$ is used to determine the spring constant k, given that F is the weight of the mass and s is the amount of stretch experienced by the spring when the mass is attached. By summing all forces acting on the mass and applying **Newton's second law of motion,** we are led to the governing DE

$$my'' + cy' + ky = F(t) \tag{66}$$

The term cy', where $c > 0$, is the **damping** force and $F(t)$ represents the external force.

In solving (66), we distinguish between several cases, depending upon whether damping effects are important and whether the external force is present. These are the following:

Case I. No Damping or External Forces

Here $c = 0$ and $F(t) \equiv 0$, and the governing equation is

$$y'' + \omega_0^2 y = 0 \tag{67}$$

where $\omega_0 = \sqrt{k/m}$, with general solution $y = C_1 \cos \omega_0 t + C_2 \sin \omega_0 t$. By defining $A = \sqrt{C_1^2 + C_2^2}$ and $\tan \phi = C_2/C_1$, the solution can always be put in the form

$$y = A \cos(\omega_0 t - \phi) \tag{68}$$

Hence, all solutions of (67) are forms of **simple harmonic motion.**

Case II. Damping But No External Forces

This time the governing DE is

$$my'' + cy' + ky = 0 \tag{69}$$

the general solution of which leads to three distinct types of motion, called **overdamped** ($c^2 > 4mk$), **critically damped** ($c^2 = 4mk$), or **underdamped** ($c^2 < 4mk$). Only the underdamped case leads to **oscillatory** type of motion. However, in each case the solution approaches its equilibrium position for increasing time owing to the damping force, that is, $y \to 0$ as $t \to \infty$.

Case III. External Force Present

When an external force is impressed on the system we must solve the full equation of motion (66). In some cases we may be able to neglect damping effects again (i.e., set $c = 0$), but either way the general solution has the form $y = y_H + y_P$, where y_H is the general solution of the associated homogeneous equation and y_P is any particular solution. If $c \neq 0$, the homogeneous solution is the **transient solution** of the system while y_P is the **steady-state solution.**

If the input force has the same frequency as the natural frequency of the system, that is, $F(t) = P \cos \omega_0 t$, and damping is neglected, a state of pure **resonance** will be achieved. Resonance can generally be controlled, however, by a sufficient amount of damping.

An RLC circuit containing a resistor, inductor, and capacitor, connected in series with an electromotive force $E(t)$, leads to equations for the charge $q(t)$ and current $i(t)$ that have the same form as (66). Thus, the analysis of simple electric circuits is basically the same as that for the spring-mass system.

Review Exercises

1. A spring for which $k = 48$ lb/ft has a 16-lb mass attached.
  (a) Find the natural frequency and period of the motion.
  (b) If the mass is released from rest 2 in. below its equilibrium position, find the subsequent motion. Neglect damping.

2. A 20-lb weight stretches a spring 6 in. from its natural length.
  (a) Find the natural frequency and period of the motion.
  (b) If the weight is initially released 3 in. below its equilibrium position with a downward velocity of 2 ft/s, find its subsequent motion.
  (c) Put the answer for (b) in the form $y = A \cos(\omega t - \phi)$.

3. A 20-g mass stretches a spring 4 cm from its natural length. In the absence of external forces and air resistance, find the position of the mass at all later times given that it was pulled 1 cm below its equilibrium position and set into motion with an upward velocity of 0.5 cm/s.

4. A 10-kg mass stretches a spring 0.7 m from its natural length. The mass is started in motion from its equilibrium position with an upward velocity of 1 m/s. If the air resistance is $90v$ N, find the subsequent motion.

5. A mass of 1/4 slug stretches a spring 1.28 ft from its natural length. The mass is started in motion from its equilibrium position with a downward velocity of 4 ft/s. If the air resistance is twice the instantaneous velocity, find the subsequent motion.

6. A mass of 1/4 slug is attached to a spring for which $k = 1$ lb/ft. The mass is started in motion 2 ft below its equilibrium position with a upward velocity of 2 ft/s. If the air resistance is equal to the instantaneous velocity, find the subsequent motion.

7. A 1-slug mass is attached to a spring for which $k = 8$ lb/ft. An external force $F(t) = 16 \cos 4t$ is applied and the mass is set into motion from its equilibrium position with zero velocity. If air resistance is 4 times the instantaneous velocity,
  (a) Find the subsequent motion.

(b) Determine the steady-state motion and find its amplitude.

8. A weight of 64 lb is attached to a spring for which $k = 50$ lb/ft. An external force $F(t) = 4 \sin 2t$ is applied to the mass, setting it into motion from rest in its equilibrium position. Neglect damping effects and find the subsequent motion.

9. A 20-lb weight stretches a spring 6 in. from its natural length. An external force $F(t) = 40 \cos 8t$ is applied and the system is set in motion from rest 2 in. below its equilibrium position. Neglect damping effects and find the subsequent motion.

10. An *RLC* series circuit has $R = 10 \, \Omega$, $C = 10^{-2}$ F, $L = 0.5$ H, and a battery source of 12 V. If the initial current and voltage in the circuit is zero,

  (a) Find the charge on the capacitor at all later times.

  (b) Find the current in the circuit at all later times.

(c) What is the steady-state charge in the circuit?

11. An *RLC* series circuit has $R = 5 \, \Omega$, $C = 4 \times 10^{-4}$ F, $L = 0.05$ H, and a battery source of 110 V. If the initial current and voltage in the circuit is zero,

  (a) Find the charge on the capacitor at all later times.

  (b) Find the current in the circuit at all later times.

  (c) What is the steady-state charge in the circuit?

12. An *RLC* series circuit has $R = 4 \, \Omega$, $C = 1/26$ F, $L = 0.5$ H, and an emf source $E(t) = 16 \cos 2t$ V. If the initial current and voltage in the circuit is zero,

  (a) Find the charge on the capacitor at all later times.

  (b) What is the steady-state charge in the circuit?

CHAPTER *6*
The Laplace Transform

The Laplace transform was originally developed as an efficient method for solving linear constant-coefficient DEs with prescribed initial conditions, that is, IVPs. Today it is an essential tool in the study of control theory, methods of design, circuit analysis, and systems of differential equations, and in solving certain integral equations and partial differential equations.

In Section 6.2 we calculate the **transforms of some elementary functions** directly from the integral definition. We also state the fundamental **existence theorem** for Laplace transforms. **Operational properties,** from which certain transforms may be determined without resorting to the defining integral, are developed in Section 6.3. We discuss methods of computing **inverse Laplace transforms** in Section 6.4, which rely mostly on **tables, operational properties,** and **partial fractions.**

Solving IVPs by the transform method is featured in Section 6.5. Some **applications** involving spring-mass systems and electric circuits are also briefly discussed. In Section 6.6 we introduce the **convolution theorem,** which is useful in finding the inverse transform of a product of functions, while **periodic functions** and **discontinuous functions** are featured in Section 6.7. **Impulse functions** are introduced in Section 6.8, which lead naturally to the **impulse response function** and to the **one-sided Green's function.** Some **historical comments** are given in Section 6.10.

6.1 Introduction

IVPs featuring DEs with constant coefficients occur in numerous applications. For example, in Chapter 5 we found that

$$my'' + cy' + ky = F(t), \qquad y(0) = y_0, \qquad y'(0) = v_0$$

and

$$Lq'' + Rq' + C^{-1}q = E(t), \qquad q(0) = q_0, \qquad q'(0) = i(0) = i_0$$

arise, respectively, in determining the motion of a spring-mass system and the charge on a series *RLC* circuit. In these applications it often happens that the input functions $F(t)$ and $E(t)$ have certain *discontinuities,* and conventional methods tend to be awkward when applied to problems of this nature. However, a powerful *operational method* was introduced in the late 1800s by 0. Heaviside[†] for solving problems such as these and others commonly occurring in engineering applications. The technique is formally equivalent to the *Laplace transform method,* so named in honor of the French mathematician P. Laplace.[‡]

The idea of transforming one function into another is commonplace in mathematics. For example, the differential operator D transforms the function $f(t)$ into the new function $D\{f(t)\} = f'(t)$. Another transformation used extensively in the calculus is that of integration. A particular transformation involving integrals that is widely used in applications is the *integral transform,* of which there are several varieties. Perhaps the best known of these integral transforms is the **Laplace transform** defined by

$$\mathscr{L}\{f(t)\} = \int_0^\infty e^{-st}f(t)\,dt \tag{1}$$

The symbol

$$\mathscr{L}\{f(t)\} = F(s) \tag{2}$$

is also commonly used to denote this transform. Thus we often write (1) in the form

$$F(s) = \int_0^\infty e^{-st}f(t)\,dt \tag{3}$$

The new function $F(s) = \mathscr{L}\{f(t)\}$ is said to be the **transform** of $f(t)$, and e^{-st} is called the **kernel** of the transformation.

> *Remark.* We use lowercase letters for functions in the t domain and the same letters capitalized for their transform.

Because the defining integral (1) is improper (owing to the infinite limit of integration), it must be evaluated through the limit process

$$\int_0^\infty e^{-st}f(t)\,dt = \lim_{b\to\infty}\int_0^b e^{-st}f(t)\,dt$$

[†] Oliver Heaviside (1850–1925) was an English engineer (see Section 6.10 for a historical account of his work).

[‡] Pierre Simon de Laplace (1749–1827). See also Section 6.10.

If the limit of the integral from 0 to b exists, we say that the integral **converges** to that limiting value; otherwise it **diverges.** For the purposes of notational abbreviation, however, we henceforth let $\Big|_0^\infty$ denote $\displaystyle\lim_{b\to\infty}(\ )\Big|_0^b$.

6.2 Laplace Transforms of Some Elementary Functions

The Laplace transform of many elementary functions can be readily obtained through routine integration of the defining integral (1). To begin, therefore, we simply use the defining integral to evaluate the transforms of some typical functions arising in practice. After that, we discuss conditions under which the Laplace transform $F(s)$ of a given function $f(t)$ is known to exist.

> **Remark.** Since the transform integral is defined only over $t \geq 0$, it makes little difference how a particular function $f(t)$ is defined for $t < 0$. In our work on Laplace transforms, therefore, we assume the definition of $f(t)$ is given only for $t \geq 0$.

■————— **EXAMPLE 1**

Find $\mathcal{L}\{1\}$.

Solution

From the defining integral (1), we have

$$\mathcal{L}\{1\} = \int_0^\infty e^{-st}\,dt = \frac{e^{-st}}{-s}\bigg|_0^\infty$$

For $s \leq 0$, the integral clearly diverges, but if we restrict $s > 0$, then $e^{-st} \to 0$ as $t \to \infty$. Hence, we deduce that

$$\mathcal{L}\{1\} = \frac{1}{s}, \qquad s > 0$$

■

■————— **EXAMPLE 2**

Find $\mathcal{L}\{e^{at}\}$.

Solution

Here we find that

$$\mathcal{L}\{e^{at}\} = \int_0^\infty e^{-st}e^{at}\,dt = \int_0^\infty e^{-(s-a)t}\,dt$$

This integral is essentially that occurring in Example 1 and hence converges for $s - a > 0$. In this case we get

$$\mathcal{L}\{e^{at}\} = \frac{1}{s-a}, \qquad s > a$$

■

Although we have not specified any range of values on the transform variable s in the definition given by (1), a restriction such as $s > c$, where c is some constant, is usually necessary for convergence of the defining integral (see Examples 1 and 2). Restrictions of this kind, however, have little effect in applications involving the transform. That is, it is generally enough to know that the transform of a given function $f(t)$ exists for some values of the transform variable s. Also, we will always treat s as a real variable in our work, but in more rigorous treatments of the Laplace transform it is common practice to treat s as a *complex* variable.[†]

EXAMPLE 3

Find $\mathcal{L}\{\sin t\}$.

Solution

Integration by parts yields

$$\mathcal{L}\{\sin t\} = \int_0^\infty e^{-st} \sin t \, dt$$

$$= -e^{-st} \cos t \Big|_0^\infty - s \int_0^\infty e^{-st} \cos t \, dt$$

$$= 1 - s \int_0^\infty e^{-st} \cos t \, dt, \qquad s > 0$$

A second integration by parts on the remaining integral now gives us

$$\mathcal{L}\{\sin t\} = 1 - s^2 \int_0^\infty e^{-st} \sin t \, dt$$

$$= 1 - s^2 \mathcal{L}\{\sin t\}$$

and solving for $\mathcal{L}\{\sin t\}$, we find

$$\mathcal{L}\{\sin t\} = \frac{1}{s^2 + 1}, \qquad s > 0 \qquad \blacksquare$$

In the evaluation of various transforms we find the following **linearity property** to be very useful.

Theorem 6.1 Linearity. *If $\mathcal{L}\{f(t)\} = F(s)$ and $\mathcal{L}\{g(t)\} = G(s)$, then for any constants C_1 and C_2,*

$$\mathcal{L}\{C_1 f(t) + C_2 g(t)\} = C_1 F(s) + C_2 G(s)$$

The proof of Theorem 6.1 follows immediately from the linearity property of integrals and is left to the exercises (see Problem 29 in Exercises 6.2).

[†] For example, see L. C. Andrews and B. K. Shivamoggi, *Integral Transforms for Engineers and Applied Mathematicians,* Macmillan, New York, 1988.

EXAMPLE 4 Evaluate $\mathscr{L}\{7 - 3e^{2t} + 5 \sin t\}$.

Solution

From the linearity property, we have

$$\mathscr{L}\{7 - 3e^{2t} + 5 \sin t\} = 7\mathscr{L}\{1\} - 3\mathscr{L}\{e^{2t}\} + 5\mathscr{L}\{\sin t\}$$

$$= \frac{7}{s} - \frac{3}{s - 2} + \frac{5}{s^2 + 1}$$

where we have used the results of Examples 1, 2, and 3. ∎

EXAMPLE 5 Find $\mathscr{L}\{\sin kt\}$ and $\mathscr{L}\{\cos kt\}$.

Solution

Here we can make use of the result of Example 2 and the linearity property to obtain both transforms at once. Based on Euler's formula (from Section 4.4.1)

$$e^{ix} = \cos x + i \sin x$$

we obtain

$$\mathscr{L}\{\cos kt\} + i\mathscr{L}\{\sin kt\} = \mathscr{L}\{e^{ikt}\}$$

$$= \frac{1}{s - ik}$$

where we have set $a = ik$ in the result of Example 2. By multiplying both numerator and denominator by $s + ik$ and separating the result into real and imaginary parts, we then get

$$\mathscr{L}\{\cos kt\} + i\mathscr{L}\{\sin kt\} = \frac{1}{s - ik} \cdot \frac{s + ik}{s + ik}$$

$$= \frac{s}{s^2 + k^2} + i\frac{k}{s^2 + k^2}$$

so upon matching up real and imaginary parts, we deduce that

$$\mathscr{L}\{\cos kt\} = \frac{s}{s^2 + k^2}$$

$$\mathscr{L}\{\sin kt\} = \frac{k}{s^2 + k^2}$$ ∎

6.2.1 Existence Theorem

Because the defining integral (1) is improper, not every function $f(t)$ will have a Laplace transform [even if $f(t)$ is continuous]. However, we can identify a rather large class of functions satisfying *two* fundamental properties such that all members of the class have a Laplace transform.

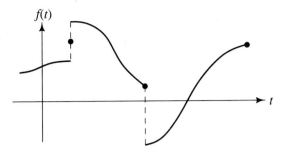

Figure 6.1 A piecewise continuous function.

The most fundamental property necessary for a function to be integrable is that it be *piecewise continuous.*

Definition 6.1 A function $f(t)$ is said to be **piecewise continuous** on the subinterval $[a, b]$ of $t \geq 0$ if

(a) $f(t)$ is defined and continuous at all but a finite number of points on $[a, b]$.
(b) The left-hand and right-hand limits exist at each point on the interval $[a, b]$.[†]

> *Remark.* The left-hand and right-hand limits are defined, respectively, by
>
> $$\lim_{\epsilon \to 0^+} f(x - \epsilon) = f(x^-) \quad \text{and} \quad \lim_{\epsilon \to 0^+} f(x + \epsilon) = f(x^+)$$
>
> Furthermore, when x is a point of continuity, $f(x^-) = f(x^+) = f(x)$.

It is not essential that a piecewise continuous function $f(t)$ be defined at every point in the interval of interest. In particular, it is often not defined at a point of discontinuity, and even when it is, it really doesn't matter what function value is assigned at these points. Also, the interval of interest may be open or closed, or open at one end and closed at the other (see Figure 6.1). Finally, we say a function is piecewise continuous on $t \geq 0$ if it is piecewise continuous on *every* finite interval of $t \geq 0$.

In addition to being piecewise continuous, we also require that the function be of *exponential order,* for which we have the following definition.

Definition 6.2 A function $f(t)$ is said to be of **exponential order** if there exist real constants c, M, and t_0 such that

$$|f(t)| < Me^{ct}, \qquad t > t_0$$

or equivalently, such that

$$\lim_{t \to \infty} f(t)e^{-ct} = 0 \ [‡]$$

[†] Only one of these limits is required at the endpoints a and b.

[‡] This limit is sometimes denoted by writing $f(t) = 0(e^{ct})$.

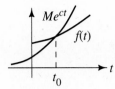

Figure 6.2a A function of exponential order.

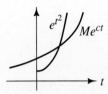

Figure 6.2b A function *not* of exponential order.

Saying a function is of exponential order means that its graph on the interval $t > t_0$ does not grow faster than the graph of Me^{ct} for appropriate values of M and c (see Figure 6.2a). For instance, the functions 1, e^{at}, and sin t are all of exponential order, whereas the function $f(t) = e^{t^2}$ is not (see Figure 6.2b), since for any value of c

$$\lim_{t \to \infty} e^{t^2} e^{-ct} \neq 0$$

Let us now state and prove our basic *existence theorem*.

Theorem 6.2 Existence Theorem.   *If $f(t)$ is piecewise continuous on $t \geq 0$ and is of exponential order as $t \to \infty$, then the Laplace transform integral $\mathscr{L}\{f(t)\} = \int_0^\infty e^{-st} f(t)\, dt$ exists.*

Proof.   *We begin by expressing the defining integral as*

$$\int_0^\infty e^{-st} f(t)\, dt = \int_0^{t_0} e^{-st} f(t)\, dt + \int_{t_0}^\infty e^{-st} f(t)\, dt$$

The first integral on the right exists since $f(t)$ is assumed to be piecewise continuous. Because $f(t)$ is also of exponential order, the second integral on the right satisfies the inequality

$$\left| \int_{t_0}^\infty e^{-st} f(t)\, dt \right| \leq \int_{t_0}^\infty e^{-st} |f(t)|\, dt$$

$$\leq M \int_{t_0}^\infty e^{-(s-c)t}\, dt$$

Hence, by direct integration of this last integral, we obtain

$$\left| \int_{t_0}^\infty e^{-st} f(t)\, dt \right| \leq \frac{Me^{-(s-c)t_0}}{s - c}, \qquad s > c$$

For $s > c$, this last expression vanishes in the limit as $t_0 \to \infty$, so we say that the integral is absolutely convergent. Thus, the Laplace transform of $f(t)$ exists.

Most functions met in practice satisfy the conditions of Theorem 6.2. However, these conditions are *sufficient* rather than necessary to ensure that a function has a Laplace transform. That is, a function may not satisfy both conditions listed in Theorem 6.2 and still have a Laplace transform. For example, $t^{-1/2}$ has an infinite discontinuity at $t = 0$ (and hence, is not piecewise continuous), but it can be shown that (see Problem 26 in Exercises 6.2)

$$\mathscr{L}\{t^{-1/2}\} = \int_0^\infty e^{-st} t^{-1/2}\, dt = \sqrt{\frac{\pi}{s}}, \qquad s > 0$$

Hence, $t^{-1/2}$ has a Laplace transform. Also, notice that if the transform of a function exists, it is *unique* since the definite integral of a function is unique.

As a final comment here, we note that evaluating the Laplace transform of a piecewise continuous function from the defining integral is quite cumbersome, owing to the necessity of evaluating integrals over each interval of definition. For example, if the discontinuous function is

$$f(t) = \begin{cases} t, & 0 < t < 2 \\ 5, & t > 2 \end{cases} \tag{4}$$

then the defining integral leads to

$$\mathscr{L}\{f(t)\} = \int_0^2 e^{-st} t\, dt + \int_2^\infty e^{-st}\, 5\, dt$$

Using integration by parts on the first integral, we obtain

$$\mathscr{L}\{f(t)\} = \left[-\frac{t}{s} e^{-st} - \frac{1}{s^2} e^{-st} \right]_0^2 + \left[-\frac{5}{s} e^{-st} \right]_2^\infty$$

which simplifies to

$$\mathscr{L}\{f(t)\} = \frac{1}{s^2} + \left(\frac{3}{s} - \frac{1}{s^2} \right) e^{-2s}, \qquad s > 0 \tag{5}$$

In Section 6.7 we discuss an alternate method for calculating such transforms that is far superior to the above procedure.

Exercises 6.2

In Problems 1 to 12, evaluate the Laplace transform of each function directly from the defining integral.

1. $f(t) = t$

2. $f(t) = t^2$

3. $f(t) = t^n,\ n = 1, 2, 3, \ldots$

4. $f(t) = \sin kt$

5. $f(t) = \cos kt$

6. $f(t) = 2e^{3t} - e^{-3t}$

7. $f(t) = e^{-at} - e^{-bt}$

■**8.** $f(t) = \sinh kt$

9. $f(t) = \cosh kt$

10. $f(t) = t^2 - 3t + 5$

11. $f(t) = \sin(at + b)$

12. $f(t) = \cos(at + b)$

In Problems 13 to 16, evaluate the Laplace transform of each function, recalling the identities $\cosh x = \frac{1}{2}(e^x + e^{-x})$ and $\sinh x = \frac{1}{2}(e^x - e^{-x})$.

13. $f(t) = \cosh^2 kt$

14. $f(t) = \sinh^2 kt$

15. $f(t) = e^{at} \cosh kt$

16. $f(t) = e^{at} \sinh kt$

In Problems 17 to 20, use integration by parts to evaluate the Laplace transform of each function.

17. $f(t) = te^{at}$

18. $f(t) = t \sin kt$

19. $f(t) = t^2 \cos kt$

*20. $f(t) = te^{-t} \cos t$

21. Using $\cos^2 x = \frac{1}{2}(1 + \cos 2x)$, evaluate $\mathcal{L}\{\cos^2 kt\}$.

■22. Using $\sin^2 x = 1 - \cos^2 x$ and Problem 21, find $\mathcal{L}\{\sin^2 kt\}$.

23. Use a trigonometric identity to evaluate $\mathcal{L}\{\sin kt \cos kt\}$.

24. Use Definition 6.2 to determine which of the following functions are of exponential order:

   (a) $f(t) = t^{100}$

   (b) $f(t) = e^{-t^2}$

   (c) $f(t) = te^t$

   (d) $f(t) = 3e^{t-t^2}$

   (e) $f(t) = 5 \sin(e^{t^2})$

   (f) $f(t) = \dfrac{\sin t}{t}$

*25. The **gamma function,** denoted by $\Gamma(x)$, is defined by the integral

$$\Gamma(x) = \int_0^\infty e^{-t} t^{x-1}\, dt, \qquad x > 0$$

Show that

   (a) $\Gamma(x + 1) = x\Gamma(x)$

   (b) $\Gamma(n + 1) = n!, \qquad n = 0, 1, 2, \ldots$

*26. Using the results of Problem 25 and the special value $\Gamma(1/2) = \sqrt{\pi}$, verify the following transform relations:

   (a) $\mathcal{L}\{t^{-1/2}\} = \sqrt{\dfrac{\pi}{s}}, \qquad s > 0$

   (b) $\mathcal{L}\{t^{1/2}\} = \dfrac{1}{2s} \sqrt{\dfrac{\pi}{s}}, \qquad s > 0$

   (c) $\mathcal{L}\{t^{5/2}\} = \dfrac{15}{8s^3} \sqrt{\dfrac{\pi}{s}}, \qquad s > 0$

   (d) $\mathcal{L}\{t^n\} = \dfrac{n!}{s^{n+1}}, \qquad n = 0, 1, 2, \ldots, \quad s > 0$

*27. Show that $\Gamma(1/2) = \sqrt{\pi}$.

   Hint. First show that

$$\Gamma(1/2) = 2 \int_0^\infty e^{-x^2}\, dx = 2 \int_0^\infty e^{-y^2}\, dy$$

then multiply these integrals, change to polar coordinates, and evaluate the resulting double integral.

*28. Given that (see Problem 25)

$$\Gamma'(x) = \int_0^\infty e^{-t}(\log t) t^{x-1}\, dt, \qquad x > 0$$

develop the Laplace transform relation

$$\mathcal{L}\{\log t\} = \frac{1}{s}\Gamma'(1) - \frac{\log s}{s}$$

29. Prove Theorem 6.1.

6.3 Operational Properties

After the transform of a function is calculated, it is usually placed in a table for reference purposes (see Table 6.1), much as we do in constructing integral tables. Once we have tabulated several such transforms, it may be possible to generate new transforms from these by the use of *operational properties*. For example, knowing the transform of $f(t)$ permits us, through the use of these properties, to directly determine the transforms of $e^{at}f(t)$, $tf(t)$, $f'(t)$, and so on. The linearity property (Theorem 6.1) plus some other useful operational properties of the transform are listed in Table 6.2.

TABLE 6.1 TABLE OF LAPLACE TRANSFORMS

$F(s) = \mathscr{L}\{f(t)\}$	$f(t) = \mathscr{L}^{-1}\{F(s)\}$
1. $\dfrac{1}{s}$	1
2. $\dfrac{1}{s^2}$	t
3. $\dfrac{1}{s^n}$ $(n = 1, 2, 3, \ldots)$	$\dfrac{t^{n-1}}{(n-1)!}$
4. $\dfrac{1}{s^{1/2}}$	$\dfrac{1}{(\pi t)^{1/2}}$
5. $\dfrac{1}{s^{3/2}}$	$2\left(\dfrac{t}{\pi}\right)^{1/2}$
6. $\dfrac{1}{s^x}$ $(x > 0)$	$\dfrac{t^{x-1}}{\Gamma(x)}$
7. $\dfrac{1}{s-a}$	e^{at}
8. $\dfrac{1}{(s-a)^2}$	te^{at}
9. $\dfrac{1}{(s-a)^n}$ $(n = 1, 2, 3, \ldots)$	$\dfrac{t^{n-1}e^{at}}{(n-1)!}$
10. $\dfrac{1}{(s-a)^x}$ $(x > 0)$	$\dfrac{t^{x-1}e^{at}}{\Gamma(x)}$
11. $\dfrac{1}{(s-a)(s-b)}$ $(a \neq b)$	$\dfrac{e^{at} - e^{bt}}{a-b}$
12. $\dfrac{s}{(s-a)(s-b)}$ $(a \neq b)$	$\dfrac{ae^{at} - be^{bt}}{a-b}$
13. $\dfrac{1}{s^2 + k^2}$	$\dfrac{1}{k}\sin kt$
14. $\dfrac{s}{s^2 + k^2}$	$\cos kt$
15. $\dfrac{1}{s^2 - k^2}$	$\dfrac{1}{k}\sinh kt$
16. $\dfrac{s}{s^2 - k^2}$	$\cosh kt$
17. $\dfrac{1}{(s-a)^2 + b^2}$	$\dfrac{1}{b}e^{at}\sin bt$
18. $\dfrac{s-a}{(s-a)^2 + b^2}$	$e^{at}\cos bt$

(*continued*)

TABLE 6.1 TABLE OF LAPLACE TRANSFORMS *(continued)*

$F(s) = \mathcal{L}\{f(t)\}$	$f(t) = \mathcal{L}^{-1}\{F(s)\}$
19. $\dfrac{1}{s(s^2 + k^2)}$	$\dfrac{1}{k^2}(1 - \cos kt)$
20. $\dfrac{1}{s^2(s^2 + k^2)}$	$\dfrac{1}{k^3}(kt - \sin kt)$
21. $\dfrac{1}{(s^2 + k^2)^2}$	$\dfrac{1}{2k^3}(\sin kt - kt \cos kt)$
22. $\dfrac{s}{(s^2 + k^2)^2}$	$\dfrac{t}{2k} \sin kt$
23. $\dfrac{s^2}{(s^2 + k^2)^2}$	$\dfrac{1}{2k}(\sin kt + kt \cos kt)$
24. $\dfrac{s}{(s^2 + a^2)(s^2 + b^2)} \quad (a^2 \neq b^2)$	$\dfrac{1}{b^2 - a^2}(\cos at - \cos bt)$
25. $\dfrac{1}{s^4 + 4k^4}$	$\dfrac{1}{4k^3}(\sin kt \cosh kt - \cos kt \sinh kt)$
26. $\dfrac{s}{s^4 + 4k^4}$	$\dfrac{1}{2k^2} \sin kt \sinh kt$
27. $\dfrac{1}{s^4 - k^4}$	$\dfrac{1}{2k^3}(\sinh kt - \sin kt)$
28. $\dfrac{s}{s^4 - k^4}$	$\dfrac{1}{2k^2}(\cosh kt - \cos kt)$

TABLE 6.2 OPERATIONAL PROPERTIES OF THE LAPLACE TRANSFORM

1. $\mathcal{L}\{C_1 f(t) + C_2 g(t)\} = C_1 F(s) + C_2 G(s)$
2. $\mathcal{L}\{e^{at} f(t)\} = F(s - a)$
3. $\mathcal{L}\{f(at)\} = (1/a)F(s/a), \quad a > 0$
4. $\mathcal{L}\{f'(t)\} = sF(s) - f(0)$
5. $\mathcal{L}\{f''(t)\} = s^2 F(s) - sf(0) - f'(0)$
6. $\mathcal{L}\{f^{(n)}(t)\} = s^n F(s) - s^{n-1} f(0) - s^{n-2} f'(0) - \cdots - f^{(n-1)}(0), \quad n = 1, 2, 3, \ldots$
7. $\mathcal{L}\{tf(t)\} = -F'(s)$
8. $\mathcal{L}\{t^n f(t)\} = (-1)^n F^{(n)}(s), \quad n = 1, 2, 3, \ldots$
9. $\mathcal{L}\left\{\displaystyle\int_0^t f(u) \, du\right\} = \dfrac{F(s)}{s}$
10. $\mathcal{L}\left\{\dfrac{f(t)}{t}\right\} = \displaystyle\int_s^\infty F(u) \, du$

6.3.1 Shift Property

Let us begin our discussion of operational properties by proving entry 2 in Table 6.2, which gives us the transform of the product $e^{at}f(t)$ from the transform of $f(t)$.

Theorem 6.3 Shifting. *If $\mathcal{L}\{f(t)\} = F(s)$, then*

$$\mathcal{L}\{e^{at}f(t)\} = F(s - a)$$

Proof. *By definition,*

$$\mathcal{L}\{e^{at}f(t)\} = \int_0^\infty e^{-st}[e^{at}f(t)]\,dt$$

$$= \int_0^\infty e^{-(s-a)t}f(t)\,dt$$

from which we deduce

$$\mathcal{L}\{e^{at}f(t)\} = F(s - a)$$

EXAMPLE 6

Evaluate $\mathcal{L}\{5e^{-2t}\cos 3t\}$.

Solution

From Example 5 and the linearity property, we know that

$$\mathcal{L}\{5\cos 3t\} = \frac{5s}{s^2 + 9}$$

and hence, through the shifting property (Theorem 6.3), it follows that

$$\mathcal{L}\{5e^{-2t}\cos 3t\} = \frac{5(s + 2)}{(s + 2)^2 + 9} = \frac{5s + 10}{s^2 + 4s + 13}$$

6.3.2 Transforms of Derivatives

The real merit of the Laplace transform is revealed by entries 4 to 6 in Table 6.2 which tell us how to transform derivatives of functions. Suppose that $f(t)$ is a continuous function with a piecewise continuous derivative $f'(t)$ on the interval $t \geq 0$. If we assume further that both $f(t)$ and $f'(t)$ are of exponential order, then integration by parts leads to

$$\mathcal{L}\{f'(t)\} = \int_0^\infty e^{-st}f'(t)\,dt$$

$$= [e^{-st}f(t)]\Big|_0^\infty + s\int_0^\infty e^{-st}f(t)\,dt$$

Because $f(t)$ is of exponential order, we have (for $s > c$) the condition $e^{-st}f(t) \to 0$ as $t \to \infty$, and consequently

$$\mathcal{L}\{f'(t)\} = sF(s) - f(0) \qquad (6)$$

where $F(s) = \mathcal{L}\{f(t)\}$.

> ***Remark.***   If $f(t)$ has a finite discontinuity at $t = 0$, we replace $f(0)$ in (6) with $f(0^+)$.

Along similar lines, suppose now that $f(t)$ and $f'(t)$ are continuous functions and that $f''(t)$ is piecewise continuous on $t \geq 0$. If all are of exponential order, then using (6) we obtain

$$\mathscr{L}\{f''(t)\} = s\mathscr{L}\{f'(t)\} - f'(0)$$
$$= s[sF(s) - f(0)] - f'(0)$$

or

$$\mathscr{L}\{f''(t)\} = s^2 F(s) - sf(0) - f'(0) \qquad (7)$$

where $F(s) = \mathscr{L}\{f(t)\}$. Generalizing (6) and (7), we have the following theorem.

> **Theorem 6.4 Differentiation.**   *If $f(t), f'(t), \ldots, f^{(n-1)}(t)$ are continuous functions and $f^{(n)}(t)$ is piecewise continuous on $t \geq 0$, and all are of exponential order, then*
>
> $$\mathscr{L}\{f^{(n)}(t)\} = s^n F(s) - s^{n-1}f(0) - s^{n-2}f'(0) - \cdots - f^{(n-1)}(0)$$
>
> *where $F(s) = \mathscr{L}\{f(t)\}$.*

EXAMPLE 7

Use Theorem 6.4 to evaluate $\mathscr{L}\{t^n\}$, $n = 0, 1, 2, \ldots$.

Solution

We could integrate by parts n times to find $\mathscr{L}\{t^n\}$, but we wish to illustrate another technique.

The function $f(t) = t^n$ and all its derivatives are continuous functions of exponential order. Also, we see that

$$f(0) = f'(0) = \cdots = f^{(n-1)}(0) = 0$$
$$f^{(n)}(t) = n!$$

The substitution of these results into the conclusion of Theorem 6.4 leads to

$$\mathscr{L}\{n!\} = s^n \mathscr{L}\{t^n\} - 0 - 0 - \cdots - 0$$

Therefore, we deduce that

$$\mathscr{L}\{t^n\} = \frac{n!\,\mathscr{L}\{1\}}{s^n} = \frac{n!}{s^{n+1}}, \qquad n = 0, 1, 2, \ldots$$

where we have used the result of Example 1. ∎

6.3.3 Derivatives of Transforms

Entries 7 and 8 in Table 6.2 enable us to readily determine the transform of products of the form $t^n f(t)$, $n = 1, 2, 3, \ldots$, from the transform of $f(t)$.

Theorem 6.5 *If $F(s) = \mathcal{L}\{f(t)\}$, then*

$$\mathcal{L}\{tf(t)\} = -F'(s)$$

whereas in general

$$\mathcal{L}\{t^n f(t)\} = (-1)^n F^{(n)}(s), \qquad n = 1, 2, 3, \ldots$$

Proof. *Let us start with the transform relation*

$$F(s) = \int_0^\infty e^{-st} f(t)\, dt$$

and formally differentiate with respect to s (under the integral sign) to obtain

$$F'(s) = \int_0^\infty (-t)e^{-st} f(t)\, dt = -\int_0^\infty e^{-st}[tf(t)]\, dt$$

Thus, we deduce that

$$\mathcal{L}\{tf(t)\} = -F'(s)$$

Repeated derivatives of the above integral give us

$$F''(s) = \int_0^\infty (-t)^2 e^{-st} f(t)\, dt$$

$$F'''(s) = \int_0^\infty (-t)^3 e^{-st} f(t)\, dt$$

while in general (for $n = 1, 2, 3, \ldots$)

$$F^{(n)}(s) = \int_0^\infty (-t)^n e^{-st} f(t)\, dt = (-1)^n \int_0^\infty e^{-st}[t^n f(t)]\, dt$$

From this last relation we finally deduce that

$$\mathcal{L}\{t^n f(t)\} = (-1)^n F^{(n)}(s), \qquad n = 1, 2, 3, \ldots$$

■
EXAMPLE 8

Evaluate $\mathcal{L}\{te^{-3t} \sin t\}$.

Solution

Based on the known transform (recall Example 3)

$$\mathcal{L}\{\sin t\} = \frac{1}{s^2 + 1}$$

and the operational property (Theorem 6.5)

$$\mathcal{L}\{tf(t)\} = -F'(s)$$

we first obtain the result

$$\mathcal{L}\{t \sin t\} = -\frac{d}{ds}\left(\frac{1}{s^2 + 1}\right) = \frac{2s}{(s^2 + 1)^2}$$

If we now apply the shifting property (Theorem 6.3), we find that

$$\mathcal{L}\{te^{-3t} \sin t\} = \frac{2(s + 3)}{[(s + 3)^2 + 1]^2} = \frac{2s + 6}{(s^2 + 6s + 10)^2}.$$

■

We found it necessary in Example 8 to use more than one operational property of the transform in conjunction with known transforms. This is common practice in evaluating the transforms of functions composed of products of other functions. If the shift property is one of the operational properties used, it is usually advantageous to apply it last.

Exercises 6.3

In Problems 1 to 12, evaluate the Laplace transform of the given function using Table 6.1 and any of the operational properties listed in Table 6.2.

1. $f(t) = 3te^{2t}$

2. $f(t) = t^3 e^{-2t}$

3. $f(t) = t^2 \sin kt$

4. $f(t) = 2e^{-t} \sin 3t$

5. $f(t) = \cosh kt \cos kt$

6. $f(t) = e^{-t}(t^2 - 2t + 7)$

7. $f(t) = 3e^{-4t}(\cos 4t - t \sin 4t)$

8. $f(t) = \dfrac{\sin t}{t}$

9. $f(t) = \dfrac{e^t - e^{-t}}{t}$

■10. $f(t) = \int_0^t (u^2 - u + e^{-u})\, du$

11. $f(t) = e^{-t} \cos^2 t$

12. $f(t) = 5te^{3t} \sin^2 t$

13. If $\mathscr{L}\{f(t)\} = F(s)$, show that
   (a) $\mathscr{L}\{f(at)\} = (1/a)F(s/a)$, $a > 0$.
   (b) Given that $\mathscr{L}\{\cos t\} = s/(s^2 + 1)$, use (a) to determine $\mathscr{L}\{\cos 4t\}$.

*14. Use entries 9 and 10 in Table 6.2 to determine the Laplace transform of the **sine integral**

$$Si(t) = \int_0^t \frac{\sin u}{u}\, du$$

15. Use Theorem 6.5 to deduce that
   (a) $\mathscr{L}\{t \cos kt\} = \dfrac{s^2 - k^2}{(s^2 + k^2)^2}$
   (b) From (a), deduce that
$$\int_0^\infty te^{-2t} \cos t\, dt = \frac{3}{25}$$

■16. Determine the Laplace transform of
$$f(t) = t^n e^{at}, \qquad n = 1, 2, 3, \ldots$$
   (a) Using Example 7 and Theorem 6.3.
   (b) Using Example 2 and Theorem 6.5.

17. By defining $g(t) = \int_0^t f(u)\, du$, use Theorem 6.5 to deduce that
$$\mathscr{L}\left\{\int_0^t f(u)\, du\right\} = \frac{F(s)}{s}$$

18. By formally integrating the transform relation
$$F(s) = \int_0^\infty e^{-st} f(t)\, dt$$
from s to ∞, deduce that
$$\mathscr{L}\left\{\frac{f(t)}{t}\right\} = \int_s^\infty F(u)\, du$$

*19. The **error function** (erf) and **complementary error function** (erfc) are defined, respectively, by
$$\text{erf}(t) = \frac{2}{\sqrt{\pi}} \int_0^t e^{-x^2}\, dx,$$
$$\text{erfc}(t) = \frac{2}{\sqrt{\pi}} \int_t^\infty e^{-x^2}\, dx$$

Show that
   (a) $\text{erf}(\infty) = 1$
      Hint. Use $\Gamma(1/2) = \sqrt{\pi}$ (see Problem 27 in Exercises 6.2).
   (b) $\text{erfc}(t) = 1 - \text{erf}(t)$

*20. Use Problem 19 to establish the following Laplace transform relations:

(a) $\mathcal{L}\{e^{-t^2/4}\} = \sqrt{\pi}e^{s^2}\,\mathrm{erf}(s)$

    Hint. Write $t^2/4 + st = (t/2 + s)^2 - s^2$

and make the change of variable $y = t/2 + s$.

(b) $\mathcal{L}\{\mathrm{erf}(t)\} = (1/s)e^{s^2/4}\,\mathrm{erfc}(s/2)$

    Hint. Change the order of integration.

6.4 Inverse Transforms

Generally, the use of Laplace transforms is effective only if we can also solve the inverse problem, that is, given $F(s)$, what is $f(t)$? In symbols we write

$$\mathcal{L}^{-1}\{F(s)\} = f(t) \tag{8}$$

called the **inverse Laplace transform** of $F(s)$. To evaluate inverse transforms, we rely primarily on three methods, namely,

  1. The use of *tables* (Table 6.1)
  2. The use of *operational properties* (Table 6.2)
  3. The method of *partial fractions*

To begin, let us recall (from Section 6.2) the transforms

$$\mathcal{L}\{1\} = \frac{1}{s}$$

$$\mathcal{L}\{e^{at}\} = \frac{1}{s - a}$$

and

$$\mathcal{L}\{\sin kt\} = \frac{k}{s^2 + k^2}$$

from which we immediately deduce the inverse transform relations

$$\mathcal{L}^{-1}\left\{\frac{1}{s}\right\} = 1$$

$$\mathcal{L}^{-1}\left\{\frac{1}{s - a}\right\} = e^{at}$$

and

$$\mathcal{L}^{-1}\left\{\frac{1}{s^2 + k^2}\right\} = \frac{1}{k}\sin kt$$

Thus, we may refer to tables (like Table 6.1) for either the transform function $F(s)$ or the inverse transform $f(t)$. If the entry we desire is not in the table, we must then resort to other methods.

When trying to find an inverse Laplace transform of a particular function, we might in general ask the question "Does the inverse Laplace transform exist?" The answer is—"not necessarily." That is, not all functions have an inverse Laplace transform. Ordinarily, the function $F(s)$ must satisfy certain continuity requirements and behave suitably as $s \to \infty$.

Theorem 6.6 *If $f(t)$ is piecewise continuous on $t \geq 0$ and of exponential order, and if $F(s) = \mathscr{L}\{f(t)\}$, then*

$$\lim_{s \to \infty} F(s) = 0$$

The proof of Theorem 6.6 follows that of Theorem 6.2 and is left to the exercises (see Problem 31 in Exercises 6.4). Although it doesn't give us conditions under which the inverse transform exists, Theorem 6.6 is significant in that it rules out certain functions as possible Laplace transforms. That is, if $F(s)$ is any function for which $\lim_{s \to \infty} F(s) \neq 0$, then it does not have any piecewise continuous function of exponential order as its inverse transform. This condition rules out functions such as polynomials in s, e^s, cos s, and so on, as possible Laplace transforms.

If the inverse Laplace transform exists, then we might also ask the question "Is it unique?" According to Theorem 6.2, a discontinuous function can have a Laplace transform. Hence, if $f(t)$ and $g(t)$ are two piecewise continuous functions that are identical everywhere except for a finite number of points, they will have the same Laplace transform, say $F(s)$. In this case, either $f(t)$ or $g(t)$ can be considered the inverse transform of $F(s)$. However, in all cases there can be only one *continuous* function that is the inverse Laplace transform.

6.4.1 Operational Properties

We previously used *operational properties* for calculating the transforms of certain functions. However, these properties are equally useful in constructing inverse transforms. For example, in terms of inverse Laplace transforms, the **linearity property** (Theorem 6.1) and **shifting property** (Theorem 6.3) become, respectively,

$$\mathscr{L}^{-1}\{C_1 F(s) + C_2 G(s)\} = C_1 f(t) + C_2 g(t) \tag{9}$$

and

$$\mathscr{L}^{-1}\{F(s - a)\} = e^{at} \mathscr{L}^{-1}\{F(s)\} \tag{10}$$

■———
EXAMPLE 9   Find $\mathscr{L}^{-1}\left\{\dfrac{3s + 7}{s^2 + 5}\right\}$.

Solution

Using the linearity property (9), we have

$$\mathscr{L}^{-1}\left\{\frac{3s + 7}{s^2 + 5}\right\} = 3\mathscr{L}^{-1}\left\{\frac{s}{s^2 + 5}\right\} + \frac{7}{\sqrt{5}}\,\mathscr{L}^{-1}\left\{\frac{\sqrt{5}}{s^2 + 5}\right\}$$

$$= 3\cos\sqrt{5}t + \frac{7}{\sqrt{5}}\sin\sqrt{5}t \qquad\qquad ■$$

EXAMPLE 10   Find $\mathscr{L}^{-1}\left\{\dfrac{s-5}{s^2+6s+13}\right\}$.

Solution

Completing the square in the denominator, we get

$$\frac{s-5}{s^2+6s+13} = \frac{s-5}{(s+3)^2+4}$$

$$= \frac{(s+3)-8}{(s+3)^2+4}$$

Thus, using (9) and (10), we obtain

$$\mathscr{L}^{-1}\left\{\frac{s-5}{s^2+6s+13}\right\} = \mathscr{L}^{-1}\left\{\frac{(s+3)-8}{(s+3)^2+4}\right\}$$

$$= e^{-3t}\mathscr{L}^{-1}\left\{\frac{s-8}{s^2+4}\right\}$$

$$= e^{-3t}\left[\mathscr{L}^{-1}\left\{\frac{s}{s^2+4}\right\} - 4\mathscr{L}^{-1}\left\{\frac{2}{s^2+4}\right\}\right]$$

or

$$\mathscr{L}^{-1}\left\{\frac{s-5}{s^2+6s+13}\right\} = e^{-3t}(\cos 2t - 4\sin 2t) \qquad \blacksquare$$

Another operational property that is sometimes useful in evaluating inverse transforms is entry 9 in Table 6.2. As an inverse transform, this property takes the form

$$\mathscr{L}^{-1}\left\{\frac{F(s)}{s}\right\} = \int_0^t f(u)\,du \qquad\qquad (11)$$

EXAMPLE 11   Find $\mathscr{L}^{-1}\left\{\dfrac{7}{s(s^2+9)}\right\}$.

Solution

By defining $F(s) = 1/(s^2+9)$, we first observe that

$$f(t) = \mathscr{L}^{-1}\left\{\frac{1}{s^2+9}\right\} = \frac{1}{3}\sin 3t$$

Hence, using (11) it follows that

$$\mathscr{L}^{-1}\left\{\frac{7}{s(s^2+9)}\right\} = \frac{7}{3}\int_0^t \sin 3u\,du$$

$$= \frac{7}{9}(1 - \cos 3t) \qquad \blacksquare$$

6.4.2 Partial Fractions

In many cases we need to find the inverse Laplace transform of a rational function, that is, a function having the form

$$F(s) = \frac{P(s)}{Q(s)}$$

where $P(s)$ and $Q(s)$ are polynomials in s. The inverse transform in such cases can often be found quite easily by representing $F(s)$ in terms of its **partial fractions.** If $P(s)$ and $Q(s)$ have no common factors and the degree of $P(s)$ is lower than that of $Q(s)$, there are *three* cases to consider which we briefly review below.

Case I. Distinct Linear Factors

If $Q(s)$ can be factored into a product of distinct linear factors

$$Q(s) = (s - a_1)(s - a_2) \cdots (s - a_n)$$

then the partial fraction expansion has the form

$$\frac{P(s)}{Q(s)} = \frac{A_1}{s - a_1} + \frac{A_2}{s - a_2} + \cdots + \frac{A_n}{s - a_n} \tag{12}$$

Case II. Repeated Linear Factors

If $Q(s)$ has a repeated factor $(s - a)^m$, among other factors, then the partial fraction expansion corresponding to this repeated factor is

$$\frac{P(s)}{Q(s)} = \frac{A_1}{s - a} + \frac{A_2}{(s - a)^2} + \cdots + \frac{A_m}{(s - a)^m} + \cdots \tag{13}$$

Case III. Quadratic Factors

If $Q(s)$ has a quadratic factor $(s^2 + as + b)$ that cannot be reduced to linear factors with real coefficients, then the partial fraction expansion corresponding to this factor is

$$\frac{P(s)}{Q(s)} = \frac{As + B}{s^2 + as + b} + \cdots \tag{14a}$$

If the quadratic factor is $(s^2 + as + b)^m$, then we assume

$$\frac{P(s)}{Q(s)} = \frac{A_1 s + B_1}{s^2 + as + b} + \frac{A_2 s + B_2}{(s^2 + as + b)^2} + \cdots$$
$$+ \frac{A_m s + B_m}{(s^2 + as + b)^m} + \cdots \tag{14b}$$

EXAMPLE 12   Find $\mathscr{L}^{-1}\left\{\dfrac{2s^2 + 5s - 1}{s^3 - s}\right\}$.

Solution

Since the factors of the denominator $s^3 - s = s(s - 1)(s + 1)$ are distinct and linear, the partial fraction expansion leads to

$$\frac{2s^2 + 5s - 1}{s^3 - s} = \frac{2s^2 + 5s - 1}{s(s - 1)(s + 1)}$$

$$= \frac{A}{s} + \frac{B}{s - 1} + \frac{C}{s + 1}$$

where A, B, and C are constants to be determined. Upon clearing fractions, this expression can be written as

$$2s^2 + 5s - 1 = A(s - 1)(s + 1) + Bs(s + 1) + Cs(s - 1)$$

which must be valid for all values of s. In particular, by setting $s = 0$, $s = 1$, and $s = -1$, respectively, we deduce that $A = 1$, $B = 3$, and $C = -2$. Hence,

$$\mathscr{L}^{-1}\left\{\frac{2s^2 + 5s - 1}{s^3 - s}\right\} = \mathscr{L}^{-1}\left\{\frac{1}{s}\right\} + 3\mathscr{L}^{-1}\left\{\frac{1}{s - 1}\right\} - 2\mathscr{L}^{-1}\left\{\frac{1}{s + 1}\right\}$$

$$= 1 + 3e^t - 2e^{-t} \qquad \blacksquare$$

EXAMPLE 13   Find $\mathscr{L}^{-1}\left\{\dfrac{1}{(s + 1)(s^2 + 4)}\right\}$.

Solution

This time the denominator has one distinct linear factor and one quadratic factor, so that the partial fraction expansion yields

$$\frac{1}{(s + 1)(s^2 + 4)} = \frac{A}{s + 1} + \frac{Bs + C}{s^2 + 4}$$

or, upon clearing fractions,

$$1 = A(s^2 + 4) + (Bs + C)(s + 1)$$

Setting $s = -1$, we find $A = 1/5$, and equating like coefficients of s^2 and s^0 yields the equations

$$s^2: \qquad 0 = A + B$$

$$s^0: \qquad 1 = 4A + C$$

From these relations we deduce that $B = -1/5$ and $C = 1/5$, and hence

$$\mathscr{L}^{-1}\left\{\frac{1}{(s+1)(s^2+1)}\right\} = \frac{1}{5}\left[\mathscr{L}^{-1}\left\{\frac{1}{s+1}\right\} - \mathscr{L}^{-1}\left\{\frac{s}{s^2+4}\right\} + \mathscr{L}^{-1}\left\{\frac{1}{s^2+4}\right\}\right]$$

$$= \frac{1}{5}\left(e^{-t} - \cos 2t + \frac{1}{2}\sin 2t\right) \qquad \blacksquare$$

Our last example contains repeated factors.

■ **EXAMPLE 14**   Find $\mathscr{L}^{-1}\left\{\dfrac{s+1}{s^2(s+2)^3}\right\}$.

Solution

Here we have repeated linear factors in the denominator so that

$$\frac{s+1}{s^2(s+2)^3} = \frac{A}{s} + \frac{B}{s^2} + \frac{C}{s+2} + \frac{D}{(s+2)^2} + \frac{E}{(s+2)^3}$$

Clearing fractions, we obtain

$$s + 1 = As(s+2)^3 + B(s+2)^3 + Cs^2(s+2)^2$$
$$+ Ds^2(s+2) + Es^2$$

By setting $s = 0$ and $s = -2$, we determine respectively that $B = 1/8$ and $E = -1/4$. Also, comparing like coefficients yields

$$s^4: \quad 0 = A + C$$
$$s^3: \quad 0 = 6A + B + 4C + D$$
$$s: \quad 1 = 8A + 12B$$

the simultaneous solution of which leads to $C = -A = 1/16$ and $D = 0$. Hence, the inverse transform we seek is

$$\mathscr{L}^{-1}\left\{\frac{s+1}{s^2(s+2)^3}\right\} = -\frac{1}{16}\mathscr{L}^{-1}\left\{\frac{1}{s}\right\} + \frac{1}{8}\mathscr{L}^{-1}\left\{\frac{1}{s^2}\right\} + \frac{1}{16}\mathscr{L}^{-1}\left\{\frac{1}{s+2}\right\}$$

$$- \frac{1}{4}\mathscr{L}^{-1}\left\{\frac{1}{(s+2)^3}\right\}$$

$$= \frac{1}{16}(-1 + 2t + e^{-2t} - 2t^2 e^{-2t}) \qquad \blacksquare$$

Exercises 6.4

In Problems 1 to 12, determine the inverse Laplace transform of each function using Tables 6.1 and 6.2 in Section 6.3.

1. $F(s) = \dfrac{1}{s^3}$

2. $F(s) = \dfrac{3}{5s^2 + 25}$

3. $F(s) = \dfrac{2}{(s - 3)^5}$

4. $F(s) = \dfrac{15}{s^2 + 4s + 13}$

5. $F(s) = \dfrac{1}{s^2 - 6s + 10}$

6. $F(s) = \dfrac{s}{s^2 - 6s + 13}$

7. $F(s) = \dfrac{13}{s^2 + 8s + 16}$

■8. $F(s) = \dfrac{s - 10}{s^2 + 6s + 13}$

9. $F(s) = \dfrac{7s - 3}{s^2 + 4s + 29}$

10. $F(s) = \dfrac{s^2}{(s + 2)^4}$

11. $F(s) = \dfrac{5s - 2}{3s^2 + 4s + 8}$

*12. $F(s) = \dfrac{3s + 1}{s(s + 1)^3}$

In Problems 13 to 24, use partial fractions or Equation (11) to evaluate the inverse Laplace transform.

13. $F(s) = \dfrac{1}{s^2 + s}$

14. $F(s) = \dfrac{1}{(s - 1)(s + 2)(s + 4)}$

15. $F(s) = \dfrac{s^2}{(s + 2)^3}$

■16. $F(s) = \dfrac{3s - 2}{s^3(s^2 + 4)}$

17. $F(s) = \dfrac{s + 1}{(s^2 - 4s)(s + 5)^2}$

18. $F(s) = \dfrac{1}{s(s^4 - 1)}$

19. $F(s) = \dfrac{s^2 + 1}{(s^2 - 1)(s^2 - 4)}$

20. $F(s) = \dfrac{4s^2 - 16}{s^3(s + 2)^3}$

21. $F(s) = \dfrac{s + 1}{s^3 + s^2 - 6s}$

*22. $F(s) = \dfrac{3s^2 - 6s + 7}{(s^2 - 2s + 5)^2}$

*23. $F(s) = \dfrac{s^2 - 3}{(s + 2)(s - 3)(s^2 + 2s + 5)}$

*24. $F(s) = \dfrac{s^3 + 16s - 24}{s^4 + 20s^2 + 64}$

25. Given that $f(t) = \mathcal{L}^{-1}\{F(s)\}$, show for constants a and b that

    (a) $\mathcal{L}^{-1}\{F(as)\} = (1/a)f(t/a), a > 0.$

    (b) $\mathcal{L}^{-1}\{F(as + b)\} = (1/a)e^{-bt/a}f(t/a), a > 0.$

■26. If $\mathcal{L}^{-1}\{s^{-1/2}e^{-1/s}\} = \cos(2\sqrt{t})/\sqrt{\pi t}$, use Problem 25 to determine $\mathcal{L}^{-1}\{s^{-1/2}e^{-a/s}\}, a > 0.$

27. Show that

$$\mathcal{L}^{-1}\left\{\frac{s}{(s + a)^2 + b^2}\right\} =$$

$$\frac{1}{b}e^{-at}(b \cos bt - a \sin bt)$$

*28. Using the operational property $f(t) = -\mathcal{L}^{-1}\{F'(s)\}/t$, show that

$$\mathcal{L}^{-1}\left\{\log\left(\frac{s + 1}{s - 1}\right)\right\} = \frac{2 \sinh t}{t}$$

*29. Let $f(t) = \int_0^\infty \dfrac{\sin tx}{x}\, dx$. By formally taking the Laplace transform of each side, show that

(a) $F(s) = \int_0^\infty \dfrac{1}{x^2 + s^2}\, dx$

(b) By integrating the integral in (a) and then finding the inverse transform of the resulting expression, deduce the value of $f(t)$.

*30. Use the technique of Problem 29 to show that

(a) $\int_0^\infty e^{-tx^2}\, dx = \dfrac{1}{2}\sqrt{\dfrac{\pi}{t}},\; t > 0$

(b) $\int_0^\infty \dfrac{\cos tx}{1 + x^2}\, dx = \dfrac{\pi}{2} e^{-t},\; t > 0$

*31. Prove Theorem 6.6.

6.5 Initial Value Problems

The Laplace transform is a powerful tool for solving *linear DEs with constant coefficients*—in particular, IVPs. The usefulness of the transform method for this class of problems rests primarily on the fact that the transform of the DE, together with prescribed initial conditions, reduces the problem to an *algebraic equation* in the transformed function. This algebraic equation is easily solved, and the inverse Laplace transform of its solution is the solution of the IVP that we seek.

In this section the unknown function will generally be denoted by $y(t)$ and its transform by $Y(s)$.

■——— **EXAMPLE 15**

Solve the IVP

$$y' + y = 3e^t, \qquad y(0) = 7$$

Solution

We first apply the Laplace transform to both sides of the DE, which formally yields

$$\mathcal{L}\{y'(t)\} + \mathcal{L}\{y(t)\} = \mathcal{L}\{3e^t\}$$

By using Theorem 6.4 (entry 4 in Table 6.2) and entry 7 in Table 6.1, we have, respectively,

$$\mathcal{L}\{y'(t)\} = sY(s) - y(0)$$
$$= sY(s) - 7$$

and

$$\mathcal{L}\{3e^t\} = \dfrac{3}{s - 1}$$

where $Y(s) = \mathcal{L}\{y(t)\}$. Thus, the given IVP is reduced to the algebraic equation

$$[sY(s) - 7] + Y(s) = \dfrac{3}{s - 1}$$

Solving for $Y(s)$, we obtain

$$Y(s) = \dfrac{7}{s + 1} + \dfrac{3}{(s - 1)(s + 1)}$$

$$= \dfrac{11/2}{s + 1} + \dfrac{3/2}{s - 1}$$

the last step of which is obtained through partial fractions. By taking the inverse Laplace transform of this last expression, we formally obtain the solution

$$y(t) = \mathcal{L}^{-1}\{Y(s)\}$$

$$= \frac{11}{2}\mathcal{L}^{-1}\left\{\frac{1}{s+1}\right\} + \frac{3}{2}\mathcal{L}^{-1}\left\{\frac{1}{s-1}\right\}$$

or, using entry 7 in Table 6.1,

$$y(t) = \tfrac{1}{2}(11e^{-t} + 3e^{t})$$

The above example, although rather simple, illustrates the basic features of the transform method. In particular, notice that the initial values are immediately incorporated into the transformed problem. Because of this, the transform method offers the advantage that the particular solution of the IVP is found directly without first producing a "general solution." Once the transform method is perfected, it is generally a very efficient technique for solving IVPs. Let us illustrate with some additional examples.

EXAMPLE 16

Solve

$$y'' + 2y' + 5y = 0, \qquad y(0) = 2, \qquad y'(0) = -4$$

Solution

Using properties of the Laplace transform of derivatives (Theorem 6.4), the transform of the DE leads to

$$\mathcal{L}\{y''(t)\} + 2\mathcal{L}\{y'(t)\} + 5\mathcal{L}\{y(t)\} = 0$$

or

$$[s^2 Y(s) - 2s + 4] + 2[sY(s) - 2] + 5Y(s) = 0$$

where the initial conditions have been incorporated. Simplifying this expression, we get

$$(s^2 + 2s + 5)Y(s) = 2s$$

which has the solution

$$Y(s) = \frac{2s}{s^2 + 2s + 5} = \frac{2(s+1) - 2}{(s+1)^2 + 4}$$

Thus, by applying the shift property (Theorem 6.3), we have

$$y(t) = \mathcal{L}^{-1}\left\{\frac{2(s+1) - 2}{(s+1)^2 + 4}\right\}$$

$$= e^{-t}\mathcal{L}^{-1}\left\{\frac{2s - 2}{s^2 + 4}\right\}$$

$$= e^{-t}\left[2\mathcal{L}^{-1}\left\{\frac{s}{s^2 + 4}\right\} - \mathcal{L}^{-1}\left\{\frac{2}{s^2 + 4}\right\}\right]$$

or

$$y(t) = e^{-t}(2\cos 2t - \sin 2t)$$

EXAMPLE 17

Solve

$$y'' - 6y' + 9y = t^2 e^{3t}, \qquad y(0) = 2, \qquad y'(0) = 6$$

Solution

Taking the transform of the DE yields

$$[s^2 Y(s) - 2s - 6] - 6[sY(s) - 2] + 9Y(s) = \frac{2}{(s-3)^3}$$

or

$$(s^2 - 6s + 9)Y(s) = 2(s - 3) + \frac{2}{(s-3)^3}$$

Since $(s^2 - 6s + 9) = (s - 3)^2$, we find

$$Y(s) = \frac{2}{s-3} + \frac{2}{(s-3)^5}$$

and upon taking the inverse transform, we get

$$y(t) = 2\mathcal{L}^{-1}\left\{\frac{1}{s-3}\right\} + \frac{2}{4!} \mathcal{L}^{-1}\left\{\frac{4!}{(s-3)^5}\right\}$$

$$= 2e^{3t} + \frac{1}{12} e^{3t}\mathcal{L}^{-1}\left\{\frac{4!}{s^5}\right\}$$

or

$$y(t) = 2e^{3t} + \frac{1}{12} t^4 e^{3t}$$

EXAMPLE 18

Solve

$$y''' + y'' - y' - y = 9e^{2t}, \qquad y(0) = 2, \qquad y'(0) = 4, \qquad y''(0) = 3$$

Solution

The term-by-term transform of the DE leads to

$$[s^3 Y(s) - 2s^2 - 4s - 3] + [s^2 Y(s) - 2s - 4] - [sY(s) - 2] - Y(s) = \frac{9}{s-2}$$

or, upon simplifying,

$$(s^3 + s^2 - s - 1)Y(s) = 2s^2 + 6s + 5 + \frac{9}{s-2}$$

Factoring, we have $s^3 + s^2 - s - 1 = (s - 1)(s + 1)^2$, and thus

$$Y(s) = \frac{2s^2 + 6s + 5}{(s-1)(s+1)^2} + \frac{9}{(s-2)(s-1)(s+1)^2}$$

$$= \frac{(2s^2 + 6s + 5)(s-2) + 9}{(s-2)(s-1)(s+1)^2}$$

Using partial fractions, we write

$$Y(s) = \frac{A}{s-2} + \frac{B}{s-1} + \frac{C}{(s+1)^2} + \frac{D}{s+1}$$

from which it is determined that $A = B = C = 1$ and $D = 0$. Hence, the inversion of this last expression yields the solution

$$y(t) = e^{2t} + e^t + te^{-t}$$ ■

6.5.1 Applications

Applications such as those discussed in Chapter 5 involving spring-mass systems and electric circuits readily lend themselves to a Laplace transform analysis. Let us illustrate with some examples. (Readers may find it useful to first briefly review Sections 5.2 to 5.4.)

EXAMPLE 19

Determine the response of a spring-mass system that is initially at rest and then at time $t = 0$ is subjected to the sinusoidal forcing function $P \cos \omega t$ (see Figure 6.3). Neglect damping effects.

Solution

The problem is characterized by

$$y'' + \omega_0^2 y = \frac{P}{m} \cos \omega t, \qquad y(0) = 0, \qquad y'(0) = 0$$

where $\omega_0 = \sqrt{k/m}$. The transform of this DE leads to

$$(s^2 + \omega_0^2)Y(s) = \frac{Ps/m}{s^2 + \omega^2}$$

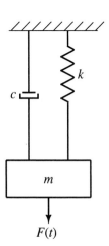

$F(t)$ **Figure 6.3** Spring-mass system.

with solution

$$Y(s) = \frac{Ps/m}{(s^2 + \omega_0^2)(s^2 + \omega^2)}$$

At this point we need to distinguish between the cases $\omega \neq \omega_0$ and $\omega = \omega_0$.

No Resonance.   If $\omega \neq \omega_0$, a partial fraction expansion yields

$$Y(s) = \frac{P}{m}\left(\frac{As + B}{s^2 + \omega^2} + \frac{Cs + D}{s^2 + \omega_0^2}\right)$$

from which we deduce that $A = -C = 1/(\omega_0^2 - \omega^2)$ and $B = D = 0$. Hence, we have

$$y(t) = \mathscr{L}^{-1}\{Y(s)\}$$

$$= \frac{P}{m(\omega_0^2 - \omega^2)}(\cos \omega t - \cos \omega_0 t)$$

which represents the superposition of two harmonic motions, corresponding to the natural frequency of the system ω_0 and the frequency ω of the forcing function.

Resonance.   When $\omega = \omega_0$, the solution of the transformed problem is

$$Y(s) = \frac{P}{m}\frac{s}{(s^2 + \omega_0^2)^2}$$

To obtain the inverse transform of this function, we first recall the similar result in Example 8 which we can express as

$$\mathscr{L}^{-1}\left\{\frac{s}{(s^2 + 1)^2}\right\} = \frac{1}{2}t \sin t$$

Now by using entry 3 in Table 6.2 written in the form

$$\mathscr{L}^{-1}\left\{\frac{1}{a}F(s/a)\right\} = f(at)$$

we see that

$$\mathscr{L}^{-1}\left\{\frac{s/a^2}{[(s/a)^2 + 1]^2}\right\} = \frac{1}{2}at \sin at$$

or, upon simplifying,

$$\mathscr{L}^{-1}\left\{\frac{s}{(s^2 + a^2)^2}\right\} = \frac{1}{2a}t \sin at$$

Finally, setting $a = \omega_0$, we are led to the result

$$y(t) = \mathscr{L}^{-1}\{Y(s)\}$$

$$= \frac{P}{2m\omega_0}t \sin \omega_0 t$$

This time the solution corresponds to *resonance* in the system since y becomes unbounded as $t \to \infty$ (recall the discussion on resonance in Section 5.3.2). ■

EXAMPLE 20

A mass of 1 kg is attached to a spring with spring constant $k = 10$ N/m. The mass initially starts moving from rest in its equilibrium position under the influence of an external force given by $1 + 2e^{-3t} \sin t$ N. If the damping constant is 6 times the instantaneous velocity, determine the position of the mass at all later times (see Figure 6.3). What is the steady-state solution?

Solution

For the given values of the input parameters, the problem is characterized by

$$y'' + 6y' + 10y = 1 + 2e^{-3t} \sin t, \qquad y(0) = 0, \qquad y'(0) = 0$$

the transform of which leads to

$$(s^2 + 6s + 10)Y(s) = \frac{1}{s} + \frac{2}{(s + 3)^2 + 1}$$

The solution of the transformed problem is readily shown to be

$$Y(s) = \frac{1}{s[(s + 3)^2 + 1]} + \frac{2}{[(s + 3)^2 + 1]^2}$$

By recognizing the inverse transform relation

$$\mathscr{L}^{-1}\left\{\frac{1}{(s + 3)^2 + 1}\right\} = e^{-3t} \sin t$$

and then using entry 9 in Table 6.2, we have the inverse transform of the first term on the right of $Y(s)$, namely,

$$\mathscr{L}^{-1}\left\{\frac{1}{s[(s + 3)^2 + 1]}\right\} = \int_0^t e^{-3u} \sin u \, du$$

$$= \frac{1}{10}\left[1 - e^{-3t}(3 \sin t + \cos t)\right]$$

To evaluate the remaining inverse transform, we first use the shift property to obtain

$$\mathscr{L}^{-1}\left\{\frac{2}{[(s + 3)^2 + 1]^2}\right\} = 2e^{-3t}\mathscr{L}^{-1}\left\{\frac{1}{(s^2 + 1)^2}\right\}$$

Next, we recall from Example 19 that

$$\mathscr{L}^{-1}\left\{\frac{s}{(s^2 + 1)^2}\right\} = \frac{1}{2}t \sin t$$

and by using entry 9 in Table 6.2 once again, we have

$$\mathscr{L}^{-1}\left\{\frac{1}{(s^2 + 1)^2}\right\} = \frac{1}{2}\int_0^t u \sin u \, du$$

$$= \frac{1}{2}(\sin t - t \cos t)$$

leading to

$$\mathscr{L}^{-1}\left\{\frac{2}{[(s+3)^2+1]^2}\right\} = 2e^{-3t}\mathscr{L}^{-1}\left\{\frac{1}{(s^2+1)^2}\right\}$$

$$= e^{-3t}(\sin t - t \cos t)$$

Upon combining the above results, we get the solution

$$y(t) = \mathscr{L}^{-1}\left\{\frac{1}{s[(s+3)^2+1]}\right\} + \mathscr{L}^{-1}\left\{\frac{2}{[(s+3)^2+1]^2}\right\}$$

$$= \frac{1}{10}[1 - e^{-3t}(3 \sin t + \cos t)] + e^{-3t}(\sin t - t \cos t)$$

or, upon simplifying,

$$y(t) = \tfrac{1}{10} + e^{-3t}[\tfrac{7}{10}\sin t - (t + \tfrac{1}{10})\cos t]$$

To find the steady-state solution, we let $t \to \infty$, which yields the constant value

$$y(t) = \tfrac{1}{10} \qquad \text{(steady-state solution)}$$

We can interpret this result as saying that the mass will eventually come to rest in a position 1/10 m below the equilibrium position of the system. ∎

The power and efficiency of the Laplace transform method is clearly illustrated by this last example. The reader should solve this problem by the conventional methods introduced in Chapter 4 to fully appreciate the benefit of the transform technique.

As our last example, let us consider a problem involving an electric circuit (see Figure 6.4).

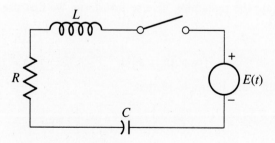

Figure 6.4 *RLC* circuit.

■ **EXAMPLE 21** The *RLC* circuit in Figure 6.4 has components $L = 0.1$ H, $R = 2\ \Omega$, $C = 1/260$ F, and $E(t) = 100 \sin 60t$ V. If $i(0) = q(0) = 0$, what is the charge on the capacitor at all later times? What is the steady-state charge?

Solution

As given, the problem can be formulated by

$$\tfrac{1}{10}q'' + 2q' + 260q = 100 \sin 60t, \qquad q(0) = 0, \qquad i(0) = q'(0) = 0$$

But, multiplying the DE by 10, we get the more convenient form

$$q'' + 20q' + 2600q = 1000 \sin 60t$$

the Laplace transform of which leads to

$$(s^2 + 20s + 2600)Q(s) = \frac{60{,}000}{s^2 + 3600}$$

Solving this transformed equation for $Q(s)$, we obtain

$$Q(s) = \frac{60{,}000}{(s^2 + 3600)[(s + 10)^2 + 2500]}$$

$$= \frac{As + B}{s^2 + 3600} + \frac{C(s + 10) + D}{(s + 10)^2 + 2500}$$

where we have anticipated use of the shift theorem in our partial fraction expansion. Observe here that, even without determining the constants A, B, C, and D, we can immediately write down the solution in the form

$$q(t) = A \cos 60t + \frac{B}{60} \sin 60t + e^{-10t}\left(C \cos 50t + \frac{D}{50} \sin 50t \right)$$

By the method of partial fractions, we find that

$$A = -\tfrac{30}{61}, \qquad B = -\tfrac{1500}{61}, \qquad C = \tfrac{30}{61}, \quad \text{and} \quad D = \tfrac{1800}{61}$$

from which we finally deduce

$$q(t) = -\tfrac{5}{61}(6 \cos 60t + 5 \sin 60t) + \tfrac{6}{61}e^{-10t}(5 \cos 50t + 6 \sin 50t)$$

Clearly, the steady-state solution is $(t \to \infty)$

$$q(t) = -\tfrac{5}{61}(6 \cos 60t + 5 \sin 60t) \qquad \blacksquare$$

Exercises 6.5

In Problems 1 to 20, use the Laplace transform method to solve the given IVP.

1. $y' = e^{2t}$, $y(0) = -1$

2. $y' - y = -e^{-t}$, $y(0) = 1$

3. $y' + y = e^t$, $y(0) = 0$

4. $y'' + y = 0$, $y(0) = 1$, $y'(0) = 0$

5. $y'' + y = 0$, $y(0) = 0$, $y'(0) = 1$

6. $y'' + y = 1$, $y(0) = 0$, $y'(0) = 0$

7. $y'' + y = 2e^t$, $y(0) = 0$, $y'(0) = 0$

8. $y'' + y' - 2y = -4$, $y(0) = 2$, $y'(0) = 3$

9. $y'' + 2y' + 2y = \sin 2t - 2 \cos 2t$, $y(0) = 0$, $y'(0) = 0$

■10. $y'' + 2y' + y = 3te^{-t}$, $y(0) = 4$, $y'(0) = 2$

11. $y'' + 4y' + 6y = 1 + e^{-t}$, $y(0) = 1$, $y'(0) = -4$

12. $y'' + 16y = \cos 4t$, $y(0) = 0$, $y'(0) = 1$

13. $y'' - 4y' + 4y = t$, $y(0) = 1$, $y'(0) = 0$

14. $y'' - y' = e^t \cos t$, $y(0) = 0$, $y'(0) = 0$

15. $y''' + 6y'' + 11y' + 6y = 0$, $y(0) = 2$, $y'(0) = 1$, $y''(0) = -1$

16. $y''' - y'' + 4y' - 4y = t$, $y(0) = 0$, $y'(0) = 0$, $y''(0) = 1$

17. $2y''' + 3y'' - 3y' - 2y = e^{-t}$, $y(0) = 0$, $y'(0) = 0$, $y''(0) = 1$

■18. $y^{(4)} - y = 0$, $y(0) = 1$, $y'(0) = 0$, $y''(0) = -1$, $y'''(0) = 0$

19. $y^{(4)} - y = t$, $y(0) = 0$, $y'(0) = 0$, $y''(0) = 0$, $y'''(0) = 0$

*20. $y^{(4)} - 4y''' + 6y'' - 4y' + y = 0$, $y(0) = 0$, $y'(0) = 1$, $y''(0) = 0$, $y'''(0) = 1$

In Problems 21 to 31, use the Laplace transform to solve the specified problem.

21. Problem 23, Exercises 5.2

22. Problem 24, Exercises 5.2

23. Problem 25, Exercises 5.2

24. Problem 30, Exercises 5.2

25. Problem 31, Exercises 5.2

26. Example 6, Section 5.3

27. Problem 7, Exercises 5.3

■28. Problem 16, Exercises 5.3

29. Problem 17, Exercises 5.3

30. Problem 8, Exercises 5.4

31. Problem 7, Exercises 5.4

*32. Show that the Laplace transform of the differential system

$$ty'' + 2(t - 1)y' - 2y = 0, \qquad y(0) = 0$$

leads to the first-order DE in the transform domain

$$(s^2 + 2s)Y'(s) + 4(s + 1)Y(s) = 0$$

Solve this DE for $Y(s)$ and invert it to find the solution $y(t)$. (This problem is one of the few variable-coefficient DEs for which the Laplace transform method proves fruitful.)

Hint. Use entry 7 in Table 6.2.

6.6 Convolution Theorem

Inverse transforms of the form $\mathscr{L}^{-1}\{F(s)G(s)\}$, where $F(s) = \mathscr{L}\{f(t)\}$ and $G(s) = \mathscr{L}\{g(t)\}$, arise quite frequently in practice. Unfortunately, it happens that

$$\mathscr{L}^{-1}\{F(s)G(s)\} \neq f(t)g(t)$$

Entry 9 in Table 6.2 (in Section 6.3), which we can also write as [recall Equation (11)]

$$\mathscr{L}^{-1}\left\{\frac{F(s)}{s}\right\} = \int_0^t f(u)\, du$$

gives us a means of finding the inverse transform of a simple product $F(s)G(s)$, where one function is $G(s) = 1/s$. Notice that, although $\mathscr{L}^{-1}\{F(s)/s\}$ is not exactly a product of inverse transforms, it is the integral of a product $f(t)g(t)$ where $g(t) = 1$. Similarly, we find in general that the inverse transform $\mathscr{L}^{-1}\{F(s)G(s)\}$ leads to an integral involving some type of product between $f(t)$ and $g(t)$.

To develop the proper relationship we seek, let us begin by writing the product of the transforms of $f(t)$ and $g(t)$ as the iterated integral

$$F(s)G(s) = \int_0^\infty e^{-sx}f(x)\,dx \cdot \int_0^\infty e^{-su}g(u)\,du$$

$$= \int_0^\infty \int_0^\infty e^{-s(x+u)}f(x)g(u)\,dx\,du$$

The change of variable $x = t - u$ then leads to

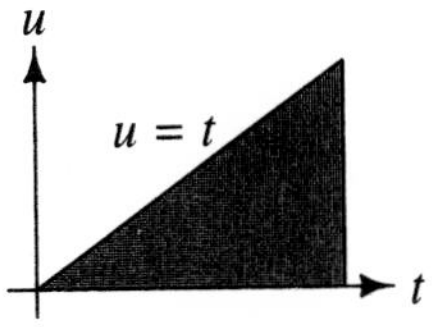

$$F(s)G(s) = \int_0^\infty \int_u^\infty e^{-st}f(t-u)g(u)\,dt\,du$$

which we can interpret as an iterated integral over the region $u \le t < \infty, 0 \le u < \infty$, as shown in Figure 6.5. If we change the order of integration, we find that the region of integration is characterized by $0 \le u \le t, 0 \le t < \infty$, and thus

$$F(s)G(s) = \int_0^\infty \int_0^t e^{-st}f(t-u)g(u)\,du\,dt$$

$$= \int_0^\infty e^{-st}\int_0^t f(t-u)g(u)\,du\,dt$$

By denoting the innermost integral by

$$(f * g)(t) = \int_0^t f(t-u)g(u)\,du \tag{15}$$

known as the **convolution integral,** we can write

$$F(s)G(s) = \mathcal{L}\{(f * g)(t)\} \tag{16}$$

Now by formally taking the inverse Laplace transform, we deduce the following important theorem, known as the *convolution theorem.*

Theorem 6.7 Convolution Theorem.   *If $f(t)$ and $g(t)$ are piecewise continuous functions on $t \ge 0$, are both of exponential order, and have Laplace transforms $F(s)$ and $G(s)$, respectively, then*

$$\mathcal{L}^{-1}\{F(s)G(s)\} = (f * g)(t) = \int_0^t f(t-u)g(u)\,du$$

■ **EXAMPLE 22**   Find $\mathcal{L}^{-1}\left\{\dfrac{1}{s^2(s^2 + k^2)}\right\}$

Solution

Let us select $F(s) = 1/s^2$ and $G(s) = 1/(s^2 + k^2)$, whose inverse transforms are, respectively,

$$\mathcal{L}^{-1}\left\{\frac{1}{s^2}\right\} = f(t) = t$$

and

$$\mathcal{L}^{-1}\left\{\frac{1}{s^2 + k^2}\right\} = g(t) = \frac{1}{k}\sin kt$$

Figure 6.5

Hence, using the convolution theorem, we write

$$\mathscr{L}^{-1}\left\{\frac{1}{s^2(s^2 + k^2)}\right\} = (f * g)(t) = \int_0^t (t - u)\frac{1}{k}\sin ku\, du$$

or

$$\mathscr{L}^{-1}\left\{\frac{1}{s^2(s^2 + k^2)}\right\} = \frac{t}{k}\int_0^t \sin ku\, du - \frac{1}{k}\int_0^t u\sin ku\, du$$

which, upon evaluation and simplification, leads to

$$\mathscr{L}^{-1}\left\{\frac{1}{s^2(s^2 + k^2)}\right\} = \frac{1}{k^3}(kt - \sin kt)$$

On the other hand, had we defined $F(s) = 1/(s^2 + k^2)$ and $G(s) = 1/s^2$, then the convolution theorem would lead to

$$(g * f)(t) = \frac{1}{k}\int_0^t u\sin[k(t - u)]\, du$$

$$= \frac{1}{k}\sin kt\int_0^t u\cos ku\, du - \frac{1}{k}\cos kt\int_0^t u\sin ku\, du$$

which simplifies to the same result, namely,

$$(g * f)(t) = \frac{1}{k^3}(kt - \sin kt) \qquad \blacksquare$$

The result $(f * g)(t) = (g * f)(t)$ in Example 22 is one of several useful properties of the convolution integral. To prove that this is a general property, let us make the change of variable $v = t - u$ in (15), which yields

$$(f * g)(t) = -\int_t^0 f(v)g(t - v)\, dv$$

$$= \int_0^t g(t - v)f(v)\, dv$$

Hence, we have derived the **commutative law**

$$(f * g)(t) = (g * f)(t) \qquad (17)$$

which is sometimes expressed as simply $f * g = g * f$.

Other properties of the convolution integral that follow from the definition are

$f * (Cg) = (Cf) * g = C(f * g)$	C constant	(18)
$f * (g + h) = f * g + f * h$	**distributive law**	(19)
$f * (g * h) = (f * g) * h$	**associative law**	(20)

the proofs of which are left to the exercises.

Exercises 6.6

In Problems 1 to 6, find the convolution $(f * g)(t)$ of the given functions and verify that $\mathcal{L}\{(f * g)(t)\} = F(s)G(s)$.

1. $f(t) = e^{2t}$, $g(t) = e^t$

2. $f(t) = t$, $g(t) = \sin 2t$

3. $f(t) = t^2$, $g(t) = e^{-2t}$

4. $f(t) = e^{-t}$, $g(t) = \cos t$

5. $f(t) = \cos t$, $g(t) = \sin t$

6. $f(t) = g(t) = \sin t$

In Problems 7 to 12, find the inverse Laplace transform of each function by using the convolution theorem.

7. $\dfrac{1}{s^2(s + 1)}$

8. $\dfrac{1}{(s^2 + 1)^2}$

9. $\dfrac{1}{s^4(s^2 + 1)}$

■10. $\dfrac{s}{(s + 1)(s^2 + 4)}$

11. $\dfrac{1}{(s + 1)^2(s^2 + 4)}$

12. $\dfrac{1}{(s^2 + a^2)(s^2 + b^2)}$, $a \neq b$

*13. Use the convolution theorem to show that

$$\mathcal{L}^{-1}\left\{\dfrac{1}{\sqrt{s}(s - 1)}\right\} = e^t \operatorname{erf}(\sqrt{t})$$

where $\operatorname{erf}(x)$ is the **error function** defined in Problem 19 in Exercises 6.3.

14. Show that the Laplace convolution integral satisfies

  (a) $f * (Cg) = (Cf) * g = C(f * g)$,

    C constant

  (b) $f * (g + h) = (f * g) + (f * h)$

*15. Show that $f * (g * h) = (f * g) * h$

16. Given $Y(s) = (s + 1)/(s^2 + 4)$, explain why we cannot set $F(s) = s + 1$ and $G(s) = 1/(s^2 + 4)$ in the convolution theorem to evaluate $\mathcal{L}^{-1}\{Y(s)\}$; that is, explain why

$$\mathcal{L}^{-1}\{Y(s)\} \neq \int_0^t f(t - u)g(u)\, du$$

where $f(t) = \mathcal{L}^{-1}\{F(s)\}$ and $g(t) = \mathcal{L}^{-1}\{G(s)\}$.

17. Given that $\mathcal{L}\{f(t)\} = 1/\sqrt{s^2 + a^2}$, evaluate

$$I = \int_0^t f(t - u)f(u)\, du$$

■18. Show that

  (a) $1 * 1 * 1 = t^2/2$

  (b) $t * t * t = t^5/5!$

  (c) $t^{m-1} * t^n = (m - 1)!n!t^{m+n}/(m + n)!$,

    $m, n = 1, 2, 3, \ldots$

19. Solve the **integral equation** for $y(t)$, where

  (a) $y(t) = 4t - 3 \displaystyle\int_0^t y(u) \sin(t - u)\, du$

  (b) $\displaystyle\int_0^t \dfrac{y(u)}{\sqrt{t - u}}\, du = \sqrt{t}$

20. Solve the **integro-differential equation** for $y(t)$, where

$$y'(t) = \int_0^t y(u) \cos(t - u)\, du, \qquad y(0) = 1$$

*21. Starting with $f(t) = \displaystyle\int_0^t u^{x-1}(t - u)^{y-1}\, du$, $x, y > 0$,

  (a) Use the convolution theorem to show that

$$F(s) = \dfrac{\Gamma(x)\Gamma(y)}{s^{x+y}}$$

    where $\Gamma(x)$ is the **gamma function** (recall Problem 25 in Exercises 6.2).

  (b) From the result in (a), establish the formula

$$\int_0^1 u^{x-1}(1 - u)^{y-1}\, du = \dfrac{\Gamma(x)\Gamma(y)}{\Gamma(x + y)},$$

$$x, y > 0$$

22. Using the convolution theorem, show that the solution of the spring-mass system

$$my'' + ky = f(t), \qquad y(0) = 0, \qquad y'(0) = 0$$

can be expressed in the form $(\omega_0 = \sqrt{k/m})$

$$y(t) = \dfrac{1}{m\omega_0} \int_0^t \sin[\omega_0(t - u)]f(u)\, du$$

23. Using the result of Problem 22, determine the response of the spring-mass system when the forcing function is given by

   (a) $f(t) = P$ (constant)
   (b) $f(t) = P \cos \omega t, \; \omega \neq \omega_0$
   (c) $f(t) = P \cos \omega_0 t$

*24. Determine the current $i(t)$ in a single-loop RLC circuit when $L = 0.1$ H, $R = 20 \, \Omega$, $C = 10^{-3}$ F,

$i(0) = 0$, and the impressed voltage is

$$E(t) = \begin{cases} 120t, & 0 \leq t < 1 \\ 0, & t > 1 \end{cases}$$

The system is characterized by

$$0.1 \frac{di}{dt} + 20i + 10^3 \int_0^t i(u)\, du = E(t),$$

$$i(0) = 0$$

6.7 Laplace Transforms of Other Functions

Other functions that are important in a variety of engineering applications are *periodic* and *discontinuous functions*. However, we limit our treatment of periodic functions to simply calculating the Laplace transform (the inverse transform problem is more complex). We also limit our treatment of discontinuous functions in this section to the *Heaviside unit function* (or *unit step function*) and discuss such notions as *impulse functions* in Section 6.8.

6.7.1 Periodic Functions

A function $f(t)$ is called **periodic** if there exists a constant $T > 0$ for which $f(t + T) = f(t)$ for all $t \geq 0$. The smallest value of T for which the property holds is called the **fundamental period** or, simply, the **period.** Familiar examples of periodic functions are $\cos t$ and $\sin t$, both of which have period 2π. Of course, we have already found the Laplace transform of each of these, but there are many other periodic functions whose definition and transform is not so easily given. Evaluation of the Laplace transform of these more general periodic functions usually involves the following theorem.

Theorem 6.8 *Let $f(t)$ be piecewise continuous on $t \geq 0$ and of exponential order. If $f(t + T) = f(t)$ for all $t \geq 0$, then*

$$\mathcal{L}\{f(t)\} = \frac{1}{1 - e^{-sT}} \int_0^T e^{-st} f(t)\, dt$$

Proof. *Let us write the Laplace transform as*

$$\mathcal{L}\{f(t)\} = \int_0^T e^{-st} f(t)\, dt + \int_T^\infty e^{-st} f(t)\, dt$$

By making the change of variable $t = u + T$ in the last integral, we obtain

$$\mathcal{L}\{f(t)\} = \int_0^T e^{-st} f(t)\, dt + \int_0^\infty e^{-s(u+T)} f(u + T)\, du$$

$$= \int_0^T e^{-st} f(t)\, dt + e^{-sT} \int_0^\infty e^{-su} f(u)\, du$$

where we are using the periodicity property of $f(t)$. The last integral is recognized as the Laplace transform of $f(t)$ again, so that

$$\mathcal{L}\{f(t)\} = \int_0^T e^{-st} f(t)\, dt + e^{-sT} \mathcal{L}\{f(t)\}$$

Finally, solving the above equation for $\mathcal{L}\{f(t)\}$ yields

$$\mathcal{L}\{f(t)\} = \frac{1}{1 - e^{-sT}} \int_0^T e^{-st} f(t)\, dt$$

EXAMPLE 23

Find the Laplace transform of the half-wave rectified sinusoidal function (see Figure 6.6)

$$f(t) = \begin{cases} \sin t, & 0 < t < \pi \\ 0, & \pi < t < 2\pi \end{cases} \qquad f(t + 2\pi) = f(t)$$

Solution

Since $T = 2\pi$, we first compute

$$\int_0^{2\pi} e^{-st} f(t)\, dt = \int_0^{\pi} e^{-st} \sin t\, dt$$

$$= \frac{1 + e^{-\pi s}}{s^2 + 1}$$

and therefore

$$\mathcal{L}\{f(t)\} = \frac{1}{1 - e^{-2\pi s}} \cdot \frac{1 + e^{-\pi s}}{s^2 + 1}$$

$$= \frac{1}{(1 - e^{-\pi s})(s^2 + 1)}$$

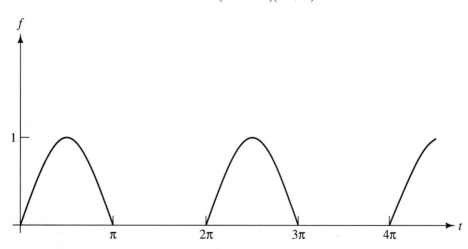

Figure 6.6

6.7.2 Discontinuous Functions

The Laplace transform has the effect of "smoothing" discontinuous functions in the transform domain. Thus, an interesting and useful application of the transform method involves solving linear DEs with discontinuous input functions. Equations of this type are commonplace in circuit analysis problems as well as in some problems involving mechanical systems.

In order to effectively deal with functions having finite jump discontinuities, it is helpful to introduce the **Heaviside unit function** (or **unit step function**). We denote this function by the symbol $h(t - a)$, and define it by

$$h(t - a) = \begin{cases} 0, & t < a \\ 1, & t \geq a \end{cases} \tag{21}$$

where $a \geq 0$ is the value where the jump discontinuity occurs (see Figure 6.7).

Figure 6.7 Heaviside unit function.

> *Remark.* How we define a discontinuous function at a point of discontinuity is of little concern here since one point (or even a finite number of points) does not affect its Laplace transform. Thus, sometimes the point $t = a$ is omitted in the definition of $h(t - a)$.

■ EXAMPLE 24

Sketch the graph of

$$f(t) = h(t - a) - h(t - b), \qquad 0 \leq a < b$$

Solution

From definition, it is clear that

$$f(t) = \begin{cases} 0 - 0 = 0, & 0 \leq t < a \\ 1 - 0 = 1, & a \leq t < b \\ 1 - 1 = 0, & t \geq b \end{cases}$$

The graph of this function is therefore the rectangle function shown in Figure 6.8.

■

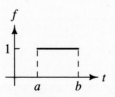

Figure 6.8

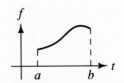

Figure 6.9

Suppose that $f(t)$ is a function that has nonzero values only on the interval $a < t < b$ (see Figure 6.9); that is, suppose that

$$f(t) = \begin{cases} f_1(t), & a < t < b \\ 0, & \text{otherwise} \end{cases} \tag{22}$$

The rectangle function introduced in Example 24 can be useful in expressing the piecewise continuous function $f(t)$ in terms of the Heaviside unit function. For example, we can write

$$f(t) = f_1(t)[h(t - a) - h(t - b)] \tag{23}$$

More generally, if (see Figure 6.10)

$$f(t) = \begin{cases} f_1(t), & 0 < t < a \\ f_2(t), & a < t < b \\ f_3(t), & t > b \end{cases} \tag{24}$$

then we can write

$$f(t) = f_1(t)[h(t) - h(t - a)] + f_2(t)[h(t - a) - h(t - b)] + f_3(t)h(t - b)$$

or, by combining like terms and using $h(t) = 1, t > 0$,

$$f(t) = f_1(t) + [f_2(t) - f_1(t)]h(t - a) + [f_3(t) - f_2(t)]h(t - b), \qquad t > 0 \tag{25}$$

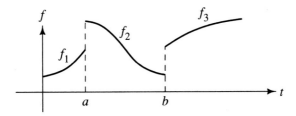

Figure 6.10

EXAMPLE 25

Express the function $f(t)$ in terms of the Heaviside unit function, where (see Figure 6.11)

$$f(t) = \begin{cases} t^2, & 0 < t < 2 \\ 3, & t > 2 \end{cases}$$

Solution

Following the above procedure, we write

$$f(t) = t^2[h(t) - h(t - 2)] + 3h(t - 2)$$
$$= t^2 + (3 - t^2)h(t - 2), \qquad t > 0 \qquad \blacksquare$$

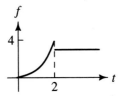

Figure 6.11

To use the Heaviside unit function in problems featuring the Laplace transform, we need to develop a formula for the Laplace transform of the product of a given function and the Heaviside unit function. First, let us observe that

$$\mathcal{L}\{h(t-a)\} = \int_0^\infty e^{-st}h(t-a)\, dt$$

$$= \int_a^\infty e^{-st}\, dt$$

which reduces to

$$\mathcal{L}\{h(t-a)\} = \frac{e^{-as}}{s}, \qquad a \geq 0, \qquad s > 0 \qquad (26)$$

Notice that when $a = 0$, Equation (26) becomes

$$\mathcal{L}\{h(t)\} = \mathcal{L}\{1\} = \frac{1}{s}$$

Theorem 6.9 *If* $\mathcal{L}\{g(t)\} = G(s)$, *then*

$$\mathcal{L}\{g(t-a)h(t-a)\} = e^{-as}G(s)$$

or, equivalently,

$$\mathcal{L}\{g(t)h(t-a)\} = e^{-as}\mathcal{L}\{g(t+a)\}$$

Proof. *From the definition,*

$$\mathcal{L}\{g(t-a)h(t-a)\} = \int_0^\infty e^{-st}g(t-a)h(t-a)\, dt$$

$$= \int_a^\infty e^{-st}g(t-a)\, dt$$

Introducing the change of variable $x = t - a$, *this expression becomes*

$$\mathcal{L}\{g(t-a)h(t-a)\} = \int_0^\infty e^{-s(x+a)}g(x)\, dx$$

$$= e^{-as}\int_0^\infty e^{-sx}g(x)\, dx$$

or

$$\mathcal{L}\{g(t-a)h(t-a)\} = e^{-as}G(s)$$

We leave it to the reader to prove the equivalence of the second relation given in the theorem.

EXAMPLE 26

Find the Laplace transform of

$$f(t) = \begin{cases} t^2, & 0 < t < 2 \\ 3, & t > 2 \end{cases}$$

Solution

Recall from Example 25 that we can write

$$f(t) = t^2 + (3 - t^2)h(t - 2)$$

The Laplace transform $f(t)$ written in this fashion leads to

$$\mathcal{L}\{f(t)\} = \mathcal{L}\{t^2\} + \mathcal{L}\{(3 - t^2)h(t - 2)\}$$

and by using the second expression in Theorem 6.8, we obtain

$$\mathcal{L}\{f(t)\} = \mathcal{L}\{t^2\} + e^{-2s}\mathcal{L}\{3 - (t + 2)^2\}$$
$$= \mathcal{L}\{t^2\} - e^{-2s}\mathcal{L}\{1 + 4t + t^2\}$$

Finally, completing the above transforms, we arrive at

$$\mathcal{L}\{f(t)\} = \frac{2}{s^3} - e^{-2s}\left(\frac{1}{s} + \frac{4}{s^2} + \frac{2}{s^3}\right)$$

To see the efficiency of the Heaviside step function, the reader should evaluate this transform by the defining integral

$$\mathcal{L}\{f(t)\} = \int_0^2 e^{-st}t^2 \, dt + 3\int_2^\infty e^{-st} \, dt \qquad \blacksquare$$

The property given in Theorem 6.9 is also useful in the development of certain inverse Laplace transforms. In this case, we write

$$\mathcal{L}^{-1}\{e^{-as}G(s)\} = g(t - a)h(t - a) \qquad (27)$$

where $g(t) = \mathcal{L}^{-1}\{G(s)\}$. Based upon this result, it is apparent that the presence of an exponential function e^{-as} in the transform domain is a warning of a discontinuity in the inverse transform function or its derivative. Moreover, the value of a gives us the location of the discontinuity.

■ **EXAMPLE 27**

Find the inverse Laplace transform of

$$F(s) = \frac{1 - 3e^{-5s}}{s^2}$$

Solution

Applying the linearity property, we find

$$\mathcal{L}^{-1}\{F(s)\} = \mathcal{L}^{-1}\left\{\frac{1}{s^2}\right\} - 3\mathcal{L}^{-1}\left\{\frac{e^{-5s}}{s^2}\right\}$$
$$= t - 3(t - 5)h(t - 5)$$

where we are using $\mathcal{L}^{-1}\{1/s^2\} = t$. Finally, we can write this inverse transform as

$$f(t) = \begin{cases} t, & 0 \leq t < 5 \\ 15 - 2t, & t \geq 5 \end{cases} \qquad \blacksquare$$

Looking back on these last two examples, we see that the first form of the property listed in Theorem 6.9 is best suited for finding inverse Laplace transforms, while the second form is more convenient when computing Laplace transforms. Let us now consider an IVP where the input function is piecewise continuous.

EXAMPLE 28

Solve the IVP

$$y'' + y = f(t), \qquad y(0) = 0, \qquad y'(0) = 0$$

where

$$f(t) = \begin{cases} \sin t, & 0 \le t < \pi/2 \\ 0, & t \ge \pi/2 \end{cases}$$

Solution

Physically, we might interpret this problem as a spring-mass system (no damping) which is at rest until time $t = 0$. It is then under the influence of the sinusoidal forcing function $\sin t$ until time $t = \pi/2$, after which the forcing function is removed.

Writing the input function as

$$f(t) = (\sin t)[1 - h(t - \pi/2)]$$

we first calculate

$$
\begin{aligned}
\mathscr{L}\{f(t)\} &= \mathscr{L}\{\sin t\} - \mathscr{L}\{(\sin t)h(t - \pi/2)\} \\
&= \mathscr{L}\{\sin t\} - e^{-\pi s/2}\mathscr{L}\{\sin(t + \pi/2)\} \\
&= \mathscr{L}\{\sin t\} - e^{-\pi s/2}\mathscr{L}\{\cos t\}
\end{aligned}
$$

which leads to

$$\mathscr{L}\{f(t)\} = \frac{1}{s^2 + 1} - \frac{se^{-\pi s/2}}{s^2 + 1}$$

Hence, the Laplace transform of the given DE yields

$$(s^2 + 1)Y(s) = \frac{1}{s^2 + 1} - \frac{se^{-\pi s/2}}{s^2 + 1}$$

with solution

$$Y(s) = \frac{1}{(s^2 + 1)^2} - \frac{se^{-\pi s/2}}{(s^2 + 1)^2}$$

Finally, by taking the inverse Laplace transform of this last expression, we find

$$y(t) = \tfrac{1}{2}(\sin t - t\cos t) - \tfrac{1}{2}[(t - \pi/2)\sin(t - \pi/2)]h(t - \pi/2)$$

which can also be written as

$$y(t) = \begin{cases} \dfrac{1}{2}(\sin t - t\cos t), & 0 \le t < \pi/2 \\[2mm] \dfrac{1}{2}\sin t - \dfrac{\pi}{4}\cos t, & t \ge \pi/2 \end{cases}$$

Observe that the solution becomes steady as soon as the external input function is removed (i.e., after $t = \pi/2$). Otherwise, owing to the term $t \cos t$, resonance would eventually take place since the input function has the natural frequency of the system.

∎

If we were to solve Example 28 by the methods of Chapter 4, we would have to solve one problem over the interval $0 \le t < \pi/2$ and a second problem over $t \ge \pi/2$. In the second problem a new initial condition would have to be prescribed, based on the solution of the first problem. However, the Laplace transform method offers the distinct advantage of not having to consider a problem for each interval separately.

Exercises 6.7

In Problems 1 to 8, use Theorem 6.8 to verify the given Laplace transform relation.

1. $\mathscr{L}\{\sin t\} = \dfrac{1}{s^2 + 1}$

2. $\mathscr{L}\{\cos kt\} = \dfrac{s}{s^2 + k^2}$

3. $\mathscr{L}\{f(t)\} = \dfrac{1}{s(1 + e^{-s})}$,   where $f(t + 2) = f(t)$ and

$$f(t) = \begin{cases} 1, & 0 \le t < 1 \\ 0, & 1 < t < 2 \end{cases}$$

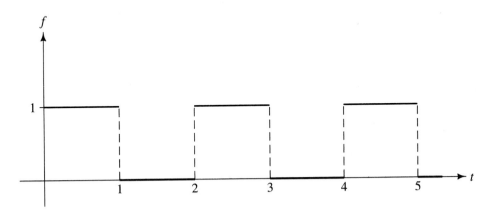

Problem 3

4. $\mathscr{L}\{f(t)\} = \dfrac{1 - (s + 1)e^{-s}}{s^2(1 - e^{-s})}$,   where $f(t + 1) = f(t)$ and

$$f(t) = t, \qquad 0 \le t < 1$$

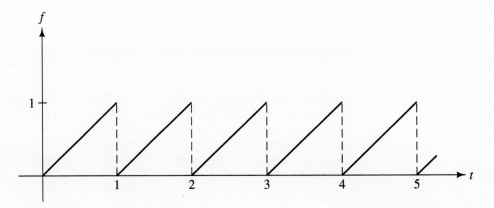

Problem 4

5. $\mathscr{L}\{f(t)\} = \dfrac{1 - (s + 1)e^{-s}}{s^2(1 - e^{-2s})}$, where $f(t + 2) = f(t)$ and

$$f(t) = \begin{cases} t, & 0 \le t < 1 \\ 0, & 1 \le t < 2 \end{cases}$$

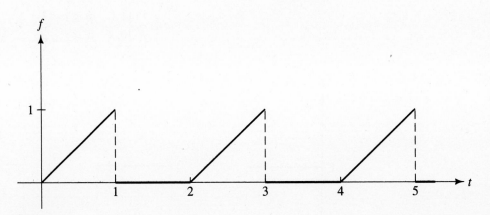

Problem 5

■**6.** $\mathscr{L}\{f(t)\} = \dfrac{1}{s} \tanh(cs/2)$, where $f(t + 2c) = f(t)$ and

$$f(t) = \begin{cases} 1, & 0 \le t \le c \\ -1, & c < t < 2c \end{cases}$$

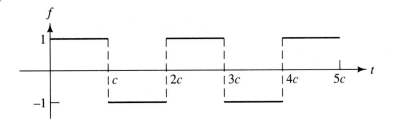

Problem 6

7. $\mathcal{L}\{f(t)\} = \dfrac{1}{s^2}\tanh(cs/2)$, where $f(t + 2c) = f(t)$ and

$$f(t) = \begin{cases} t, & 0 \le t \le c \\ 2c - t, & c < t < 2c \end{cases}$$

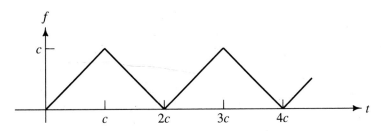

Problem 7

8. $\mathcal{L}\{|\sin t|\} = \dfrac{1}{s^2 + 1}\coth(\pi s/2)$

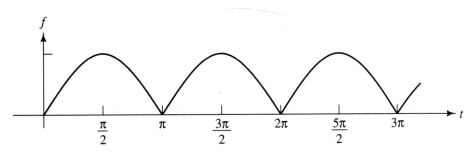

Problem 8

In Problems 9 to 14, sketch the graph of the given function for $t \geq 0$.

9. $f(t) = t^2 h(t) + (5 - t^2)h(t - 3)$

10. $f(t) = h(t - 1) + 2h(t - 3) - 6h(t - 4)$

11. $f(t) = h(t) - h(t - \pi) + (\sin t)h(t - 2\pi)$

12. $f(t) = g(t - 2)h(t - 2)$, where $g(t) = t^2$

13. $f(t) = t^2 - (t^2 - 2t + 4)h(t - 2)$

14. $f(t) = t^2 - t^2 h(t - 2)$

In Problems 15 to 22, express each function in terms of the Heaviside unit function and find its Laplace transform.

15. $f(t) = \begin{cases} 2, & 0 \leq t \leq 1 \\ t, & t > 1 \end{cases}$

16. $f(t) = \begin{cases} t^2, & 0 \leq t < 3 \\ 4, & t > 3 \end{cases}$

17. $f(t) = \begin{cases} \sin t, & 0 \leq t < \pi \\ 0, & t > \pi \end{cases}$

18. $f(t) = \begin{cases} \sin 3t, & 0 \leq t \leq \dfrac{\pi}{2} \\ \\ 0, & t > \dfrac{\pi}{2} \end{cases}$

19. $f(t) = \begin{cases} 0, & 0 \leq t < 3 \\ e^{-t}, & t > 3 \end{cases}$

20. $f(t) = \begin{cases} e^{-t}, & 0 \leq t < 3 \\ 0, & t > 3 \end{cases}$

21. $f(t) = \begin{cases} t^2, & 0 \leq t < 1 \\ 3, & 1 < t < 4 \\ 0, & t > 4 \end{cases}$

■22. $f(t) = \begin{cases} 1, & 0 \leq t < 1 \\ 0, & 1 < t < 2 \\ 1, & 2 < t < 3 \\ 0, & t > 3 \end{cases}$

In Problems 23 to 30, find the inverse Laplace transform.

23. $F(s) = \dfrac{5e^{-3s} - e^{-s}}{s}$

24. $F(s) = \dfrac{e^{-s}}{s^2}$

25. $F(s) = \dfrac{3e^{-2s} - 1}{s^2}$

26. $F(s) = \dfrac{e^{-3s}}{s + 2}$

27. $F(s) = \dfrac{e^{-3s}}{(s + 2)^3}$

■28. $F(s) = \dfrac{1 - e^{-\pi s}}{s^2 + 4}$

29. $F(s) = \dfrac{s(1 + e^{-\pi s})}{s^2 + 4}$

30. $F(s) = \dfrac{(s - 2)e^{-s}}{s^2 - 4s + 3}$

In Problems 31 to 36, solve the given IVP.

31. $y' + y = f(t)$, $y(0) = 1$, where

$$f(t) = \begin{cases} 0, & 0 \leq t < 1 \\ 1, & t > 1 \end{cases}$$

32. $y' + y = f(t)$, $y(0) = 0$, where

$$f(t) = \begin{cases} 1, & 0 \leq t < 1 \\ 0, & t > 1 \end{cases}$$

33. $y'' + 4y = f(t)$, $y(0) = 1$, $y'(0) = 0$, where

$$f(t) = \begin{cases} 4t, & 0 \leq t < 1 \\ 4, & t > 1 \end{cases}$$

34. $y'' + y = f(t)$, $y(0) = 0$, $y'(0) = 0$, where

$$f(t) = \begin{cases} 4, & 0 \leq t < 2 \\ t + 2, & t > 2 \end{cases}$$

35. $y'' + 4y = f(t)$, $y(0) = 0$, $y'(0) = 1$, where

$$f(t) = \begin{cases} \cos 4t, & 0 \leq t < \pi \\ 0, & t > \pi \end{cases}$$

36. $y'' - 5y' + 6y = h(t - 1)$, $y(0) = 0$, $y'(0) = 1$

In Problems 37 to 40, solve the IVP for the spring-mass system

$$my'' + cy' + ky = f(t), \qquad y(0) = y_0, \qquad y'(0) = v_0$$

using the prescribed values of the parameters.

37. $m = 1$, $c = 0$, $k = 4$, $y_0 = v_0 = 0$,

$$f(t) = \begin{cases} 0, & 0 < t < 3 \\ 1, & t \ge 3 \end{cases}$$

■ **38.** $m = 1$, $c = 0$, $k = 4$, $y_0 = v_0 = 0$,

$$f(t) = \begin{cases} \sin t, & 0 < t < 2\pi \\ 0, & t \ge 2\pi \end{cases}$$

39. $m = 1$, $c = 3$, $k = 2$, $y_0 = 2$, $v_0 = -3$,

$$f(t) = \begin{cases} 0, & 0 \le t < 2 \\ e^{-t}, & t \ge 2 \end{cases}$$

40. $m = 1$, $c = 5$, $k = 6$, $y_0 = 0$, $v_0 = 1$,

$$f(t) = \begin{cases} 0, & 0 \le t < 2 \\ t, & t \ge 2 \end{cases}$$

[0] 6.8 Impulse Functions

In certain applications it is convenient to introduce the concept of an impulse, which is the result of a sudden excitation administered to a system, such as a sharp blow or voltage surge. Let us imagine that the sudden excitation, which we denote by $d_\epsilon(t - a)$, has a nonzero value over the short interval of time $a - \epsilon < t < a + \epsilon$, but is otherwise zero. The total **impulse** (force times duration) imparted to the system is thus defined by

$$I = \int_{-\infty}^{\infty} d_\epsilon(t - a)\, dt = \int_{a-\epsilon}^{a+\epsilon} d_\epsilon(t - a)\, dt, \qquad (\epsilon > 0) \tag{28}$$

The value of I is a *measure of the strength of the sudden excitation*.

In order to provide a mathematical model of the function $d_\epsilon(t - a)$, it is convenient to think of it as having a constant value over the interval $a - \epsilon < t < a + \epsilon$ (see Figure 6.12). Furthermore, we wish to choose this constant value in such a way that the total impulse given by (28) is unity. Hence, we write

$$d_\epsilon(t - a) = \begin{cases} \dfrac{1}{2\epsilon}, & a - \epsilon < t < a + \epsilon \\ 0, & \text{otherwise} \end{cases} \tag{29}$$

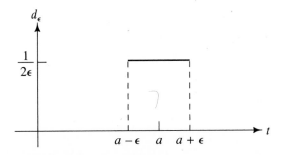

Figure 6.12 Impulse function.

Now let us idealize the function $d_\epsilon(t - a)$ by requiring it to act over shorter and shorter intervals of time by allowing $\epsilon \to 0$. Although the interval about $t = a$ is shrinking to zero, we still want $I = 1$, that is,

$$\lim_{\epsilon \to 0} I = \lim_{\epsilon \to 0} \int_{-\infty}^{\infty} d_\epsilon(t - a) \, dt = 1 \qquad (30)$$

We can use the result of this limit process to define an "idealized" **unit impulse function**, $\delta(t - a)$, which has the property of imparting an impulse of magnitude 1 to the system at time $t = a$ but being zero for all other values of t. The defining properties of this function are therefore

$$\delta(t - a) = 0, \qquad t \neq a$$

$$\int_{-\infty}^{\infty} \delta(t - a) \, dt = 1 \qquad\qquad (31)$$

By a similar kind of limit process, it is possible to define the integral of a product of the unit impulse function and any continuous and bounded function $f(t)$; that is,

$$\int_{-\infty}^{\infty} \delta(t - a)f(t) \, dt = \lim_{\epsilon \to 0} \int_{-\infty}^{\infty} d_\epsilon(t - a)f(t) \, dt$$

$$= \lim_{\epsilon \to 0} \frac{1}{2\epsilon} \int_{a-\epsilon}^{a+\epsilon} f(t) \, dt$$

Recalling that

$$\int_{a}^{b} f(t) \, dt = f(\xi)(b - a), \qquad a < \xi < b \qquad (32)$$

which is the **mean value theorem** of the integral calculus, we find that

$$\int_{-\infty}^{\infty} \delta(t - a)f(t) \, dt = \lim_{\epsilon \to 0} \frac{1}{2\epsilon} \cdot f(\xi) \cdot 2\epsilon$$

for some ξ in the interval $a - \epsilon < \xi < a + \epsilon$. Consequently, in the limit we see that $\xi \to a$, and deduce that[†]

$$\int_{-\infty}^{\infty} \delta(t - a)f(t) \, dt = f(a) \qquad (33)$$

Equation (33) is called the **sifting property** of $\delta(t - a)$ but actually serves as its definition in practice.

Obviously the "function" $\delta(t - a)$, also known as the **Dirac delta function**,[‡] is not a function in the usual sense of the word. It has significance only as part of an integrand. In dealing with this function, it is usually best to avoid the idea of assigning "function values" and instead refer to its integral property (33), even though it has no meaning as an ordinary integral.

[†] Note that $\xi = \xi(\epsilon)$, so that $\lim_{\epsilon \to 0} f[\xi(\epsilon)] = f[\lim_{\epsilon \to 0} \xi(\epsilon)] = f(a)$.

[‡] Named after Paul A. M. Dirac (1902–1984), who was awarded the Nobel Prize (with E. Schrödinger) in 1933 for his work in quantum mechanics.

Using the integral property (33), we see that

$$\mathcal{L}\{\delta(t-a)\} = \int_0^\infty e^{-st}\delta(t-a)\,dt$$

$$= e^{-as}, \qquad a \geq 0 \tag{34}$$

Hence, in spite of its peculiar nature, the transform of $\delta(t-a)$ is a perfectly normal function. Also observe that $\mathcal{L}\{\delta(t)\} = 1$.

EXAMPLE 29

Solve the IVP

$$y'' + y = \delta(t-\pi), \qquad y(0) = 0, \qquad y'(0) = 0$$

Solution

By formally taking the Laplace transform of each side of the DE, we obtain

$$s^2 Y(s) + Y(s) = e^{-\pi s}$$

with solution

$$Y(s) = \frac{e^{-\pi s}}{s^2 + 1}$$

Now, using the result

$$\mathcal{L}^{-1}\left\{\frac{1}{s^2 + 1}\right\} = \sin t$$

we see that

$$y(t) = \mathcal{L}^{-1}\left\{\frac{e^{-\pi s}}{s^2 + 1}\right\} = [\sin(t-\pi)]h(t-\pi)$$

or (see Figure 6.13)

$$y(t) = \begin{cases} 0, & 0 \leq t < \pi \\ \sin(t-\pi), & t \geq \pi \end{cases}$$

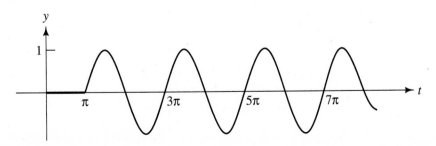

Figure 6.13

We might physically interpret this problem as representing an electric circuit to which an instantaneous unit voltage impulse is applied at time $t = \pi$. Hence, there is no response until time $t = \pi$ (owing to the homogeneous initial conditions), but the impulse voltage produces a response that lasts indefinitely (there is no resistor for damping effects). Interestingly, the solution is a continuous function for all $t \geq 0$ in spite of the singular nature of the impulse function. ■

6.8.1 Impulse Response Function

In the study of linear systems, such as electric circuits, the systems engineer characterizes the system under study by the **transfer function** $G(s)$, defined as the ratio of the Laplace transform of the output function $y(t)$ to the Laplace transform of the input (or forcing) function $f(t)$, assuming that all initial conditions are zero. That is, we write $G(s) = Y(s)/F(s)$, where $Y(s) = \mathscr{L}\{y(t)\}$ and $F(s) = \mathscr{L}\{f(t)\}$.

Let us assume the linear system is governed by the IVP

$$ay'' + by' + cy = f(t), \qquad y(0) = 0, \qquad y'(0) = 0 \tag{35}$$

where a, b, and c are constants. The Laplace transform of (35) leads to

$$(as^2 + bs + c)Y(s) = F(s)$$

from which we deduce the transfer function

$$G(s) = \frac{Y(s)}{F(s)} = \frac{1}{as^2 + bs + c} \tag{36}$$

By defining the inverse Laplace transform

$$g(t) = \mathscr{L}^{-1}\{G(s)\} = \mathscr{L}^{-1}\left\{\frac{1}{as^2 + bs + 1}\right\} \tag{37}$$

we can use the convolution theorem to write the solution of (35) in the form

$$y(t) = \int_0^t g(t-u)f(u)\,du = \int_0^t g(u)f(t-u)\,du \tag{38}$$

We see, therefore, that knowledge of the transfer function or its inverse Laplace transform allows us to solve (35) for any input function $f(t)$. In particular, if the input function is the unit impulse, that is, if $f(t) = \delta(t)$, then (38) reduces to simply

$$y(t) = \int_0^t g(t-u)\delta(u)\,du = g(t) \tag{39}$$

In words, $g(t)$ is the *response of the linear system at rest that is subject to a unit impulse input at time $t = 0$*. For this reason, the function $g(t)$ is known as the **impulse response function** of the system described by (35).

■─────────
EXAMPLE 30

Use the Laplace transform to construct the impulse response function associated with the system described by

$$M[y] = f(t), \qquad y(0) = \alpha, \qquad y'(0) = \beta$$

where M is the differential operator

$$M = 2D^2 - 4D + 10$$

Also, determine the response of the system when an arbitrary input function $f(t)$ is applied and the initial values are both zero.

Solution

By writing $M = 2(D^2 - 2D + 5)$, we first determine

$$g(t) = \frac{1}{2} \mathscr{L}^{-1} \left\{ \frac{1}{s^2 - 2s + 5} \right\}$$

$$= \frac{1}{2} \mathscr{L}^{-1} \left\{ \frac{1}{(s-1)^2 + 4} \right\}$$

which, through the shift property, leads immediately to the impulse response function

$$g(t) = \tfrac{1}{4} e^t \sin 2t$$

The response of the system to an arbitrary input function $f(t)$ is the solution of the IVP

$$2y'' + 4y' + 10y = f(t), \qquad y(0) = 0, \qquad y'(0) = 0$$

Using the impulse response function $g(t)$, we can write

$$y(t) = \frac{1}{4} \int_0^t e^{t-u} \sin 2(t-u) f(u) \, du$$

or, equivalently

$$y(t) = \frac{1}{4} \int_0^t e^u \sin 2u f(t-u) \, du \qquad \blacksquare$$

6.8.2 One-Sided Green's Function

The impulse response function is directly related to the *one-sided Green's function* that we introduced in Section 5.5. To see this, let us first rewrite the IVP (35) as

$$y'' + By' + Cy = \frac{1}{a} f(t), \qquad y(0) = 0, \qquad y'(0) = 0 \qquad (40)$$

where the DE is in *normal form* ($B = b/a$ and $C = c/a$). By the method of the one-sided Green's function, (40) has the formal solution

$$y(t) = \frac{1}{a} \int_0^t g_1(t, u) f(u) \, du \qquad (41)$$

where $g_1(t, u)$ is the one-sided Green's function and we are using u as the dummy variable here rather than τ. A comparison of (41) with (38) suggests that $g(t - u) = (1/a)g_1(t, u)$. In other words, the impulse response function and the one-sided Green's function differ at most by a multiplicative constant. Hence, it makes little difference in practice which of these functions we use in solving a given IVP.

▬▬▬▬▬▬▬▬
EXAMPLE 31

Determine the one-sided Green's function for the operator given in Example 30.

Solution

Given that the impulse response function is

$$g(t) = \tfrac{1}{4}e^t \sin 2t$$

the one-sided Green's function is

$$g_1(t, u) = 2g(t - u) = \tfrac{1}{2}e^{t-u} \sin 2(t - u)$$ ■

Comparison of the convolution theorem of the Laplace transform and the Green's function method shows that they are equivalent for constant-coefficient equations. In fact, it now follows that the one-sided Green's function for a constant-coefficient DE can always be expressed as a function of the variable $t - u$. The method of Green's function is more general than that of the Laplace transform, however, since it can be applied to any variable-coefficient equation (at least in theory) and is more readily adapted to problems when the initial data are prescribed at a point other than $t = 0$.

Exercises 6.8

In Problems 1 to 6, solve the given IVP.

1. $y'' + 4y = \delta(t - 2\pi)$, $y(0) = 1$, $y'(0) = 0$

2. $y'' + 2y' + 2y = \delta(t - \pi)$, $y(0) = 0$, $y'(0) = 0$

3. $y'' + 2y' + 2y = \delta(t - \pi)$, $y(0) = 1$, $y'(0) = 0$

4. $y'' + 4y = \delta(t - \pi) - \delta(t - 2\pi)$, $y(0) = 0$, $y'(0) = 0$

5. $y'' + 2y' + y = \delta(t) + h(t - 2\pi)$, $y(0) = 0$, $y'(0) = 1$

■**6.** $y'' + y = \delta(t - \pi) \cos t$, $y(0) = 0$, $y'(0) = 1$

7. A spring-mass system receiving an impulse p_0 at time $t = 0$ is governed by the IVP

$$my'' + ky = p_0\delta(t), \qquad y(0) = 0, \qquad y'(0) = 0$$

Show that the solution of this problem is the same as that for the IVP

$$my'' + ky = 0, \qquad y(0) = 0, \qquad y'(0) = v_0$$

illustrating that the input $p_0\delta(t)$ does indeed impart an initial *momentum* $p_0 = mv_0$ to the system.

8. A spring-mass system at rest until time $t = 0$ is then subjected to the sinusoidal force $f(t) = P \sin \omega t$. At time $t = \pi$ seconds, the mass is given a sharp blow from below that instantaneously imparts an upward impulse of 5 units to the system. Neglect damping effects and describe the motion of the system.

9. A simple LC circuit with inductance $L = 2$ H and capacitance $C = 0.02$ F is connected to a 24-V battery. If the switch is alternately closed and open at times $t = 0$, $\pi/5$, $2\pi/5$, $3\pi/5$, ..., the current $i(t)$ satisfies the IVP

$$i'' + 100i = 12 \sum_{n=0}^{\infty} (-1)^n \delta(t - n\pi/5),$$

$$i(0) = 0, \qquad i'(0) = 0$$

Show that if $n\pi/5 < t < (n + 1)\pi/5$, then

$$i(t) = \begin{cases} \tfrac{6}{5} \sin 10t, & \text{if } n \text{ is even} \\ 0, & \text{if } n \text{ is odd} \end{cases}$$

We can interpret this solution as the output of a *half-wave rectifier* that clips (sets to zero) the negative portion of a sine wave (recall Example 23).

In Problems 10 to 16, find the transfer function $G(s)$ and the impulse response function $g(t)$ associated with the given DE. Also, write out the response of the system to the general input function $f(t)$.

10. $y'' + 9y = f(t)$

11. $2y'' - 5y' - 3y = f(t)$

■**12.** $y'' + 2y' - 15y = f(t)$

13. $9y'' + 6y' + y = f(t)$

14. $9y'' - 12y' + 4y = f(t)$

15. $y'' - 6y' + 25y = f(t)$

16. $2y'' - 3y' + 10y = f(t)$

In Problems 17 to 24, use the Laplace transform method to determine the one-sided Green's function for the given differential operator.

17. $M = D^2$

18. $M = (D - a)^2$

19. $M = D^2 + 2D + 5$

20. $M = (D - a)(D - b)$, $a \neq b$

21. $M = D^n$, $n = 2, 3, 4, \ldots$

■**22.** $M = D^2(D^2 - 1)$

23. $M = D^4 - 1$

24. $M = D^3 - 6D^2 + 11D - 6$

*** 25.** By assuming that the fundamental theorem of differential calculus applies to the function

$$f(t) = \int_{-\infty}^{t} \delta(u - a) \, du$$

formally show that

$$h'(t - a) = \delta(t - a)$$

6.9 Chapter Summary

In this chapter we have introduced the **Laplace transform,** defined by

$$\mathscr{L}\{f(t)\} = F(s) = \int_0^\infty e^{-st} f(t) \, dt \tag{42}$$

as a tool for solving IVPs with constant coefficients. The Laplace transform of many commonly occurring functions can be obtained directly by evaluating the defining integral, whereupon they are placed in a table like Table 6.1. Other transforms may often be determined from these tabulated entries by the use of certain **operational properties** (see Table 6.2).

Given the transform function $F(s)$, we then say the original function is the **inverse transform** and write $f(t) = \mathscr{L}^{-1}\{F(s)\}$. The tables are useful in finding inverse transforms of many functions but, in addition, we may also use operational properties or partial fraction expansions.

The Laplace transform applied to the IVP

$$ay'' + by' + cy = f(t), \qquad y(0) = \alpha, \qquad y'(0) = \beta \tag{43}$$

reduces it to the algebraic equation

$$(as^2 + bs + c)Y(s) - a(\alpha s + \beta) - b\beta = F(s) \tag{44}$$

Upon solving this simple equation for $Y(s)$, we then evaluate its inverse transform to recover the desired solution $y(t)$. It is the last step of evaluating $\mathscr{L}^{-1}\{Y(s)\}$ that is usually the most difficult part of the process. However, the Laplace transform technique is generally very efficient in that it eliminates the need to first find a general solution of the DE and then impose the initial conditions to determine the values of the arbitrary constants.

Review Exercises

In Problems 1 to 6, find the Laplace transform of the given function through the use of tables and/or operational properties.

1. $f(t) = \sinh at - \sin at$

2. $f(t) = t^3 e^{4t}$

3. $f(t) = t^2 \sin kt$

4. $f(t) = te^{2t} \sin t$

5. $f(t) = \begin{cases} 5, & 0 < t < 3 \\ t, & t > 3 \end{cases}$

6. $f(t) = t \cosh kt$

In Problems 7 to 12, find the inverse transform of the given function.

7. $F(s) = \dfrac{s + 3}{s^2 + 5}$

8. $F(s) = \dfrac{s - 3}{s^2 - 6s + 25}$

9. $F(s) = \dfrac{6s - 4}{s^2 - 4s + 20}$

10. $F(s) = \dfrac{se^{-\pi s}}{s^2 + 4}$

11. $F(s) = \dfrac{4}{(s + 1)(s^2 + 1)}$

12. $F(s) = \dfrac{5s^2 - 15s - 11}{(s + 1)(s - 2)^3}$

In Problems 13 to 20, use the Laplace transform to solve the given IVP.

13. $y'' + 9y = 0$, $y(0) = 5$, $y'(0) = -3$

14. $y'' - 3y' + 4y = 0$, $y(0) = 1$, $y'(0) = 5$

15. $y'' - y' - 2y = 4t^2$, $y(0) = 6$, $y'(0) = 5$

16. $y'' + y' - 2y = 4 \cos t$, $y(0) = 0$, $y'(0) = 0$

17. $y'' + 20y' + 200y = 0$, $y(0) = 0$, $y'(0) = 24$

18. $y'' - 3y' + 2y = 4e^{2t}$, $y(0) = -3$, $y'(0) = 5$

19. $y'' + 2y' + 5y = e^{-t} \sin t$, $y(0) = 0$, $y'(0) = 1$

20. $y''' + y' = e^t$, $y(0) = 0$, $y'(0) = 0$, $y''(0) = 0$

6.10 Historical Comments

The **Laplace transform** dates back to P. S. Laplace, who made use of the transform integral in his work on probability theory in the 1780s. Poisson also knew of the integral in the 1820s, and it occurred in the famous 1811 paper of Joseph Fourier (1768–1830) on heat conduction (see also Chapter 11). Nonetheless, it was Heaviside who popularized the use of the Laplace transform as a computational tool in elementary differential equations and electrical engineering.

Heaviside was an English electrical engineer who set himself apart from the established scientists of the day by studying mostly alone. He was a self-made man, without academic credentials, but he made significant contributions to the development and application of electromagnetic theory. He called mathematicians "woodenheaded" when his powerful new mathematical methods perplexed them, and he also alienated his fellow electrical engineers who didn't know what to make of him. Because of poor relations with his fellow scientists, much of his work did not receive the credit he deserved. Even today his name is often not mentioned in connection with some of his most important contributions to science, such as his pioneering work on discontinuous functions using operational methods and his mathematical solution of the distortionless transmission line.

CHAPTER 7
Power Series Methods

The methods of solution presented in Chapters 4 to 6 are generally limited to either constant-coefficient DEs or Cauchy–Euler equations. We find that **variable-coefficient DEs** must usually be solved by some **power series method** wherein the solution is expressed in the form of an infinite series.

The **general theory** of power series is reviewed in Section 7.2 while in Section 7.3 we present the general power series method for second-order DEs with expansions about **ordinary points.** For such points, we can always produce two linearly independent solutions in the form of standard power series. Points that are not ordinary are classified as **singular points,** which are further classified as **regular** or **irregular.**

The power series method is modified in Sections 7.4 and 7.5 to account for expansions about **regular singular points.** This modification of the power series method, known as the **Frobenius method,** leads to three possible cases of solution depending upon the nature of the roots of the **indicial equation.** Finding solutions about irregular singular points is not considered at all since there is no general theory for this case.

In Sections 7.6 and 7.7 we develop solutions of two special variable-coefficient DEs, called **Legendre's equation** and **Bessel's equation,** that appear in a variety of applications. Some historical comments appear in Section 7.9.

7.1 Introduction

In Chapter 4 we found that solving homogeneous linear DEs with *constant coefficients* can be reduced to the algebraic problem of finding the roots of its auxiliary equation. Even the class of Cauchy–Euler equations, which have *variable coefficients*, can be solved by algebraic methods. However, this is not the case for most other variable-coefficient DEs because their solutions typically involve *nonelementary* functions.[†] The methods to be discussed in this chapter produce solutions in the form of power series and, for that reason, the general procedure is referred to as the **power series method.** Although it can be extended to higher-order DEs, we restrict the presentation primarily to homogeneous second-order equations with polynomial coefficients. Nonetheless, the general procedure is applicable to nonhomogeneous DEs and to those with coefficients more general than polynomials.

7.2 Review of Power Series

We assume the reader is familiar with the basic properties of power series, but a short review of such properties may be useful for reference purposes.

A **power series** is an infinite series of the form

$$\sum_{n=0}^{\infty} c_n(x - x_0)^n = c_0 + c_1(x - x_0) + c_2(x - x_0)^2 + \cdots \tag{1}$$

where $c_0, c_1, c_2, \ldots, c_n, \ldots$ are called the coefficients of the series and x_0 is the center of the series or the point about which the series is expanded. When $x = x_0$, the series has the sum c_0, but generally we are interested in whether the series also has a sum for other values of x. We say that (1) **converges** at the point x if the limit of partial sums

$$\lim_{N \to \infty} \sum_{n=0}^{N} c_n(x - x_0)^n$$

exists (as a finite value); otherwise the series is said to **diverge.** We say the power series (1) **converges absolutely** at the point x if $\sum_{n=0}^{\infty} |c_n| |x - x_0|^n$ converges.

Theorem 7.1 *To every power series $\sum_{n=0}^{\infty} c_n(x - x_0)^n$ is a number R, $0 \le R < \infty$, called the **radius of convergence**, with the property that the series converges absolutely for $|x - x_0| < R$ and diverges for $|x - x_0| > R$. If the series converges for all values of x, then $R \to \infty$. However, if the series converges only at x_0, then $R = 0$.*

Notice that Theorem 7.1 does not include the values of x for which $|x - x_0| = R$. In general we find that, once R has been determined, these endpoints must be analyzed

---

[†] Functions such as Legendre polynomials, Bessel functions, and so on.

separately to determine convergence of the series (1). In any case, the radius of convergence R is usually found by means of the **ratio test.**

Theorem 7.2 Ratio Test.   *Given the series* $\sum_{n=0}^{\infty} c_n(x - x_0)^n$ *and*

$$\lim_{n \to \infty} \left| \frac{c_{n+1}(x - x_0)^{n+1}}{c_n(x - x_0)^n} \right| = |x - x_0| \lim_{n \to \infty} \left| \frac{c_{n+1}}{c_n} \right| = L$$

then

(a) *If* $L < 1$, *the series* **converges.**
(b) *If* $L > 1$, *the series* **diverges.**
(c) *If* $L = 1$, *we cannot make a conclusion—the series may converge or diverge.*

EXAMPLE 1

Find the complete interval of convergence of

$$\sum_{n=0}^{\infty} \frac{(-1)^n(x - 2)^n}{3^n(n + 1)}$$

Solution

Application of the ratio test yields

$$\lim_{n \to \infty} \left| \frac{(-1)^{n+1}(x - 2)^{n+1}}{3^{n+1}(n + 2)} \cdot \frac{3^n(n + 1)}{(-1)^n(x - 2)^n} \right| = \frac{|x - 2|}{3} \lim_{n \to \infty} \frac{n + 1}{n + 2} = \frac{|x - 2|}{3}$$

Thus, the series converges for $|x - 2| < 3$, which identifies the *radius of convergence* as $R = 3$. We deduce, therefore, that the given series converges (at least) for $-1 < x < 5$ and diverges for $x < -1$ and $x > 5$.

Let us now check the endpoints $x = -1$ and $x = 5$. We first observe that at $x = -1$ the series reduces to

$$\sum_{n=0}^{\infty} \frac{(-1)^n(-3)^n}{3^n(n + 1)} = \sum_{n=0}^{\infty} \frac{1}{n + 1} = 1 + \frac{1}{2} + \frac{1}{3} + \frac{1}{4} + \cdots$$

which is the *harmonic series.* From the calculus we know this series is *divergent.* At the other endpoint $x = 5$, we see that

$$\sum_{n=0}^{\infty} \frac{(-1)^n 3^n}{3^n(n + 1)} = \sum_{n=0}^{\infty} \frac{(-1)^n}{n + 1} = 1 - \frac{1}{2} + \frac{1}{3} - \frac{1}{4} + \cdots$$

called the *alternating harmonic series* and it *converges conditionally.*[†] Thus, the given series converges on $-1 < x \leq 5$. ∎

A power series about x_0 with a positive radius of convergence R converges to some function $f(x)$ in the interval $|x - x_0| < R$. In this case, we say that $f(x)$ is **analytic** at x_0. Polynomials, e^x, and $\sin x$ are examples of functions that are analytic everywhere, whereas $\tan x$ is analytic everywhere except where $\cos x = 0$.

[†] If the series $\sum u_n$ converges but the related series $\sum |u_n|$ diverges, we say that $\sum u_n$ *converges conditionally.*

Hence, if $f(x)$ is analytic at x_0, it has the representation

$$f(x) = \sum_{n=0}^{\infty} c_n(x - x_0)^n, \qquad |x - x_0| < R \tag{2}$$

where $c_n = f^{(n)}(x_0)/n!$, $n = 0, 1, 2, \ldots$. The series (2) is called a **Taylor series,** or a **Maclaurin series** if $x_0 = 0$.[†] Familiar examples of power series are:

$$\frac{1}{1 - x} = \sum_{n=0}^{\infty} x^n, \qquad -1 < x < 1 \tag{3}$$

$$e^x = \sum_{n=0}^{\infty} \frac{x^n}{n!}, \qquad -\infty < x < \infty \tag{4}$$

$$\sin x = \sum_{n=0}^{\infty} (-1)^n \frac{x^{2n+1}}{(2n + 1)!}, \qquad -\infty < x < \infty \tag{5}$$

$$\cos x = \sum_{n=0}^{\infty} (-1)^n \frac{x^{2n}}{(2n)!}, \qquad -\infty < x < \infty \tag{6}$$

$$\log x = \sum_{n=1}^{\infty} \frac{(-1)^{n-1}}{n} (x - 1)^n, \qquad 0 < x \le 2 \tag{7}$$

Taylor series like (2) have several important properties that are significant in the study of power series solutions of DEs. In particular, if $R > 0$, we can say the following:

1. The function $f(x)$ is a continuous function on the interval $|x - x_0| < R$.
2. The series (2) can be differentiated termwise to produce the power series for $f'(x)$, that is,

$$f'(x) = \sum_{n=1}^{\infty} nc_n(x - x_0)^{n-1}, \qquad |x - x_0| < R$$

3. If a and b are within the interval of convergence, the series (2) can be integrated termwise to yield the convergent series

$$\int_a^b f(t)\, dt = \sum_{n=0}^{\infty} \frac{c_n}{n + 1} (b^{n+1} - a^{n+1})$$

4. Convergent power series with a common interval of convergence may be added by summing corresponding terms of the series, that is,

$$\sum_{n=0}^{\infty} b_n(x - x_0)^n + \sum_{n=0}^{\infty} c_n(x - x_0)^n = \sum_{n=0}^{\infty} (b_n + c_n)(x - x_0)^n$$

5. If $\sum_{n=0}^{\infty} c_n(x - x_0)^n = 0$, then $c_n = 0$, $n = 0, 1, 2, \ldots$.

Many of the power series of interest have centers at $x = 0$ as given by (3) to (6) above, but not all such as (7). Observe, however, that by making the change of variable

---

[†] Named after Brook Taylor (1685–1731) and Colin Maclaurin (1698–1746). See the historical comments in Section 7.9.

$t = x - 1$ in (7), we obtain the power series

$$\log(1 + t) = \sum_{n=1}^{\infty} \frac{(-1)^{n-1}}{n} t^n, \qquad -1 < t \le 1 \tag{8}$$

which has center at $t = 0$. Because we can always make such changes of variable, much of the discussion in the following sections will be confined to power series for which $x_0 = 0$ without loss of generality.

7.2.1 Shift of Index of Summation

In solving DEs by the power series method it can be useful at times to *shift* the index of summation by one or more values. For instance, the replacement of n by $n - 1$ in (5) above leads to

$$\sum_{\substack{n=0 \\ \big\downarrow \\ n \to n-1}}^{\infty} (-1)^n \frac{x^{2n+1}}{(2n+1)!} = \sum_{n=1}^{\infty} (-1)^{n-1} \frac{x^{2n-1}}{(2n-1)!} \tag{9}$$

The equivalence of these expressions is made clear by writing out the first few terms of each series.

> **Remark.**   When shifting the summation index, it is sometimes convenient to first introduce a new index variable k, say let $n = k - 1$, and then change k to n.

EXAMPLE 2

Make an appropriate index shift to rewrite the series

$$\sum_{n=0}^{\infty} n(n-1)c_n x^n \text{ as a series in powers of } n - 2.$$

Solution

Clearly the change $n \to n - 2$ will lead to a series in x^{n-2}. Hence,

$$\sum_{\substack{n=0 \\ \big\downarrow \\ n \to n-2}}^{\infty} n(n-1)c_n x^n = \sum_{n=2}^{\infty} (n-2)(n-3)c_{n-2} x^{n-2} \qquad ■$$

Exercises 7.2

In Problems 1 to 8, find the interval of convergence of the given series.

1. $\displaystyle\sum_{n=1}^{\infty} \frac{x^n}{n}$

2. $\displaystyle\sum_{n=0}^{\infty} nx^n$

3. $\displaystyle\sum_{n=0}^{\infty} \frac{n!}{2^n} x^n$

4. $\displaystyle\sum_{n=0}^{\infty} \frac{2^n}{n!} x^n$

5. $\displaystyle\sum_{n=0}^{\infty} \frac{(-1)^n n^2 (x-5)^n}{3^n}$

7. $\displaystyle\sum_{n=0}^{\infty} \frac{(-1)^n (x/2)^{2n}}{(n!)^2}$

*6. $\displaystyle\sum_{n=1}^{\infty} \frac{1 \cdot 3 \cdot 5 \cdots (2n-1)}{2 \cdot 5 \cdot 8 \cdots (3n-1)} (x-1)^n$

■8. $\displaystyle\sum_{n=0}^{\infty} \frac{(n!)^2}{(2n)!} x^{2n+1}$

In Problems 9 to 12, find a Taylor series about x_0 for the given function.

9. $f(x) = e^{-x}, x_0 = 0$

11. $f(x) = \sin x, x_0 = \pi/2$

12. $f(x) = \cos x, x_0 = \pi$

10. $f(x) = \dfrac{1}{1+x}, x_0 = 0$

In Problems 13 to 16, make an appropriate index shift to rewrite the series as directed.

13. $\displaystyle\sum_{n=2}^{\infty} \frac{n(n+3)}{n!} x^{n-2}$ as a series in x^n

14. $\displaystyle\sum_{n=0}^{\infty} \frac{3(n+1)}{n!} x^{n+1}$ as a series in x^n

15. $\displaystyle\sum_{n=0}^{\infty} (-1)^n \frac{x^{2n}}{(2n)!}$ as a series in x^{2n-2}

■16. $\displaystyle\sum_{n=1}^{\infty} \frac{(-1)^n}{2 \cdot 4 \cdot 6 \cdots (2n)} x^{n-1}$ as a series in x^{n+1}

17. Using termwise integration, show that

$$\int_0^1 \left(\sum_{n=0}^{\infty} \frac{x^n}{n!} \right) dx = e - 1$$

*18. Starting with the geometric series

$$\frac{1}{1-x} = \sum_{n=0}^{\infty} x^n, \qquad -1 < x < 1$$

(a) Make a change of variable to derive the Taylor series for

$$f(x) = \frac{1}{1+x^2}$$

(b) Use the answer in (a) to determine the Taylor series about $x_0 = 0$ for

$$g(x) = \tan^{-1} x$$

7.3 Solutions Near an Ordinary Point

From the calculus we know that an analytic function can be expanded in a power series like (3) to (7) above. Because the solution y of a DE must be continuous and satisfy certain differentiability requirements, it may be reasonable to assume that it is analytic and can thus be expressed in a power series of the form

$$y = \sum_{n=0}^{\infty} c_n x^n, \qquad |x| < R \tag{10}$$

Given that this is the case, it follows that power series for $y', y'', \ldots$ can be found by termwise differentiation of (10), that is,

$$y' = \sum_{n=0}^{\infty} n c_n x^{n-1} \tag{11a}$$

$$y'' = \sum_{n=0}^{\infty} n(n-1) c_n x^{n-2} \tag{11b}$$

and so on. Using these series to solve a particular DE is somewhat like the method of undetermined coefficients (see Section 4.7.1). That is, (10) is the solution we seek where only the constants $c_0, c_1, c_2, \ldots, c_n, \ldots$ need to be determined.

Let us illustrate the basic procedure by applying it to the simple equation

$$y' - y = 0 \tag{12}$$

Of course, we know from previous techniques that the general solution of this DE is $y = Ce^x$, where C is an arbitrary constant, but now we wish to find it by the method of power series.

To begin, let us substitute the series (10) and (11a) for y and y' directly into the DE. This action leads to

$$y' - y = \sum_{n=0}^{\infty} nc_n x^{n-1} - \sum_{n=0}^{\infty} c_n x^n = 0$$

To combine these two series, we must have the same exponent for x in each corresponding term, and both summation indices must start at the same value. Therefore, let us replace the index n by $n - 1$ in the second sum, which leads to

$$\sum_{n=0}^{\infty} nc_n x^{n-1} - \underset{\underset{n \to n-1}{\Downarrow}}{\sum_{n=0}^{\infty}} c_n x^n = \sum_{n=0}^{\infty} nc_n x^{n-1} - \sum_{n=1}^{\infty} c_{n-1} x^{n-1} = 0$$

Next, we write the first term of the first series (which in this case is zero) outside the sum to obtain

$$0 \cdot c_0 x^{-1} + \sum_{n=1}^{\infty} nc_n x^{n-1} - \sum_{n=1}^{\infty} c_{n-1} x^{n-1} = 0$$

or, now adding the two series termwise,

$$0 \cdot c_0 x^{-1} + \sum_{n=1}^{\infty} (nc_n - c_{n-1}) x^{n-1} = 0 \tag{13}$$

Because x can take on various values, the coefficient of x^{n-1} must vanish in order to have (13) identically zero; that is,

$$nc_n - c_{n-1} = 0, \qquad n = 1, 2, 3, \ldots \tag{14}$$

whereas c_0 can remain *arbitrary* since the term in which it appears is already zero. The relationship (14) is called a **recurrence formula** for the unknown coefficients, because c_n can be determined once c_{n-1} is known. And since $n \neq 0$, we find it convenient to rewrite (14) as

$$c_n = \frac{c_{n-1}}{n}, \qquad n = 1, 2, 3, \ldots \tag{15}$$

Successively setting $n = 1, 2, 3, \ldots$ in (15) leads to

$$c_1 = c_0$$

$$c_2 = \frac{c_1}{2} = \frac{c_0}{2}$$

$$c_3 = \frac{c_2}{3} = \frac{c_0}{3 \cdot 2} = \frac{c_0}{3!}$$

$$c_4 = \frac{c_3}{4} = \frac{c_0}{4!}$$

and so on. The substitution of these results into (10) then yields

$$y = c_0 + c_1 x + c_2 x^2 + c_3 x^3 + c_4 x^4 + \cdots$$

$$= c_0 + c_0 x + \frac{c_0}{2!} x^2 + \frac{c_0}{3!} x^3 + \frac{c_0}{4!} x^4 + \cdots$$

or

$$y = c_0 \left(1 + x + \frac{x^2}{2!} + \frac{x^3}{3!} + \frac{x^4}{4!} + \cdots \right) \tag{16}$$

Because c_0 is an arbitrary constant, we claim that (16) is the *general solution* of (12). Moreover, the general term of the series is clearly $x^n/n!$, and thus we can write (16) as

$$y = c_0 \sum_{n=0}^{\infty} \frac{x^n}{n!} \tag{17}$$

We recognize the power series in (17) as that of e^x, and therefore we have deduced that the general solution of (12) is

$$y = c_0 e^x \tag{18}$$

in agreement with our previous result.

Although the above example was quite elementary, it still illustrates the basic manipulations required to solve a DE by the power series method. Even when the coefficients of the DE are polynomials in x, rather than constants, the manipulations required differ little from those used in this simple example. Basically, there are two rules of operation that we always use in the power series method.

Rule 1 Adjust all exponents on x, by making appropriate index shifts, to equal the **smallest** one among the various series.

Rule 2 When combining series under one summation sign, start all series with the **largest** of all the beginning values. Terms preceding the first value of the summations must then be added outside the summation sign.

The power series method illustrated above works for many variable-coefficient DEs appearing in practice, but not for all. For this reason it is helpful to identify a special class of DEs for which we know in advance that the power series method works. To do so involves the notion of ordinary and singular points.

7.3.1 Ordinary and Singular Points: Polynomial Coefficients

The method of solution of the homogeneous DE

$$A_2(x)y'' + A_1(x)y' + A_0(x)y = 0 \tag{19}$$

about a point x_0 will depend on whether this point is considered an *ordinary* or *singular point* of the DE. In order to decide between these choices, let us first consider the special case in which the coefficients $A_0(x)$, $A_1(x)$, and $A_2(x)$ are all *polynomials* with no common factors. We then say that x_0 is an **ordinary point** of the DE (19) provided that $A_2(x_0) \neq 0$. If $A_2(x_0) = 0$, we say that x_0 is a **singular point** of the DE.

When classifying points as singular or ordinary, we usually do so by identifying all singular points first. Therefore, in the case of polynomial coefficients the singular points are simply solutions of $A_2(x) = 0$. However, we must include all *complex solutions* in addition to any real solutions of this polynomial equation. All points that are not singular are then ordinary points.

EXAMPLE 3

Identify the singular points of the DEs:
  (a) $(1 - x^2)y'' - 2xy' + 2y = 0$
  (b) $xy'' + y' + xy = 0$
  (c) $(x^2 + 4)y'' + xy' - 2y = 0$

Solution

   (a) Here $A_2(x) = 1 - x^2$. By setting $1 - x^2 = 0$, we see that $x = \pm 1$ are *singular points*. All other points, real and complex, are *ordinary points*.
   (b) Since $A_2(x) = x$, it follows that $x = 0$ is the only *singular point*. All other points are *ordinary*.
   (c) By setting $A_2(x) = x^2 + 4 = 0$, we find that the *singular points* in this case are complex, namely, $x = \pm 2i$. Again, all other points are *ordinary*. ∎

7.3.2 Ordinary and Singular Points: General Coefficients

While many DEs that commonly occur in practice have polynomial coefficients, this is not true of all DEs. Therefore, we need to modify our definitions of singular and ordinary points to include the more general case

$$A_2(x)y'' + A_1(x)y' + A_0(x)y = 0 \tag{20}$$

where the coefficients $A_0(x)$, $A_1(x)$, and $A_2(x)$ are not restricted to polynomials. For discussion purposes here, we divide (20) by $A_2(x)$ to put the DE in *normal form*, that is,

$$y'' + a(x)y' + b(x)y = 0 \tag{21}$$

where $a(x) = A_1(x)/A_2(x)$ and $b(x) = A_0(x)/A_2(x)$. We now define ordinary and singular points in all cases by the following definition.

Definition 7.1 We say that x_0 is an **ordinary point** of the DE

$$y'' + a(x)y' + b(x)y = 0$$

if both $a(x)$ and $b(x)$ have convergent power series about x_0 with positive radii of convergence.[†] If one or both of these functions fails to have such a series about x_0, we say that x_0 is a **singular point** of the DE.

EXAMPLE 4

Identify the singular points of
(a) $2xy'' + (\sin x)y = 0$
(b) $y'' - (\cot x)y = 0$

Solution

(a) Writing the equation in normal form

$$y'' + \left(\frac{\sin x}{2x}\right)y = 0$$

we see that *all* points are ordinary points, including $x = 0$, since

$$\frac{\sin x}{2x} = \frac{1}{2}\left(1 - \frac{x^2}{3!} + \frac{x^4}{5!} - \frac{x^6}{7!} + \cdots\right), \qquad -\infty < x < \infty^{‡}$$

(b) Because $\cot x$ fails to be defined where $\sin x = 0$, it cannot have a power series about $x = 0, \pm\pi, \pm 2\pi, \pm 3\pi, \ldots$, which therefore denote the singular points. ■

7.3.3 General Method

We now consider the general problem of solving a DE in the neighborhood of an *ordinary point* of the DE, starting with the following important theorem.

Theorem 7.3 *If x_0 is an ordinary point of the DE*

$$A_2(x)y'' + A_1(x)y' + A_0(x)y = 0$$

then the general solution is given by

$$y = \sum_{n=0}^{\infty} c_n(x - x_0)^n = c_0 y_1(x) + c_1 y_2(x)$$

where c_0 and c_1 are arbitrary constants. Moreover, the series for $y_1(x)$ and $y_2(x)$ will converge at least in the interval $|x - x_0| < R$, where R is the distance from x_0 to the nearest singular point (real or complex).

[†] Recall from Section 7.2 that such functions are called *analytic*.

[‡] Actually, the function $(\sin x)/2x$ has what is called a *removable singularity* at $x = 0$.

For most problems we are concerned with whether the point $x = 0$ is an ordinary point of the DE or not. This is so because most DEs are formulated in such a way that the origin is a special point on the interval of interest; thus, we seek series expansions about it.

EXAMPLE 5

Based on Theorem 7.3, find the radius of convergence of the power series solutions about $x = 0$ of the DE

$$(x^2 - 2x + 5)y'' - 3xy' + (x + 1)^2 y = 0$$

Solution

Since $A_2(x) = x^2 - 2x + 5$, we find the singular points by setting

$$x^2 - 2x + 5 = 0$$

the solutions of which are $x = 1 \pm 2i$. Hence, based on Theorem 7.3, $R = |1 \pm 2i| = \sqrt{5}$.[†] That is, the power series solutions about $x = 0$ of the given DE will converge at least in the interval $|x| < \sqrt{5}$. ∎

EXAMPLE 6

Solve

$$(1 - x^2)y'' - 2xy' + 2y = 0^‡$$

Solution

The coefficients are polynomials so the singularities occur where $1 - x^2 = 0$, or $x = \pm 1$; hence, $x = 0$ is an ordinary point. Consequently, we look for solutions of the form

$$y = \sum_{n=0}^{\infty} c_n x^n, \qquad |x| < 1$$

where the interval of convergence is at least $|x| < 1$. By differentiating, we formally obtain

$$y' = \sum_{n=0}^{\infty} nc_n x^{n-1}, \qquad y'' = \sum_{n=0}^{\infty} n(n-1)c_n x^{n-2}$$

Replacing y, y', and y'' in the DE with their series, we have

$$(1 - x^2) \sum_{n=0}^{\infty} n(n-1)c_n x^{n-2} - 2x \sum_{n=0}^{\infty} nc_n x^{n-1} + 2 \sum_{n=0}^{\infty} c_n x^n = 0$$

[†] Recall that if $z_1 = x_1 + iy_1$ and $z_2 = x_2 + iy_2$ are complex numbers, then

$$|z_2 - z_1| = \sqrt{(x_2 - x_1)^2 + (y_2 - y_1)^2}.$$

[‡] This DE is a special case of *Legendre's equation* (see Section 7.6).

or, by multiplying by the polynomial coefficients,

$$\sum_{n=0}^{\infty} n(n-1)c_n x^{n-2} - \sum_{n=0}^{\infty} n(n-1)c_n x^n - \sum_{n=0}^{\infty} 2nc_n x^n + \sum_{n=0}^{\infty} 2c_n x^n = 0$$
$$\Downarrow_{n \to n-2} \qquad\qquad \Downarrow_{n \to n-2} \qquad \Downarrow_{n \to n-2}$$

Now let us replace n by $n-2$ in the last three sums to get

$$\sum_{n=0}^{\infty} n(n-1)c_n x^{n-2} - \sum_{n=2}^{\infty} (n-2)(n-3)c_{n-2} x^{n-2}$$

$$- \sum_{n=2}^{\infty} 2(n-2)c_{n-2} x^{n-2} + \sum_{n=2}^{\infty} 2c_{n-2} x^{n-2} = 0$$

or

$$0 \cdot c_0 x^{-2} + 0 \cdot c_1 x^{-1} + \sum_{n=2}^{\infty} [n(n-1)c_n - (n-2)(n-3)c_{n-2}$$

$$- 2(n-2)c_{n-2} + 2c_{n-2}]x^{n-2} = 0$$

Clearly, c_0 and c_1 are arbitrary, but the remaining constants must satisfy the recurrence relation

$$n(n-1)c_n - [(n-2)(n-3) + 2(n-2) - 2]c_{n-2} = 0, \qquad n = 2, 3, 4, \ldots$$

which can be simplified to

$$c_n = \left(\frac{n-3}{n-1}\right)c_{n-2} \quad n = 2, 3, 4, \ldots$$

Setting $n = 2, 3, 4, \ldots$ successively into the above recurrence relation yields

$$c_2 = -c_0$$
$$c_3 = 0$$
$$c_4 = \tfrac{1}{3}c_2 = -\tfrac{1}{3}c_0$$
$$c_5 = \tfrac{2}{4}c_3 = 0$$
$$c_6 = \tfrac{3}{5}c_4 = -\tfrac{1}{5}c_0$$
$$c_7 = \tfrac{4}{6}c_5 = 0$$
$$c_8 = \tfrac{5}{7}c_6 = -\tfrac{1}{7}c_0$$
$$\cdots\cdots\cdots\cdots$$

Thus, our solution becomes

$$y = c_0 + c_1 x + c_2 x^2 + c_3 x^3 + c_4 x^4 + c_5 x^5 + \cdots$$

$$= c_0 + c_1 x - c_0 x^2 - \frac{c_0}{3}x^4 - \frac{c_0}{5}x^6 - \frac{c_0}{7}x^8 + \cdots$$

which, after combining all terms containing c_0, leads to

$$y = c_0(1 - x^2 - \tfrac{1}{3}x^4 - \tfrac{1}{5}x^6 - \tfrac{1}{7}x^8 - \cdots) + c_1 x$$

$$= c_0 y_1(x) + c_1 y_2(x), \qquad |x| < 1$$

where

$$y_1(x) = 1 - x^2 - \tfrac{1}{3}x^4 - \tfrac{1}{5}x^6 - \tfrac{1}{7}x^8 - \cdots$$

and

$$y_2(x) = x$$

Lastly we can say that, based on Theorem 7.3, $y_1(x)$ and $y_2(x)$ form a fundamental set of solutions on the interval $|x| < 1$.   ■

Sometimes the pattern of the general term in a series is clear, and then it can be useful to write that series in summation notation. This is particularly helpful for estimating the sum of the series for some x in the interval of convergence. In the above example, for instance, we see that

$$y_1(x) = 1 - x^2 - \tfrac{1}{3}x^4 - \tfrac{1}{5}x^6 - \tfrac{1}{7}x^8 - \cdots$$

$$= 1 - \sum_{n=1}^{\infty} \frac{x^{2n}}{2n - 1}$$

EXAMPLE 7

Solve the IVP

$$y'' + xy' + y = 0, \qquad y(0) = 3, \qquad y'(0) = -7$$

Solution

Again the coefficients are polynomials and since $A_2(x) = 1 \neq 0$, the given DE has no singular points. Thus, $x = 0$ is an ordinary point and the series solutions about this point will converge for all x. By substituting the series representations for y, y', and y'' into the DE and simplifying, we obtain

$$\sum_{n=0}^{\infty} n(n - 1)c_n x^{n-2} + \sum_{n=0}^{\infty} nc_n x^n + \sum_{n=0}^{\infty} c_n x^n = 0$$

$$\Downarrow \qquad\qquad \Downarrow$$
$$n \to n - 2 \qquad n \to n - 2$$

After changing indices as shown, we get

$$\sum_{n=0}^{\infty} n(n - 1)c_n x^{n-2} + \sum_{n=2}^{\infty} (n - 2)c_{n-2} x^{n-2} + \sum_{n=2}^{\infty} c_{n-2} x^{n-2} = 0$$

or

$$0 \cdot c_0 x^{-2} + 0 \cdot c_1 x^{-1} + \sum_{n=2}^{\infty} [n(n - 1)c_n + (n - 2)c_{n-2} + c_{n-2}]x^{n-2} = 0$$

Clearly, both c_0 and c_1 are arbitrary, and the remaining constants are solutions of the recurrence relation

$$n(n - 1)c_n + (n - 2)c_{n-2} + c_{n-2} = 0, \qquad n = 2, 3, 4, \ldots$$

Upon simplification, we see that

$$c_n = -\frac{1}{n}c_{n-2}, \qquad n = 2, 3, 4, \ldots$$

from which we deduce

$$c_2 = -\tfrac{1}{2}c_0$$

$$c_3 = -\tfrac{1}{3}c_1$$

$$c_4 = -\tfrac{1}{4}c_2 = \frac{c_0}{2 \cdot 4}$$

$$c_5 = -\tfrac{1}{5}c_3 = \frac{c_1}{3 \cdot 5}$$

$$c_6 = -\tfrac{1}{6}c_4 = -\frac{c_0}{2 \cdot 4 \cdot 6}$$

$$c_7 = -\tfrac{1}{7}c_5 = -\frac{c_1}{3 \cdot 5 \cdot 7},$$

. . . .

Hence, the general solution is given by

$$y(x) = c_0 + c_1 x + c_2 x^2 + c_3 x^3 + c_4 x^4 + \cdots$$

$$= c_0\left(1 - \frac{x^2}{2} + \frac{x^4}{2 \cdot 4} - \frac{x^6}{2 \cdot 4 \cdot 6} + \cdots\right)$$

$$+ c_1\left(x - \frac{x^3}{3} + \frac{x^5}{3 \cdot 5} - \frac{x^7}{3 \cdot 5 \cdot 7} + \cdots\right)$$

Before imposing the initial conditions on this series solution, we first observe that the derivative is

$$y'(x) = c_0\left(-x + \frac{x^3}{2} - \cdots\right) + c_1\left(1 - x^2 + \cdots\right)$$

Now setting $x = 0$ in $y(x)$ and $y'(x)$, it follows that

$$y(0) = c_0 = 3$$

$$y'(0) = c_1 = -7$$

The solution of the IVP is therefore given by

$$y = 3\left(1 - \frac{x^2}{2} + \frac{x^4}{2 \cdot 4} - \frac{x^6}{2 \cdot 4 \cdot 6} + \cdots\right)$$

$$- 7\left(x - \frac{x^3}{3} + \frac{x^5}{3 \cdot 5} - \frac{x^7}{3 \cdot 5 \cdot 7} + \cdots\right)$$

The general term of both series solutions above have well-defined patterns. For example, the first series leads to

$$y_1(x) = 1 - \frac{x^2}{2} + \frac{x^4}{2 \cdot 4} - \frac{x^6}{2 \cdot 4 \cdot 6} + \cdots$$

$$= 1 + \left(-\frac{x^2}{2}\right) + \frac{1}{2!}\left(-\frac{x^2}{2}\right)^2 + \frac{1}{3!}\left(-\frac{x^2}{2}\right)^3 + \cdots$$

$$= \sum_{n=0}^{\infty} \frac{(-1)^n}{n!}\left(\frac{x^2}{2}\right)^n$$

which we recognize as

$$y_1(x) = e^{-x^2/2}$$

The second series can likewise be expressed in the form

$$y_2(x) = x - \frac{x^3}{3} + \frac{x^5}{3 \cdot 5} - \frac{x^7}{3 \cdot 5 \cdot 7} + \cdots$$

$$= x - \frac{2x^3}{3!} + \frac{2^2 2! x^5}{5!} - \frac{2^3 3! x^7}{7!} + \cdots$$

or

$$y_2(x) = \sum_{n=0}^{\infty} \frac{(-1)^n 2^n n!}{(2n+1)!} x^{2n+1}$$

Thus, our solution becomes

$$y = 3e^{-x^2/2} - 7\sum_{n=0}^{\infty} \frac{(-1)^n 2^n n!}{(2n+1)!} x^{2n+1}, \qquad |x| < \infty \qquad \blacksquare$$

Remark. Notice in Example 7 that $c_0 = y(0)$ and $c_1 = y'(0)$. It is easily shown that in general this will be the case. (Proof?)

Although there is the natural tendency to immediately perform the indicated products that occur in the coefficients of the solution, as in the above example, it is usually best to initially refrain from doing this. That is, if there is a recognizable pattern from which the general term of the series can be identified, it is usually easier to recognize this pattern by leaving the coefficients expressed as products.

If, in solving an IVP, the initial values of y and y' are prescribed at a point other than $x = 0$, the solution technique is slightly altered. For instance, if the initial conditions in Example 7 were prescribed at $x = 1$, that is, if

$$y'' + xy' + y = 0, \qquad y(1) = 3, \qquad y'(1) = -7 \qquad (22)$$

it would be best to seek solutions of the form

$$y = \sum_{n=0}^{\infty} c_n(x-1)^n \qquad (23)$$

rather than in powers of x. Then once again we would find $c_0 = y(1) = 3$ and $c_1 = y'(1) = -7$. The alternative is to set $t = x - 1$ in (22), which leads to

$$y'' + (t + 1)y' + y = 0, \qquad y(0) = 3, \qquad y'(0) = -7 \tag{24}$$

where the primes now denote differentiation with respect to t. Setting $t = x - 1$ also in (23), we can then proceed as before.

■————————
EXAMPLE 8

Solve the IVP

$$y'' + (x + 1)^2 y' - 4(x + 1)y = 0, \qquad y(-1) = 5, \qquad y'(-1) = 2$$

Solution

Because the initial conditions are prescribed at $x = -1$, rather than at $x = 0$, we first make the change of variable $t = x + 1$. In terms of t, therefore, the given IVP becomes

$$y'' + t^2 y' - 4ty = 0, \qquad y(0) = 5, \qquad y'(0) = 2$$

where the primes now denote differentiation with respect to t. Clearly, all points of this DE are ordinary; in particular, $t = 0$ is an ordinary point. Hence, we seek solutions of the form

$$y = \sum_{n=0}^{\infty} c_n t^n$$

which, according to Theorem 7.3, converge for all values of t (and consequently for all values of x).

By substituting this series for y (and the corresponding series for y' and y'') into the above DE, we obtain

$$\sum_{n=0}^{\infty} n(n-1)c_n t^{n-2} + \underset{n \to n-3}{\sum_{n=0}^{\infty} nc_n t^{n+1}} - \underset{n \to n-3}{\sum_{n=0}^{\infty} 4c_n t^{n+1}} = 0$$

Making the appropriate shifts in indices and combining terms, we are then led to

$$0 \cdot c_0 t^{-2} + 0 \cdot c_1 t^{-1} + 2c_2 t^0 + \sum_{n=3}^{\infty} [n(n-1)c_n + (n-3)c_{n-3} - 4c_{n-3}]t^{n-2} = 0$$

Once again we see that c_0 and c_1 are arbitrary, but here we must set $c_2 = 0$. The remaining constants are then determined from the recurrence formula

$$n(n-1)c_n + (n-7)c_{n-3} = 0$$

or

$$c_n = -\frac{n-7}{n(n-1)} c_{n-3}, \qquad n = 3, 4, 5, \ldots$$

Hence, other than c_0 and c_1, we see that

$$c_2 = 0$$

$$c_3 = -\frac{(-4)}{3 \cdot 2} c_0 = \frac{4}{3 \cdot 2} c_0$$

$$c_4 = -\frac{(-3)}{4 \cdot 3} c_1 = \frac{3}{4 \cdot 3} c_1$$

$$c_5 = -\frac{(-2)}{5 \cdot 4} c_2 = 0$$

$$c_6 = -\frac{(-1)}{6 \cdot 5} c_3 = \frac{4}{6 \cdot 5 \cdot 3 \cdot 2} c_0$$

$$c_7 = 0$$

$$c_8 = 0$$

$$c_9 = -\frac{2}{9 \cdot 8} c_6 = -\frac{2 \cdot 4}{9 \cdot 8 \cdot 6 \cdot 5 \cdot 3 \cdot 2} c_0$$

$$c_{10} = -\frac{3}{10 \cdot 9} c_7 = 0$$

$$c_{11} = 0$$

$$\dots\dots\dots\dots\dots\dots\dots\dots\dots\dots\dots\dots$$

Thus, the solution becomes

$$y(t) = c_0 + c_1 t + c_2 t^2 + c_3 t^3 + c_4 t^4 + \cdots$$

$$= c_0 \left(1 + \frac{4}{3 \cdot 2} t^3 + \frac{4}{6 \cdot 5 \cdot 3 \cdot 2} t^6 - \frac{2 \cdot 4}{9 \cdot 8 \cdot 6 \cdot 5 \cdot 3 \cdot 2} t^9 + \cdots \right) + c_1 \left(t + \frac{1}{4} t^4 \right)$$

Applying the initial conditions, we see that $c_0 = 5$ and $c_1 = 2$. Finally, setting $t = x + 1$, we get the desired solution

$$y(x) = 5 \left[1 + \frac{4}{3 \cdot 2} (x + 1)^3 + \frac{4}{6 \cdot 5 \cdot 3 \cdot 2} (x + 1)^6 \right.$$

$$\left. - \frac{2 \cdot 4}{9 \cdot 8 \cdot 6 \cdot 5 \cdot 3 \cdot 2} (x + 1)^9 + \cdots \right] + 2 \left[(x + 1) + \frac{1}{4} (x + 1)^4 \right]$$

$$= 5 \left[1 + \frac{2}{3} (x + 1)^3 + \frac{1}{45} (x + 1)^6 - \frac{1}{1620} (x + 1)^9 + \cdots \right]$$

$$+ 2 \left[(x + 1) + \frac{1}{4} (x + 1)^4 \right]$$

In some cases the recurrence formula may involve more than two terms in the unknown coefficients. Consider the next example.

EXAMPLE 9

Solve

$$y'' + (1 + x)y = 0$$

Solution

This DE has no singularities and thus we are sure of obtaining two series solutions that converge for all x. Because of the similarity in this DE and that in the last example, we might let $t = 1 + x$ and find a solution about $t = 0$. This would lead to a series in x about $x = -1$. However, since initial conditions are not prescribed, this is unnecessary. Also, as we previously mentioned, many DEs are formulated in such a way that $x = 0$ is a special point of interest. In such cases it is often desirable to obtain a series expansion about the point $x = 0$, rather than some other point, since it may reveal more information about the solution in the vicinity of the origin.

Suppose we seek a series solution about $x = 0$. The substitution of the appropriate series for y, y', and y'' into the DE yields

$$\sum_{n=0}^{\infty} n(n-1)c_n x^{n-2} + \underset{n \to n-2}{\sum_{n=0}^{\infty} c_n x^n} + \underset{n \to n-3}{\sum_{n=0}^{\infty} c_n x^{n+1}} = 0$$

or, changing indices as shown,

$$\sum_{n=0}^{\infty} n(n-1)c_n x^{n-2} + \sum_{n=2}^{\infty} c_{n-2} x^{n-2} + \sum_{n=3}^{\infty} c_{n-3} x^{n-2} = 0$$

Next, starting all summations at $n = 3$, we have

$$0 \cdot c_0 x^{-2} + 0 \cdot c_1 x^{-1} + (2c_2 + c_0)x^0 + \sum_{n=3}^{\infty} [n(n-1)c_n + c_{n-2} + c_{n-3}]x^{n-2} = 0$$

Thus, c_0 and c_1 are arbitrary while

$$c_2 = -\frac{c_0}{2}$$

$$c_n = -\frac{c_{n-2} + c_{n-3}}{n(n-1)}, \qquad n = 3, 4, 5, \ldots$$

To simplify matters here and still obtain two linearly independent solutions, let us first set $c_0 = 1$ and $c_1 = 0$ (since they are arbitrary constants, we can equate them to any value). This will produce one solution function $y_1(x)$. Then, we set $c_0 = 0$ and $c_1 = 1$, which will produce a second solution function $y_2(x)$ that is linearly independent of $y_1(x)$.

Setting $c_0 = 1$ and $c_1 = 0$, we find that

$$c_2 = -\frac{1}{2} c_0 = -\frac{1}{2}$$

$$c_3 = -\frac{c_1 + c_0}{3 \cdot 2} = -\frac{1}{3!}$$

$$c_4 = -\frac{c_2 + c_1}{4 \cdot 3} = \frac{1}{4!}$$

$$c_5 = -\frac{c_3 + c_2}{5 \cdot 4} = \frac{4}{5!}$$

· · · · · · · · · · · · · · · · ·

In this case, the first solution is

$$y_1(x) = 1 - \frac{x^2}{2!} - \frac{x^3}{3!} + \frac{x^4}{4!} + \frac{4x^5}{5!} + \cdots$$

Similarly, if we now set $c_0 = 0$ and $c_1 = 1$, we obtain

$$c_2 = -\frac{1}{2} c_0 = 0$$

$$c_3 = -\frac{c_1 + c_0}{3 \cdot 2} = -\frac{1}{3!}$$

$$c_4 = -\frac{c_2 + c_1}{4 \cdot 3} = -\frac{2}{4!}$$

$$c_5 = -\frac{c_3 + c_2}{5 \cdot 4} = \frac{1}{5!}$$

· · · · · · · · · · · · · · · · ·

providing us with the second solution

$$y_2(x) = x - \frac{x^3}{3!} - \frac{2x^4}{4!} + \frac{x^5}{5!} + \cdots$$

The general solution is therefore given by the linear combination

$$y = Ay_1(x) + By_2(x)$$

$$= A\left(1 - \frac{x^2}{2!} - \frac{x^3}{3!} + \frac{x^4}{4!} + \frac{4x^5}{5!} + \cdots\right)$$

$$+ B\left(x - \frac{x^3}{3!} - \frac{2x^4}{4!} + \frac{x^5}{5!} + \cdots\right)$$

where A and B are arbitrary constants.   ∎

> **Remark.** Setting one of the arbitrary constants to zero and finding a corresponding solution (as in the above example) is not required but merely a gimmick to perhaps simplify some of the algebraic manipulations in obtaining the constants.

Exercises 7.3

In Problems 1 to 8, list all singular points (real and complex) of the given DE.

1. $(1 + x)y' + y = 0$

2. $x(3 + x)y'' - (3 + x)y' + 2xy = 0$

3. $5y'' + 6xy' - x^2y = 0$

4. $y'' - \dfrac{4}{x(1 + 4x^2)} y' + \dfrac{1}{x^2(1 + 4x^2)} y = 0$

5. $(x^2 - x - 2)y'' + xy' - (x + 1)y = 0$

■**6.** $(\cos x)y'' - (\sin x)y = 0$

7. $e^{-x}y'' - x^2y' + (x^3 - 1)y = 0$

8. $[\log(x - 3)]y'' - e^xy' = 0$

In Problems 9 to 12, find a power series solution about $x = 0$ for the given first-order DE.

9. $(1 + x)y' + y = 0$

10. $y' + 2xy = 0$

11. $y' + y = x$

***12.** $y' + (\sin x)y = 0$

   Hint. In Problem 12 expand $\sin x$ in a power series about $x = 0$ and find only the first three nonzero terms in the general solution.

In Problems 13 to 22, find at least the first *four* nonzero terms in the power series expansion about $x = 0$ of the solutions. Also, state the interval of convergence as predicted by Theorem 7.3.

13. $y'' + 4y = 0$

14. $y'' - 4y = 0$

15. $(1 + 4x^2)y'' - 8y = 0$

16. $y'' + 2xy' + 5y = 0$

17. $(1 - x^2)y'' - 2xy' + 12y = 0$

18. $(x^2 + 4)y'' + 2xy' - 12y = 0$

19. $y'' + xy' + (x^2 + 2)y = 0$

■**20.** $y'' + x^2y' + 3xy = 0$

21. $(x - 1)y'' + y' = 0$

22. $(x + 3)y'' + (x + 2)y' + y = 0$

In Problems 23 to 26, find at least the first *four* nonzero terms in the power series expansion about $x = 0$ of the solution to the given IVP.

23. $y'' - xy' - y = 0$, $y(0) = 1$, $y'(0) = 0$

■**24.** $y'' + xy' - 2y = 0$, $y(0) = 0$, $y'(0) = 1$

25. $(x^2 - 1)y'' + 3xy' + xy = 0$, $y(0) = 4$, $y'(0) = 6$

***26.** $3y'' - y' + (x + 1)y = 1$, $y(0) = 0$, $y'(0) = 0$

In Problems 27 to 29, find at least the first *four* nonzero terms in the power series expansions about $x = 1$ of the solutions of the given DE and state the interval of convergence.

27. $y'' + (x - 1)y = 0$

***28.** $xy'' + y' + xy = 0$, $y(1) = 0$, $y'(1) = -1$

29. $(x^2 - 2x + 2)y'' - 4(x - 1)y' + 6y = 0$

***30.** Find at least the first *four* nonzero terms in the power series expansion about $x = 0$ of the

solutions of

$$y'' - 4xy' - 4y = e^x$$

Hint. Expand e^x in a power series about $x = 0$.

In Problems 31 and 32, find at least the first *four* nonzero terms of the power series expansion about $x = 0$ of the solutions of the given DE.

*31. $y''' - 3xy' - y = 0$

*32. $y''' + x^2y'' + 5xy' + 3y = 0$

*33. Hermite's equation[†] is given by

$$y'' - 2xy' + 2ny = 0, \qquad n \geq 0$$

Obtain two linearly independent solutions for the case when

(a) $n = 1$

(b) $n = 4$

(c) Show that when n is any nonnegative integer, one solution of Hermite's DE is always a polynomial of degree n.

*34. Chebyshëv's equation[‡] is given by

$$(1 - x^2)y'' - xy' + n^2y = 0, \qquad n \geq 0$$

Obtain two linearly independent solutions for the case when

(a) $n = 1$

(b) $n = 4$

(c) Show that when n is any nonnegative integer, one solution of Chebyshëv's DE is always a polynomial of degree n.

7.4 Solutions Near a Singular Point—Part I

In the neighborhood of a singular point the behavior of the solution of a DE, and hence that of the physical system described by the DE, is usually quite different from that near an ordinary point. Thus, solutions in the neighborhood of singular points are often of special interest. Unfortunately, when x_0 is a singular point of the DE, it may not be possible to find a power series solution of the form

$$y = \sum_{n=0}^{\infty} c_n(x - x_0)^n$$

Of course, we could always choose an ordinary point about which to form a series expansion, but such solutions are often not valid at the singular points and thus do not provide us with the information we seek.

Even though a standard power series solution about the singular point may not exist, we may be able to find a solution of the slightly different form

$$y = (x - x_0)^s \sum_{n=0}^{\infty} c_n(x - x_0)^n \tag{25}$$

where s is an unknown parameter to be determined along with the c's.

[†] Named after the French mathematician Charles Hermite (1822–1901). Hermite showed in 1858 that the general fifth-degree equation can be solved by elliptic functions and proved in 1873 that e is a transcendental number (i.e., a type of irrational number). He made contributions to the fields of number theory and analysis.

[‡] Named after the Russian mathematician Pafnuti Lvovich Chebyshëv (1821–1894). Much of his work involved prime numbers, but he is also known for his work in probability, the theory of numbers, and the approximation of functions. Other spellings of his name include *Tchebysheff* and *Tchebycheff*, among others.

In general, a singular point x_0 of the DE

$$y'' + a(x)y' + b(x)y = 0 \qquad (26)$$

is one for which either $\lim_{x \to x_0} a(x) \to \infty$ or $\lim_{x \to x_0} b(x) \to \infty$ (or both). In other words, either (or both) $a(x)$ or $b(x)$ fails to have a power series about a point if it is a singular point. A singular point is further classified as *regular* or *irregular* according to the following definition.

Definition 7.2 If x_0 is a singular point of (26), it is classified as a **regular singular point** (RSP) if both

$$(x - x_0)a(x) \quad \text{and} \quad (x - x_0)^2 b(x)$$

have power series about x_0; otherwise, we say that x_0 is an **irregular singular point** (ISP).

In the special case when $a(x)$ and $b(x)$ in (26) are both *rational functions,* an RSP is a singular point for which both $(x - x_0)a(x)$ and $(x - x_0)^2 b(x)$ have *finite limits* as $x \to x_0$. Thus, in this case the denominator of $a(x)$ cannot contain a factor $(x - x_0)$ to a power higher than 1, and the denominator of $b(x)$ cannot contain a factor $(x - x_0)$ to a power higher than 2.

EXAMPLE 10

Classify the singular points of
(a) $xy'' + y' + xy = 0$
(b) $x(x - 1)^2(x + 3)y'' + 5x^2 y' + (2x^3 + 1)y = 0$

Solution

(a) Clearly, $x = 0$ is the only singular point. Dividing the DE by x puts it in the normal form

$$y'' + \frac{1}{x} y' + y = 0$$

where we identify

$$a(x) = \frac{1}{x}, \qquad b(x) = 1$$

Because $a(x)$ and $b(x)$ are both rational functions and

$$\lim_{x \to 0} xa(x) = 1, \qquad \lim_{x \to 0} x^2 b(x) = 0$$

we conclude that $x = 0$ is an RSP.

(b) The singular points are $x = 0, 1, -3$. Putting the DE in normal form, we see that

$$a(x) = \frac{5x}{(x - 1)^2(x + 3)}, \qquad b(x) = \frac{2x^3 + 1}{x(x - 1)^2(x + 3)}$$

Thus, for $x = 0$, we find

$$\lim_{x \to 0} xa(x) = \lim_{x \to 0} \frac{5x^2}{(x-1)^2(x+3)} = 0$$

$$\lim_{x \to 0} x^2 b(x) = \lim_{x \to 0} \frac{x(2x^3 + 1)}{(x-1)^2(x+3)} = 0$$

Clearly, $x = 0$ is an RSP. Similarly, at $x = 1$ and $x = -3$, we find, respectively,

$$\lim_{x \to 1} (x-1)a(x) = \infty, \qquad \lim_{x \to 1} (x-1)^2 b(x) = \frac{3}{4}$$

and

$$\lim_{x \to -3} (x+3)a(x) = -\frac{15}{16}, \qquad \lim_{x \to -3} (x+3)^2 b(x) = 0$$

Thus, $x = 1$ is an ISP while $x = -3$ is an RSP. ■

In the remainder of the chapter we concern ourselves only with the theory of solutions near *regular* singular points.

7.4.1 Regular Singular Points: Method of Frobenius

In 1873, a method was published by the German mathematician G. Frobenius[†] for finding a solution of a DE about a regular singular point. The method is based upon the following theorem, and although it can be generalized to any RSP, we present the theorem and method only for the case when $x = 0$ is the RSP. Once again this is simply a mathematical convenience and causes no loss of generality. For example, if $x = x_0$ is the singular point of interest, we can transform the DE to one for which zero is the singular point by setting $t = x - x_0$.

Theorem 7.4 *If $x = 0$ is an RSP of the DE*

$$A_2(x)y'' + A_1(x)y' + A_0(x)y = 0$$

then there exists at least one solution of the form

$$y = x^s \sum_{n=0}^{\infty} c_n x^n, \qquad c_0 \neq 0, \qquad x > 0$$

Moreover, the series will converge at least in the interval $0 \leq x < R$, where R is the distance from $x = 0$ to the nearest other singular point.

In finding a solution of the form suggested in Theorem 7.4, it is necessary to determine the values of the parameter s in addition to the coefficients $c_0, c_1, c_2, \ldots,$

[†] F. Georg Frobenius (1849–1917) is known for his research in algebra and analysis (see also Section 7.9).

c_n, ... The restriction $x > 0$ is required to prevent complex solutions that might arise for $x < 0$ and certain values of s. If we need a solution that is valid for $x < 0$, we can make the simple change of variable $t = -x$ and solve the resulting DE for $t > 0$.

If we formally differentiate the series

$$y = x^s \sum_{n=0}^{\infty} c_n x^n = \sum_{n=0}^{\infty} c_n x^{n+s}, \qquad x > 0 \tag{27}$$

then

$$y' = \sum_{n=0}^{\infty} (n+s)c_n x^{n+s-1} \tag{28a}$$

$$y'' = \sum_{n=0}^{\infty} (n+s)(n+s-1)c_n x^{n+s-2} \tag{28b}$$

and so on. The substitution of these series for y, y', and y'' into the second-order homogeneous DE (in normal form)

$$y'' + a(x)y' + b(x)y = 0 \tag{29}$$

leads to

$$\sum_{n=0}^{\infty} (n+s)(n+s-1)c_n x^{n+s-2} + a(x) \sum_{n=0}^{\infty} (n+s)c_n x^{n+s-1} + b(x) \sum_{n=0}^{\infty} c_n x^{n+s} = 0$$

or, upon combining the series,

$$\sum_{n=0}^{\infty} [(n+s)(n+s-1) + xa(x)(n+s) + x^2 b(x)]c_n x^{n+s-2} = 0 \tag{30}$$

Assuming that $x = 0$ is an RSP of (29), it follows that $xa(x)$ and $x^2 b(x)$ have power series expansions of the form

$$\begin{aligned} xa(x) &= \alpha_0 + \alpha_1 x + \alpha_2 x^2 + \cdots \\ x^2 b(x) &= \beta_0 + \beta_1 x + \beta_2 x^2 + \cdots \end{aligned} \tag{31}$$

Now let us substitute these expressions into (30) and group the resulting terms in like powers of x. This action yields the rather cumbersome expression

$$\begin{aligned} [s(s-1) + \alpha_0 s + \beta_0]c_0 x^{s-2} &+ [(s+1)sc_1 + (s+1)\alpha_0 c_1 \\ &+ s\alpha_1 c_0 + \beta_0 c_1 + \beta_1 c_0]x^{s-1} + \cdots = 0 \end{aligned} \tag{32}$$

As before, this last equation can be satisfied only if the coefficients of all powers of x vanish independently. In particular, the first term involving x^{s-2} can vanish ($x \neq 0$) if and only if

$$[s(s-1) + \alpha_0 s + \beta_0]c_0 = 0$$

Because we assume in Theorem 7.4 that $c_0 \neq 0$, it follows that the term in brackets must vanish, that is,

$$s(s-1) + \alpha_0 s + \beta_0 = 0 \tag{33}$$

This important quadratic equation in s is called the **indicial equation** of the DE (29). The two roots of the indicial equation, s_1 and s_2, are referred to as the **exponents** of the singularity. It can be shown that one solution of the form given in Theorem 7.4 always exists, corresponding to the *larger* root of the indicial equation. Finding a second linearly independent solution, however, will depend upon the nature of the two roots s_1 and s_2. Specifically, there are three separate cases to consider:

Case I—Roots differing by a noninteger

Case II—Equal roots

Case III—Roots differing by a nonzero integer

We consider the above three cases separately under the assumption that s_1 and s_2 are *real* roots of the indicial equation (33). Complex roots are also possible, but we will not discuss such cases.

EXAMPLE 11

Find the indicial equation and its solutions at the singularity $x = 0$ of

$$2xy'' + (1 - 2x)y' - y = 0$$

Solution

In normal form the DE becomes

$$y'' + \left(\frac{1}{2x} - 1\right)y' - \frac{1}{2x}y = 0$$

from which we identify

$$a(x) = \frac{1}{2x} - 1, \qquad b(x) = -\frac{1}{2x}$$

Hence, it follows that

$$\alpha_0 = \lim_{x \to 0} xa(x) = \lim_{x \to 0}\left(\frac{1}{2} - x\right) = \frac{1}{2}$$

$$\beta_0 = \lim_{x \to 0} x^2 b(x) = \lim_{x \to 0}\left(-\frac{x}{2}\right) = 0$$

Using $\alpha_0 = 1/2$ and $\beta_0 = 0$, the indicial equation (33) reduces to

$$s(s - 1) + \tfrac{1}{2}s = 0$$

or

$$s^2 - s/2 = s(s - 1/2) = 0$$

The solutions of this equation are clearly given by

$$s_1 = 0, \qquad s_2 = \tfrac{1}{2}$$

which belong to Case I above.

■

7.4.2 Case I: Roots Differing by a Noninteger

The case where $s_1 - s_2 \neq N$, $N = 0, \pm 1, \pm 2, \ldots$, always leads to two linearly independent solutions of the form given in Theorem 7.4. Specifically, we have the following theorem.

Theorem 7.5 *If $x = 0$ is an RSP of the DE*

$$A_2(x)y'' + A_1(x)y' + A_0(x)y = 0$$

and s_1 and s_2 are distinct roots of the indicial equation whose difference is not integral, then the general solution is given by

$$y = Ay_1(x) + By_2(x), \qquad x > 0$$

where A and B are arbitrary constants and where

$$y_1(x) = x^{s_1} \sum_{n=0}^{\infty} c_n(s_1)x^n, \qquad c_0(s_1) = 1$$

and

$$y_2(x) = x^{s_2} \sum_{n=0}^{\infty} c_n(s_2)x^n, \qquad c_0(s_2) = 1$$

The two sets of coefficients $c_n(s_1)$ and $c_n(s_2)$, $n = 0, 1, 2, \ldots$, occurring in Theorem 7.5 are obtained independently by replacing s by s_1 and s by s_2, respectively, in the recurrence formula. Let us illustrate with the following example.

EXAMPLE 12

Find the general solution of

$$2xy'' + (1 - 2x)y' - y = 0, \qquad x > 0$$

Solution

Since the coefficients are polynomials, we see immediately that $x = 0$ is the only singular point of the DE. Furthermore, it is easy to verify that it is an RSP.

By assuming a series solution of the form

$$y = x^s \sum_{n=0}^{\infty} c_n x^n = \sum_{n=0}^{\infty} c_n x^{n+s}$$

and substituting this expression and its derivatives into the DE, we obtain

$$\sum_{n=0}^{\infty} 2(n + s)(n + s - 1)c_n x^{n+s-1} + \sum_{n=0}^{\infty} (n + s)c_n x^{n+s-1}$$

$$- \underset{\underset{n \to n-1}{\Downarrow}}{\sum_{n=0}^{\infty}} 2(n + s)c_n x^{n+s} - \underset{\underset{n \to n-1}{\Downarrow}}{\sum_{n=0}^{\infty}} c_n x^{n+s} = 0$$

Making the appropriate change of indices as shown, we are led to

$$\sum_{n=0}^{\infty} 2(n+s)(n+s-1)c_n x^{n+s-1} + \sum_{n=0}^{\infty} (n+s)c_n x^{n+s-1}$$

$$-\sum_{n=1}^{\infty} 2(n+s-1)c_{n-1}x^{n+s-1} - \sum_{n=1}^{\infty} c_{n-1}x^{n+s-1} = 0$$

and by combining all terms under one summation sign, we have

$$[2s(s-1)+s]c_0 x^{s-1} + \sum_{n=1}^{\infty} [2(n+s)(n+s-1)c_n + (n+s)c_n$$

$$-2(n+s-1)c_{n-1} - c_{n-1}]x^{n+s-1} = 0$$

The coefficient of c_0 equated to zero gives us the indicial equation

$$2s^2 - s = s(2s-1) = 0$$

(which is equivalent to that found in Example 11). The roots are $s_1 = 0$ and $s_2 = 1/2$, and thus we can use Theorem 7.5 to find the general solution. The recurrence formula (for all s) is

$$(n+s)[2(n+s)-1]c_n(s) = [2(n+s)-1]c_{n-1}(s), \qquad n = 1, 2, 3, \ldots$$

where we are emphasizing the dependency of the c's upon the parameter s. But since $2(n+s)-1 \neq 0$ for $s = 0$ or $s = 1/2$ and $n = 1, 2, 3, \ldots$, the recurrence formula reduces to

$$c_n(s) = \frac{c_{n-1}(s)}{n+s}, \qquad n = 1, 2, 3, \ldots$$

Now putting $s = s_1 = 0$ in this general recurrence formula and setting $c_0(0) = 1$ for mathematical convenience, we obtain the particular recurrence relation $s_1 = 0$:

$$c_n = \frac{c_{n-1}}{n}, \qquad n = 1, 2, 3, \ldots$$

Hence, we find successively that

$$c_0 = 1$$

$$c_1 = c_0 = 1$$

$$c_2 = \frac{c_1}{2} = \frac{1}{2}$$

$$c_3 = \frac{c_2}{3} = \frac{1}{3!}$$

$$c_4 = \frac{c_3}{4} = \frac{1}{4!}$$

and so on. The first solution is therefore

$$y_1(x) = x^0(c_0 + c_1 x + c_2 x^2 + c_3 x^3 + \cdots)$$

$$= 1 + x + \frac{x^2}{2!} + \frac{x^3}{3!} + \frac{x^4}{4!} + \cdots$$

or

$$y_1(x) = e^x$$

Similarly, for $s = s_2 = 1/2$, the recurrence formula becomes $s_2 = \frac{1}{2}$:

$$c_n = \frac{2c_{n-1}}{2n + 1}, \qquad n = 1, 2, 3, \ldots$$

from which we find [here again we set $c_0(1/2) = 1$]

$$c_0 = 1$$

$$c_1 = \frac{2c_0}{3} = \frac{2}{3}$$

$$c_2 = \frac{2c_1}{5} = \frac{2^2}{3 \cdot 5}$$

$$c_3 = \frac{2c_2}{7} = \frac{2^3}{3 \cdot 5 \cdot 7}$$

and so forth. A second linearly independent solution is therefore given by

$$y_2(x) = x^{1/2}(c_0 + c_1 x + c_2 x^2 + c_3 x^3 + \cdots)$$

$$= x^{1/2}\left(1 + \frac{2x}{3} + \frac{2^2 x^2}{3 \cdot 5} + \frac{2^3 x^3}{3 \cdot 5 \cdot 7} + \cdots\right)$$

or

$$y_2(x) = x^{1/2}\left(1 + \frac{2x}{3} + \frac{4x^2}{15} + \frac{8x^3}{105} + \cdots\right)$$

Finally, the general solution we seek is

$$y = Ay_1(x) + By_2(x)$$

$$= Ae^x + Bx^{1/2}\left(1 + \frac{2x}{3} + \frac{4x^2}{15} + \frac{8x^3}{105} + \cdots\right)$$

where A and B are arbitrary constants. In this case it can be shown both $y_1(x)$ and $y_2(x)$ converge on the interval $|x| < \infty$. ■

The choice of setting $c_0(s) = 1$ for both $s = s_1$ and $s = s_2$ is one simply for convenience. That is, we do not need arbitrary constants in $y_1(x)$ and $y_2(x)$ provided we later introduce arbitrary constants A and B as given in Theorem 7.5.

> **Remark.** As a final comment here we note that if the roots s_1 and s_2 are complex, this case can be handled like Case I above except that the solutions corresponding to s_1 and s_2 will be complex functions. Real solutions can be obtained, however, by taking appropriate linear combinations of the complex solutions, similar to the technique used in Section 4.4.1 in obtaining trigonometric solutions from complex exponentials.

Exercises 7.4

In Problems 1 to 8, locate and classify the singular points of the DE.

1. $x^3(x^2 - 1)y'' - x(x + 1)y' + (x - 1)y = 0$

2. $(x^4 - 1)y'' + xy' + y = 0$

3. $x^4(x^2 + 1)(x - 1)^2 y'' + 4x^3(x - 1)y'$
    $+ (x + 1)y = 0$

4. $y'' + xy = 0$

5. $x^2(x - 4)^2 y'' + 3xy' - (x - 4)y = 0$

6. $x^2(x + 1)^2 y'' + (x^2 - 1)y' + 2y = 0$

7. $xy'' + (x - 3)^{-2}y = 0$

■**8.** $(x^5 + x^4 - 6x^3)y'' + 3x^2 y' + (x - 2)y = 0$

In Problems 9 to 14, find the roots of the indicial equation belonging to the singular point $x = 0$ and identify them as belonging to Case I, II, or III.

9. $xy'' + y' - 4y = 0$

10. $4x^2 y'' + (1 - 2x)y = 0$

11. $2x^2 y'' - xy' + (1 + x)y = 0$

12. $4xy'' + 3y' + 3y = 0$

13. $x^2 y'' + x^2 y' - 2y = 0$

■**14.** $x^2 y'' + xy' + (x^2 - 1)y = 0$

In Problems 15 to 24, use the method of Frobenius to obtain two linearly independent series solutions about the point $x = 0$. Find at least the first *three* nonzero terms of each series.

15. $2xy'' + (1 + x)y' - 2y = 0$

16. $4xy'' + 3y' + 3y = 0$

17. $2xy'' + 5(1 + 2x)y' + 5y = 0$

18. $x^2 y'' + x(x - \frac{1}{2})y' + \frac{1}{2}y = 0$

19. $2x^2 y'' + 3xy' - y = 0$

20. $2x^2 y'' + xy' - y = 0$

21. $2x^2 y'' - xy' + (1 + x)y = 0$

22. $2x^2 y'' - xy' + (x - 5)y = 0$

23. $2x^2 y'' + xy' + (x^2 - 3)y = 0$

■**24.** $2xy'' + (1 + 2x)y' - 5y = 0$

*25. The DE

$$x(1 - x)y'' + [c - (a + b + 1)x]y' - aby = 0$$

where a, b, and c are constants, is called the **hypergeometric equation,** and its solutions are called **hypergeometric functions.**

(a) Show that $x = 0$ and $x = 1$ are RSPs of the DE.

(b) Assuming $c \neq 0, -1, -2, \ldots$, show that one solution of this DE is

$$y_1(x) = F(a, b; c; x)$$

$$= 1 + \frac{ab}{c}x + \frac{a(a + 1)b(b + 1)}{c(c + 1)}\frac{x^2}{2!} + \cdots$$

*26. When $a = 1$ and $c = b$ in Problem 25, show that the series in (b) reduces to the geometric series; that is, show that

$$y_1(x) = F(1, b; b; x) = \frac{1}{1 - x}$$

*27. Assuming that $1 - c$ is not an integer, show that

$$y_2(x) = x^{1-c}F(1 - c + a, 1 - c + b; 2 - c; x)$$

is a second linearly independent solution of Problem 25.

*28. The DE

$$xy'' + (c - x)y' - ay = 0$$

where a and c are constants, is called the **confluent hypergeometric equation,** and its solutions are called **confluent hypergeometric functions.**

(a) Show that $x = 0$ is an RSP of the DE.

(b) Assuming that $c \neq 0, -1, -2, \ldots$, show that one solution of this DE is

$$y_1(x) = M(a; c; x)$$

$$= 1 + \frac{a}{c}x + \frac{a(a + 1)}{c(c + 1)} \frac{x^2}{2!} + \cdots$$

*29. When $c = a$ in Problem 28, show that the solution in (b) reduces to

$$y_1(x) = M(a; a; x) = e^x$$

*30. Assuming that $1 - c$ is not an integer, show that

$$y_2(x) = x^{1-c}M(a + 1 - c; 2 - c; x)$$

is a second linearly independent solution of Problem 28.

7.5 Solutions Near a Singular Point—Part II

When $s_1 - s_2 = N$, $N = 0, \pm 1, \pm 2, \ldots$, we are not always able to find *two* solutions of the general form

$$y = x^s \sum_{n=0}^{\infty} c_n x^n$$

In fact, we find that in general one of the solutions contains a *logarithmic* term. For discussion purposes, we consider separately the cases $s_1 = s_2$ and $s_1 - s_2 = N$, $N = \pm 1, \pm 2, \ldots$.

7.5.1 Case II: Equal Roots

When $s_1 = s_2$, the procedure used in Section 7.4.2 will yield only one solution. Nonetheless, two linearly independent solutions can be shown to exist in this case corresponding to

$$
\begin{aligned}
y_1(x) &= y(x, s_1) \\
&= x^{s_1}\left[1 + \sum_{n=1}^{\infty} c_n(s_1)x^n\right], \qquad c_0(s_1) = 1
\end{aligned}
\tag{34}
$$

and

$$
\begin{aligned}
y_2(x) &= \frac{\partial y(x, s)}{\partial s}\bigg|_{s=s_1} \\
&= y_1(x) \log x + x^{s_1} \sum_{n=1}^{\infty} c_n'(s_1)x^n
\end{aligned}
\tag{35}
$$

where

$$
y(x, s) = x^s\left[1 + \sum_{n=1}^{\infty} c_n(s)x^n\right], \qquad c_0(s) = 1
\tag{36}
$$

Remark. By (36), we simply mean the function obtained for arbitrary s which becomes a solution y_1 upon setting $s = s_1$.

To see why $y_2(x)$ can be generated from the derivative $\partial y(x, s)/\partial s$, let us consider the Cauchy–Euler equation

$$4x^2 y'' + y = 0 \tag{37}$$

whose indicial equation has the double root $s_1 = s_2 = 1/2$. (Recall from Example 31 in Section 4.9 that the general solution is $y = C_1 x^{1/2} + C_2 x^{1/2} \log x, x > 0$). In normal form, (37) becomes

$$N[y] = 0$$

where $N[y] = y'' + y/4x^2$. By seeking a solution of the form $y(x, s) = x^s$, we obtain

$$N[y(x, s)] = [s(s - 1) + \tfrac{1}{4}]x^{s-2}$$

$$= (s - \tfrac{1}{2})^2 x^{s-2} \tag{38}$$

Clearly, the right-hand side of (38) goes to zero if we set $s = 1/2$, giving us the solution $y_1(x) = y(x, 1/2) = x^{1/2}$. Also, by formally differentiating (38) with respect to s, we obtain

$$\frac{\partial}{\partial s} N[y(x, s)] = N\left[\frac{\partial y(x, s)}{\partial s}\right]$$

$$= [2(s - \tfrac{1}{2}) + (s - \tfrac{1}{2})^2 \log x]x^{s-2}$$

where we have reversed the order of differentiation with respect to s and x. If we now set $s = 1/2$, this last expression reduces to

$$N\left[\frac{\partial y(x, s)}{\partial s}\right]\Bigg|_{s=1/2} = 0$$

from which, using $y(x, s) = x^s$, we deduce that a second linearly independent solution is given by

$$y_2(x) = \frac{\partial y(x, s)}{\partial s}\Bigg|_{s=1/2} = x^{1/2} \log x$$

In the more general case, we have the DE

$$N[y] \equiv y'' + a(x)y' + b(x)y = 0 \tag{39}$$

The substitution of (36) into (39) leads to

$$N[y(x, s)] = (s - s_1)^2 x^{s-2} \tag{40}$$

where all remaining terms of the series vanish by appropriately selecting the c's. The repeated factor $(s - s_1)^2$, of course, is a consequence of the double root of the indicial equation. From this point the analysis is the same as that above for the Cauchy–Euler equation (37).

Finally, we observe that obtaining $y_2(x)$ from $\partial y(x, s)/\partial s$ requires us to find the derivatives $c'_n(s)$, which are usually expressed in the form of a finite product of functions. In such cases it is generally easier to use the technique of *logarithmic differentiation,* briefly reviewed in the following example.

EXAMPLE 13

Find the derivative $y'(s)$, where

$$y(s) = \frac{s^2(s + 2)}{(4s + 1)^3(5s - 3)^6}$$

Solution

To begin, we take the logarithm of each side to obtain

$$\log y(s) = 2 \log s + \log(s + 2) - 3 \log(4s + 1) - 6 \log(5s - 3)$$

The derivative of this expression leads to

$$\frac{y'(s)}{y(s)} = \frac{2}{s} + \frac{1}{s + 2} - \frac{12}{4s + 1} - \frac{30}{5s - 3}$$

from which we deduce

$$y'(s) = \left(\frac{2}{s} + \frac{1}{s + 2} - \frac{12}{4s + 1} - \frac{30}{5s - 3}\right) y(s) \qquad\blacksquare$$

Another way of obtaining a solution $y_2(x)$ in the case of equal roots is to use the method of Section 4.3. That is, if $y_1(x) = y(x, s_1)$ is a solution of (39), a second linearly independent solution is given by the expression

$$y_2(x) = y_1(x) \int \frac{\exp\left(-\int a(x)\, dx\right)}{[y_1(x)]^2}\, dx \qquad (41)$$

However, using (41) to generate $y_2(x)$ is quite tedious when $y_1(x)$ is an infinite series.

EXAMPLE 14

Find the general solution of

$$xy'' + y' + xy = 0, \qquad x > 0^\dagger$$

Solution

The assumption $y = \sum_{n=0}^{\infty} c_n x^{n+s}$ leads to

$$\sum_{n=0}^{\infty} c_n(n + s)(n + s - 1)x^{n+s-1} + \sum_{n=0}^{\infty} c_n(n + s)x^{n+s-1} + \sum_{\substack{n=0 \\ \Downarrow \\ n \to n-2}}^{\infty} c_n x^{n+s+1} = 0$$

---

† This DE is a special case of *Bessel's equation* (see Section 7.7).

After a change of index in the last sum, we obtain

$$[s(s - 1) + s]c_0 x^{s-1} + [(s + 1)s + s + 1]c_1 x^s$$

$$+ \sum_{n=2}^{\infty} [(n + s)(n + s - 1)c_n + (n + s)c_n + c_{n-2}]x^{n+s-1} = 0$$

Next, by setting the coefficient of c_0 to zero we have

$$s(s - 1) + s = 0$$

from which we deduce $s_1 = s_2 = 0$. Hence, c_0 is arbitrary but the remaining constants must satisfy

$$[(s + 1)s + s + 1]c_1 = 0$$

$$[(n + s)(n + s - 1) + (n + s)]c_n + c_{n-2} = 0, \qquad n = 2, 3, 4, \ldots$$

For $s = 0$, the coefficient of c_1 does not vanish so we are forced to set $c_1 = 0$. The remaining recurrence formula simplifies to

$$(n + s)^2 c_n + c_{n-2} = 0$$

or

$$c_n(s) = -\frac{c_{n-2}(s)}{(s + n)^2}, \qquad n = 2, 3, 4, \ldots$$

In this last form of the recurrence formula we are emphasizing the dependence of these constants on the parameter s. Since the solution formula (35) requires derivatives of the c's with respect to s, we will not substitute the value $s = 0$ into the recurrence formula until later. By setting $c_0(s) = 1$ for convenience, the above recurrence formula leads to

$$c_2(s) = -\frac{c_0(s)}{(s + 2)^2} = -\frac{1}{(s + 2)^2}$$

$$c_3(s) = -\frac{c_1(s)}{(s + 3)^2} = 0$$

$$c_4(s) = -\frac{c_2(s)}{(s + 4)^2} = \frac{1}{(s + 2)^2(s + 4)^2}$$

$$c_5(s) = 0$$

$$c_6(s) = -\frac{c_4(s)}{(s + 6)^2} = -\frac{1}{(s + 2)^2(s + 4)^2(s + 6)^2}$$

. . . .

In this case the general pattern for $c_n(s)$ is quite clear. For instance, we see that $c_n(s) = 0$, $n = 1, 3, 5, \ldots$, whereas for even values of n we find it convenient to write

$$c_n(s) \equiv c_{2m}(s) = \frac{(-1)^m}{(s + 2)^2(s + 4)^2 \cdots (s + 2m)^2}, \qquad m = 1, 2, 3, \ldots$$

To obtain $y_1(x)$, we set $s = 0$ in the above formula, finding that

$$c_{2m}(0) = \frac{(-1)^m}{(2^2)(4^2)\cdots(2^2 m^2)} = \frac{(-1)^m}{2^{2m}(m!)^2}, \quad m = 1, 2, 3, \ldots$$

Hence, the solution we seek becomes

$$y_1(x) = c_0 + \sum_{m=1}^{\infty} c_{2m}(0)x^{2m}$$

$$= 1 + \sum_{m=1}^{\infty} \frac{(-1)^m}{2^{2m}(m!)^2} x^{2m}$$

Recognizing that $c_{2m}(0) = 1$ when $m = 0$, we can write y_1 in the more compact form

$$y_1(x) = J_0(x) = \sum_{m=0}^{\infty} \frac{(-1)^m}{(m!)^2}\left(\frac{x}{2}\right)^{2m}$$

where the symbol $J_0(x)$ denotes the **Bessel function of the first kind** and order zero (see also Section 7.7).

Next, in order to calculate the derivatives $c'_{2m}(s)$, we use logarithmic differentiation, from which we deduce

$$c'_{2m}(s) = \left(-\frac{2}{s+2} - \frac{2}{s+4} - \cdots - \frac{2}{s+2m}\right)c_{2m}(s)$$

When $s = 0$, we get

$$c'_{2m}(0) = \frac{(-1)^{m-1}}{2^{2m}(m!)^2}\left(1 + \frac{1}{2} + \frac{1}{3} + \cdots + \frac{1}{m}\right)$$

It is a common practice to introduce the notation

$$H_m = 1 + \frac{1}{2} + \frac{1}{3} + \cdots + \frac{1}{m}$$

for the partial sum of the harmonic series. Thus, based on (35), we obtain the second solution

$$y_2(x) = y_1(x)\log x + \sum_{m=1}^{\infty} \frac{(-1)^{m-1}H_m}{(m!)^2}\left(\frac{x}{2}\right)^{2m}$$

The general solution, therefore, is

$$y = Ay_1(x) + By_2(x)$$

$$= (A + B\log x)\sum_{m=0}^{\infty} \frac{(-1)^m}{(m!)^2}\left(\frac{x}{2}\right)^{2m} + B\sum_{m=1}^{\infty} \frac{(-1)^{m-1}H_m}{(m!)^2}\left(\frac{x}{2}\right)^{2m}, \quad x > 0$$

where A and B are arbitrary constants. ∎

Rather than relying on (35) to find the second solution in Example 14 we could use (41), which in this case becomes

$$y_2(x) = y_1(x)\int \frac{dx}{x[y_1(x)]^2} \tag{42}$$

Performing some lengthy calculations, we have

$$\frac{1}{[y_1(x)]^2} = \frac{1}{[1 - (x^2/4) + (x^4/64) - (x^6/2304) + \cdots]^2}$$

$$= \frac{1}{1 - (x^2/2) + (3x^4/32) - (5x^6/576) + \cdots}$$

$$= 1 + \frac{x^2}{2} + \frac{5x^4}{32} + \frac{23x^6}{576} + \cdots$$

this last step resulting from long division. Thus, from (42) we obtain

$$y_2(x) = y_1(x) \int \left(\frac{1}{x} + \frac{x}{2} + \frac{5x^3}{32} + \frac{23x^5}{576} + \cdots \right) dx$$

$$= y_1(x) \left(\log x + \frac{x^2}{4} + \frac{5x^4}{128} + \frac{23x^6}{3456} + \cdots \right)$$

$$= y_1(x) \log x + \left(1 - \frac{x^2}{4} + \frac{x^4}{64} - \cdots \right) \left(\frac{x^2}{4} + \frac{5x^4}{128} + \frac{23x^6}{3456} + \cdots \right)$$

or, upon completing the product,

$$y_2(x) = y_1(x) \log x + \frac{x^2}{4} - \frac{3x^4}{128} + \frac{11x^6}{13,824} + \cdots \qquad (43)$$

Although not immediately obvious, this is the same solution found in Example 14 for $y_2(x)$, the verification of which we leave to the reader.

7.5.2 Case III: Roots Differing by a Nonzero Integer

Let us suppose that $s_1 > s_2$ and that $s_1 - s_2 = N$, where $N = 1, 2, 3, \ldots$. In this case we are always assured of one solution, corresponding to the larger root s_1, which has the form

$$y_1(x) = y(x, s_1)$$

$$= x^{s_1} \left[1 + \sum_{n=1}^{\infty} c_n(s_1)x^n \right], \qquad c_0(s_1) = 1 \qquad (44)$$

where

$$y(x, s) = x^s \left[1 + \sum_{n=1}^{\infty} c_n(s)x^n \right], \qquad c_0(s) = 1 \qquad (45)$$

A second solution $y_2(x)$ can then be constructed by using either (41) or a procedure similar to that used in Case II in Section 7.5.1 leading to a logarithmic type of solution. However, it turns out that both $y_1(x)$ and $y_2(x)$ can also be obtained by using the *smaller* root s_2 of the indicial equation. To see this, let us assume the DE is in normal form

$$y'' + a(x)y' + b(x)y = 0 \qquad (46)$$

and introduce the normal operator

$$N[y] = y'' + a(x)y' + b(x)y \tag{47}$$

If we assume that $y(x, s)$ is defined by (45), then it can be shown that the substitution of $y(x, s)$ into (47) will eventually lead to

$$N[y(x, s)] = (s - s_1)(s - s_2)c_0(s)x^{s-2} \tag{48}$$

where all remaining terms on the right-hand side vanish by properly selecting the c's. Using the larger root s_1, only one solution is possible. However, if we set $c_0(s) = s - s_2$, then (48) becomes

$$N[y(x, s)] = (s - s_1)(s - s_2)^2 x^{s-2} \tag{49}$$

and two solutions corresponding to $s = s_2$ are then available to us, using $y(x, s)$ and $\partial y(x, s)/\partial s$ and following the technique of equal roots discussed in Section 7.5.1.[†] In particular, it can be shown that

$$y_1(x) = y(x, s_2)$$

$$= x^{s_2} \sum_{n=1}^{\infty} c_n(s_2)x^n \tag{50}$$

which we claim is equivalent to calculating $y_1(x)$ by (44). The solution $y_2(x)$ then assumes the form

$$y_2(x) = \frac{\partial y(x, s)}{\partial s}\bigg|_{s=s_2}$$

$$= Ky_1(x)\log x + x^{s_2} \sum_{n=0}^{\infty} c_n'(s_2)x^n \tag{51}$$

where K is a constant that in some instances is zero and

$$y(x, s) = x^s\left[(s - s_2) + \sum_{n=1}^{\infty} c_n(s)x^n\right] \tag{52}$$

where we have written $c_0(s) = s - s_2$.

In some special cases the technique of using the smaller root of the indicial equation may produce two solutions, neither of which involves a logarithmic term. Typically, this will happen when s_1 is a positive integer and both c_0 and c_{s_1} turn out to be arbitrary. This special case is called the **nonlog case,** and when it arises, it is the simplest of all cases to handle. To test for this case, we usually substitute $s = s_2$ directly into the recurrence formula in the hopes that two arbitrary constants turn up (usually c_0 and c_{s_1}). If not, we must construct $y(x, s)$ as defined by (52) and then proceed as described above. Let us illustrate the simple case first.

[†] Although it may seem that we could also set $c_0(s) = s - s_1$ and obtain two solutions corresponding to the larger root $s = s_1$, this turns out not to be the case.

<table>
<tr><td>**EXAMPLE 15**
Nonlog Case</td><td>Find the general solution of</td></tr>
</table>

$$xy'' - (4 + x)y' + 2y = 0, \qquad x > 0$$

Solution

The substitution of $y = \sum_{n=0}^{\infty} c_n x^{n+s}$ into the given DE leads to

$$\sum_{n=0}^{\infty} (n + s)(n + s - 1)c_n x^{n+s-1} - \sum_{n=0}^{\infty} 4(n + s)c_n x^{n+s-1}$$

$$- \sum_{\substack{n=0 \\ \Downarrow \\ n \to n-1}}^{\infty} (n + s)c_n x^{n+s} + \sum_{\substack{n=0 \\ \Downarrow \\ n \to n-1}}^{\infty} 2c_n x^{n+s} = 0$$

which simplifies to

$$[s(s - 1) - 4s]c_0 x^{s-1} + \sum_{n=1}^{\infty} [(n + s)(n + s - 1)c_n - 4(n + s)c_n$$

$$- (n + s - 1)c_{n-1} + 2c_{n-1}]x^{n+s-1} = 0$$

Therefore, the indicial equation

$$s(s - 1) - 4s = s(s - 5) = 0$$

gives $s_1 = 5$ and $s_2 = 0$ as roots. Simplifying the expression under the summation, we obtain the recurrence formula

$$(n + s)(n + s - 5)c_n = (n + s - 3)c_{n-1}, \qquad n = 1, 2, 3, \ldots$$

which, for the smaller root $s = 0$, reduces to

$$n(n - 5)c_n = (n - 3)c_{n-1}, \qquad n = 1, 2, 3, \ldots$$

Since division by $(n - 5)$ is not permitted when $n = 5$, it is best to write out separate relations for the c's through $n = 5$. Thus,

$n = 1$: $\qquad\qquad\qquad -4c_1 = -2c_0$

$n = 2$: $\qquad\qquad\qquad -6c_2 = -c_1$

$n = 3$: $\qquad\qquad\qquad -6c_3 = 0 \cdot c_2 = 0$

$n = 4$: $\qquad\qquad\qquad -4c_4 = c_3 = 0$

$n = 5$: $\qquad\qquad\qquad 0 \cdot c_5 = 2c_4 = 0$

from which we deduce

$$c_1 = \tfrac{1}{2}c_0, \qquad c_2 = \tfrac{1}{6}c_1 = \tfrac{1}{12}c_0, \qquad c_3 = c_4 = 0$$

and since $0 \cdot c_5 = 0$ is satisfied for any value of c_5, we see that c_5 is arbitrary. For $n > 5$, it follows that

$$c_n = \frac{(n - 3)c_{n-1}}{n(n - 5)}, \qquad n = 6, 7, 8, \ldots$$

Proceeding now as usual, we find

$$c_6 = \frac{3}{6 \cdot 1} c_5$$

$$c_7 = \frac{4}{7 \cdot 2} c_6 = \frac{3 \cdot 4}{7 \cdot 6 \cdot 2 \cdot 1} c_5$$

$$c_8 = \frac{5}{8 \cdot 3} c_7 = \frac{3 \cdot 4 \cdot 5}{8 \cdot 7 \cdot 6 \cdot 3 \cdot 2 \cdot 1} c_5$$

. . . .

Collecting terms, we obtain the general solution

$$y = x^0(c_0 + c_1 x + c_2 x^2 + c_3 x^3 + c_4 x^4 + c_5 x^5 + \cdots)$$

$$= c_0\left(1 + \frac{1}{2}x + \frac{1}{12}x^2\right)$$

$$+ c_5\left(x^5 + \frac{3}{6 \cdot 1}x^6 + \frac{3 \cdot 4}{7 \cdot 6 \cdot 2 \cdot 1}x^7 + \frac{3 \cdot 4 \cdot 5}{8 \cdot 7 \cdot 6 \cdot 3 \cdot 2 \cdot 1}x^8 + \cdots\right)$$

or finally

$$y = c_0(1 + \tfrac{1}{2}x + \tfrac{1}{12}x^2)$$

$$+ c_5 x^5(1 + \tfrac{1}{2}x + \tfrac{1}{7}x^2 + \tfrac{5}{168}x^3 + \cdots), \qquad x > 0 \qquad \blacksquare$$

EXAMPLE 16
Nonlog Case

Find the general solution of

$$x^2 y'' + xy' + (x^2 - \tfrac{1}{4})y = 0, \qquad x > 0^\dagger$$

Solution

Proceeding as before, the substitution of $y = \sum_{n=0}^{\infty} c_n x^{n+s}$ into the DE leads to

$$\sum_{n=0}^{\infty} (n+s)(n+s-1)c_n x^{n+s} + \sum_{n=0}^{\infty} (n+s)c_n x^{n+s} + \sum_{\substack{n=0 \\ \Downarrow \\ n \to n-2}}^{\infty} c_n x^{n+s+2} - \sum_{n=0}^{\infty} \frac{1}{4}c_n x^{n+s} = 0$$

or

$$\left(s^2 - \frac{1}{4}\right)c_0 x^s + \left(s + \frac{3}{2}\right)\left(s + \frac{1}{2}\right)c_1 x^{s+1}$$

$$+ \sum_{n=2}^{\infty}\left[(n+s)(n+s-1)c_n + (n+s)c_n + c_{n-2} - \frac{1}{4}\right]x^{n+s} = 0$$

The indicial equation $s^2 - 1/4 = 0$ has roots $s_1 = 1/2$ and $s_2 = -1/2$. Using the smaller root $s_2 = -1/2$ leads to both c_0 and c_1 arbitrary in this instance since the

† This DE is a special case of *Bessel's equation* (see Section 7.7).

term above containing c_1 vanishes for this value of s. Setting $s = -1/2$ in the terms under the summation sign, the remaining coefficients are then determined by the recurrence formula

$$c_n = -\frac{c_{n-2}}{n(n-1)}, \qquad n = 2, 3, 4, \ldots$$

We obtain successively

$$c_2 = -\frac{c_0}{2 \cdot 1} = -\frac{c_0}{2!}$$

$$c_3 = -\frac{c_1}{3 \cdot 2} = -\frac{c_1}{3!}$$

$$c_4 = -\frac{c_2}{4 \cdot 3} = \frac{c_0}{4!}$$

$$c_5 = -\frac{c_3}{5 \cdot 4} = \frac{c_1}{5!}$$

. . . .

and thus

$$y = x^{-1/2}(c_0 + c_1 x + c_2 x^2 + c_3 x^3 + c_4 x^4 + \cdots)$$

$$= x^{-1/2}c_0\left(1 - \frac{x^2}{2!} + \frac{x^4}{4!} - \frac{x^6}{6!} + \cdots\right)$$

$$+ x^{-1/2}c_1\left(x - \frac{x^3}{3!} + \frac{x^5}{5!} - \frac{x^7}{7!} + \cdots\right)$$

In this case, however, we recognize each of the above series as that of $\cos x$ and $\sin x$, respectively, and so we can express our solution in the more convenient form

$$y = c_0 \frac{\cos x}{\sqrt{x}} + c_1 \frac{\sin x}{\sqrt{x}}, \qquad x > 0$$ ∎

We conclude this section with an example involving the logarithmic case.

EXAMPLE 17
Log Case

Find the general solution of

$$xy'' + 3y' - y = 0, \qquad x > 0$$

Solution

By substituting $y = \sum_{n=0}^{\infty} c_n x^{n+s}$ into the DE, we get

$$\sum_{n=0}^{\infty} (n+s)(n+s-1)c_n x^{n+s-1} + \sum_{n=0}^{\infty} 3(n+s)c_n x^{n+s-1} - \sum_{n=0}^{\infty} c_n x^{n+s} = 0$$

$$\underset{n \to n-1}{\Downarrow}$$

which simplifies to

$$[s(s - 1) + 3s]c_0 x^{s-1} + \sum_{n=1}^{\infty} [(n + s)(n + s - 1)c_n + 3(n + s)c_n - c_{n-1}]x^{n+s-1} = 0$$

Here we find that the coefficient of c_0 set to zero gives

$$s^2 + 2s = s(s + 2) = 0$$

or $s_1 = 0$ and $s_2 = -2$. Making the substitution $s = -2$ in the recurrence formula

$$(n + s)(n + s - 1)c_n + 3(n + s)c_n = c_{n-1}$$

yields

$$n(n - 2)c_n = c_{n-1}, \qquad n = 1, 2, 3, \ldots$$

Again we do not divide by the factor $n(n - 2)$ until $n > 2$. Thus,

$n = 1$: $\qquad\qquad\qquad\qquad\qquad -c_1 = c_0$

$n = 2$: $\qquad\qquad\qquad\qquad\qquad 0 \cdot c_2 = c_1$

and

$$c_n = \frac{c_{n-1}}{n(n - 2)}, \qquad n = 3, 4, 5, \ldots$$

The first two relations can be satisfied only for $c_1 = c_0 = 0$, which contradicts our assumption that $c_0 \neq 0$. Therefore, we have failed to produce two solutions this time that do not involve logarithmic terms. (It may at first seem that the way to proceed is to set $c_0 = c_1 = 0$ and leave c_2 arbitrary. Doing so, however, will only produce one solution, which will then be reobtained automatically when solving the DE as shown below.)

We now return to the recurrence formula and express it as a function of s; that is, we write

$$c_0(s) = s + 2$$

$$c_n(s) = \frac{c_{n-1}(s)}{(s + n)(s + n + 2)}, \qquad n = 1, 2, 3, \ldots$$

Hence,

$$c_1(s) = \frac{c_0(s)}{(s + 1)(s + 3)} = \frac{s + 2}{(s + 1)(s + 3)}$$

$$c_2(s) = \frac{c_1(s)}{(s + 2)(s + 4)} = \frac{1}{(s + 1)(s + 3)(s + 4)}$$

$$c_3(s) = \frac{c_2(s)}{(s + 3)(s + 5)} = \frac{1}{(s + 1)(s + 3)^2(s + 4)(s + 5)}$$

$$c_4(s) = \frac{c_3(s)}{(s + 4)(s + 6)}$$

$$= \frac{1}{(s + 1)(s + 3)^2(s + 4)^2(s + 5)(s + 6)}$$

. . . .

Based on Equation (52), we have

$$y(x, s) = x^s\left[(s + 2) + \sum_{n=1}^{\infty} c_n(s)x^n\right]$$

$$= (s + 2)x^s + \frac{(s + 2)x^{s+1}}{(s + 1)(s + 3)} + \frac{x^{s+2}}{(s + 1)(s + 3)(s + 4)}$$

$$+ \frac{x^{s+3}}{(s + 1)(s + 3)^2(s + 4)(s + 5)}$$

$$+ \frac{x^{s+4}}{(s + 1)(s + 3)^2(s + 4)^2(s + 5)(s + 6)} + \cdots$$

from which we can immediately generate $y_1(x)$ by setting $s = -2$, that is,

$$y_1(x) = y(x, -2)$$

$$= 0 \cdot x^{-2} + 0 \cdot x^{-1} - \frac{1}{2!} - \frac{x}{3!} - \frac{x^2}{2!4!} - \frac{x^3}{3!5!} - \cdots$$

or

$$y_1(x) = -\sum_{n=0}^{\infty} \frac{x^n}{n!(n + 2)!}$$

To generate $y_2(x)$, we first need to calculate the derivative

$$\frac{\partial y(x, s)}{\partial s} = y(x, s) \log x + x^s + \frac{(s + 2)x^{s+1}}{(s + 1)(s + 3)}\left(\frac{1}{s + 2} - \frac{1}{s + 1} - \frac{1}{s + 3}\right)$$

$$+ \frac{x^{s+2}}{(s + 1)(s + 3)(s + 4)}\left(-\frac{1}{s + 1} - \frac{1}{s + 3} - \frac{1}{s + 4}\right)$$

$$+ \frac{x^{s+3}}{(s + 1)(s + 3)^2(s + 4)(s + 5)}\left(-\frac{1}{s + 1} - \frac{2}{s + 3} - \frac{1}{s + 4} - \frac{1}{s + 5}\right)$$

$$+ \frac{x^{s+4}}{(s + 1)(s + 3)^2(s + 4)^2(s + 5)(s + 6)}$$

$$\times \left(-\frac{1}{s + 1} - \frac{2}{s + 3} - \frac{2}{s + 4} - \frac{1}{s + 5} - \frac{1}{s + 6}\right) + \cdots$$

Now setting $s = -2$, we obtain

$$y_2(x) = \frac{\partial y(x, s)}{\partial s}\bigg|_{s=-2}$$

$$= y_1(x) \log x + x^{-2} - x^{-1} + \frac{1}{4} + \frac{11}{36}x + \frac{31}{576}x^2 + \cdots$$

The general solution, therefore, is given by

$$y = Ay_1(x) + By_2(x), \qquad x > 0$$

where A and B are arbitrary constants.

Exercises 7.5

In Problems 1 to 4, use logarithmic differentiation to find the derivative of the given function.

1. $y(s) = \dfrac{s + n}{s(s + 1)(s + 2) \cdots (s + n - 1)}$

2. $y(s) = \dfrac{s^4}{(s + 1)^3(s + 2)^6}$

3. $y(s) = \dfrac{(s + n)^3}{[(s + 1)(s + 2) \cdots (s + n - 1)]^2}$

■4. $y(s) = \dfrac{s^5(s - 3)^2}{(s + 1)^2[(s + 2)(s + 3) \cdots (s + n)]^3}$

In Problems 5 to 14, show that the roots of the indicial equation are equal and obtain at least the first *three* nonzero terms of the series solutions by the method of Frobenius about the point $x = 0$.

5. $xy'' + y' - 4y = 0$

6. $x^2y'' + 3xy' + (1 - 2x)y = 0$

7. $x^2y'' - x(1 + x)y' + y = 0$

8. $4x^2y'' + (1 - 2x)y = 0$

9. $x^2y'' + 5xy' + 4y = 0$

10. $x^2y'' + 3xy' + (1 + 4x^2)y = 0$

11. $x^2y'' + x(x - 1)y' + (1 - x)y = 0$

■12. $xy'' + (1 - x)y' - y = 0$

13. $4x^2y'' + y = 0$

14. $x^2y'' - xy' + y = 0$

In Problems 15 to 25, show that the roots of the indicial equation differ by a nonzero integer and obtain at least the first *three* nonzero terms of the series solutions by the method of Frobenius about the point $x = 0$.

Nonlog Cases

15. $x^2y'' + 2x(x - 2)y' + 2(2 - 3x)y = 0$

16. $xy'' + (x - 6)y' - 3y = 0$

17. $xy'' - (x + 3)y' + 2y = 0$

18. $x^2y'' + x^2y' - 2y = 0$

19. $x^2y'' + x^2y' + (x - 2)y = 0$

Log Cases

20. $xy'' + (3 - 2x)y' + 8y = 0$

21. $xy'' + y = 0$

■22. $x^2y'' + (x^2 - 3x)y' + 3y = 0$

23. $x^2y'' + x(1 - x)y' - (1 + 3x)y = 0$

24. $x^2y'' + xy' + (x^2 - 1)y = 0$

25. $2xy'' + 6y' + y = 0$

[0] 7.6 Legendre's Equation

Several variable-coefficient DEs occur so frequently in practice that they have been named and their solutions, called **special functions,** have been extensively studied. One such equation that is of special importance is **Legendre's equation**[†] of order α, given by

$$(1 - x^2)y'' - 2xy' + \alpha(\alpha + 1)y = 0 \qquad (53)$$

where α is a constant. This DE arises in various applications involving spherical symmetry, as in finding the steady-state temperature within a solid spherical ball whose surface temperature is prescribed.

[†] See the historical comments on Legendre in Section 7.9.

Legendre's equation (53) has *singularities* at $x = \pm 1$ since $A_2(x) = 1 - x^2$ vanishes at these points. All other points are ordinary points. In particular, $x = 0$ is an ordinary point, and thus we seek a power series solution of the form

$$y = \sum_{k=0}^{\infty} c_k x^k \qquad (54)$$

which, based on Theorem 7.3, will converge at least in the open interval $-1 < x < 1$.

By substituting the series (54) and its derivatives into Equation (53), we obtain

$$(1 - x^2) \sum_{k=0}^{\infty} k(k-1)c_k x^{k-2} - 2x \sum_{k=0}^{\infty} kc_k x^{k-1} + \alpha(\alpha+1) \sum_{k=0}^{\infty} c_k x^k = 0$$

Making appropriate shifts of indices and combining all terms under one summation sign, we finally get

$$0 \cdot c_0 x^{-2} + 0 \cdot c_1 x^{-1} + \sum_{k=2}^{\infty} [k(k-1)c_k$$
$$- (k-2)(k-3)c_{k-2} - 2(k-2)c_{k-2} + \alpha(\alpha+1)c_{k-2}]x^{k-2} = 0 \qquad (55)$$

Consequently, both c_0 and c_1 are arbitrary, and the remaining constants can be determined by equating to zero the coefficient of x^{k-2}; that is,

$$k(k-1)c_k - [(k-2)(k-1) - \alpha(\alpha+1)]c_{k-2} = 0, \qquad k = 2, 3, 4, \ldots \qquad (56)$$

which is the recurrence formula. After rearranging terms, we can write the recurrence relation in the more useful form

$$c_k = \frac{(k-\alpha-2)(k+\alpha-1)}{k(k-1)} c_{k-2}, \qquad k = 2, 3, 4, \ldots \qquad (57)$$

This recurrence formula gives each coefficient in terms of the second one preceding it, except for c_0 and c_1 which are arbitrary. We find successively,

$$c_2 = -\frac{\alpha(\alpha+1)}{2!} c_0$$

$$c_3 = -\frac{(\alpha-1)(\alpha+2)}{3!} c_1$$

$$c_4 = -\frac{(\alpha-2)(\alpha+3)}{4 \cdot 3} c_2 = \frac{(\alpha-2)\alpha(\alpha+1)(\alpha+3)}{4!} c_0$$

$$c_5 = -\frac{(\alpha-3)(\alpha+4)}{5 \cdot 4} c_3 = \frac{(\alpha-3)(\alpha-1)(\alpha+2)(\alpha+4)}{5!} c_1$$

and so forth. By inserting these coefficients back into (54), we obtain the general solution

$$y = c_0 y_1(x) + c_1 y_2(x) \qquad (58)$$

where

$$y_1(x) = 1 - \frac{\alpha(\alpha+1)}{2!} x^2 + \frac{(\alpha-2)\alpha(\alpha+1)(\alpha+3)}{4!} x^4 - \cdots \qquad (59)$$

and

$$y_2(x) = x - \frac{(\alpha - 1)(\alpha + 2)}{3!} x^3 + \frac{(\alpha - 3)(\alpha - 1)(\alpha + 2)(\alpha + 4)}{5!} x^5 - \cdots \tag{60}$$

7.6.1 Legendre Polynomials

The convergence of (59) and (60) is guaranteed by the power series method only for the *open* interval $-1 < x < 1$. In fact, standard convergence tests reveal that, for general α, the above two series diverge at the endpoints $x = \pm 1$. Unfortunately, in most applications we require that the solution be *bounded* throughout the interval—including the endpoints. To remedy this situation, we can select the parameter α in such a way that one of the series (59) and (60) truncates. That is, if $\alpha = n$, where $n = 0, 1, 2, \ldots$, then either y_1 or y_2 will reduce to a *polynomial* of degree n, which of course is bounded everywhere. If n is even, y_1 reduces to a polynomial, and if n is odd, y_2 is a polynomial. That multiple of the polynomial of degree n that has the value unity when $x = 1$ is called the nth **Legendre polynomial** and is denoted by the symbol $P_n(x)$. Thus, the only bounded solutions of Legendre's equation (53) are multiples of

$$y = P_n(x), \qquad n = 0, 1, 2, \ldots \tag{61}$$

where

$$P_n(x) = \begin{cases} y_1(x)/y_1(1) & n \text{ even} \\ y_2(x)/y_2(1) & n \text{ odd} \end{cases} \tag{62}$$

The first few Legendre polynomials obtained from (62) are

$$P_0(x) = 1$$
$$P_1(x) = x$$
$$P_2(x) = \tfrac{1}{2}(3x^2 - 1) \tag{63}$$
$$P_3(x) = \tfrac{1}{2}(5x^3 - 3x)$$
$$P_4(x) = \tfrac{1}{8}(35x^4 - 30x^2 + 3)$$

while in general it can be shown that

$$P_n(x) = \sum_{k=0}^{[n/2]} \frac{(-1)^k (2n - 2k)! x^{n-2k}}{2^n k! (n - k)! (n - 2k)!}, \qquad n = 0, 1, 2, \ldots \tag{64}$$

where

$$[n/2] = \begin{cases} n/2, & n \text{ even} \\ (n - 1)/2, & n \text{ odd} \end{cases} \tag{65}$$

The first few Legendre polynomials are sketched over the interval $-1 \leq x \leq 1$ in Figure 7.1.

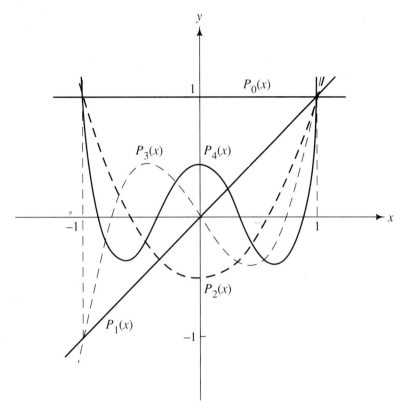

Figure 7.1 Graph of $P_n(x)$, $n = 0, 1, 2, 3, 4$.

> *Remark.*   A second linearly independent solution of Legendre's equation
> (53) can be obtained by retaining the remaining infinite series above when
> α is an integer. Multiples of these series form what are called **Legendre
> functions of the second kind.** For further information, see L. C. Andrews,
> *Special Functions for Engineers and Applied Mathematicians.* Macmillan,
> New York, 1985.

Exercises 7.6

In Problems 1 to 4, determine which Legendre polynomial $P_n(x)$ is a solution of the given DE.

1. $(1 - x^2)y'' - 2xy' = 0$

2. $(1 - x^2)y'' - 2xy' + 2y = 0$

3. $(1 - x^2)y'' - 2xy' + 12y = 0$

■**4.** $(1 - x^2)y'' - 2xy' + 30y = 0$

5. To within a multiplicative constant, determine the Wronskian of Legendre's equation (53).

■**6.** Show that a second solution of Legendre's equation for $\alpha = n$, $n = 0, 1, 2, \ldots$, is given by

$$y_2 = P_n(x) \int \frac{dx}{(1 - x^2)[P_n(x)]^2}$$

7. Using the result of Problem 6, show that a second solution of Legendre's equation for

$\alpha = n$, $n = 0, 1, 2, \ldots$, is given by

(a) $y_2 = \dfrac{1}{2} \log \dfrac{1+x}{1-x}$ when $n = 0$

(b) $y_2 = \dfrac{x}{2} \log \dfrac{1+x}{1-x} - 1$ when $n = 1$

*8. For the case $\alpha = n$, $n = 0, 1, 2, \ldots$, show that one solution of Legendre's equation (53) about the singular point $x = 1$ is given by

$$y = \sum_{k=0}^{n} \frac{(n+k)!}{2^k (k!)^2 (n-k)!} (x-1)^k$$

*9. (*Orthogonality*) Show that Legendre's equation for $\alpha = n$, $n = 0, 1, 2, \ldots$, can be expressed in the form

$$\frac{d}{dx}\left[(1-x^2)y'\right] + n(n+1)y = 0$$

and use this result to prove that $P_n(x)$ and $P_k(x)$ are **orthogonal** on $-1 \le x \le 1$; that is, show that

$$\int_{-1}^{1} P_n(x)P_k(x)\, dx = 0, \qquad k \ne n$$

Hint. Note that $P_n(x)$ and $P_k(x)$ satisfy, respectively,

$$\frac{d}{dx}\left[(1-x^2)P_n'(x)\right] + n(n+1)P_n(x) = 0$$

$$\frac{d}{dx}\left[(1-x^2)P_k'(x)\right] + k(k+1)P_k(x) = 0$$

Multiply the first DE by $P_k(x)$ and the second by $P_n(x)$, subtract the results, and integrate from -1 to 1.

*10. Use the orthogonality property in Problem 9 to show that

(a) $\displaystyle\int_{-1}^{1} P_n(x)\, dx = 0, \qquad n = 1, 2, 3, \ldots$

(b) $\displaystyle\int_{-1}^{1} xP_n(x)\, dx = 0, \qquad n = 0, 2, 3, 4, \ldots$

*11. (*Laguerre polynomials*) Show that one solution of **Laguerre's equation**[†]

$$xy'' + (1-x)y' + ny = 0, \qquad n = 0, 1, 2, \ldots$$

is given by the **Laguerre polynomials**

$$y_1 = L_n(x) = \sum_{k=0}^{n} \frac{(-1)^k n! x^k}{(n-k)!(k!)^2}$$

*12. (*Hermite polynomials*) The **Hermite polynomials**[‡] are defined by

$$H_n(x) = \sum_{k=0}^{[n/2]} \frac{(-1)^k n!(2x)^{n-2k}}{k!(n-2k)!}, \qquad n = 0, 1, 2, \ldots$$

(a) Show that $y = H_n(x)$ satisfies **Hermite's equation**

$$y'' - 2xy' + 2ny = 0$$

(b) If $u = H_n(x)e^{-x^2/2}$ is a solution of **Weber's equation**

$$u'' + (\lambda - x^2)u = 0$$

deduce that $\lambda = 2n - 1$, $n = 0, 1, 2, \ldots$.

[0] 7.7 Bessel's Equation

Solutions of **Bessel's equation**

$$x^2 y'' + xy' + (x^2 - v^2)y = 0, \qquad x > 0 \tag{66}$$

are called **Bessel functions** of order v, where we assume $v \ge 0$. The equation is named in honor of the German astronomer F. W. Bessel,[§] who investigated its solutions in

---

[†] Named after the French mathematician Edmond Laguerre (1834–1886).

[‡] Named after Charles Hermite.

[§] See the historical comments on Bessel in Section 7.9.

connection with his studies of planetary motion. Bessel's equation also occurs in the study of free vibrations of a circular membrane, finding the temperature distribution in a solid cylinder, and in electromagnetic theory. Other areas of application are far too numerous to mention. In fact, Bessel functions occur so frequently in engineering and physics applications that they are undoubtedly the most important functions beyond the study of elementary functions in the calculus.

7.7.1 The Gamma Function

Solutions of Bessel's equation (66) for general values of the parameter v are usually expressed in terms of the **gamma function,** defined by (recall Problems 25 to 28 in Exercises 6.2)

$$\Gamma(x) = \int_0^\infty e^{-t} t^{x-1} \, dt, \qquad x > 0 \tag{67}$$

The most important property of this function is its **recurrence formula**

$$\Gamma(x + 1) = x\Gamma(x) \tag{68}$$

obtained by replacing x with $x + 1$ in (67) and integrating by parts (see Problem 1 in Exercises 7.7).

For $x = 1$, it follows immediately from definition that

$$\Gamma(1) = \int_0^\infty e^{-t} \, dt = 1 \tag{69}$$

The gamma function $\Gamma(x)$ evaluated at other integer values can also be obtained through further integration by parts (see Problem 2 in Exercises 7.7). However, using $\Gamma(1) = 1$ and repeated use of the recurrence formula (68), it also follows that

$$\Gamma(2) = 1 \cdot \Gamma(1) = 1$$

$$\Gamma(3) = 2 \cdot \Gamma(2) = 2 \cdot 1$$

$$\Gamma(4) = 3 \cdot \Gamma(3) = 3 \cdot 2 \cdot 1$$

so that, in general, we deduce the interesting formula

$$\Gamma(n + 1) = n!, \qquad n = 0, 1, 2, \ldots \tag{70}$$

Thus, the gamma function, which is not restricted to integer values, is a generalization of the factorial function $n!$. In particular, we have now proved that $0! = 1$, a result that is usually puzzling to algebra students (since it is defined, but not derived).

The gamma function is not defined for $x < 0$ by the integral definition (67). However, by writing the recurrence formula (68) in the form

$$\Gamma(x) = \frac{\Gamma(x + 1)}{x}, \qquad x \neq 0 \tag{71}$$

we can, by its repeated use, extend the definition of $\Gamma(x)$ to all $x < 0$ except for $x = -n$, $n = 0, 1, 2, \ldots$. At the nonpositive integers it can be shown that $|\Gamma(-n)| = \infty$, $n = 0, 1, 2, \ldots$ (see Problem 5 in Exercises 7.7).

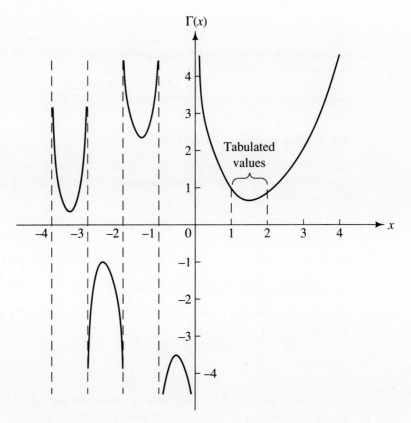

Figure 7.2 Graph of $\Gamma(x)$.

The graph of the gamma function for $x < 0$ as well as $x > 0$ is shown in Figure 7.2.

7.7.2 Bessel Functions of the First Kind

In Examples 14 and 16 we solved Bessel's equation for the special cases where $v = 0$ and $v = 1/2$, respectively. Here now we wish to consider the general case for arbitrary v.

It is easily verified that the point $x = 0$ is a *regular singular point* of Equation (66). Thus, we seek a series solution of the form

$$y = x^s \sum_{k=0}^{\infty} c_k x^k = \sum_{k=0}^{\infty} c_k x^{k+s} \tag{72}$$

The substitution of (72) into (66), followed by some algebraic manipulation, leads to

$$\sum_{k=0}^{\infty} [(k+s)(k+s-1)c_k + (k+s)c_k - v^2 c_k]x^{k+s} + \sum_{k=0}^{\infty} c_k x^{k+s+2} = 0$$

which can be expressed in the equivalent form

$$c_0(s^2 - v^2)x^s + c_1[(s + 1)^2 - v^2]x^{s+1}$$

$$+ \sum_{k=2}^{\infty} \{[(k + s)^2 - v^2]c_k + c_{k-2}\}x^{k+s} = 0 \qquad (73)$$

The values of s leading to solutions of the form (72) are found by setting the coefficient of c_0 to zero; thus, we see that $s = \pm v$. This choice of s leaves c_0 arbitrary, and the remaining constants must satisfy

$$c_1[(s + 1)^2 - v^2] = 0$$

$$[(k + s)^2 - v^2]c_k + c_{k-2} = 0, \qquad k = 2, 3, 4, \ldots \qquad (74)$$

For $s = \pm v$, however, it follows that $c_1 = 0$ and

$$c_k = -\frac{c_{k-2}}{k^2 \pm 2kv}, \qquad k = 2, 3, 4, \ldots \qquad (75)$$

Because $c_1 = 0$, it is clear from (75) that all c_k with odd index also vanish.

Unless we know the value of v, we do not know which of the three cases of the method of Frobenius we are dealing with. Therefore, to be precise and to produce at least one solution of Bessel's equation, we select $s = +v$, and then (75) for even values of k can be written as

$$c_k = c_{2m} = -\frac{c_{2m-2}}{2^2 m(m + v)}, \qquad m = 1, 2, 3, \ldots$$

By repeated use of this recurrence formula, it follows that

$$c_{2m} = \frac{(-1)^m c_0}{2^{2m} m!(v + 1)(v + 2) \cdots (v + m)} \qquad (76)$$

Next, using properties of the gamma function, it can be shown that (see Problem 4 in Exercises 7.7)

$$\Gamma(v + m + 1) = \Gamma(v + 1)(v + 1)(v + 2) \cdots (v + m)$$

which allows us to write the general term of the series as

$$c_{2m} = \frac{(-1)^m \Gamma(v + 1)c_0}{2^{2m} m! \Gamma(v + m + 1)}, \qquad m = 1, 2, 3, \ldots \qquad (77)$$

Thus, we obtain the solution

$$y_1 = c_0 x^v + c_0 x^v \sum_{m=1}^{\infty} \frac{(-1)^m \Gamma(v + 1)(x/2)^{2m}}{m! \Gamma(v + m + 1)}$$

$$= c_0 x^v \sum_{m=0}^{\infty} \frac{(-1)^m \Gamma(v + 1)(x/2)^{2m}}{m! \Gamma(v + m + 1)} \qquad (78)$$

It is customary to set

$$c_0 = \frac{1}{2^v \Gamma(v + 1)}$$

and denote the resulting infinite series in (78) by the symbol

$$J_v(x) = \sum_{m=0}^{\infty} \frac{(-1)^m (x/2)^{2m+v}}{m!\,\Gamma(m+v+1)} \tag{79}$$

known as the **Bessel function of the first kind** and order v.

The most commonly occurring Bessel functions are those for which $v = n$, $n = 0, 1, 2, \ldots$. In this case, (79) becomes

$$J_n(x) = \sum_{m=0}^{\infty} \frac{(-1)^m (x/2)^{2m+n}}{m!\,(m+n)!} \tag{80}$$

For example, when $n = 0$ we obtain the series (also recall Example 14)

$$J_0(x) = \sum_{m=0}^{\infty} \frac{(-1)^m (x/2)^{2m}}{(m!)^2}$$

$$= 1 - \frac{x^2}{2^2} + \frac{x^4}{2^4 2!\,2!} - \frac{x^6}{2^6 3!\,3!} + \cdots$$

The graphs of $J_0(x)$, $J_1(x)$, and $J_2(x)$ are shown in Figure 7.3. These functions exhibit an oscillatory behavior somewhat like that of the sinusoidal functions, except that the amplitude of the Bessel functions diminishes as x increases and the zeros of these functions are not evenly spaced. (For a further discussion of the oscillatory nature of Bessel functions, see Section 5.6.)

For $s_1 = v \geq 0$, we have shown that $y_1 = J_v(x)$ is a solution of (66). If we set $s_2 = -v$ and find that $s_1 - s_2 \neq N$, $N = 0, \pm 1, \pm 2, \ldots$, then clearly

$$y_2(x) = J_{-v}(x) = \sum_{m=0}^{\infty} \frac{(-1)^m (x/2)^{2m-v}}{m!\,\Gamma(m-v+1)} \tag{81}$$

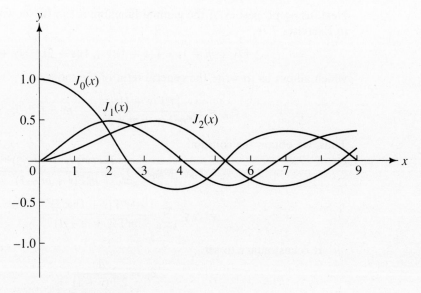

Figure 7.3 Graph of $J_n(x)$, $n = 0, 1, 2$.

is another linearly independent solution of (66), obtained by the formal replacement of v with $-v$ in (79). In fact, if v is not an integer, it can be shown that y_2 given by (81) is always linearly independent of $y_1 = J_v(x)$. In general, therefore, if v is not an integer it follows that

$$y = C_1 J_v(x) + C_2 J_{-v}(x), \qquad v \neq 0, 1, 2, \ldots \qquad (82)$$

is a *general solution* of (66). From its series definition (81), it follows that $J_{-v}(x)$, $v \neq 0, 1, 2, \ldots$, is *not bounded* at $x = 0$. Hence, in applications where we want bounded solutions at $x = 0$, we must discard this solution by setting $C_2 = 0$.

EXAMPLE 18

Express the solution to Example 16 in terms of Bessel functions; that is, solve

$$x^2 y'' + xy' + (x^2 - \tfrac{1}{4})y = 0, \qquad x > 0$$

Solution

We previously found the general solution

$$y = C_1 \frac{\cos x}{\sqrt{x}} + C_2 \frac{\sin x}{\sqrt{x}}$$

Here now we recognize the DE as Bessel's equation (66) with $v = 1/2$. Thus, the general solution can also be expressed as

$$y = C_1 J_{1/2}(x) + C_2 J_{-1/2}(x), \qquad x > 0$$

suggesting a possible connection between Bessel functions and the circular functions (see Problem 9 in Exercises 7.7). ■

7.7.3 Bessel Functions of the Second Kind

When $v = 0$, we initially obtain only one solution of Bessel's equation, namely, $y_1 = J_0(x)$. Since this situation corresponds to Case II in Section 7.5, we know that a second linearly independent solution exists of the form

$$y_2(x) = J_0(x) \log x + \sum_{k=0}^{\infty} b_k x^k$$

In fact, in Example 14 we found that this solution was given by

$$y_2(x) = J_0(x) \log x + \sum_{m=1}^{\infty} \frac{(-1)^{m-1} H_m}{(m!)^2} \left(\frac{x}{2}\right)^{2m}$$

It is customary, however, to choose a linear combination of $J_0(x)$ and $y_2(x)$ as the "second" solution of the DE rather than to simply take $y_2(x)$. This special combination is defined by

$$Y_0(x) = \frac{2}{\pi} \left[y_2(x) + (\gamma - \log 2) J_0(x) \right]$$

$$= \frac{2}{\pi} J_0(x)[\log(x/2) + \gamma] + \frac{2}{\pi} \sum_{m=1}^{\infty} \frac{(-1)^{m-1} H_m}{(m!)^2} \left(\frac{x}{2}\right)^{2m} \qquad (83)$$

called the **Bessel function of the second kind** and order zero, and chosen so that $Y_0(x)$ satisfies some of the same recurrence formulas as $J_0(x)$. Here γ is *Euler's constant* defined by

$$\gamma = \lim_{n \to \infty} (H_n - \log n) \cong 0.5772$$

Using both $J_0(x)$ and $Y_0(x)$, the general solution of Bessel's equation of order zero can be written as

$$y = C_1 J_0(x) + C_2 Y_0(x) \tag{84}$$

When $v = n$, $n = 1, 2, 3, \ldots$, the special case of $J_{-v}(x)$ defined by (81) that arises is given by

$$J_{-n}(x) = \sum_{m=0}^{\infty} \frac{(-1)^m (x/2)^{2m-n}}{m! \Gamma(m - n + 1)}$$

However, through properties of the gamma function and a change of index, it can be shown that $J_{-n}(x)$ can also be expressed in the form (see Problem 10 in Exercises 7.7)

$$J_{-n}(x) = (-1)^n \sum_{m=0}^{\infty} \frac{(-1)^m (x/2)^{2m+n}}{m!(m + n)!}$$

from which we deduce

$$J_{-n}(x) = (-1)^n J_n(x), \qquad n = 0, 1, 2, \ldots \tag{85}$$

Hence, although $J_{-n}(x)$ is indeed a solution of Bessel's equation, it is not linearly independent of $J_n(x)$. For this reason it cannot be used directly in the construction of a general solution.

To produce a second solution of Bessel's equation that is linearly independent of $J_n(x)$ when $v = n$, $n = 1, 2, 3, \ldots$, we start by introducing the function

$$Y_v(x) = \frac{J_v(x) \cos v\pi - J_{-v}(x)}{\sin v\pi} \tag{86}$$

called the **Bessel function of the second kind** and order v. For v not an integer, the function $Y_v(x)$ is well defined by (86), but for integral values of v it reduces to the indeterminate form 0/0. In these cases we must define

$$Y_n(x) = \lim_{v \to n} Y_v(x), \qquad n = 0, 1, 2, \ldots \tag{87}$$

the evaluation of which involves L'Hôpital's rule.[†] The particular function $Y_0(x)$ defined by (87) is in fact the same as that defined by (83). The graphs of $Y_0(x)$, $Y_1(x)$, and $Y_2(x)$ are all shown in Figure 7.4.

[†] See, for example, Chapter 6 in L. C. Andrews, *Special Functions for Engineers and Applied Mathematicians,* Macmillan, New York, 1985.

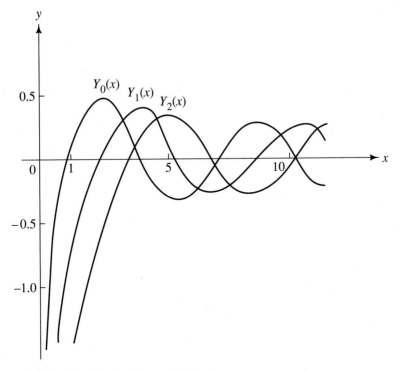

Figure 7.4 Graph of $Y_n(x)$, $n = 0, 1, 2$.

Since both $J_v(x)$ and $J_{-v}(x)$ are solutions of (66) for all $v \geq 0$, it follows from the linearity property that $Y_v(x)$ is also a solution of Bessel's equation. Moreover, it can be shown that it is linearly independent of $J_v(x)$, even for $v = n$, $n = 0, 1, 2, \ldots$. Hence, we conclude that the *general solution* of Bessel's equation (66) for all $v \geq 0$ is given by

$$y = C_1 J_v(x) + C_2 Y_v(x), \qquad v \geq 0 \tag{88}$$

However, like $J_{-v}(x)$, the function $Y_v(x)$ is not bounded at $x = 0$. Thus, in many applications this solution must be discarded by selecting $C_2 = 0$ in (88).

Exercises 7.7

1. Through integration by parts on the expression

$$\Gamma(x + 1) = \int_0^\infty e^{-t} t^x \, dt$$

deduce the recurrence formula $\Gamma(x + 1) = x\Gamma(x)$.

2. Directly from the integral definition,
   (a) Show that $\Gamma(2) = 1$.
   (b) Use integration by parts n times to deduce that

$$\Gamma(n + 1) = n!, \qquad n = 0, 1, 2, \ldots$$

*3. Establish that

$$\Gamma(1/2) = 2 \int_0^\infty e^{-x^2}\, dx = 2 \int_0^\infty e^{-y^2}\, dy$$

and use this result to deduce that $\Gamma(1/2) = \sqrt{\pi}$. *Hint.* Multiply the integrals to obtain $[\Gamma(1/2)]^2$ and then evaluate the resulting double integral by changing to polar coordinates.

4. By the recurrence formula show that

$$\Gamma(m + v + 1) = (m + v)\Gamma(m + v)$$

and more generally that (by repeated use)

$$\Gamma(m + v + 1) = \Gamma(v + 1)(v + 1)(v + 2)\cdots(v + m)$$

5. By repeated use of the recurrence formula,
(a) Deduce that

$$\Gamma(x) = \frac{\Gamma(x + m)}{x(x + 1)(x + 2)\cdots(x + m - 1)}$$

where $m = 1, 2, 3, \ldots$.

(b) From (a), deduce that

$$\frac{1}{\Gamma(-n)} = 0, \qquad n = 0, 1, 2, \ldots$$

6. Write out the first four terms of the series representations for both $J_1(x)$ and $J_2(x)$.

7. Use the series representation to show that
(a) $J_0(0) = 1$ and $J_v(0) = 0$ for $v > 0$
(b) $J_1'(0) = 1/2$ and $J_v'(0) = 0$ for $v > 1$

■8. Show that $J_n(-x) = (-1)^n J_n(x)$, $n = 0, 1, 2, \ldots$.

9. Establish the following identities ($x > 0$)

(a) $J_{1/2}(x) = \sqrt{\dfrac{2}{\pi x}} \sin x$

(b) $J_{-1/2}(x) = \sqrt{\dfrac{2}{\pi x}} \cos x$

Hint. Compare series representations and use $\Gamma(1/2) = \sqrt{\pi}$ and $\Gamma(3/2) = \sqrt{\pi}/2$.

*10. Show that

$$J_{-n}(x) = \sum_{m=n}^\infty \frac{(-1)^m (x/2)^{2m-n}}{m!\,\Gamma(m - n + 1)}$$

and by making the change of index $k = m - n$, deduce that

$$J_{-n}(x) = (-1)^n J_n(x), \qquad n = 0, 1, 2, \ldots$$

In Problems 11 to 14, express the general solution of the DE in terms of Bessel functions.

11. $x^2 y'' + xy' + (x^2 - 1)y = 0$
■12. $xy'' + y' + xy = 0$
13. $4x^2 y'' + 4xy' + (4x^2 - 1)y = 0$
*14. $x^2 y'' + xy' + (4x^2 - 9)y = 0$
Hint. Let $t = 2x$.

15. To within a multiplicative constant, determine the Wronskian of Bessel's equation

$$x^2 y'' + xy' + (x^2 - v^2)y = 0, \qquad v \ge 0$$

16. Show directly that the series for $J_n(x)$, $n = 0, 1, 2, \ldots$, converges absolutely for all x.

17. By differentiating the series termwise, show that

$$J_0'(x) = -J_1(x)$$

18. Verify directly that $y = J_0(kx)$, where k is a constant, is a solution of

$$xy'' + y' + k^2 xy = 0$$

*19. (*Orthogonality*) By writing the DE in Problem 18 in the form

$$\frac{d}{dx}(xy') + k^2 xy = 0$$

show that $J_0(k_m x)$ and $J_0(k_n x)$ are **orthogonal** on the interval $0 \le x \le 1$ provided $J_0(k_n) = 0$, $n = 1, 2, 3, \ldots$; that is, show that

$$\int_0^1 x J_0(k_m x) J_0(k_n x)\, dx = 0, \qquad m \ne n$$

Hint. See the hint given in Problem 9 in Exercises 7.6.

■20. Show that the change of variable $y = u(x)/\sqrt{x}$ reduces Bessel's equation (66) to the form

$$u'' + \left(1 + \frac{\frac{1}{4} - v^2}{x^2}\right)u = 0$$

and use this result to find a general solution, in terms of trigonometric functions, of Bessel's equation when $v = 1/2$.

In Problems 21 to 24, obtain the general solution of the given DE, using the fact that $y = x^a[C_1 J_v(bx^c) + C_2 Y_v(bx^c)]$ is the general solution of

$$x^2 y'' + (1 - 2a)xy' + [b^2 c^2 x^{2c}(a^2 - c^2 v^2)]y = 0, \qquad v \geq 0, \qquad b > 0$$

*21. $x^2 y'' + 5xy' + (9x^2 - 12)y = 0$

*22. $y'' + xy = 0$

    Hint. Multiply the DE by x^2.

*23. $4x^2 y'' + (1 + 4x)y = 0$

*24. $x^2 y'' - 5xy' + (64x^8 + 5)y = 0$

7.8 Chapter Summary

In this chapter we discussed the **power series method** for solving variable coefficient DEs of the general form

$$A_2(x)y'' + A_1(x)y' + A_0(x)y = 0 \tag{89}$$

where $A_0(x)$, $A_1(x)$, and $A_2(x)$ are polynomials with no common factors for all. Those points for which $A_2(x) = 0$ are called **singular points** of the DE, while all others are called **ordinary points.** Generally, we are concerned with whether the point $x = 0$ is an ordinary point or singular point. If $x = 0$ is an ordinary point, then linearly independent solutions y_1 and y_2 *always* exist as power series of the form

$$y = \sum_{n=0}^{\infty} c_n x^n \tag{90}$$

from which we can construct the general solution of (89). Both series converge at least in the interval $|x| < R$, where R is the distance between $x = 0$ and the nearest singularity. To find series for y_1 and y_2, we substitute the series (90) directly into the DE and group like terms. We then set all coefficients of such terms to zero, which leads to the **recurrence formula** for the determination of the c's.

If $x = 0$ is a singular point, then we further investigate to decide whether it is **regular** or **irregular.** To make this last decision, we first put the DE in normal form

$$y'' + a(x)y' + b(x)y = 0 \tag{91}$$

Then, if $\lim_{x \to 0} xa(x)$ and $\lim_{x \to 0} x^2 b(x)$ both exist, we say that $x = 0$ is a regular singular point (RSP); otherwise it is an irregular singular point (ISP). If $x = 0$ is an RSP, then there always exists one solution of the form

$$y = x^s \sum_{n=0}^{\infty} c_n x^n = \sum_{n=0}^{\infty} c_n x^{n+s} \tag{92}$$

Solving the **indicial equation** for s, we find two roots, namely, s_1 and s_2. Depending upon the nature of s_1 and s_2, there are three cases to consider for the determination of the second solution of (91). When the difference $s_1 - s_2$ is *not* an integer, there are two solutions of the form (92). However, when $s_1 - s_2$ is equal to an integer (possibly zero), then the second solution usually involves a logarithmic term.

Review Exercises

In Problems 1 to 4, find the singular points and classify them as regular (RSP) or irregular (ISP).

1. $y'' - xy' + 2y = 0$

2. $x^2 y'' + 5x(x + 3)y' - y = 0$

3. $x^2(x^2 - 4)^2 y'' - 3(x^2 - 4)y' + 8y = 0$

4. $x(x^2 + 1)^2 y'' + 4x^2 y' - 3y = 0$

In Problems 5 to 10, find at least the first four nonzero terms in the power series expansion about $x = 0$ of the solutions. Also, state the recurrence relation and the interval of convergence.

5. $y'' + y = 0$

6. $y'' - xy' + 2y = 0$

7. $(x^2 - 1)y'' + xy' - y = 0$

8. $y'' - xy = 0$

9. $y'' - x^2 y' - y = 0$, $y(0) = 2$, $y'(0) = 3$

10. $(x^2 + 1)y'' + xy' - y = 0$, $y(0) = 5$, $y'(0) = -1$

In Problems 11 to 14, find the roots of the indicial equation and classify them as belonging to Case I, II, or III.

11. $2x^2 y'' + 7x(x + 1)y' - 3y = 0$

12. $3x^2 y'' - xy' + y = 0$

13. $x^2 y'' + xy' + x^2 y = 0$

14. $x^2 y'' + x(x - 2)y' + 2y = 0$

In Problems 15 to 20, use the method of Frobenius to find two linearly independent solutions about the point $x = 0$. Find at least the first three nonzero terms of each series.

15. $3x^2 y'' - xy' + y = 0$

16. $3xy'' + 2y' + x^2 y = 0$

17. $xy'' + y' + y = 0$

18. $x^2 y'' - xy' + (1 - x)y = 0$

19. $xy'' + 4y' - xy = 0$

20. $x(1 - x)y'' - 3y' + 2y = 0$

7.9 Historical Comments

In the early eighteenth century, **power series** expansions for functions like e^x, $\sin x$, and arctan x were published by the Scottish mathematician C. Maclaurin—a child prodigy who became professor of mathematics at the University of Aberdeen, Scotland, at the age of 19. The British mathematician B. Taylor generalized this work in 1715 by publishing the famous theorem now bearing his name.

Series solutions of differential equations have been used since the time of Newton, but often without regard to such details as convergence. In 1732, D. Bernoulli derived a series solution to a problem involving the oscillations of a heavy hanging chain. Today we recognize his solution as that of a zero-order **Bessel function,** which he obtained a century before Bessel's pioneering work on Bessel functions. Around 1750, Euler began to make fairly extensive use of power series in solving differential equations and he also obtained a series solution to Bessel's equation prior to the time of Bessel.

Euler also started a development of the **Frobenius method** of solution, but this was formalized by G. Frobenius, who published his famous paper in 1873 on finding an infinite series solution at a regular singular point. Frobenius also made substantial contributions to the theory of groups, and in 1896 he developed the theory of group characteristics for non-Abelian groups, now important in the study of quantum mechanics.

The use of series became more widespread in the nineteenth century with the development of **special functions.** In 1785, A. M. Legendre developed the gravitational potential associated with a point mass into a series of two distance variables whereby he noticed that the coefficients in this series were polynomials (later called **Legendre polynomials**) that exhibited interesting properties. He is perhaps best remembered, however, for spending forty years systematizing and extending the theory of elliptic integrals. Legendre was educated at the College Mazarin in Paris and was appointed professor of mathematics at the École Militaire at the age of 25. The first systematic study of **Bessel's equation** and its solutions was carried out by F. W. Bessel, a German astronomer. He derived various expressions for $J_n(x)$ as well as certain recurrence relations and results concerning the zeros of **Bessel functions.** While serving as first director of the observatory at Königsberg, Bessel measured the positions of approximately 60,000 stars and was the first to measure the distance of a fixed star from the sun. Bessel received the Lalande prize of the Paris Institute for his calculations on the comet of 1807 in addition to many civil honors. The major development of the theory of **hypergeometric functions** was carried out in 1812 by Carl Friedrich Gauss (1777–1855), and in 1836 a similar analysis on the **confluent hypergeometric function** was carried out by E. E. Kummer (1810–1893).

CHAPTER 8
Linear Systems of Equations

In this chapter we consider problems for which there is more than one unknown function, which leads us into **linear systems of equations.** We discuss some of the theory of **first-order systems** of equations in Section 8.2, primarily for systems of two equations. In Section 8.3 we illustrate the solution of a linear system of equations by the **operator method,** followed in Section 8.4 by some elementary **applications.** We follow this in Section 8.5 with the **method of Laplace transforms,** which reduces the differential system of equations to an algebraic system.

Matrix methods permit a more systematic treatment of solving linear systems of first-order DEs. A brief review of **matrix operations** is presented in Section 8.6, and in Section 8.7 we again discuss the general theory of **first-order linear systems** using a **matrix formulation.** In solving **homogeneous systems with constant coefficients** in Section 8.8 we are led to the concept of **eigenvalues** of a matrix **A** and their associated **eigenvectors.** The eigenvalues and eigenvectors are used in developing the **fundamental solution set** from which we obtain the **general solution of the homogeneous system.** We solve **nonhomogeneous systems** by the method of **variation of parameters** in Section 8.9, while in Section 8.10 we illustrate the concept of the **exponential matrix function** in solving first-order systems of equations.

8.1 Introduction

Many applications of differential equations involve not one but several dependent variables, each of which is a function of a single independent variable, usually time. The formulation of such problems leads to a *system of differential equations,* rather than a single equation as we have studied thus far.

As an elementary example leading to a system of equations, let us consider the motion of a single particle in space governed by Newton's second law of motion. If the mass of the particle is denoted by m and its position vector by $\mathbf{r} = (x, y, z)$, then Newton's law of motion becomes

$$m \frac{d^2 \mathbf{r}}{dt^2} = \mathbf{F} \tag{1}$$

where $\mathbf{F} = (F_1, F_2, F_3)$ is the sum of external forces acting on the particle. By equating like components of (1), we obtain the equivalent **system of differential equations**

$$\begin{aligned} mx'' &= F_1 \\ my'' &= F_2 \\ mz'' &= F_3 \end{aligned} \tag{2}$$

where the primes denote differentiation with respect to time t. Systems of equations are likewise used to describe the motions of coupled spring-mass systems and *RLC* circuits connected in parallel. Examples of such systems will be discussed later.

Systems of equations also arise in the classical ecological problem of the prey and the predator, which involves the struggle for survival among different species of animals living in the same environment. One kind of animal eats the other as a means of survival, while the other develops methods of evasion in order to avoid being eaten. Problems of this nature were independently modeled by A. J. Lotka (1880–1949) in 1925 and by V. Volterra (1860–1940) in 1926. They suggested that the instantaneous populations $x(t)$ and $y(t)$ of both species are solutions of the system of equations

$$\begin{aligned} x' &= ax - bxy \\ y' &= -cy + dxy \end{aligned} \tag{3}$$

where a, b, c, and d are all positive constants. The constants a and c represent the growth rate of the prey and the death rate of the predator, respectively, whereas b and d are measures of the effect of the interaction between the species. These equations are now widely known as the **Lotka-Volterra equations.**

Finally, another way in which a system of equations can arise is to convert an nth-order DE to a system of first-order equations. For instance, given the nth-order DE (not necessarily linear)

$$y^{(n)} = F\left[t, y, y', \ldots, y^{(n-1)}\right] \tag{4}$$

let us introduce n new variables $y_1, y_2, \ldots, y_n$ defined by

$$y_1 = y, \qquad y_2 = y', \qquad y_3 = y'', \qquad \ldots, \qquad y_n = y^{(n-1)}$$

In this case we see that (4) is equivalent to the first-order system of equations

$$y_1' = y_2$$
$$y_2' = y_3$$
$$\cdots\cdots\cdots\cdots\cdots\cdots\cdots$$
$$y_{n-1}' = y_n$$
$$y_n' = F(t, y_1, y_2, \ldots, y_n)$$

(5)

We consider only *linear* systems in this chapter, many of which are expressed as a *first-order* system like (5). Although much of the theory is applicable to systems with variable coefficients, we solve only those with *constant coefficients*. Also, we consider only the case where the number of unknowns and the number of equations are the same.

8.2 Theory of First-Order Systems

First-order systems involving only *two unknowns* can always be expressed in the form

$$x' = a_{11}(t)x + a_{12}(t)y + f_1(t)$$
$$y' = a_{21}(t)x + a_{22}(t)y + f_2(t)$$

(6)

called the **canonical form.** The generalization to systems of n equations follows in a natural sort of way. For notational simplicity, however, we find it convenient at this point to discuss primarily systems featuring only two unknowns. Also, our intention here is to simply present some of the basic theory. Solution techniques will be discussed in subsequent sections.

If $f_1(t) \equiv f_2(t) \equiv 0$, we say the system (6) is **homogeneous** and write

$$x' = a_{11}(t)x + a_{12}(t)y$$
$$y' = a_{21}(t)x + a_{22}(t)y$$

(7)

Otherwise we say that (6) is **nonhomogeneous.** The point of view and terminology that we use in solving (6) or (7) emphasize the similarities between such systems and the linear second-order DEs discussed in Chapter 4. That such similarities exist, of course, is a consequence of the fact that all second-order linear DEs can be expressed as a system of two first-order linear equations.

EXAMPLE 1

Reduce the following second-order DE to a system of first-order DEs:

$$my'' + cy' + ky = F(t)$$

Solution

To start, we rewrite the DE in the form

$$y'' = -\frac{c}{m}y' - \frac{k}{m}y + \frac{F(t)}{m}$$

and then let $y = y_1$ and $y' = y_2$. Thus, since $y'_1 = y' = y_2$, it follows that

$$y'_1 = y_2$$

$$y'_2 = -\frac{c}{m} y_2 - \frac{k}{m} y_1 + \frac{F(t)}{m}$$

■

Linear systems of equations that involve some DEs of order higher than first-order can likewise be rewritten as a system of equations involving only first-order DEs. However, the new system of first-order equations will then consist of more equations than the original system. The general procedure for accomplishing this is to solve for the highest-order derivative in each of the unknowns before introducing new variables. Let us illustrate with an example.

EXAMPLE 2

Reduce the following system of equations to a system involving only first-order DEs:

$$x'' - 2x' - y' = -e^{2t}$$

$$-6x + y' = t$$

Solution

Let us rewrite the system in the form

$$x'' = 2x' + y' - e^{2t}$$

$$y' = 6x + t$$

and introduce the new variables $y_1 = x$, $y_2 = y$, and $y_3 = x'$. Thus, $y'_1 = y_3$ and consequently

$$y'_1 = 2y_3 + y'_2 - e^{2t}$$

$$y'_2 = 6y_1 + t$$

or

$$y'_1 = y_3$$

$$y'_2 = 6y_1 + t$$

$$y'_3 = 6y_1 + 2y_3 + t - e^{2t}$$

■

A **solution** of the linear system (6) is a set of functions $\{x(t), y(t)\}$ with continuous first derivatives that simultaneously satisfy both equations of the system identically on some open interval I. The following theorem, fundamental to the solution of systems of DEs with initial conditions, is analogous to Theorem 5.1 concerning the existence and uniqueness of the solution of a second-order DE satisfying initial conditions.

Theorem 8.1 Existence-Uniqueness. *If the functions* $a_{11}(t)$, $a_{12}(t)$, $a_{21}(t)$, $a_{22}(t)$, $f_1(t)$, *and* $f_2(t)$ *are continuous on some interval* I *containing the point* t_0, *then there exists exactly one solution pair* $\{x(t), y(t)\}$ *of the system*

$$x' = a_{11}(t)x + a_{12}(t)y + f_1(t)$$

$$y' = a_{21}(t)x + a_{22}(t)y + f_2(t)$$

satisfying the initial conditions $x(t_0) = x_0$, $y(t_0) = y_0$.

One of the reasons we are interested in first-order systems is that they lend themselves more readily to the development of the theory, which corresponds closely with the theory of first-order linear DEs discussed in Chapter 2. Also, most computational algorithms in numerical techniques are established for first-order equations (see Chapter 10), and in practical applications the problem often requires some numerical calculations.

8.2.1 Fundamental Solution Sets

To begin, let us first address the *homogeneous* system described by (7). For such systems, the *superposition principle* once again plays an important role.

Theorem 8.2 Superposition Principle. *If* $\{x_1(t), y_1(t)\}$ *and* $\{x_2(t), y_2(t)\}$ *are both solution sets of the homogeneous system*

$$x' = a_{11}(t)x + a_{12}(t)y$$

$$y' = a_{21}(t)x + a_{22}(t)y$$

on some interval I, *then the pair of functions* $\{x(t), y(t)\}$, *where*

$$x(t) = C_1 x_1(t) + C_2 x_2(t)$$

$$y(t) = C_1 y_1(t) + C_2 y_2(t)$$

is also a solution set on I *for any constants* C_1 *and* C_2.

The proof of Theorem 8.2, which is similar to that of Theorem 4.1, is left to the exercises (see Problem 15 in Exercises 8.2).

EXAMPLE 3

The pairs of functions $\{e^{-t}, -e^{-t}\}$ and $\{e^{4t}, \frac{3}{2}e^{4t}\}$ are each solution sets of the homogeneous system

$$x' = x + 2y$$

$$y' = 3x + 2y$$

Verify directly that the set $\{x(t), y(t)\}$ is also a solution, for any constants C_1 and C_2, where

$$x(t) = C_1 e^{-t} + C_2 e^{4t}$$

$$y(t) = -C_1 e^{-t} + \tfrac{3}{2}C_2 e^{4t}$$

Solution

Direct substitution of $x(t)$ and $y(t)$ into the first DE yields

$$x' - x - 2y = -C_1 e^{-t} + 4C_2 e^{4t} - C_1 e^{-t} - C_2 e^{4t} + 2C_1 e^{-t} - 3C_2 e^{4t}$$
$$= 0$$

and, similarly, for the second DE,

$$y' - 3x - 2y = C_1 e^{-t} + 6C_2 e^{4t} - 3C_1 e^{-t} - 3C_2 e^{4t} + 2C_1 e^{-t} - 3C_2 e^{4t}$$
$$= 0 \qquad \blacksquare$$

In order to define what we mean by a *general solution* of a homogeneous system of equations, we have to introduce the notion of "linearly independent solution sets." For the problems considered here involving only two equations, we can say that $\{x_1(t), y_1(t)\}$ and $\{x_2(t), y_2(t)\}$ are **linearly independent** if neither is a constant multiple of the other. If one is a constant multiple of the other, we say the sets are **linearly dependent.** More generally we define these terms according to the following definition, which can be extended to linear systems of any size.

> **Definition 8.1** Two solution sets $\{x_1(t), y_1(t)\}$ and $\{x_2(t), y_2(t)\}$ are said to be **linearly dependent** on some interval I if and only if there exist constants C_1 and C_2, not both zero, such that
>
> $$C_1 x_1(t) + C_2 x_2(t) = 0$$
> $$C_1 y_1(t) + C_2 y_2(t) = 0$$
>
> for every t in I. If this is true only for $C_1 = C_2 = 0$, we say the solution sets are **linearly independent.**

■ EXAMPLE 4

Show that the sets $\{e^{3t}, 2e^t\}$ and $\{-4e^{3t}, -8e^t\}$ are *linearly dependent* on every interval, while $\{e^{3t}, 2e^t\}$ and $\{-4e^{3t}, e^t\}$ are *linearly independent* on every interval.

Solution

In the first case we see that the sets are proportional since the second set is a multiple of the first set, that is,

$$\{-4e^{3t}, -8e^t\} = -4\{e^{3t}, 2e^t\}$$

Hence, we deduce that they are *linearly dependent* on every interval. The remaining sets of functions are clearly not proportional, and thus they must be *linearly independent*. Also, using Definition 8.1 in the latter case, we see that

$$C_1 e^{3t} - 4C_2 e^{3t} = 0$$
$$2C_1 e^t + C_2 e^t = 0$$

is satisfied only for $C_1 = C_2 = 0$. (To see this more clearly, take $t = 0$ and try to solve for C_1 and C_2.) $\qquad \blacksquare$

If $\{x_1(t), y_1(t)\}$ and $\{x_2(t), y_2(t)\}$ are linearly independent solution sets of the *homogeneous* system of equations

$$x' = a_{11}(t)x + a_{12}(t)y$$

$$y' = a_{21}(t)x + a_{22}(t)y$$

on some interval I, we say they form a **fundamental set of solutions** on this interval. A **general solution** of the system on I, therefore, is given by the set $\{x_H(t), y_H(t)\}$, where C_1 and C_2 are arbitrary constants and

$$x_H(t) = C_1 x_1(t) + C_2 x_2(t)$$

$$y_H(t) = C_1 y_1(t) + C_2 y_2(t)$$

(8)

8.2.2 Nonhomogeneous Systems

Any solution set $\{x_P(t), y_P(t)\}$ satisfying the *nonhomogeneous* system

$$x' = a_{11}(t)x + a_{12}(t)y + f_1(t)$$

$$y' = a_{21}(t)x + a_{22}(t)y + f_2(t)$$

is called a **particular solution.** The following relevant theorem is analogous to Theorem 4.7, the proof of which is left to the exercises (see Problem 16 in Exercises 8.2).

Theorem 8.3 *If $\{x_P(t), y_P(t)\}$ is any particular solution set of the nonhomogeneous system of equations*

$$x' = a_{11}(t)x + a_{12}(t)y + f_1(t)$$

$$y' = a_{21}(t)x + a_{22}(t)y + f_2(t)$$

on an interval I and $\{x_H(t), y_H(t)\}$ is the general solution of the associated homogeneous system $[f_1(t) \equiv f_2(t) \equiv 0]$ on I, then the general solution of the nonhomogeneous system on I is the solution set $\{x(t), y(t)\}$, where

$$x(t) = x_H(t) + x_P(t)$$

$$= C_1 x_1(t) + C_2 x_2(t) + x_P(t)$$

$$y(t) = y_H(t) + y_P(t)$$

$$= C_1 y_1(t) + C_2 y_2(t) + y_P(t)$$

Exercises 8.2

In Problems 1 to 10, rewrite the given DEs as a system of first-order equations.

1. $y'' + k^2 y = P \sin \omega t$

2. $4y''' + y' = e^t$

3. $y^{(4)} - k^4 y = f(t)$

4. $t^2 y'' + ty' - y = t^2 \log t$

5. $x'' - x - y' = e^t$
$x + y' = t - 10$

■**6.** $2x'' + x - 2y' = 7$
  $x' - y' = e^{3t}$

7. $m_1 y_1'' + (k_1 + k_2)y_1 - k_2 y_2 = 0$
  $m_2 y_2'' - k_1 y_1 + k_2 y_2 = 0$

8. $x'' - x' + 5x + 2y'' = e^t$
  $-2x + y'' + 2y = 3t^2$

9. $x' + 4x - y' = 3t$
  $x' + y' - 2y = 5t$

10. $(D - 1)x - 2Dy = t^2$
  $x + Dy = 3t - 2$

In Problems 11 to 14, the given sets of functions are solutions of some linear system. Determine whether they form linearly *dependent* or *independent* sets of solutions.

11. $\{e^{5t}, 2e^{3t}\}, \{-3e^{5t}, -6e^{3t}\}$

■**12.** $\{e^{5t}, 2e^{3t}\}, \{-3e^{5t}, -e^{3t}\}$

13. $\{e^t, 2e^t + 8te^t\}, \{-e^t, 6e^t - 8te^t\}$

*__14.__ $\{1 + t, -2 + 2t, 4 + 2t\}, \{1, -2, 4\},$
  $\{3 + 2t, -6 + 4t, 12 + 4t\}$

*__15.__ Prove Theorem 8.2.

*__16.__ Prove Theorem 8.3.

8.3 The Operator Method

In this section we illustrate a *solution technique* that theoretically can be applied to all linear systems with *constant coefficients*.

To begin our solution procedure, let us start with the simple system of equations

$$x' - y = 0$$
$$y' + 4x = 0 \tag{9}$$

or, arranging the unknowns sequentially in both equations with x first and using the operator notation $D = d/dt$,

$$Dx - y = 0$$
$$4x + Dy = 0 \tag{10}$$

To solve (10), we eliminate one of the variables, reducing it to a single DE in one unknown. Suppose we want to eliminate y. This can be done by "operating" on the first equation with D to obtain

$$D^2 x - Dy = 0$$
$$4x + Dy = 0$$

and then adding the resulting two equations. This action leads to the single DE

$$(D^2 + 4)x = 0 \tag{11}$$

the general solution of which is

$$x(t) = C_1 \cos 2t + C_2 \sin 2t \tag{12}$$

In a similar manner, if we multiply the first equation in (10) by 4 and operate on the second equation by D, we obtain

$$4Dx - 4y = 0$$
$$4Dx + D^2 y = 0$$

Subtracting these last equations eliminates x to give us

$$(D^2 + 4)y = 0 \tag{13}$$

with general solution

$$y(t) = C_3 \cos 2t + C_4 \sin 2t \tag{14}$$

Although (12) and (14) represent solutions of the original system (10), they are not solutions for every choice of constants $C_1, C_2, C_3,$ and C_4. That is, since each equation in (10) is a first-order DE, we expect at most one arbitrary constant from each. Therefore, it is necessary to find a proper relationship between the constants $C_1, C_2, C_3,$ and C_4, which we do by substituting both (12) and (14) into one of the original equations in (10), say the first. This yields

$$-2C_1 \sin 2t + 2C_2 \cos 2t - C_3 \cos 2t - C_4 \sin 2t = 0$$

or

$$(2C_2 - C_3) \cos 2t - (2C_1 + C_4) \sin 2t = 0 \tag{15}$$

Hence, (15) is satisfied only if $2C_2 - C_3 = 0$ and $2C_1 + C_4 = 0$, or $C_3 = 2C_2$ and $C_4 = -2C_1$. In this case we claim that the **general solution** of (9) [or (10)] is given by

$$\begin{aligned} x(t) &= C_1 \cos 2t + C_2 \sin 2t \\ y(t) &= 2C_2 \cos 2t - 2C_1 \sin 2t \end{aligned} \tag{16}$$

where C_1 and C_2 are arbitrary constants. [The reader should verify that (16) also satisfies the second equation in (9).]

In order to systematize the above procedure used to solve (9), let us consider the general system of equations

$$\begin{aligned} L_1[x] + L_2[y] &= f_1(t) \\ L_3[x] + L_4[y] &= f_2(t) \end{aligned} \tag{17}$$

where $L_1, L_2, L_3,$ and L_4 are **linear differential operators with constant coefficients,** but *not* restricted to first-order operators. When using general operators we use the bracket notation shown in (17), but we generally omit it in the case of simple operators like D or D^2. Operating on the first equation by L_4 and the second equation by L_2, we obtain

$$\begin{aligned} L_4 L_1[x] + L_4 L_2[y] &= L_4[f_1(t)] \\ L_2 L_3[x] + L_2 L_4[y] &= L_2[f_2(t)] \end{aligned}$$

Since all operators have constant coefficients it follows that $L_4 L_2[y] = L_2 L_4[y]$. Hence, by subtracting the above results, we can eliminate y from the system to obtain

$$(L_1 L_4 - L_2 L_3)[x] = g_1(t) \tag{18a}$$

where $g_1(t) = L_4[f_1(t)] - L_2[f_2(t)]$. In a similar fashion, the elimination of x gives us

$$(L_1 L_4 - L_2 L_3)[y] = g_2(t) \tag{18b}$$

where $g_2(t) = L_1[f_2(t)] - L_3[f_1(t)]$. Equations (18a) and (18b) can now be solved independently, but their solutions must be substituted back into one of the original equations in (17) to determine the relation between the "arbitrary" constants.

8.3.1 Determinant Formulation

The above procedure is basically the same as that used to solve simultaneous *linear algebraic equations*. Solving a system of algebraic equations can be greatly systematized by using the **method of determinants** known as **Cramer's rule.** That is, if we wish to solve the linear algebraic system

$$a_1 x + b_1 y = c_1$$

$$a_2 x + b_2 y = c_2$$

then Cramer's rule states that x satisfies the determinant relation[†]

$$\begin{vmatrix} a_1 & b_1 \\ a_2 & b_2 \end{vmatrix} x = \begin{vmatrix} c_1 & b_1 \\ c_2 & b_2 \end{vmatrix}$$

and y satisfies

$$\begin{vmatrix} a_1 & b_1 \\ a_2 & b_2 \end{vmatrix} y = \begin{vmatrix} a_1 & c_1 \\ a_2 & c_2 \end{vmatrix}$$

Thus, the solutions x and y can be expressed as simply the ratio of two determinants.

Following the above example, we find that the determinant formulations of Equations (18a) and (18b) are

$$\begin{vmatrix} L_1 & L_2 \\ L_3 & L_4 \end{vmatrix} x = \begin{vmatrix} f_1(t) & L_2 \\ f_2(t) & L_4 \end{vmatrix}$$

$$= L_4[f_1(t)] - L_2[f_2(t)] \tag{19a}$$

and

$$\begin{vmatrix} L_1 & L_2 \\ L_3 & L_4 \end{vmatrix} y = \begin{vmatrix} L_1 & f_1(t) \\ L_3 & f_2(t) \end{vmatrix}$$

$$= L_1[f_2(t)] - L_3[f_1(t)] \tag{19b}$$

respectively. It is important to keep in mind that the determinants on the right-hand side of each equation must be interpreted so that when "evaluated," they produce the functions $L_4[f_1(t)] - L_2[f_2(t)]$ and $L_1[f_2(t)] - L_3[f_1(t)]$, respectively. Also, we see that the left-hand sides of (19a) and (19b) are the same. This means that the homogeneous solutions of these equations will involve the same linearly independent

---

[†] Recall that the value of a 2×2 determinant is given by

$$\begin{vmatrix} a_1 & b_1 \\ a_2 & b_2 \end{vmatrix} = a_1 b_2 - a_2 b_1$$

functions, but *not* the same arbitrary constants. The number of independent arbitrary constants in the final solutions should be equal to the order of the operator determinant $L_1L_4 - L_2L_3$.

> **Remark.**   If $L_1L_4 - L_2L_3 = 0$, the system either has **infinitely many solutions** or **no solutions,** depending upon whether the determinants on the right-hand side of (19a) and (19b) vanish.

The use of Cramer's rule to solve equations can be applied in general to systems of n equations, regardless of the order of the operators $L_1, L_2, \ldots, L_n, \ldots, L_{2n}$. From a practical point of view it works best on systems of two or three equations with at most second-order derivatives since the amount of computation becomes unwieldy for larger systems.

■──────────
EXAMPLE 5

Solve the system of equations

$$x' = 2x + y + t$$
$$y' = x + 2y + t^2$$

Solution

In operator notation the system of equations becomes

$$(D - 2)x - y = t$$
$$-x + (D - 2)y = t^2$$

which leads to the determinant formulation

$$\begin{vmatrix} D-2 & -1 \\ -1 & D-2 \end{vmatrix} x = \begin{vmatrix} t & -1 \\ t^2 & D-2 \end{vmatrix}$$

and

$$\begin{vmatrix} D-2 & -1 \\ -1 & D-2 \end{vmatrix} y = \begin{vmatrix} D-2 & t \\ -1 & t^2 \end{vmatrix}$$

After expansion of the determinants, we have

$$(D^2 - 4D + 3)x = t^2 - 2t + 1$$
$$(D^2 - 4D + 3)y = -2t^2 + 3t$$

The homogeneous solution for x is

$$x_H(t) = C_1e^t + C_2e^{3t}$$

while a particular solution is found to be

$$x_P(t) = \tfrac{1}{3}t^2 + \tfrac{2}{9}t + \tfrac{11}{27}$$

Hence,

$$x(t) = x_H(t) + x_P(t)$$
$$= C_1e^t + C_2e^{3t} + \tfrac{1}{3}t^2 + \tfrac{2}{9}t + \tfrac{11}{27}$$

Rather than solve the nonhomogeneous DE above for y, we merely observe that

$$y = x' - 2x - t$$

obtained from the first equation of the system. Therefore, it follows that

$$y(t) = C_1 e^t + 3C_2 e^{3t} + \tfrac{2}{3}t + \tfrac{2}{9} - 2(C_1 e^t + C_2 e^{3t}$$
$$+ \tfrac{1}{3}t^2 + \tfrac{2}{9}t + \tfrac{11}{27}) - t$$
$$= -C_1 e^t + C_2 e^{3t} - \tfrac{2}{3}t^2 - \tfrac{7}{9}t - \tfrac{16}{27} \qquad \blacksquare$$

8.3.2 Homogeneous Equations of First Order

In this section we illustrate a more efficient method of solving *homogeneous equations of first order* of the form

$$x' = a_{11}x + a_{12}y$$
$$y' = a_{21}x + a_{22}y \qquad (20)$$

Knowing that (20) is equivalent to a homogeneous second-order DE for either x or y, we can assume a solution exists of the form $x(t) = Ae^{\lambda t}$, $y(t) = Be^{\lambda t}$, where A, B, and λ are constants to be determined. The direct substitution of these trial solutions into (20) yields

$$A\lambda e^{\lambda t} = a_{11}Ae^{\lambda t} + a_{12}Be^{\lambda t}$$
$$B\lambda e^{\lambda t} = a_{21}Ae^{\lambda t} + a_{22}Be^{\lambda t}$$

After dividing by the nonzero factor $e^{\lambda t}$, and rearranging the remaining terms, we are left with the algebraic system of equations

$$(a_{11} - \lambda)A + a_{12}B = 0$$
$$a_{21}A + (a_{22} - \lambda)B = 0 \qquad (21)$$

We wish to find values of λ such that (21) has nonzero solutions for A and B. From the theory of equations, this will happen if and only if the coefficient determinant

$$\Delta(\lambda) = \begin{vmatrix} a_{11} - \lambda & a_{12} \\ a_{21} & a_{22} - \lambda \end{vmatrix} = 0 \qquad (22)$$

Evaluating the determinant, we obtain the quadratic equation

$$\lambda^2 - (a_{11} + a_{22})\lambda + (a_{11}a_{22} - a_{12}a_{21}) = 0 \qquad (23)$$

called the **auxiliary equation,** or **characteristic equation,** of (20). It is interesting that (23) is exactly the same auxiliary equation as that associated with the second-order DEs $(L_1 L_4 - L_2 L_3)[x] = 0$ and $(L_1 L_4 - L_2 L_3)[y] = 0$. Thus, once we know the roots of (23), λ_1 and λ_2, we can say the general solution for x has the form $x(t) = C_1 e^{\lambda_1 t} + C_2 e^{\lambda_2 t}$, where C_1 and C_2 are arbitrary constants. The solution for y can then be determined from the first equation in (20). Of course, we could just as easily have said the general solution for y is $y(t) = C_1 e^{\lambda_1 t} + C_2 e^{\lambda_2 t}$, and then we would use the second equation in (20) to determine x. As in Chapter 4, we need to distinguish between cases when λ_1 and λ_2 are *real and distinct, equal,* or *complex conjugates.*

EXAMPLE 6

Solve the system of equations

$$x' = -4x - y$$
$$y' = x - 2y$$

Solution

Here $a_{11} = -4$, $a_{12} = -1$, $a_{21} = 1$, and $a_{22} = -2$, so that Equation (22) becomes

$$\Delta(\lambda) = \begin{vmatrix} -4 - \lambda & -1 \\ 1 & -2 - \lambda \end{vmatrix} = \lambda^2 + 6\lambda + 9 = 0$$

or

$$(\lambda + 3)^2 = 0$$

Hence, $\lambda_1 = \lambda_2 = -3$. The general solution for x, for example, is therefore

$$x(t) = (C_1 + C_2 t)e^{-3t}$$

To solve for y, we simply recognize from the first equation of the system that $y = -4x - x'$, from which we obtain

$$y(t) = -4(C_1 + C_2 t)e^{-3t} + 3(C_1 + C_2 t)e^{-3t} - C_2 e^{-3t}$$
$$= -(C_1 + C_2)e^{-3t} - C_2 t e^{-3t}$$

Thus, the general solution is

$$x(t) = C_1 e^{-3t} + C_2 t e^{-3t}$$
$$y(t) = -(C_1 + C_2)e^{-3t} - C_2 t e^{-3t} \qquad ■$$

EXAMPLE 7

Solve the system of equations

$$x' = 6x - y$$
$$y' = 5x + 4y$$

Solution

The auxiliary or characteristic equation is

$$\Delta(\lambda) = \begin{vmatrix} 6 - \lambda & -1 \\ 5 & 4 - \lambda \end{vmatrix} = \lambda^2 - 10\lambda + 29 = 0$$

with complex roots $\lambda = 5 \pm 2i$. Hence, the general solution for x is

$$x(t) = e^{5t}(C_1 \cos 2t + C_2 \sin 2t)$$

From the first equation, we have $y = 6x - x'$, or

$$y(t) = 6e^{5t}(C_1 \cos 2t + C_2 \sin 2t) - 5e^{5t}(C_1 \cos 2t + C_2 \sin 2t)$$
$$- e^{5t}(-2C_2 \sin 2t + 2C_1 \cos 2t)$$

from which we deduce the general solution

$$x(t) = e^{5t}(C_1 \cos 2t + C_2 \sin 2t)$$

$$y(t) = e^{5t}[(C_1 - 2C_2) \cos 2t + (2C_1 + C_2) \sin 2t]$$

Exercises 8.3

In Problems 1 to 10, find the general solution of the system of equations.

1. $x' = 2x - y$
$y' = 3x - 2y$

2. $x' = 4x - 3y$
$y' = 5x - 4y$

3. $x' = 4x - 3y$
$y' = 8x - 6y$

4. $x' = 3x - 18y$
$y' = 2x - 9y$

5. $x' = x - 4y$
$y' = x + y$

■6. $x' = 3x - 2y$
$y' = 2x + 3y$

7. $x' = 6x - 5y$
$y' = x + 2y$

8. $x' = 3x + 2y$
$y' = -5x + y$

*9. $x' = x + 2y + z$
$y' = 6x - y$
$z' = -x - 2y - z$

*10. $x' = -x + z$
$y' = -y + z$
$z' = -x + y$

11. Given that $b \neq 0$, find the general solution of

$$x' = ax + by$$

$$y' = bx + ay$$

In Problems 12 to 28, solve the given system of equations, making sure the proper number of constants appears in the general solution. When initial conditions are prescribed, find the particular solution.

12. $x' - 2x + 5y = -\sin 2t, x(0) = 0$
$y' - x + 2y = t, y(0) = 1$

13. $x' + y' + 2y = 0$
$x' - 3x - 2y = 0$

14. $x' - 2x + y = t$
$y' - 3x + 2y = 2t$

15. $x' - 2x - y = e^t, x(0) = 1$
$y' - 4x + y = -e^t, y(0) = -1$

16. $x'' + 5x - 2y = 0$
$y'' + 2y - 2x = 0$

17. $x'' + x' - x + y'' - 3y' + 2y = 0$
$x' + 2x + 2y' - 4y = 0$

18. $(D^2 - 3D + 2)x + (D - 1)y = 0$
$(D - 2)x + (D + 1)y = 0$

19. $(D^2 - 4D + 4)x + (D^2 + 2D)y = 0$
$(D^2 - 2D)x + (D^2 + 4D + 4)y = 0$

■20. $x' + x + y' - y = 2$
$3x + y' + 2y = -1$

21. $2x' - 3x - 2y' = t$
$2x' + 3x + 2y' + 8y = 2$

22. $x' - 2x + y' - 4y = e^t$
$x' + y' - y = e^{4t}$

23. $x' + y' + 2y = \sin t$
$x' - x + y' - y = 0$

24. $(2D - 1)x + (D - 1)y = 1$
$(D + 2)x + (D - 1)y = t$

25. $(D^2 - 4D + 4)x + 3Dy = 1$
$(D - 2)x + (D + 2)y = 0$

*26. $2x' + y' - y + z' + 2z = 0$
$x' + 2x + 2y' - 3y - z' + 6z = 0$

*27. $x' - 6y = 0$
$x - y' + z = 0$
$x + y - z' = 0$

*28. $x' + z = e^t$
$x' - x + y' + z' = 0$
$x + 2y + z' = e^t$

In Problems 29 to 31, show that each system of equations is *degenerate* in that it has either no solution or infinitely many linearly independent solutions.

29. $x' + y' + y = e^t$
     $x' + y' + y = e^t + 3$

31. $x'' + 3x' + 2x + y'' + 2y' = 0$
     $x' + x + y' = 0$

■ 30. $x'' - x + y' - y = \sin t$
     $x' + x + y = 2e^t$

8.4 Applications

Linear systems of differential equations arise in numerous types of applications. In particular, they are associated with mechanical systems such as *spring-mass systems*, certain *electrical networks*, and *mixing problems* involving more than one compartment. In this section we look briefly at some simple problems of this nature.

8.4.1 Coupled Spring-Mass Systems

Suppose two masses m_1 and m_2 are connected to two springs having spring constants k_1 and k_2, respectively (see Figure 8.1). Let y_1 and y_2 denote the vertical displacements of the masses from their equilibrium positions. When the system is in motion, the stretching of the lower spring is $y_2 - y_1$, exerting a force of $k_2(y_2 - y_1)$ on the upper mass m_1 (Hooke's law). Also, the upper spring exerts the force $-k_1 y_1$ on this mass, and so, invoking Newton's second law of motion (See Section 5.2), the small motions of the mass m_1 are governed by

$$m_1 y_1'' = -k_1 y_1 + k_2(y_2 - y_1) \tag{24}$$

In a similar manner, it follows that the equation of motion for the lower mass m_2 is

$$m_2 y_2'' = -k_2(y_2 - y_1) \tag{25}$$

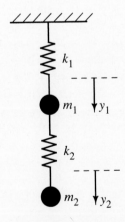

Figure 8.1 Coupled spring-mass system.

Hence, the **coupled spring-mass system** in Figure 8.1, in the absence of damping effects and external forces, is governed by the system of second-order DEs

$$m_1 y_1'' + (k_1 + k_2)y_1 - k_2 y_2 = 0$$
$$m_2 y_2'' - k_2 y_1 + k_2 y_2 = 0$$

(26)

EXAMPLE 8

Determine the free motions of a double spring-mass system composed of two unit masses and springs with constants $k_1 = 6$ lb/ft and $k_2 = 4$ lb/ft. Assume that both masses start from their equilibrium positions, but that m_1 has a downward initial velocity of 5 ft/s and m_2 an upward initial velocity of 5 ft/s. Neglect damping.

Solution

The system here is described by Equations (26), which, for the specific parameter values given above, reduces to

$$y_1'' + 10y_1 - 4y_2 = 0$$
$$y_2'' - 4y_1 + 4y_2 = 0$$

or, in operator notation,

$$(D^2 + 10)y_1 - 4y_2 = 0$$
$$-4y_1 + (D^2 + 4)y_2 = 0$$

The prescribed initial conditions are

$$y_1(0) = 0, \qquad y_1'(0) = 5, \qquad y_2(0) = 0, \qquad y_2'(0) = -5$$

In the determinant formulation, we obtain

$$\begin{vmatrix} D^2 + 10 & -4 \\ -4 & D^2 + 4 \end{vmatrix} y = (D^4 + 14D^2 + 24)y = 0$$

which is the same DE for both y_1 and y_2. Let us consider the case first where $y = y_1$, that is,

$$(D^4 + 14D^2 + 24)y_1 = (D^2 + 2)(D^2 + 12)y_1 = 0$$

The roots of the auxiliary polynomial are $m = \pm\sqrt{2}i, \pm 2\sqrt{3}i$, leading to the general solution

$$y_1(t) = C_1 \cos \sqrt{2}t + C_2 \sin \sqrt{2}t + C_3 \cos 2\sqrt{3}t + C_4 \sin 2\sqrt{3}t.$$

Solving the first DE above for y_2, we see that $y_2 = \frac{1}{4}(y_1'' + 10y_1)$, or

$$y_2(t) = 2C_1 \cos \sqrt{2}t + 2C_2 \sin \sqrt{2}t - \frac{1}{2}(C_3 \cos 2\sqrt{3}t + C_4 \sin 2\sqrt{3}t)$$

If we now impose the prescribed initial conditions, it follows that

$$y_1(0) = C_1 + C_3 = 0$$
$$y_1'(0) = \sqrt{2}C_2 + 2\sqrt{3}C_4 = 5$$
$$y_2(0) = 2C_1 - \frac{1}{2}C_3 = 0$$
$$y_2'(0) = 2\sqrt{2}C_2 - \sqrt{3}C_4 = -5$$

the simultaneous solution of which yields

$$C_1 = C_3 = 0, \qquad C_2 = -\frac{\sqrt{2}}{2}, \qquad C_4 = \sqrt{3}$$

Hence, the solution we seek is

$$y_1(t) = -\frac{\sqrt{2}}{2}\sin\sqrt{2}t + \sqrt{3}\sin 2\sqrt{3}t$$

$$y_2(t) = -\sqrt{2}\sin\sqrt{2}t - \frac{\sqrt{3}}{2}\sin 2\sqrt{3}t$$

The motion of each mass, therefore, is the superposition of two harmonic motions. ■

8.4.2 Electrical Networks

An **electrical network** composed of components connected in parallel gives rise to simultaneous DEs. For example, the *RLC* circuit shown in Figure 8.2 has two loops. By application of Kirchhoff's voltage law (Section 5.4), we obtain

$$Li_1' + Ri_2 = E(t)$$

$$\frac{1}{C}i_3 - Ri_2' = 0 \tag{27}$$

where i_1, i_2, and i_3 denote the current in each part of the circuit. However, by Kirchhoff's current law, we see that

$$i_1 = i_2 + i_3$$

and thus we can eliminate i_3 from (24) by writing $i_3 = i_1 - i_2$. In this case, (27) becomes

$$Li_1' + Ri_2 = E(t)$$

$$i_2' + \frac{1}{RC}(i_2 - i_1) = 0 \tag{28}$$

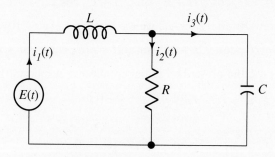

Figure 8.2 *RLC* circuit.

EXAMPLE 9

Let $R = 100\ \Omega$, $L = 4$ H, $C = 10^{-4}$ F, and $E(t) = 100$ V in the network of Figure 8.2. If initially the currents i_1 and i_2 are both zero, find the currents at all later times.

Solution

Substituting the appropriate values of the parameters into (28), we have

$$4i_1' + 100i_2 = 100, \qquad i_1(0) = 0$$

$$i_2' + 100(i_2 - i_1) = 0, \qquad i_2(0) = 0$$

or, upon simplifying and putting into determinant form,

$$\begin{vmatrix} D & 25 \\ -100 & D + 100 \end{vmatrix} i_1 = \begin{vmatrix} 25 & 25 \\ 0 & D + 100 \end{vmatrix}$$

$$\begin{vmatrix} D & 25 \\ -100 & D + 100 \end{vmatrix} i_2 = \begin{vmatrix} D & 25 \\ -100 & 0 \end{vmatrix}$$

Expanding these determinants leads to

$$(D + 50)^2 i_1 = 2500$$

$$(D + 50)^2 i_2 = 2500$$

Solving the first DE for i_1, we find

$$i_1(t) = (C_1 + C_2 t)e^{-50t} + 1$$

and by substituting this into the first of the original system of equations, it can be shown that

$$i_2(t) = (2C_1 - \tfrac{1}{25}C_2)e^{-50t} + 2C_2 t e^{-50t} + 1$$

Finally, the initial conditions require that

$$i_1(0) = C_1 + 1 = 0$$

$$i_2(0) = 2C_1 - \tfrac{1}{25}C_2 + 1 = 0$$

or $C_1 = -1$ and $C_2 = -25$. Thus,

$$i_1(t) = 1 - e^{-50t} - 25t e^{-50t}$$

$$i_2(t) = 1 - e^{-50t} - 50t e^{-50t}$$

and

$$i_3(t) = i_1(t) - i_2(t)$$

$$= 25t e^{-50t}$$

Observe that the steady-state currents ($t \to \infty$) are

$$i_1(t) = i_2(t) = 1, \qquad i_3(t) = 0$$

8.4.3 Mixing Problems

As our last representative area of application, let us consider an example of a mixing problem involving two interconnected tanks (see Section 3.5 for a discussion of mixing problems involving only one tank). Problems in this general category are often referred to as **multiple-compartment problems.**

EXAMPLE 10   Tanks A and B, each holding 100 gal of liquid, are interconnected by pipes with liquid freely flowing between the tanks (see Figure 8.3). A brine solution with a concentration of 2 lb/gal of salt flows into tank A from an outside source at the rate of 6 gal/min. The diluted solution (kept well stirred) flows out of tank A into tank B at the rate of 3 gal/min and out of the system at the rate of 4 gal/min. In addition, the liquid in tank B flows back into tank A at the rate of 1 gal/min and out of the system at the rate of 2 gal/min. If, initially, tank A contains only water and tank B contains a brine solution with 200 lb of salt, what is the amount of salt in each tank at some later time t?

Solution

Let $x(t)$ and $y(t)$ denote the amounts of salt in tanks A and B, respectively, at time t. First we recognize that the volume of liquid in tank A remains constant since liquid flows into and out of tank A at the rate of 7 gal/min. The same is true of tank B since liquid flows into and out of it at the rate of 3 gal/min.

The rate of change of salt in tank A satisfies the relation

$$\frac{dx}{dt} = (\text{rate of salt in}) - (\text{rate of salt out})$$

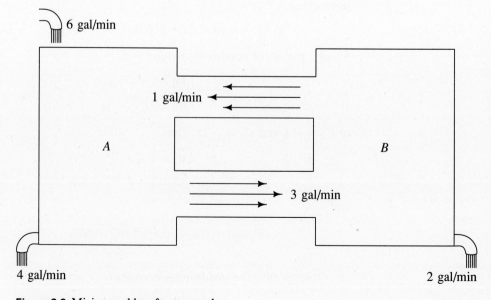

Figure 8.3 Mixing problem for two tanks.

Salt enters tank A from the outside at a rate of 12 lb/min (i.e., 6 gal/min with 2 lb/gal). Salt also enters tank A from tank B at the rate of $y/100$ lb/min. Thus, for tank A we have

$$\text{Rate of salt in} = 12 + \frac{y}{100}$$

Salt leaves tank A to the outside at the rate of $4x/100$ lb/min and leaves tank A to enter tank B at the rate of $3x/100$ lb/min. Hence, for tank A

$$\text{Rate of salt out} = \frac{4x}{100} + \frac{3x}{100} = \frac{7x}{100}$$

which leads to

$$\frac{dx}{dt} = 12 + \frac{y}{100} - \frac{7x}{100}$$

Along similar lines, it can be shown that the rate of change of salt in tank B is governed by

$$\frac{dy}{dt} = \frac{3x}{100} - \frac{3y}{100}$$

The initial conditions are $x(0) = 0$ and $y(0) = 200$, so that rewriting the above DEs in operator notation and including the initial conditions, we obtain

$$(D + \tfrac{7}{100})x - \tfrac{1}{100}y = 12, \qquad x(0) = 0$$

$$-\tfrac{3}{100}x + (D + \tfrac{3}{100})y = 0, \qquad y(0) = 200$$

By eliminating the variable y, we find that

$$\left(D^2 + \frac{1}{10}D + \frac{18}{10{,}000}\right)x = \frac{36}{100}$$

The auxiliary equation $m^2 + \tfrac{1}{10}m + 18/10{,}000 = 0$ has roots $\tfrac{1}{100}(-5 \pm \sqrt{7})$, which we approximate by -0.0235 and -0.0765. Hence, the general solution is

$$x(t) \cong C_1 e^{-0.0235t} + C_2 e^{-0.0765} + 200$$

To determine $y(t)$, we write the first equation above in the form

$$y = 100\left(\frac{dx}{dt} + \frac{7}{100}x - 12\right)$$

which yields

$$y(t) \cong 4.646C_1 e^{-0.0235t} - 0.646C_2 e^{-0.0765t} + 200$$

Finally, by imposing the prescribed initial conditions we see that $C_1 \cong -24.4$ and $C_2 \cong -175.6$, from which we obtain

$$x(t) \cong 200 - 24.4e^{-0.0235t} - 175.6e^{-0.0765t}$$

and

$$y(t) \cong 200 - 113.4e^{-0.0235t} + 113.4e^{-0.0765t}$$

Observe that both $x(t)$ and $y(t)$ approach 200 lb salt as $t \to \infty$. Thus, the concentration of salt in each tank approaches 2 lb/gal, which is that of the salt brine mixture entering tank A. ∎

Exercises 8.4

1. Consider the coupled spring-mass system described by (26) when $m_1 = m_2 = 1$, $k_1 = 3$, and $k_2 = 2$.
   (a) What are the natural frequencies of the system?
   (b) Find the positions of the masses at any time t, given the initial conditions $y_1(0) = 0$, $y_1'(0) = 1$, $y_2(0) = 1$, $y_2'(0) = 0$.

∎2. Consider the coupled spring-mass system described by (26) when $m_1 = 4$, $m_2 = 2$, $k_1 = 8$, and $k_2 = 4$.
   (a) What are the natural frequencies of the system?
   (b) Find the positions of the masses at any time t, given the initial conditions $y_1(0) = 0$, $y_1'(0) = 0$, $y_2(0) = 0$, $y_2'(0) = -2$.

3. Solve Problem 2(b) if a force $F(t) = 40 \sin 3t$ is applied to mass m_1 for $t \geq 0$.

4. Solve Problem 2(b) if a force $F(t) = 40 \sin 3t$ is applied to mass m_2 for $t \geq 0$.

5. Show that the spring-mass system in the accompanying figure is governed by the system of equations

$$y_1'' + 2ky_1 - ky_2 = 0$$
$$y_2'' + 2ky_2 - ky_1 = 0$$

6. Solve the system of equations in Problem 5 if $k = 3$ and the initial conditions are $y_1(0) = 1$, $y_1'(0) = 3$, $y_2(0) = 1$, $y_2'(0) = -3$.

7. Find the currents i_1, i_2, and i_3 in the network in Figure 8.2 when $R = 100\ \Omega$, $L = 1$ H, $C = 10^{-4}$ F, and $E(t) = 100$ V. Assume that i_1 and i_2 are initially zero.

∎8. Find the currents i_1, i_2, and i_3 in the network in Figure 8.2 when $R = 100\ \Omega$, $L = 4$ H, $C =$

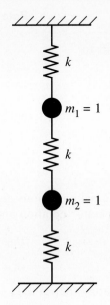

Problem 5

10^{-4} F, and $E(t) = 100e^{-1000t}$ V. Assume that i_1 and i_2 are initially zero.

9. Solve Problem 8 when all conditions are the same except that $E(t) = 100 \sin 60\pi t$ V.

10. Show that the network illustrated in the accompanying figure is governed by the system of equations

$$R_1 i_1 + L i_2' + R_2 i_2 = E(t)$$
$$R_1 i_1 + L_2 i_3' = E(t)$$

Eliminate i_3 and rewrite the system of equations in terms of i_1 and i_2 alone.

11. Solve the system of equations in Problem 10 when $R_1 = 6\ \Omega$, $R_2 = 5\ \Omega$, $L_1 = L_2 = 1$ H, $E(t) = 50 \sin t$ V, and $i_1(0) = i_2(0) = 0$.

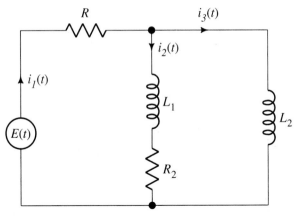

Problem 10

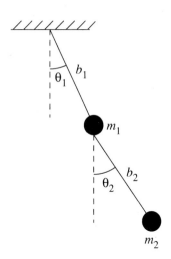

Problem 12

*12. A double pendulum oscillates in a vertical plane under the influence of gravity alone (see accompanying figure). For small displacements, it can be shown that the equations of motion are

$$(m_1 + m_2)b_1^2\theta_1'' + m_2b_1b_2\theta_2'' + (m_1 + m_2)b_1g\theta_1 = 0$$

$$m_2b_2^2\theta_2'' + m_2b_1b_2\theta_1'' + m_2b_2g\theta_2 = 0$$

Show that both θ_1 and θ_2 are solutions of the fourth-order homogeneous DE

$$[m_1m_2b_1^2b_2^2D^4 + m_2b_1b_2g(m_1 + m_2)(b_1 + b_2)D^2$$
$$+ m_2b_1b_2g^2(m_1 + m_2)]\theta = 0$$

*13. Find the natural frequencies of the system in Problem 12 when (assume $g = 32$)

(a) $m_1 = m_2 = 1$, $b_1 = b_2 = 1$
(b) $m_1 = 2$, $m_2 = 1$, $b_1 = b_2 = 1$
(c) $m_1 = 1$, $m_2 = 2$, $b_1 = b_2 = 1$

(d) $m_1 = m_2 = 1$, $b_1 = 1$, $b_2 = 2$
(e) $m_1 = m_2 = 1$, $b_1 = 2$, $b_2 = 1$
Hint. Recall that $f = \omega/2\pi$.

*14. Solve the system of equations in Problem 12 when $m_1 = 3$, $m_2 = 1$, $b_1 = b_2 = 16$, and the initial conditions are (assume $g = 32$)

$$\theta_1(0) = 0, \quad \theta_1'(0) = 0, \quad \theta_2(0) = \tfrac{1}{2}, \quad \theta_2'(0) = 0$$

15. Solve Example 10 under the conditions that 5 gal/min of liquid flows from tank A into tank B, 3 gal/min flows from tank B into tank A, and all other conditions remain the same.

16. Solve Example 10 under the conditions that 1 gal/min of liquid flows from tank A into tank B, 4 gal/min of brine solution flows into tank A from the outside, no liquid flows out of the system from tank B, and all other conditions remain the same.

8.5 The Method of Laplace Transforms

In practice, initial conditions are almost always prescribed along with the system of differential equations. In such cases the **method of Laplace transforms** is generally more convenient to use than the operator method of Section 8.3. (This section can be omitted if Chapter 6 has not been covered.)

The Laplace transform applied to a differential system reduces it to an *algebraic* system in the transformed functions. Standard solution techniques can then be applied to solve the algebraic system, such as Cramer's rule if the number of equations is small. The inverse Laplace transform of the solution of the algebraic system will then produce the solution functions of the original differential system.

■──────────
EXAMPLE 11
──────────

Solve

$$x'' - x + 5y' = t$$
$$y'' - 4y - 2x' = -2$$

subject to the initial conditions

$$x(0) = 0, \qquad x'(0) = 0, \qquad y(0) = 0, \qquad y'(0) = 0$$

Solution

If we let $X(s) = \mathcal{L}\{x(t)\}$ and $Y(s) = \mathcal{L}\{y(t)\}$, then application of the Laplace transform to each given equation results in the pair of algebraic equations

$$s^2 X(s) - X(s) + 5sY(s) = \frac{1}{s^2}$$

$$s^2 Y(s) - 4Y(s) - 2sX(s) = -\frac{2}{s}$$

which incorporates the prescribed initial conditions. Upon simplification, we have

$$(s^2 - 1)X(s) + 5sY(s) = \frac{1}{s^2}$$

$$-2sX(s) + (s^2 - 4)Y(s) = -\frac{2}{s}$$

the simultaneous solution of which is given by

$$X(s) = \frac{11s^2 - 4}{s^2(s^2 + 1)(s^2 + 4)}$$

$$Y(s) = \frac{-2s^2 + 4}{s(s^2 + 1)(s^2 + 4)}$$

In order to invert these expressions, we first expand the right-hand members in terms of their partial fractions, which yields

$$X(s) = -\frac{1}{s^2} + \frac{5}{s^2 + 1} - \frac{4}{s^2 + 4}$$

$$Y(s) = \frac{1}{s} - \frac{2s}{s^2 + 1} + \frac{s}{s^2 + 4}$$

Hence, using tables of Laplace transforms, it follows that

$$x(t) = \mathscr{L}^{-1}\{X(s)\}$$

$$= \mathscr{L}^{-1}\left\{\frac{1}{s^2}\right\} + 5\mathscr{L}^{-1}\left\{\frac{1}{s^2+1}\right\} - 2\mathscr{L}^{-1}\left\{\frac{2}{s^2+4}\right\}$$

or

$$x(t) = -t + 5\sin t - 2\sin 2t$$

Similarly, we find that

$$y(t) = 1 - 2\cos t + \cos 2t \qquad \blacksquare$$

EXAMPLE 12   Solve

$$y_1'' + 10y_1 - 4y_2 = 0$$

$$y_2'' - 4y_1 + 4y_2 = 0$$

subject to the initial conditions

$$y_1(0) = 0, \qquad y_1'(0) = 5, \qquad y_2(0) = 0, \qquad y_2'(0) = -5$$

Solution

(This system is that corresponding to the double spring-mass system described in Example 8 in Section 8.3.)

Using the notation $Y_1(s) = \mathscr{L}\{y_1(t)\}$ and $Y_2(s) = \mathscr{L}\{y_2(t)\}$, the Laplace transform of the differential system reduces to

$$s^2 Y_1(s) - 5 + 10Y_1(s) - 4Y_2(s) = 0$$

$$s^2 Y_2(s) + 5 - 4Y_1(s) + 4Y_2(s) = 0$$

or

$$(s^2 + 10)Y_1(s) - 4Y_2(s) = 5$$

$$-4Y_1(s) + (s^2 + 4)Y_2(s) = -5$$

Solving these equations for $Y_1(s)$ and $Y_2(s)$ gives us

$$Y_1(s) = \frac{5s^2}{(s^2+2)(s^2+12)}$$

$$Y_2(s) = -\frac{5(s^2+6)}{(s^2+2)(s^2+12)}$$

The partial fraction expansion of these functions yields

$$Y_1(s) = -\frac{1}{s^2+2} + \frac{6}{s^2+12}$$

$$Y_2(s) = -\frac{2}{s^2+2} - \frac{3}{s^2+12}$$

the inverse Laplace transform of which leads to

$$y_1(t) = -\frac{\sqrt{2}}{2} \sin \sqrt{2}t + \sqrt{3} \sin 2\sqrt{3}t$$

$$y_2(t) = -\sqrt{2} \sin \sqrt{2}t - \frac{\sqrt{3}}{2} \sin 2\sqrt{3}t \qquad \blacksquare$$

Exercises 8.5

In Problems 1 to 12, solve the system of equations by use of the Laplace transform.

1. $x' = -x + y, x(0) = 0$
  $y' = 2x, y(0) = 1$

2. $x' = y, x(0) = 1$
  $y' = x, y(0) = 0$

3. $x' = x - 2y, x(0) = -1$
  $y' = 5x - y, y(0) = 2$

4. $x' = 4x - 2y, x(0) = 2$
  $y' = 5x + 2y, y(0) = -2$

5. $x' = x - y + e^t, x(0) = 1$
  $y' = 2x + 3y + e^{-t}, y(0) = 0$

■6. $x' - 4x + 2y = e^t, x(0) = 1$
  $y' - 5x - 2y = -t, y(0) = 0$

7. $2x' + y' - y = t, x(0) = 1$
  $x' + y' = t^2, y(0) = 0$

8. $2x' + y' - 2x = 1, x(0) = 0$
  $x' + y' - 3x - 3y = 2, y(0) = 0$

9. $x'' + x - y = 0, x(0) = 0, x'(0) = -2$
  $y'' + y - x = 0, y(0) = 0, y'(0) = 1$

10. $x' - 4x + y''' = 6 \sin t, x(0) = 0$
  $x' - 2y''' + 2x = 0, y(0) = 0, y'(0) = 0,$
  $y''(0) = 0$

11. $x'' + 2x - y' = 2t + 5, x(0) = 3, x'(0) = 0$
  $x' - x + y' + y = -2t - 1, y(0) = -3$

*12. $x'' - x + 5y' = f(t), x(0) = 0, x'(0) = 0$
  $y'' - 4y - 2x' = 0, y(0) = 0, y'(0) = 0$
  where $f(t) = \begin{cases} 6t, & 0 \le t < 2 \\ 12, & t > 2 \end{cases}$

In Problems 13 to 20, use Laplace transforms to solve the referenced problem.

13. Problem 1 in Exercises 8.4

14. Problem 2 in Exercises 8.4

15. Problem 3 in Exercises 8.4

16. Problem 4 in Exercises 8.4

17. Problem 7 in Exercises 8.4

■18. Problem 8 in Exercises 8.4

19. Problem 15 in Exercises 8.4

20. Problem 16 in Exercises 8.4

[0] 8.6 Review of Matrices

The study of systems of linear DEs can be greatly systematized by the use of matrix methods. The matrix formulation of these problems is particularly useful when the number of equations is greater than two. In this section we provide a brief review of matrix and vector concepts to aid us in applying this powerful technique in Sections 8.7 to 8.9. The reader already familiar with matrices and vectors may go directly to Section 8.7.

Matrices and vectors are denoted by bold letters such as **A, B, X, Y, ϕ,** .... A **matrix** is a rectangular array of numbers, or **elements,** arranged in m rows and n columns; that is,

$$\mathbf{A} = \begin{pmatrix} a_{11} & a_{12} & \cdots & a_{1n} \\ a_{21} & a_{22} & \cdots & a_{2n} \\ \vdots & \vdots & & \vdots \\ a_{m1} & a_{m2} & \cdots & a_{mn} \end{pmatrix} \tag{29}$$

We call **A** an $m \times n$ matrix with elements a_{ij} ($i = 1, 2, \ldots, m; j = 1, 2, \ldots, n$). The first subscript denotes the row while the second subscript identifies the column. It is also customary to write $\mathbf{A} = [a_{ij}]$.

If $m = n$, we say that **A** is a **square matrix.** For example,

$$\mathbf{A} = \begin{pmatrix} a_{11} & a_{12} \\ a_{21} & a_{22} \end{pmatrix}$$

is a 2×2 square matrix. A matrix with only one column is called a **column vector.** A 3×1 column vector, for instance, appears like

$$\mathbf{Y} = \begin{pmatrix} y_1 \\ y_2 \\ y_3 \end{pmatrix}$$

Similarly, if the matrix has only one row, it is called a **row vector.** In some cases the elements of a matrix may be functions, such as

$$\mathbf{A}(t) = \begin{pmatrix} a_{11}(t) & a_{12}(t) \\ a_{21}(t) & a_{22}(t) \end{pmatrix}$$

and

$$\mathbf{Y}(t) = \begin{pmatrix} y_1(t) \\ y_2(t) \\ y_3(t) \end{pmatrix}$$

Finally, the symbol **0**, called the **zero matrix,** will denote any matrix or vector with all elements equal to zero.

8.6.1 Algebraic Properties

Various algebraic properties can be identified with certain matrices related to each other primarily through their size. These properties are listed below.

   1. *Equality.*   Two $m \times n$ matrices $\mathbf{A} = [a_{ij}]$ and $\mathbf{B} = [b_{ij}]$ are said to be equal if and only if

$$a_{ij} = b_{ij} \tag{30}$$

for each i and j. In this case we write $\mathbf{A} = \mathbf{B}$.

2. *Addition.*   The sum of two $m \times n$ matrices $\mathbf{A} = [a_{ij}]$ and $\mathbf{B} = [b_{ij}]$ is the $m \times n$ matrix $\mathbf{C} = [c_{ij}]$, where

$$a_{ij} + b_{ij} = c_{ij} \tag{31}$$

for each i and j. Thus, we write $\mathbf{A} + \mathbf{B} = \mathbf{C}$. For example, if

$$\mathbf{A} = \begin{pmatrix} 1 & -4 & 5 \\ 6 & 2 & 0 \end{pmatrix}, \qquad \mathbf{B} = \begin{pmatrix} -5 & 4 & 1 \\ 0 & -1 & 3 \end{pmatrix}$$

then

$$\mathbf{A} + \mathbf{B} = \begin{pmatrix} 1-5 & -4+4 & 5+1 \\ 6+0 & 2-1 & 0+3 \end{pmatrix} = \begin{pmatrix} -4 & 0 & 6 \\ 6 & 1 & 3 \end{pmatrix}$$

Matrix addition enjoys the following properties which are analogous to those of addition of real numbers:

$$\mathbf{A} + \mathbf{B} = \mathbf{B} + \mathbf{A} \tag{32}$$

$$\mathbf{A} + (\mathbf{B} + \mathbf{C}) = (\mathbf{A} + \mathbf{B}) + \mathbf{C} \tag{33}$$

$$\mathbf{A} + \mathbf{0} = \mathbf{A} \tag{34}$$

3. *Multiplication by a number.*   The product of an $m \times n$ matrix $\mathbf{A} = [a_{ij}]$ and a (real or complex) number c is defined by

$$c\mathbf{A} = c[a_{ij}] = [ca_{ij}] \tag{35}$$

(That is, each element a_{ij} is multiplied by the constant c.) In particular if $c = -1$, we have

$$(-1)\mathbf{A} = -\mathbf{A} \tag{36}$$

If c and k are any numbers, it follows that

$$c(\mathbf{A} + \mathbf{B}) = c\mathbf{A} + c\mathbf{B} \tag{37}$$

$$(c + k)\mathbf{A} = c\mathbf{A} + k\mathbf{A} \tag{38}$$

$$c(k\mathbf{A}) = (ck)\mathbf{A} \tag{39}$$

and

$$1\mathbf{A} = \mathbf{A} \tag{40}$$

4. *Multiplication.*   Let $\mathbf{A} = [a_{ij}]$ be an $m \times n$ matrix and $\mathbf{B} = [b_{ij}]$ be an $n \times r$ matrix. The product $\mathbf{AB} = \mathbf{C}$ is an $m \times r$ matrix $\mathbf{C} = [c_{ij}]$, where

$$c_{ij} = \sum_{k=1}^{n} a_{ik}b_{kj} \tag{41}$$

Essentially, matrix multiplication consists of scalar products of the *row* vectors of $\mathbf{A}$ by the *column* vectors of $\mathbf{B}$. For example, if

$$\mathbf{A} = \begin{pmatrix} 1 & 6 \\ -4 & 2 \\ 5 & 0 \end{pmatrix}, \qquad \mathbf{B} = \begin{pmatrix} -5 & 4 & 1 \\ 0 & -1 & 3 \end{pmatrix}$$

then

$$\mathbf{AB} = \begin{pmatrix} 1 & 6 \\ -4 & 2 \\ 5 & 0 \end{pmatrix} \begin{pmatrix} -5 & 4 & 1 \\ 0 & -1 & 3 \end{pmatrix}$$

$$= \begin{pmatrix} (1)(-5) + (6)(0) & (1)(4) + (6)(-1) & (1)(1) + (6)(3) \\ (-4)(-5) + (2)(0) & (-4)(4) + (2)(-1) & (-4)(1) + (2)(3) \\ (5)(-5) + (0)(0) & (5)(4) + (0)(-1) & (5)(1) + (0)(3) \end{pmatrix}$$

$$= \begin{pmatrix} -5 & -2 & 19 \\ 20 & -18 & 2 \\ -25 & 20 & 5 \end{pmatrix}$$

Remark. In order that the product **AB** may be defined, it is necessary that the number of columns in **A** be equal to the number of rows in **B**.

Matrix multiplication, when defined, obeys the following properties:

$$(\mathbf{AB})\mathbf{C} = \mathbf{A}(\mathbf{BC}) \tag{42}$$

$$\mathbf{A}(\mathbf{B} + \mathbf{C}) = \mathbf{AB} + \mathbf{AC} \tag{43}$$

$$\mathbf{AB} \neq \mathbf{BA}, \quad \text{in general} \tag{44}$$

In order to illustrate property (44), consider the following example.

EXAMPLE 13

Calculate both **AB** and **BA**, where

$$\mathbf{A} = \begin{pmatrix} 2 & 1 \\ 3 & 4 \end{pmatrix}, \qquad \mathbf{B} = \begin{pmatrix} 1 & -2 \\ 5 & 3 \end{pmatrix}$$

Solution

According to definition,

$$\mathbf{AB} = \begin{pmatrix} 2 & 1 \\ 3 & 4 \end{pmatrix} \begin{pmatrix} 1 & -2 \\ 5 & 3 \end{pmatrix}$$

$$= \begin{pmatrix} (2)(1) + (1)(5) & (2)(-2) + (1)(3) \\ (3)(1) + (4)(5) & (3)(-2) + (4)(3) \end{pmatrix}$$

or, upon simplifying,

$$\mathbf{AB} = \begin{pmatrix} 7 & -1 \\ 23 & 6 \end{pmatrix}$$

Similarly,

$$\mathbf{BA} = \begin{pmatrix} 1 & -2 \\ 5 & 3 \end{pmatrix} \begin{pmatrix} 2 & 1 \\ 3 & 4 \end{pmatrix}$$

$$= \begin{pmatrix} (1)(2) + (-2)(3) & (1)(1) + (-2)(4) \\ (5)(2) + (3)(3) & (5)(1) + (3)(4) \end{pmatrix}$$

or

$$\mathbf{BA} = \begin{pmatrix} -4 & -7 \\ 19 & 17 \end{pmatrix}$$

Clearly, $\mathbf{AB} \neq \mathbf{BA}$.   ∎

5. *Transpose.*   Associated with each matrix $\mathbf{A}$ is the matrix $\mathbf{A}^T$, known as the transpose of $\mathbf{A}$, and obtained from $\mathbf{A}$ by interchanging the rows and columns of $\mathbf{A}$. For example, if $\mathbf{A} = [a_{ij}]$, then $\mathbf{A}^T = [a_{ji}]$. More specifically, if

$$\mathbf{A} = \begin{pmatrix} 1 & -4 & 5 \\ 6 & 2 & 0 \end{pmatrix}$$

then

$$\mathbf{A}^T = \begin{pmatrix} 1 & 6 \\ -4 & 2 \\ 5 & 0 \end{pmatrix}$$

If $\mathbf{A}$ is a square matrix and $\mathbf{A}^T = \mathbf{A}$, we say that $\mathbf{A}$ is a **symmetric matrix.** Such matrices have important and useful properties, some of which we see in Section 8.8.

6. *Identity.*   The identity matrix, defined by

$$\mathbf{I} = \begin{pmatrix} 1 & 0 \\ 0 & 1 \end{pmatrix}, \qquad \mathbf{I} = \begin{pmatrix} 1 & 0 & 0 \\ 0 & 1 & 0 \\ 0 & 0 & 1 \end{pmatrix}, \qquad \cdots \tag{45}$$

is an $n \times n$ matrix $\mathbf{I} = [a_{ij}]$ for which $a_{ij} = 1$ for $i = j$ and $a_{ij} = 0$ for $i \neq j$ ($i = 1, 2, \ldots, n; j = 1, 2, \ldots, n$). The significance of this matrix is the property

$$\mathbf{AI} = \mathbf{IA} = \mathbf{A} \tag{46}$$

where both $\mathbf{A}$ and $\mathbf{I}$ are $n \times n$ square matrices.

7. *Inverse.*   Division by matrices is not defined, but the equivalent of this operation can be accomplished by multiplication by what is called a matrix inverse. That is, given a square matrix $\mathbf{A}$, we wish to find another square matrix $\mathbf{B}$ such that $\mathbf{AB} = \mathbf{I}$, where $\mathbf{I}$ is the identity. If $\mathbf{B}$ exists, we say that $\mathbf{B}$ is the inverse of $\mathbf{A}$ and write $\mathbf{B} = \mathbf{A}^{-1}$. A given matrix $\mathbf{A}$ and its inverse therefore satisfy the property

$$\mathbf{AA}^{-1} = \mathbf{A}^{-1}\mathbf{A} = \mathbf{I} \tag{47}$$

If $\mathbf{A}^{-1}$ exists, we say that $\mathbf{A}$ is **nonsingular;** otherwise it is **singular.** It can be shown that a square matrix $\mathbf{A}$ is nonsingular if and only if $\det(\mathbf{A}) \neq 0$, where $\det(\mathbf{A})$ denotes the *determinant* of $\mathbf{A}$. For example, if $\mathbf{A}$ is the 2×2 matrix

$$\mathbf{A} = \begin{pmatrix} a_{11} & a_{12} \\ a_{21} & a_{22} \end{pmatrix} \tag{48}$$

then its inverse is given by

$$\mathbf{A}^{-1} = \frac{1}{\det(\mathbf{A})} \begin{pmatrix} a_{22} & -a_{12} \\ -a_{21} & a_{11} \end{pmatrix}, \qquad \det(\mathbf{A}) \neq 0 \tag{49}$$

In the case of an $n \times n$ nonsingular matrix $\mathbf{A}$, its inverse is given by

$$\mathbf{A}^{-1} = \frac{1}{\det(\mathbf{A})} [A_{ij}]^T \tag{50}$$

where A_{ij} is the *cofactor* of a_{ij} in $\det(\mathbf{A})$. That is, if M_{ij} is the $(n-1) \times (n-1)$ determinant obtained by deleting the ith row and jth column from $\det(\mathbf{A})$, then $A_{ij} = (-1)^{i+j} M_{ij}$. For example, the inverse of the 3×3 matrix

$$\mathbf{A} = \begin{pmatrix} a_{11} & a_{12} & a_{13} \\ a_{21} & a_{22} & a_{23} \\ a_{31} & a_{32} & a_{33} \end{pmatrix} \tag{51}$$

is

$$\mathbf{A}^{-1} = \frac{1}{\det(\mathbf{A})} \begin{pmatrix} A_{11} & A_{21} & A_{31} \\ A_{12} & A_{22} & A_{32} \\ A_{13} & A_{23} & A_{33} \end{pmatrix} \tag{52}$$

where

$$A_{11} = \begin{vmatrix} a_{22} & a_{23} \\ a_{32} & a_{33} \end{vmatrix}$$

$$A_{12} = - \begin{vmatrix} a_{21} & a_{23} \\ a_{31} & a_{33} \end{vmatrix}$$

and so on.

EXAMPLE 14

Find the matrix inverse $\mathbf{A}^{-1}$, where

$$\mathbf{A} = \begin{pmatrix} 1 & 2 \\ -4 & 3 \end{pmatrix}$$

Solution

We first calculate

$$\det(\mathbf{A}) = \begin{vmatrix} 1 & 2 \\ -4 & 3 \end{vmatrix} = (1)(3) - (2)(-4)$$

or

$$\det(\mathbf{A}) = 11 \neq 0$$

Hence, using (49), we immediately deduce that

$$\mathbf{A}^{-1} = \frac{1}{11} \begin{pmatrix} 3 & -2 \\ 4 & 1 \end{pmatrix} = \begin{pmatrix} \frac{3}{11} & -\frac{2}{11} \\ \frac{4}{11} & \frac{1}{11} \end{pmatrix}$$

Checking our result, we see that

$$\mathbf{A}^{-1}\mathbf{A} = \frac{1}{11} \begin{pmatrix} 3 & -2 \\ 4 & 1 \end{pmatrix} \begin{pmatrix} 1 & 2 \\ -4 & 3 \end{pmatrix} = \frac{1}{11} \begin{pmatrix} 11 & 0 \\ 0 & 11 \end{pmatrix} = \begin{pmatrix} 1 & 0 \\ 0 & 1 \end{pmatrix}$$

and

$$\mathbf{AA}^{-1} = \begin{pmatrix} 1 & 2 \\ -4 & 3 \end{pmatrix} \frac{1}{11} \begin{pmatrix} 3 & -2 \\ 4 & 1 \end{pmatrix} = \frac{1}{11} \begin{pmatrix} 11 & 0 \\ 0 & 11 \end{pmatrix} = \begin{pmatrix} 1 & 0 \\ 0 & 1 \end{pmatrix}$$ ∎

■ **EXAMPLE 15**

Find the matrix inverse of

$$\mathbf{A} = \begin{pmatrix} 4 & -2 \\ -2 & 1 \end{pmatrix}$$

Solution

Since $\det(\mathbf{A}) = 0$, this matrix is *singular* and hence has no inverse. ∎

■ **EXAMPLE 16**

Find the matrix inverse of

$$\mathbf{A} = \begin{pmatrix} 2 & 0 & -1 \\ 1 & 1 & 0 \\ 0 & -3 & 1 \end{pmatrix}$$

Solution

Here $\det(\mathbf{A}) = 5$ so the given matrix is nonsingular. The nine cofactors are

$$A_{11} = \begin{vmatrix} 1 & 0 \\ -3 & 1 \end{vmatrix} = 1, \qquad A_{12} = -\begin{vmatrix} 1 & 0 \\ 0 & 1 \end{vmatrix} = -1, \qquad A_{13} = \begin{vmatrix} 1 & 1 \\ 0 & -3 \end{vmatrix} = -3$$

$$A_{21} = -\begin{vmatrix} 0 & -1 \\ -3 & 1 \end{vmatrix} = -3, \qquad A_{22} = \begin{vmatrix} 2 & -1 \\ 0 & 1 \end{vmatrix} = 2, \qquad A_{23} = -\begin{vmatrix} 2 & 0 \\ 0 & -3 \end{vmatrix} = 6$$

$$A_{31} = \begin{vmatrix} 0 & -1 \\ 1 & 0 \end{vmatrix} = 1, \qquad A_{32} = -\begin{vmatrix} 0 & -1 \\ -3 & 1 \end{vmatrix} = 3, \qquad A_{33} = \begin{vmatrix} 2 & 0 \\ 1 & 1 \end{vmatrix} = 2$$

Hence, from (52) it follows that

$$\mathbf{A}^{-1} = \frac{1}{5} \begin{pmatrix} 1 & -3 & 1 \\ -1 & 2 & 3 \\ -3 & 6 & 2 \end{pmatrix} = \begin{pmatrix} \frac{1}{5} & -\frac{3}{5} & \frac{1}{5} \\ -\frac{1}{5} & \frac{2}{5} & \frac{3}{5} \\ -\frac{3}{5} & \frac{6}{5} & \frac{2}{5} \end{pmatrix}$$

We leave it to the reader to verify that $\mathbf{A}^{-1}\mathbf{A} = \mathbf{AA}^{-1} = \mathbf{I}$. ∎

In the case of matrices larger than 3×3 the inverse is usually found by the use of elementary row operations, called **row reduction.** Since our discussion in the following sections will usually involve either 2×2 or 3×3 matrices, we refer the interested reader to any standard text on matrix theory for a discussion of row reduction.

8.6.2 Matrix Functions

Sometimes we need to perform calculus operations on a matrix function $\mathbf{A}(t)$. The derivative and integral, respectively, of $\mathbf{A}(t) = [a_{ij}(t)]$ are defined by

$$\mathbf{A}'(t) = [a'_{ij}(t)] \tag{53}$$

and

$$\int_{t_0}^{t} \mathbf{A}(s) \, ds = \left[\int_{t_0}^{t} a_{ij}(s) \, ds \right] \tag{54}$$

For example, if

$$\mathbf{A}(t) = \begin{pmatrix} \cos 3t \\ e^{2t} \\ 3t^2 - 5 \end{pmatrix}$$

then

$$\mathbf{A}'(t) = \begin{pmatrix} -3 \sin 3t \\ 2e^{2t} \\ 6t \end{pmatrix}$$

Exercises 8.6

In Problems 1 to 4, use the given matrices to calculate (a) $\mathbf{A} + \mathbf{B}$, (b) $\mathbf{A} - \mathbf{B}$, and (c) $3\mathbf{B} - 2\mathbf{A}$.

1. $\mathbf{A} = \begin{pmatrix} 3 & 4 \\ 5 & 6 \end{pmatrix}$, $\mathbf{B} = \begin{pmatrix} -1 & 2 \\ 0 & 3 \end{pmatrix}$

2. $\mathbf{A} = \begin{pmatrix} -2 & 2 & 4 \\ 1 & 1 & 3 \end{pmatrix}$, $\mathbf{B} = \begin{pmatrix} 1 & 0 & 2 \\ -3 & 1 & 4 \end{pmatrix}$

3. $\mathbf{A} = \begin{pmatrix} -2 & 0 \\ 4 & 1 \\ 7 & 3 \end{pmatrix}$, $\mathbf{B} = \begin{pmatrix} 3 & -1 \\ 0 & 2 \\ -4 & -2 \end{pmatrix}$

4. $\mathbf{A} = \begin{pmatrix} 1 & -3 & 4 \\ 2 & 5 & -1 \\ 0 & -4 & -2 \end{pmatrix}$, $\mathbf{B} = \begin{pmatrix} 0 & 1 & 1 \\ 2 & 1 & 0 \\ -1 & 2 & 3 \end{pmatrix}$

In Problems 5 to 10, find all products $\mathbf{AB}$ and $\mathbf{BA}$ that are defined.

5. $\mathbf{A} = \begin{pmatrix} 3 & 4 \\ 5 & 6 \end{pmatrix}$, $\mathbf{B} = \begin{pmatrix} -1 & 2 \\ 0 & 3 \end{pmatrix}$

6. $\mathbf{A} = \begin{pmatrix} -2 & 5 \\ 1 & 4 \end{pmatrix}$, $\mathbf{B} = \begin{pmatrix} 3 & 2 \\ -3 & 4 \end{pmatrix}$

7. $\mathbf{A} = \begin{pmatrix} 1 & 2 \\ -1 & 3 \end{pmatrix}$, $\mathbf{B} = \begin{pmatrix} 3 & 2 & 1 \\ 2 & 1 & 4 \end{pmatrix}$

■8. $\mathbf{A} = \begin{pmatrix} 2 & 1 & 0 \\ -1 & -2 & 2 \end{pmatrix}$, $\mathbf{B} = \begin{pmatrix} 2 & 4 \\ 1 & -1 \\ 3 & 1 \end{pmatrix}$

9. $\mathbf{A} = (6 \quad -2 \quad 1)$, $\mathbf{B} = \begin{pmatrix} 1 \\ 3 \\ -2 \end{pmatrix}$

10. $\mathbf{A} = (3 \quad -1 \quad 1)$, $\mathbf{B} = \begin{pmatrix} 2 & 0 \\ 1 & 4 \\ -2 & 1 \end{pmatrix}$

In Problems 11 to 18, determine whether the given matrix is singular or nonsingular. If nonsingular, find its inverse $\mathbf{A}^{-1}$ and show that $\mathbf{A}^{-1}\mathbf{A} = \mathbf{AA}^{-1} = \mathbf{I}$.

11. $\mathbf{A} = \begin{pmatrix} 3 & -6 \\ 2 & -4 \end{pmatrix}$

12. $\mathbf{A} = \begin{pmatrix} 1 & 4 \\ 2 & 5 \end{pmatrix}$

13. $\mathbf{A} = \begin{pmatrix} -4 & -8 \\ 3 & 5 \end{pmatrix}$

14. $\mathbf{A} = \begin{pmatrix} 2 & 2 \\ 7 & 10 \end{pmatrix}$

15. $\mathbf{A} = \begin{pmatrix} 2 & 2 & 0 \\ -2 & 1 & 1 \\ 3 & 0 & 1 \end{pmatrix}$

■**16.** $\mathbf{A} = \begin{pmatrix} 2 & 1 & 0 \\ -1 & 2 & 1 \\ 1 & 2 & 1 \end{pmatrix}$

17. $\mathbf{A} = \begin{pmatrix} 0 & 1 & 0 \\ 2 & 0 & -1 \\ 1 & 3 & 1 \end{pmatrix}$

18. $\mathbf{A} = \begin{pmatrix} 1 & 0 & 2 \\ 2 & -3 & 4 \\ 0 & 2 & 1 \end{pmatrix}$

19. If $\mathbf{A}$ is nonsingular, show that the inverse of $\mathbf{A}^{-1}$ is $\mathbf{A}$.

****20.** If $\mathbf{A} = \mathbf{A}^{-1}$, show that $\det(\mathbf{A}) = \pm 1$.

21. Show that $\mathbf{A}(t)$ is nonsingular for all t and find its inverse $\mathbf{A}^{-1}(t)$, where

$$\mathbf{A}(t) = \begin{pmatrix} e^{-t} & 2e^{4t} \\ -e^{-t} & 3e^{4t} \end{pmatrix}$$

22. Show that $\mathbf{A}(t)$ is nonsingular for all t and find its inverse $\mathbf{A}^{-1}(t)$, where

$$\mathbf{A}(t) = \begin{pmatrix} e^{-3t} & (t+1)e^{-3t} \\ -e^{-3t} & -(t+2)e^{-3t} \end{pmatrix}$$

23. Compute $\mathbf{A}'(0)$, where $\mathbf{A}(t)$ is the matrix given in Problem 21.

24. Compute $\mathbf{A}'(0)$, where $\mathbf{A}(t)$ is the matrix given in Problem 22.

25. Compute $\int_0^t \mathbf{A}(s)\, ds$, where $\mathbf{A}(t)$ is the matrix given in Problem 21.

■**26.** Compute $\int_0^t \mathbf{A}(s)\, ds$, where $\mathbf{A}(t)$ is the matrix given in Problem 22.

****27.** Given the matrices

$$\mathbf{Y}(t) = \begin{pmatrix} x(t) \\ y(t) \end{pmatrix}, \qquad \mathbf{A}(t) = \begin{pmatrix} 1 & 2 \\ 3 & 2 \end{pmatrix}$$

show that the matrix equation

$$\mathbf{Y}'(t) = \mathbf{A}(t)\mathbf{Y}(t)$$

is equivalent to the system of equations

$$x' = x + 2y$$
$$y' = 3x + 2y$$

****28.** Given the matrices

$$\mathbf{Y}(t) = \begin{pmatrix} x(t) \\ y(t) \end{pmatrix}, \qquad \mathbf{A}(t) = \begin{pmatrix} -4 & 2 \\ 2 & -1 \end{pmatrix},$$

$$\mathbf{F}(t) = \begin{pmatrix} 3e^t \\ t \end{pmatrix}$$

show that the matrix equation

$$\mathbf{Y}'(t) = \mathbf{A}(t)\mathbf{Y}(t) + \mathbf{F}(t)$$

is equivalent to the system of equations

$$x' = -4x + 2y + 3e^t$$
$$y' = 2x - y + t$$

8.7 First-Order Systems—Matrix Formulation

The theory and solution techniques associated with solving linear systems of equations are well suited for the use of matrix methods. This is because first-order linear systems of DEs can always be formulated as a single matrix equation and then solved by application of standard matrix operations. For example, the system of n equations

$$y_1' = a_{11}(t)y_1 + a_{12}(t)y_2 + \cdots + a_{1n}(t)y_n + f_1(t)$$
$$y_2' = a_{21}(t)y_1 + a_{22}(t)y_2 + \cdots + a_{2n}(t)y_n + f_2(t) \tag{55}$$
$$\cdots\cdots\cdots\cdots\cdots\cdots\cdots\cdots\cdots\cdots\cdots\cdots\cdots\cdots\cdots$$
$$y_n' = a_{n1}(t)y_1 + a_{n2}(t)y_2 + \cdots + a_{nn}(t)y_n + f_n(t)$$

can be represented by the single *matrix equation*

$$\mathbf{Y}'(t) = \mathbf{A}(t)\mathbf{Y}(t) + \mathbf{F}(t) \tag{56}$$

where $\mathbf{A}(t)$ is the $n \times n$ **coefficient matrix** defined by

$$\mathbf{A}(t) = \begin{pmatrix} a_{11}(t) & a_{12}(t) & \cdots & a_{1n}(t) \\ a_{21}(t) & a_{22}(t) & \cdots & a_{2n}(t) \\ \vdots & \vdots & & \vdots \\ a_{n1}(t) & a_{n2}(t) & \cdots & a_{nn}(t) \end{pmatrix} \tag{57}$$

and $\mathbf{Y}(t)$ and $\mathbf{F}(t)$ are column vectors defined, respectively, by

$$\mathbf{Y}(t) = \begin{pmatrix} y_1(t) \\ y_2(t) \\ \vdots \\ y_n(t) \end{pmatrix}, \qquad \mathbf{F}(t) = \begin{pmatrix} f_1(t) \\ f_2(t) \\ \vdots \\ f_n(t) \end{pmatrix} \tag{58}$$

In the discussion to follow, we usually write the matrix equation (56) more simply as

$$\mathbf{Y}' = \mathbf{A}(t)\mathbf{Y} + \mathbf{F}(t)$$

where the dependency on t is not explicitly shown for $\mathbf{Y}$. The important observation to make is that (56) assumes the same form regardless of the number of equations n in the system. The use of vectors and matrices is not only notationally expedient, but the theory associated with the matrix equation more clearly exemplifies the similarity between solving (56) and the first-order linear DEs discussed in Chapter 2.

EXAMPLE 17

Formulate the following system of equations as a single matrix equation:

$$x' = -4x + 2y + \frac{1}{t}$$

$$y' = 2x - y + 4 + \frac{2}{t}$$

Solution

By introducing the matrices

$$\mathbf{A}(t) = \begin{pmatrix} -4 & 2 \\ 2 & -1 \end{pmatrix}, \qquad \mathbf{Y}(t) = \begin{pmatrix} x(t) \\ y(t) \end{pmatrix}, \qquad \mathbf{F}(t) = \begin{pmatrix} \dfrac{1}{t} \\ 4 + \dfrac{2}{t} \end{pmatrix}$$

we find the given system of equations is equivalent to

$$\begin{pmatrix} x'(t) \\ y'(t) \end{pmatrix} = \begin{pmatrix} -4 & 2 \\ 2 & -1 \end{pmatrix} \begin{pmatrix} x(t) \\ y(t) \end{pmatrix} + \begin{pmatrix} \dfrac{1}{t} \\ 4 + \dfrac{2}{t} \end{pmatrix}$$

which we may also write as

$$\mathbf{Y}' = \begin{pmatrix} -4 & 2 \\ 2 & -1 \end{pmatrix} \mathbf{Y} + \begin{pmatrix} \dfrac{1}{t} \\ 4 + \dfrac{2}{t} \end{pmatrix}$$ ∎

Equation (56) is called a **homogeneous** equation when the column vector $\mathbf{F}(t) = 0$ for all t in some interval I;[†] otherwise it is said to be **nonhomogeneous**. We say that the vector $\mathbf{Y}$ is a **solution** of (56) if it has a continuous derivative on the interval I and identically satisfies (56) on this interval. Analogous to Theorem 8.1, we have the following *existence-uniqueness theorem* for IVPs involving systems of any order n.

Theorem 8.4 Existence-Uniqueness. *If* $\mathbf{A}(t)$ *is an* $n \times n$ *matrix and* $\mathbf{F}(t)$ *is a column vector with n elements, and both matrices have continuous matrix elements on the interval I containing the point t_0, then there exists a unique solution on I of the IVP*

$$\mathbf{Y}' = \mathbf{A}(t)\mathbf{Y} + \mathbf{F}(t), \qquad \mathbf{Y}(t_0) = \mathbf{Y}_0$$

8.7.1 Homogeneous Systems

To study the nature of the solutions of the general nonhomogeneous system (56), we first investigate the solutions of the associated **homogeneous equation**

$$\mathbf{Y}' = \mathbf{A}(t)\mathbf{Y} \tag{59}$$

obtained from (56) by setting $\mathbf{F}(t) \equiv 0$. Suppose we know that $\mathbf{Y}^{(1)}(t)$ and $\mathbf{Y}^{(2)}(t)$[‡] are both solutions of (59). From the **superposition principle** it follows that

$$\mathbf{Y} = C_1 \mathbf{Y}^{(1)}(t) + C_2 \mathbf{Y}^{(2)}(t) \tag{60}$$

is also a solution of (59) for any constants C_1 and C_2. More generally, we have the following theorem analogous to Theorem 8.2.

Theorem 8.5 Superposition Principle. *If* $\mathbf{Y}^{(1)}(t)$, $\mathbf{Y}^{(2)}(t)$, ..., $\mathbf{Y}^{(n)}(t)$ *are all solutions of the nth-order homogeneous system*

$$\mathbf{Y}' = \mathbf{A}(t)\mathbf{Y}$$

on some interval I, then the linear combination

$$\mathbf{Y} = C_1 \mathbf{Y}^{(1)}(t) + C_2 \mathbf{Y}^{(2)}(t) + \cdots + C_n \mathbf{Y}^{(n)}(t)$$

where $C_1, C_2, \ldots, C_n$ *are arbitrary constants, is also a solution on the interval I.*

[†] The interval I should not be confused with the identity matrix $\mathbf{I}$.

[‡] The superscripts in parentheses on column vectors do not refer to derivatives here, but merely play the same role as subscripts.

To illustrate Theorem 8.5, first observe that

$$\mathbf{Y}^{(1)}(t) = \begin{pmatrix} e^{-t} \\ -e^{-t} \end{pmatrix} = \begin{pmatrix} 1 \\ -1 \end{pmatrix} e^{-t}, \qquad \mathbf{Y}^{(2)}(t) = \begin{pmatrix} 2e^{4t} \\ 3e^{4t} \end{pmatrix} = \begin{pmatrix} 2 \\ 3 \end{pmatrix} e^{4t}$$

are both solutions of the homogeneous system (see Example 18 below)

$$\mathbf{Y}' = \begin{pmatrix} 1 & 2 \\ 3 & 2 \end{pmatrix} \mathbf{Y}$$

Based on Theorem 8.5, it follows that

$$\mathbf{Y} = C_1 \mathbf{Y}^{(1)}(t) + C_2 \mathbf{Y}^{(2)}(t)$$

$$= C_1 \begin{pmatrix} 1 \\ -1 \end{pmatrix} e^{-t} + C_2 \begin{pmatrix} 2 \\ 3 \end{pmatrix} e^{4t}$$

is also a solution for any constants C_1 and C_2.

We say that $\mathbf{Y}^{(1)}(t)$ and $\mathbf{Y}^{(2)}(t)$ are **linearly dependent** if one is a constant multiple of the other; otherwise, they are **linearly independent**. In the case of n solutions we have the following definition.

Definition 8.2 The vector functions $\mathbf{Y}^{(1)}(t), \mathbf{Y}^{(2)}(t), \dots, \mathbf{Y}^{(n)}(t)$ are said to be **linearly dependent** on some interval I provided there exist constants $C_1, C_2, \dots, C_n$, not all zero, such that

$$C_1 \mathbf{Y}^{(1)}(t) + C_2 \mathbf{Y}^{(2)}(t) + \cdots + C_n \mathbf{Y}^{(n)}(t) = 0$$

for every t in the interval. If this is true only for $C_1 = C_2 = \cdots = C_n = 0$, we say the vector functions are **linearly independent**.

Just as in the case of a single nth-order linear DE (Chapter 4), there is also a *Wronskian* determinant from which the linear independence or dependence of a set of solution vectors can be resolved. For example, in the case of only two functions, denoted by $\mathbf{Y}^{(1)}(t) = \begin{pmatrix} x_1(t) \\ y_1(t) \end{pmatrix}$ and $\mathbf{Y}^{(2)}(t) = \begin{pmatrix} x_2(t) \\ y_2(t) \end{pmatrix}$, their Wronskian is defined by

$$W(t) = \begin{vmatrix} x_1(t) & x_2(t) \\ y_1(t) & y_2(t) \end{vmatrix} \tag{61}$$

In the general case, we introduce the notation

$$\mathbf{Y}^{(j)}(t) = \begin{pmatrix} y_{1j}(t) \\ \vdots \\ y_{ij}(t) \\ \vdots \\ y_{nj}(t) \end{pmatrix} \tag{62}$$

so that $y_{ij}(t)$ denotes the ith component of the jth vector $\mathbf{Y}^{(j)}(t)$. In terms of this notation, the **Wronskian** of n vectors is the $n \times n$ determinant

$$W(t) = \begin{vmatrix} y_{11}(t) & y_{12}(t) & \cdots & y_{1n}(t) \\ y_{21}(t) & y_{22}(t) & \cdots & y_{2n}(t) \\ \vdots & \vdots & & \vdots \\ y_{n1}(t) & y_{n2}(t) & \cdots & y_{nn}(t) \end{vmatrix} \tag{63}$$

Notice that the columns of (63) form the solution vectors $\mathbf{Y}^{(1)}(t), \mathbf{Y}^{(2)}(t), \ldots, \mathbf{Y}^{(n)}(t)$. We state the following theorem without proof.

Theorem 8.6 *Let* $\mathbf{Y}^{(1)}(t), \mathbf{Y}^{(2)}(t), \ldots, \mathbf{Y}^{(n)}(t)$ *be n solution vectors of the homogeneous system*

$$\mathbf{Y}' = \mathbf{A}(t)\mathbf{Y}$$

on some interval I. A necessary and sufficient condition that the set of solutions be linearly independent is that $W(t) \neq 0$ *for every t in the interval, where* $W(t)$ *is defined by (63).*

Since the vectors $\mathbf{Y}^{(1)}(t), \mathbf{Y}^{(2)}(t), \ldots, \mathbf{Y}^{(n)}(t)$ are solutions of a homogeneous system of DEs, it can be shown that either $W(t) \neq 0$ or $W(t) = 0$ for every t in the interval. Thus, for example, if $W(t) \neq 0$ for some t_0 in the interval, the solutions are linearly independent on the interval.

EXAMPLE 18

Show that $\mathbf{Y}^{(1)}(t) = \begin{pmatrix} e^{-t} \\ -e^{-t} \end{pmatrix}$ and $\mathbf{Y}^{(2)}(t) = \begin{pmatrix} 2e^{4t} \\ 3e^{4t} \end{pmatrix}$ are both solutions of the matrix equation

$$\mathbf{Y}' = \begin{pmatrix} 1 & 2 \\ 3 & 2 \end{pmatrix}\mathbf{Y}$$

and use the Wronskian to show that they are linearly independent solutions.

Solution

Clearly, $\dfrac{d\mathbf{Y}^{(1)}}{dt} = \begin{pmatrix} -e^{-t} \\ e^{-t} \end{pmatrix}$ and $\dfrac{d\mathbf{Y}^{(2)}}{dt} = \begin{pmatrix} 8e^{4t} \\ 12e^{4t} \end{pmatrix}$, so that by substituting these expressions into the above system written in matrix notation, we see that

$$\begin{pmatrix} -e^{-t} \\ e^{-t} \end{pmatrix} = \begin{pmatrix} 1 & 2 \\ 3 & 2 \end{pmatrix}\begin{pmatrix} e^{-t} \\ -e^{-t} \end{pmatrix} = \begin{pmatrix} -e^{-t} \\ e^{-t} \end{pmatrix}$$

and

$$\begin{pmatrix} 8e^{4t} \\ 12e^{4t} \end{pmatrix} = \begin{pmatrix} 1 & 2 \\ 3 & 2 \end{pmatrix}\begin{pmatrix} 2e^{4t} \\ 3e^{4t} \end{pmatrix} = \begin{pmatrix} 8e^{4t} \\ 12e^{4t} \end{pmatrix}$$

proving that $Y^{(1)}(t)$ and $Y^{(2)}(t)$ are indeed solution vectors. Then, using (61), we see that

$$W(t) = \begin{vmatrix} e^{-t} & 2e^{4t} \\ -e^{-t} & 3e^{4t} \end{vmatrix} = 5e^{3t} \neq 0$$

from which we deduce that the solution vectors are linearly independent. ∎

A set of n linearly independent solution vectors $Y^{(1)}(t)$, $Y^{(2)}(t), \ldots, Y^{(n)}(t)$ of a system of n homogeneous equations on some interval I is said to form a **fundamental set of solution vectors** on this interval. In this case the **general solution** of the homogeneous system $Y' = AY$ is given by

$$Y = C_1 Y^{(1)}(t) + C_2 Y^{(2)}(t) + \cdots + C_n Y^{(n)}(t) \tag{64}$$

where $C_1, C_2, \ldots, C_n$ are arbitrary constants.

8.7.2 Fundamental Matrix

If $Y^{(1)}(t)$, $Y^{(2)}(t), \ldots, Y^{(n)}(t)$ form a fundamental set of solutions of the matrix equation

$$Y' = A(t)Y \tag{65}$$

then the related matrix

$$\phi(t) = \begin{pmatrix} y_{11}(t) & y_{12}(t) & \cdots & y_{1n}(t) \\ y_{21}(t) & y_{22}(t) & \cdots & y_{2n}(t) \\ \vdots & \vdots & & \vdots \\ y_{n1}(t) & y_{n2}(t) & \cdots & y_{nn}(t) \end{pmatrix} \tag{66}$$

whose columns are simply the vectors $Y^{(1)}(t)$, $Y^{(2)}(t), \ldots, Y^{(n)}(t)$, is called the **fundamental matrix** of the system of equations (65).

For instance, we see that

$$Y = C_1 \begin{pmatrix} e^{-t} \\ -e^{-t} \end{pmatrix} + C_2 \begin{pmatrix} 2e^{4t} \\ 3e^{4t} \end{pmatrix}$$

is a general solution of the system in Example 18 and its fundamental matrix is therefore

$$\phi(t) = \begin{pmatrix} e^{-t} & 2e^{4t} \\ -e^{-t} & 3e^{4t} \end{pmatrix}$$

Observe that the determinant of the fundamental matrix (66) is the Wronskian defined by (63), that is, $\det[\phi(t)] = W(t)$. Thus, since $W(t) \neq 0$ on the interval of interest, it follows that $\phi(t)$ is *nonsingular* and therefore has a unique inverse $\phi^{-1}(t)$. The significance of this observation becomes clear by considering the IVP

$$Y' = A(t)Y, \qquad Y(0) = Y_0 \tag{67}$$

where $\mathbf{Y}_0$ is a prescribed vector. In terms of the fundamental matrix $\phi(t)$, the general solution of the DE in (67) is given by

$$\mathbf{Y} = \phi(t)\mathbf{C} \tag{68}$$

where $\mathbf{C}$ is the arbitrary column vector

$$\mathbf{C} = \begin{pmatrix} C_1 \\ C_2 \\ \vdots \\ C_n \end{pmatrix} \tag{69}$$

Imposing the initial condition in (67) on the general solution (68), we obtain

$$\mathbf{Y}(0) = \phi(0)\mathbf{C} = \mathbf{Y}_0$$

from which we deduce

$$\mathbf{C} = \phi^{-1}(0)\mathbf{Y}_0 \tag{70}$$

Hence, the solution of the IVP (67) is formally given by

$$\mathbf{Y} = \phi(t)\phi^{-1}(0)\mathbf{Y}_0 \tag{71}$$

EXAMPLE 19   Given that $\mathbf{Y}^{(1)}(t) = \begin{pmatrix} 1 \\ -1 \end{pmatrix} e^{-2t}$ and $\mathbf{Y}^{(2)}(t) = \begin{pmatrix} 3 \\ 5 \end{pmatrix} e^{6t}$ are solution vectors of

$$\mathbf{Y}' = \begin{pmatrix} 1 & 3 \\ 5 & 3 \end{pmatrix}\mathbf{Y},$$

find the fundamental matrix $\phi(t)$ and use it to find a solution satisfying the initial condition $\mathbf{Y}(0) = \begin{pmatrix} 0 \\ 1 \end{pmatrix}$.

Solution

It follows immediately that the fundamental matrix is

$$\phi(t) = \begin{pmatrix} e^{-2t} & 3e^{6t} \\ -e^{-2t} & 5e^{6t} \end{pmatrix}$$

By setting $t = 0$, we find $\phi(0) = \begin{pmatrix} 1 & 3 \\ -1 & 5 \end{pmatrix}$, the determinant of which is $\det[\phi(0)] = (1)(5) - (3)(-1) = 8$; hence,

$$\phi^{-1}(0) = \frac{1}{8}\begin{pmatrix} 5 & -3 \\ 1 & 1 \end{pmatrix}$$

From (71), we now obtain

$$\mathbf{Y} = \frac{1}{8}\begin{pmatrix} e^{-2t} & 3e^{6t} \\ -e^{-2t} & 5e^{6t} \end{pmatrix}\begin{pmatrix} 5 & -3 \\ 1 & 1 \end{pmatrix}\begin{pmatrix} 0 \\ 1 \end{pmatrix} = \frac{1}{8}\begin{pmatrix} e^{-2t} & 3e^{6t} \\ -e^{-2t} & 5e^{6t} \end{pmatrix}\begin{pmatrix} -3 \\ 1 \end{pmatrix}$$

or

$$\mathbf{Y} = \frac{1}{8}\begin{pmatrix} -3e^{-2t} + 3e^{6t} \\ 3e^{-2t} + 5e^{6t} \end{pmatrix} = \frac{1}{8}\left[\begin{pmatrix} -3 \\ 3 \end{pmatrix}e^{-2t} + \begin{pmatrix} 3 \\ 5 \end{pmatrix}e^{6t} \right]$$

■

8.7.3 Nonhomogeneous Systems

Finally, with respect to the general *nonhomogeneous* system

$$\mathbf{Y}' = \mathbf{A}(t)\mathbf{Y} + \mathbf{F}(t)$$

we have the following theorem analogous to Theorem 4.7. The proof also parallels that of Theorem 4.7 and is left to the exercises (see Problem 27 in Exercises 8.7).

Theorem 8.7 *If* $\mathbf{Y}_P$ *is any particular solution of the nth-order system of equations*

$$\mathbf{Y}' = \mathbf{A}(t)\mathbf{Y} + \mathbf{F}(t)$$

on some interval I and

$$\mathbf{Y}_H = C_1\mathbf{Y}^{(1)}(t) + C_2\mathbf{Y}^{(2)}(t) + \cdots + C_n\mathbf{Y}^{(n)}(t)$$

is the general solution on I of the associated homogeneous system $\mathbf{Y}' = \mathbf{A}(t)\mathbf{Y}$, *then the general solution of the nonhomogeneous system is* $\mathbf{Y} = \mathbf{Y}_H + \mathbf{Y}_P$.

■ **EXAMPLE 20**

Verify that $\mathbf{Y}_P = \begin{pmatrix} 4 \\ 3 \end{pmatrix}e^{-3t}$ is a particular solution of the nonhomogeneous system

$$\mathbf{Y}' = \begin{pmatrix} -10 & 6 \\ -12 & 7 \end{pmatrix}\mathbf{Y} + \begin{pmatrix} 10 \\ 18 \end{pmatrix}e^{-3t}$$

Solution

First, we observe that $\mathbf{Y}'_P = -3\begin{pmatrix} 4 \\ 3 \end{pmatrix}e^{-3t} = \begin{pmatrix} -12 \\ -9 \end{pmatrix}e^{-3t}$, and thus the direct substitution of $\mathbf{Y}_P$ and its derivative $\mathbf{Y}'_P$ into the equation yields

$$\begin{pmatrix} -12 \\ -9 \end{pmatrix}e^{-3t} = \begin{pmatrix} -10 & 6 \\ -12 & 7 \end{pmatrix}\begin{pmatrix} 4 \\ 3 \end{pmatrix}e^{-3t} + \begin{pmatrix} 10 \\ 18 \end{pmatrix}e^{-3t}$$

$$= \begin{pmatrix} -40 + 18 \\ -48 + 21 \end{pmatrix}e^{-3t} + \begin{pmatrix} 10 \\ 18 \end{pmatrix}e^{-3t}$$

$$= \begin{pmatrix} -22 \\ -27 \end{pmatrix}e^{-3t} + \begin{pmatrix} 10 \\ 18 \end{pmatrix}e^{-3t} = \begin{pmatrix} -12 \\ -9 \end{pmatrix}e^{-3t}$$

which proves our result.

■

Exercises 8.7

In Problems 1 to 10, write the given system of equations in matrix notation.

1. $x' = 2x - y$
   $y' = 3x - 2y$

2. $x' = 4x - 3y$
   $y' = 5x - 4y$

3. $x' = 4x - 3y$
   $y' = 8x - 6y$

4. $x' = 3x + 2y$
   $y' = 6x - y$

5. $x' = x + z$
   $y' = x + y$
   $z' = -2x - z$

6. $x' = x + y + z$
   $y' = 2x + y - z$
   $z' = -3x + 2y + 4z$

7. $x' = 2x - 5y - \sin 2t,\ x(0) = 0$
   $y' = x - 2y + t,\ y(0) = 1$

■8. $x' = 2x - y + z + t,\ x(0) = 1$
   $y' = 3x - 2y + 2t,\ y(0) = 0$
   $z' = -5y + z + t^2 - 1,\ z(0) = -3$

9. $x' = x - y + 3z + e^t,\ x(0) = 1$
   $y' = 2x + 3y - 4z + e^{-t},\ y(0) = -5$
   $z' = 3x - y - z - 1,\ z(0) = 2$

*10. $2x' + y' - y = t,\ x(0) = 1$
   $x' + y' = t^2,\ y(0) = 0$

In Problems 11 to 14, rewrite the given matrix equation without the use of matrices.

11. $\mathbf{Y}' = \begin{pmatrix} -10 & 6 \\ -12 & 7 \end{pmatrix}\mathbf{Y} + \begin{pmatrix} 10e^{-3t} \\ 18e^{-3t} \end{pmatrix},\ \mathbf{Y}(0) = \begin{pmatrix} -1 \\ 2 \end{pmatrix}$

12. $\mathbf{Y}' = \begin{pmatrix} 1 & -1 \\ 1 & 1 \end{pmatrix}\mathbf{Y} + \begin{pmatrix} 3e^t \\ 3e^t \end{pmatrix},\ \mathbf{Y}(0) = \begin{pmatrix} 3 \\ 1 \end{pmatrix}$

13. $\mathbf{Y}' = \begin{pmatrix} 1 & 2 & 1 \\ 6 & -1 & 0 \\ -1 & -2 & -1 \end{pmatrix}\mathbf{Y} + \begin{pmatrix} -1 \\ 3 \\ 0 \end{pmatrix}e^{2t},$

$\mathbf{Y}(0) = \begin{pmatrix} 0 \\ -1 \\ 5 \end{pmatrix}$

14. $\mathbf{Y}' = \begin{pmatrix} 5 & 2 & -2 \\ 7 & 0 & -2 \\ 11 & 1 & -1 \end{pmatrix}\mathbf{Y} + \begin{pmatrix} 3 \\ 7 \\ -2 \end{pmatrix}t + \begin{pmatrix} 4 \\ 0 \\ 1 \end{pmatrix}\cos t,$

$\mathbf{Y}(0) = \begin{pmatrix} 1 \\ 0 \\ -1 \end{pmatrix}$

In Problems 15 to 20, show that the given vectors are solutions and further that they form a fundamental solution set of the given homogeneous system.

15. $\mathbf{Y}' = \begin{pmatrix} 4 & -3 \\ 8 & -6 \end{pmatrix}\mathbf{Y},\ \mathbf{Y}^{(1)} = \begin{pmatrix} 1 \\ \frac{4}{3} \end{pmatrix},$

$\mathbf{Y}^{(2)} = \begin{pmatrix} 1 \\ 2 \end{pmatrix}e^{-t}$

16. $\mathbf{Y}' = \begin{pmatrix} 3 & -18 \\ 2 & -9 \end{pmatrix}\mathbf{Y},\ \mathbf{Y}^{(1)} = \begin{pmatrix} 18 \\ 6 \end{pmatrix}e^{-3t},$

$\mathbf{Y}^{(2)} = \begin{pmatrix} 18t \\ 6t - 1 \end{pmatrix}e^{-3t}$

17. $\mathbf{Y}' = \begin{pmatrix} 3 & 2 \\ -5 & 1 \end{pmatrix}\mathbf{Y},$

$\mathbf{Y}^{(1)} = \begin{pmatrix} 2\cos 3t \\ -\cos 3t - 3\sin 3t \end{pmatrix}e^{2t},$

$\mathbf{Y}^{(2)} = \begin{pmatrix} 2\sin 3t \\ 3\cos 3t - \sin 3t \end{pmatrix}e^{2t}$

■18. $\mathbf{Y}' = \begin{pmatrix} 1 & 3 \\ 1 & -1 \end{pmatrix}\mathbf{Y},\ \mathbf{Y}^{(1)} = \begin{pmatrix} 3 \\ 1 \end{pmatrix}e^{2t},$

$\mathbf{Y}^{(2)} = \begin{pmatrix} 1 \\ -1 \end{pmatrix}e^{-2t}$

19. $\mathbf{Y}' = \begin{pmatrix} 1 & 2 & 1 \\ 6 & -1 & 0 \\ -1 & -2 & -1 \end{pmatrix} \mathbf{Y}, \ \mathbf{Y}^{(1)} = \begin{pmatrix} 1 \\ 6 \\ -13 \end{pmatrix},$

$\mathbf{Y}^{(2)} = \begin{pmatrix} 1 \\ -2 \\ -1 \end{pmatrix} e^{-4t}, \ \mathbf{Y}^{(3)} = \begin{pmatrix} 2 \\ 3 \\ -2 \end{pmatrix} e^{3t}$

20. $\mathbf{Y}' = \begin{pmatrix} 5 & 2 & -2 \\ 7 & 0 & -2 \\ 11 & 1 & -1 \end{pmatrix} \mathbf{Y}, \ \mathbf{Y}^{(1)} = \begin{pmatrix} 2 \\ 2 \\ 6 \end{pmatrix} e^t,$

$\mathbf{Y}^{(2)} = \begin{pmatrix} 3 \\ 3 \\ 6 \end{pmatrix} e^{3t}, \ \mathbf{Y}^{(3)} = \begin{pmatrix} 0 \\ 1 \\ 2 \end{pmatrix} e^{-2t}$

In Problems 21 to 26, use the given solution vectors to form the fundamental matrix $\phi(t)$, and then in Problems 21 to 25 find the solution satisfying the initial condition $\mathbf{Y}(0) = \begin{pmatrix} 1 \\ 2 \end{pmatrix}$. In Problem 26, assume that

$$\mathbf{Y}(0) = \begin{pmatrix} 1 \\ 2 \\ 0 \end{pmatrix}.$$

21. See Problem 15.
22. See Problem 16.
23. See Problem 17.
■**24.** See Problem 18.
25. See Problem 19.
26. See Problem 20.
*****27.** Prove Theorem 8.7.

28. Show that $\mathbf{Y}_P = \begin{pmatrix} -3 \\ 3 \end{pmatrix} e^t$ is a particular solution of the equation in Problem 12.

29. Show that $\mathbf{Y}_P = -\dfrac{1}{6} \begin{pmatrix} 1 \\ 5 \end{pmatrix} e^t$ is a particular solution of the equation

$$\mathbf{Y}' = \begin{pmatrix} 2 & 1 \\ 4 & -1 \end{pmatrix} \mathbf{Y} + \begin{pmatrix} e^t \\ -e^t \end{pmatrix}$$

*****30.** Show that

$$\mathbf{Y}_P = \begin{pmatrix} -2\cos t \, \log|\sec t + \tan t| \\ -1 + \sin t \, \log|\sec t + \tan t| \end{pmatrix} e^t$$

is a particular solution of the equation

$$\mathbf{Y}' = \begin{pmatrix} 1 & 2 \\ -\frac{1}{2} & 1 \end{pmatrix} \mathbf{Y} + \begin{pmatrix} 0 \\ e^t \tan t \end{pmatrix}$$

8.8 Homogeneous Systems with Constant Coefficients

We now wish to illustrate the power of the matrix method in solving *homogeneous first-order systems with constant coefficients,* such as

$$\begin{aligned} y_1' &= a_{11}y_1 + a_{12}y_2 + \cdots + a_{1n}y_n \\ y_2' &= a_{21}y_1 + a_{22}y_2 + \cdots + a_{2n}y_n \\ &\qquad \cdots \cdots \cdots \cdots \cdots \cdots \cdots \\ y_n' &= a_{n1}y_1 + a_{n2}y_2 + \cdots + a_{nn}y_n \end{aligned} \tag{72}$$

Formulated as a *matrix equation,* (72) becomes

$$\mathbf{Y}' = \mathbf{AY} \tag{73}$$

where $\mathbf{A}$ is the $n \times n$ matrix

$$\mathbf{A} = \begin{pmatrix} a_{11} & a_{12} & \cdots & a_{1n} \\ a_{21} & a_{22} & \cdots & a_{2n} \\ \vdots & \vdots & & \vdots \\ a_{n1} & a_{n2} & \cdots & a_{nn} \end{pmatrix} \tag{74}$$

whose elements are all *constants*. By analogy with our solution treatment of nth-order DEs with constant coefficients (recall Section 4.4), we seek solutions of (72) that are of the general form

$$y_i = k_i e^{\lambda t}, \qquad i = 1, 2, \ldots, n$$

where $\lambda, k_1, k_2, \ldots, k_n$ are scalar constants to be determined. In matrix form we write this as

$$\mathbf{Y} = \begin{pmatrix} k_1 \\ k_2 \\ k_3 \\ \vdots \\ k_n \end{pmatrix} e^{\lambda t} = \mathbf{K} e^{\lambda t} \tag{75}$$

Since $\mathbf{Y}' = \lambda \mathbf{K} e^{\lambda t}$, the substitution of (75) into (73) leads to

$$\lambda \mathbf{K} e^{\lambda t} = \mathbf{A} \mathbf{K} e^{\lambda t}$$

which, after canceling the common factor $e^{\lambda t}$ and turning the equation around, becomes

$$\mathbf{A}\mathbf{K} = \lambda \mathbf{K} \tag{76}$$

Hence, we see that $\mathbf{Y} = \mathbf{K} e^{\lambda t}$ is a nontrivial solution of (73) provided the product $\mathbf{A}\mathbf{K}$ is a scalar multiple of $\mathbf{K}$. It is customary here to introduce the identity matrix $\mathbf{I}$ and rearrange (76) in the form

$$(\mathbf{A} - \lambda \mathbf{I})\mathbf{K} = \mathbf{0} \tag{77}$$

We interpret (77) as a homogeneous system of n algebraic equations that can have a nontrivial solution if and only if its coefficient determinant is zero, that is, if and only if

$$\det(\mathbf{A} - \lambda \mathbf{I}) = \Delta(\lambda) = 0 \tag{78}$$

What we have shown here is that to find a solution vector of the homogeneous system $\mathbf{Y}' = \mathbf{A}\mathbf{Y}$, we must find a value of λ satisfying (78) and then use it in (77) to determine the vector $\mathbf{K}$.

A value of λ satisfying $\det(\mathbf{A} - \lambda I) = 0$ is called an **eigenvalue** of the matrix $\mathbf{A}$, and the *nonzero* vector $\mathbf{K}$ satisfying $\mathbf{A}\mathbf{K} = \lambda \mathbf{K}$ is called its associated **eigenvector.** In general, we have the following theorem.

> **Theorem 8.6** *For each eigenvalue λ of the constant-coefficient square matrix* **A** *and each eigenvector* **K** *belonging to λ, the vector function*
>
> $$\mathbf{Y} = \mathbf{K}e^{\lambda t}$$
>
> *is a nontrivial solution of the homogeneous matrix equation*
>
> $$\mathbf{Y}' = \mathbf{A}\mathbf{Y}$$

Writing (78) out in detail we have

$$\Delta(\lambda) = \begin{vmatrix} a_{11} - \lambda & a_{12} & \cdots & a_{1n} \\ a_{12} & a_{22} - \lambda & \cdots & a_{2n} \\ \vdots & \vdots & & \vdots \\ a_{n1} & a_{n2} & \cdots & a_{nn} - \lambda \end{vmatrix} = 0 \tag{79}$$

and upon expanding the determinant, we get the nth-degree polynomial

$$(-1)^n \lambda^n + b_{n-1} \lambda^{n-1} + \cdots + b_1 \lambda + b_0 = 0 \tag{80}$$

Equation (80) is called the **characteristic equation of the matrix A.** Based on the fundamental theorem of algebra, we know that (80) has exactly n roots—some of which may be complex or repeated—and thus we see that the $n \times n$ matrix **A** has exactly n eigenvalues.

Once we know the eigenvalues $\lambda_1, \lambda_2, \ldots, \lambda_n$ we attempt to find n *linearly independent* eigenvectors $\mathbf{K}^{(1)}, \mathbf{K}^{(2)}, \ldots, \mathbf{K}^{(n)}$ such that we generate n linearly independent solution vectors

$$\mathbf{Y}^{(1)}(t) = \mathbf{K}^{(1)}e^{\lambda_1 t}, \qquad \mathbf{Y}^{(2)}(t) = \mathbf{K}^{(2)}e^{\lambda_2 t}, \qquad \ldots, \qquad \mathbf{Y}^{(n)}(t) = \mathbf{K}^{(n)}e^{\lambda_n t}$$

Doing so, the general solution of $\mathbf{Y}' = \mathbf{A}\mathbf{Y}$ assumes the form

$$\mathbf{Y} = C_1 \mathbf{K}^{(1)}e^{\lambda_1 t} + C_2 \mathbf{K}^{(2)}e^{\lambda_2 t} + \cdots + C_n \mathbf{K}^{(n)}e^{\lambda_n t} \tag{81}$$

where $C_1, C_2, \ldots, C_n$ are arbitrary constants.

Before discussing this problem any further in the general sense, however, it is instructive to first consider only the case of two equations and two unknowns, corresponding to the case of a second-order linear DE. This we do in Section 8.8.1, and then we briefly investigate the procedure for higher-order systems in Section 8.8.2.

8.8.1 Systems of Two Equations

Here we restrict our attention to the linear system

$$\mathbf{Y}' = \mathbf{A}\mathbf{Y} \tag{82}$$

where **A** is a 2×2 matrix with *constant* elements, that is,

$$\mathbf{A} = \begin{pmatrix} a_{11} & a_{12} \\ a_{21} & a_{22} \end{pmatrix}$$

For the simple case of a system of only two equations, the determinant $\Delta(\lambda)$ given by (79) takes the explicit form

$$\Delta(\lambda) = \begin{vmatrix} a_{11} - \lambda & a_{12} \\ a_{21} & a_{22} - \lambda \end{vmatrix} = 0 \tag{83}$$

Expanding this determinant, we are led at once to the characteristic equation

$$\lambda^2 - (a_{11} + a_{22})\lambda + (a_{11}a_{22} - a_{12}a_{21}) = 0 \tag{84}$$

Since (84) is quadratic, it clearly has two solutions λ_1, λ_2 that may be found by using the quadratic formula. To obtain one solution of (82) corresponding to, say, $\lambda = \lambda_1$, we substitute this value back into the matrix equation

$$(\mathbf{A} - \lambda \mathbf{I})\mathbf{K} = \mathbf{0}$$

to find nonzero values for k_1 and k_2, which form the vector components of $\mathbf{K}^{(1)} = \begin{pmatrix} k_1 \\ k_2 \end{pmatrix}$.[†] Once found, we obtain the solution vector

$$\mathbf{Y}^{(1)}(t) = \begin{pmatrix} k_1 \\ k_2 \end{pmatrix} e^{\lambda_1 t} \tag{85}$$

Our method of deriving a second linearly independent solution set corresponding to $\lambda = \lambda_2$ will vary depending upon whether the characteristic roots λ_1 and λ_2 are **real and distinct, real and equal,** or **complex conjugates.** We consider each of these cases separately.

Case I. Real and Distinct Roots

When the roots λ_1 and λ_2 of the characteristic equation (84) are real and distinct, there are two *linearly independent* eigenvectors $\mathbf{K}^{(1)}$ and $\mathbf{K}^{(2)}$ belonging, respectively, to λ_1 and λ_2. Eigenvectors $\mathbf{K}^{(1)}$ and $\mathbf{K}^{(2)}$ are both found in the same fashion by substituting eigenvalues λ_1 and λ_2, respectively, into the matrix equation $(\mathbf{A}\lambda - \mathbf{I})\mathbf{K} = \mathbf{0}$. The fundamental set of solution vectors in this case is given by $\mathbf{Y}^{(1)}(t) = \mathbf{K}^{(1)}e^{\lambda_1 t}$ and $\mathbf{Y}^{(2)}(t) = \mathbf{K}^{(2)}e^{\lambda_2 t}$, leading to the *general solution*

$$\mathbf{Y} = C_1 \mathbf{K}^{(1)}e^{\lambda_1 t} + C_2 \mathbf{K}^{(2)}e^{\lambda_2 t} \tag{86}$$

■―――――――
EXAMPLE 21

Use matrix methods to solve the system

$$x' = x + 2y$$
$$y' = 3x + 2y$$

Solution

In matrix notation, this system becomes

$$\mathbf{Y}' = \begin{pmatrix} 1 & 2 \\ 3 & 2 \end{pmatrix} \mathbf{Y}$$

―――――――
[†] For simplicity of notation here we use k_i instead of k_{ij} for the elements of the eigenvector $\mathbf{K}^{(j)}$.

The characteristic equation of the coefficient matrix is

$$\begin{vmatrix} 1-\lambda & 2 \\ 3 & 2-\lambda \end{vmatrix} = (1-\lambda)(2-\lambda) - 6 = 0$$

or

$$\lambda^2 - 3\lambda - 4 = (\lambda - 4)(\lambda + 1) = 0$$

Hence, the eigenvalues are $\lambda_1 = -1$ and $\lambda_2 = 4$.

The equation $(\mathbf{A}\lambda - \mathbf{I})\mathbf{K} = 0$ from which the eigenvectors $\mathbf{K}$ are determined assumes the form

$$\begin{pmatrix} 1-\lambda & 2 \\ 3 & 2-\lambda \end{pmatrix}\begin{pmatrix} k_1 \\ k_2 \end{pmatrix} = \begin{pmatrix} 0 \\ 0 \end{pmatrix}$$

Thus, for $\lambda_1 = -1$, we obtain the algebraic equations

$$2k_1 + 2k_2 = 0$$
$$3k_1 + 3k_2 = 0$$

both of which are equivalent. (This will always be the case.) Thus, $k_2 = -k_1$, so that by choosing $k_1 = 1$ it follows that $k_2 = -1$. The eigenvector corresponding to the eigenvalue $\lambda_1 = -1$ can then be represented by any constant multiple of[†]

$$\mathbf{K}^{(1)} = \begin{pmatrix} 1 \\ -1 \end{pmatrix}$$

In the same way, corresponding to $\lambda_2 = 4$, we find the equation

$$-3k_1 + 2k_2 = 0$$

Among other choices, we select $k_1 = 2$ and $k_2 = 3$, leading to the eigenvector

$$\mathbf{K}^{(2)} = \begin{pmatrix} 2 \\ 3 \end{pmatrix}$$

Based upon these results, we obtain two solution vectors given by

$$\mathbf{Y}^{(1)}(t) = \mathbf{K}^{(1)}e^{-t}, \qquad \mathbf{Y}^{(2)}(t) = \mathbf{K}^{(2)}e^{4t}$$

which are clearly not proportional, and hence form a fundamental solution set. The general solution is therefore

$$\mathbf{Y} = C_1\mathbf{Y}^{(1)}(t) + C_2\mathbf{Y}^{(2)}(t)$$

$$= C_1\begin{pmatrix} 1 \\ -1 \end{pmatrix}e^{-t} + C_2\begin{pmatrix} 2 \\ 3 \end{pmatrix}e^{4t}$$

[†] Eigenvectors are unique only to within a multiplicative constant.

where C_1 and C_2 are any constants. In terms of the fundamental matrix, we can also express the general solution as

$$\mathbf{Y} = \begin{pmatrix} e^{-t} & 2e^{4t} \\ -e^{-t} & 3e^{4t} \end{pmatrix} \begin{pmatrix} C_1 \\ C_2 \end{pmatrix}$$

while in scalar form the general solution is

$$x(t) = C_1 e^{-t} + 2C_2 e^{4t}$$
$$y(t) = -C_1 e^{-t} + 3C_2 e^{4t}$$ ■

Although we won't prove it here, it can be shown that if the matrix $\mathbf{A}$ in the matrix equation $\mathbf{Y}' = \mathbf{AY}$ is *symmetric*, all eigenvalues and eigenvectors will be real. Moreover, a full set of linearly independent eigenvectors *always* exists, even in the case of repeated eigenvalues (see Case II below). However, when $\mathbf{A}$ is not symmetric there is no guarantee of these properties. In particular, the eigenvalues and eigenvectors can be complex when $\mathbf{A}$ fails to be symmetric (see Case III below).

■ EXAMPLE 22

Use the fundamental matrix to solve the IVP

$$\mathbf{Y}' = \begin{pmatrix} 1 & 2 \\ 3 & 2 \end{pmatrix} \mathbf{Y}, \qquad \mathbf{Y}(0) = \begin{pmatrix} 5 \\ 0 \end{pmatrix}$$

Solution

We recognize the matrix equation as being the same as that in Example 21, except here an initial condition is also prescribed. Hence, the fundamental matrix is

$$\boldsymbol{\phi}(t) = \begin{pmatrix} e^{-t} & 2e^{4t} \\ -e^{-t} & 3e^{4t} \end{pmatrix}$$

Based upon Equation (71), the particular solution we seek has the form

$$\mathbf{Y} = \boldsymbol{\phi}(t)\boldsymbol{\phi}^{-1}(0)\mathbf{Y}_0$$

where $\mathbf{Y}_0 = \begin{pmatrix} 5 \\ 0 \end{pmatrix}$ is the prescribed initial vector. By setting $t = 0$ in the fundamental matrix, we find

$$\boldsymbol{\phi}(0) = \begin{pmatrix} 1 & 2 \\ -1 & 3 \end{pmatrix}, \qquad \det[\boldsymbol{\phi}(0)] = 5$$

from which we compute

$$\boldsymbol{\phi}^{-1}(0) = \frac{1}{5}\begin{pmatrix} 3 & -2 \\ 1 & 1 \end{pmatrix}$$

Therefore, it follows that

$$\mathbf{Y} = \begin{pmatrix} e^{-t} & 2e^{4t} \\ -e^{-t} & 3e^{4t} \end{pmatrix} \frac{1}{5} \begin{pmatrix} 3 & -2 \\ 1 & 1 \end{pmatrix} \begin{pmatrix} 5 \\ 0 \end{pmatrix}$$

$$= \begin{pmatrix} e^{-t} & 2e^{4t} \\ -e^{-t} & 3e^{4t} \end{pmatrix} \begin{pmatrix} 3 \\ 1 \end{pmatrix}$$

$$= \begin{pmatrix} 3e^{-t} + 2e^{4t} \\ -3e^{-t} + 3e^{4t} \end{pmatrix} = \begin{pmatrix} 3 \\ -3 \end{pmatrix} e^{-t} + \begin{pmatrix} 2 \\ 3 \end{pmatrix} e^{4t}$$

Once again, in scalar form this becomes

$$x(t) = 3e^{-t} + 2e^{4t}$$

$$y(t) = -3e^{-t} + 3e^{4t}$$ ∎

Case II. Equal Roots

If the two roots λ_1 and λ_2 of the characteristic equation are real and equal, then setting $\lambda_1 = \lambda_2 = \lambda$ produces only one solution vector of the system of DEs, namely, $\mathbf{Y}^{(1)}(t) = \mathbf{K}^{(1)}e^{\lambda t}$. By analogy with second-order DEs, we might expect to find a second linearly independent solution vector of the form

$$\mathbf{Y}^{(2)}(t) = \mathbf{K}^{(2)}te^{\lambda t}$$

Unfortunately, this need not be the case. Instead, it turns out that we must seek a second solution vector of the slightly more general form

$$\mathbf{Y}^{(2)}(t) = [\mathbf{K}^{(2)}t + \mathbf{L}^{(2)}]e^{\lambda t} \tag{87}$$

where $\mathbf{K}^{(2)} = \begin{pmatrix} k_1 \\ k_2 \end{pmatrix}$ and $\mathbf{L}^{(2)} = \begin{pmatrix} l_1 \\ l_2 \end{pmatrix}$ are both vectors that must be determined. The substitution of (87) into the equation $\mathbf{Y}' = \mathbf{AY}$ yields the condition

$$[\lambda\mathbf{K}^{(2)}t + \lambda\mathbf{L}^{(2)} + \mathbf{K}^{(2)}]e^{\lambda t} = [\mathbf{AK}^{(2)}t + \mathbf{AL}^{(2)}]e^{\lambda t}$$

Equating terms in $te^{\lambda t}$ and $e^{\lambda t}$, we obtain the relations

$$\lambda\mathbf{K}^{(2)} = \mathbf{AK}^{(2)}$$

$$\lambda\mathbf{L}^{(2)} + \mathbf{K}^{(2)} = \mathbf{AL}^{(2)}$$

which may be rearranged as

$$(\mathbf{A} - \lambda\mathbf{I})\mathbf{K}^{(2)} = \mathbf{0} \tag{88}$$

$$(\mathbf{A} - \lambda\mathbf{I})\mathbf{L}^{(2)} = \mathbf{K}^{(2)} \tag{89}$$

The first of these equations shows that $\mathbf{K}^{(2)}$ is an *eigenvector* of $\mathbf{A}$. Hence, we may choose to write $\mathbf{K}^{(2)} = \mathbf{K}^{(1)}$, although in some cases it may be better to choose another eigenvector (when possible) so that the *nonhomogeneous* equation (89) can be solved for $\mathbf{L}^{(2)}$. The *general solution* in this case has the form

$$\mathbf{Y} = C_1\mathbf{K}^{(1)}e^{\lambda t} + C_2[\mathbf{K}^{(2)}t + \mathbf{L}^{(2)}]e^{\lambda t} \tag{90}$$

EXAMPLE 23   Solve the homogeneous system

$$x' = -4x - y$$
$$y' = x - 2y$$

Solution

The characteristic equation for the coefficient matrix of this system of equations is

$$\Delta(\lambda) = \begin{vmatrix} -(4 + \lambda) & -1 \\ 1 & -(2 + \lambda) \end{vmatrix} = \lambda^2 + 6\lambda + 9 = 0$$

which gives us the double root $\lambda_1 = \lambda_2 = -3$. Setting $\lambda = -3$ in the matrix equation

$$\begin{pmatrix} -(4 + \lambda) & -1 \\ 1 & -(2 + \lambda) \end{pmatrix}\begin{pmatrix} k_1 \\ k_2 \end{pmatrix} = \begin{pmatrix} 0 \\ 0 \end{pmatrix}$$

yields the relation

$$-k_1 - k_2 = 0$$

or $k_2 = -k_1$. Choosing $k_1 = 1$, it follows that $k_2 = -1$ and thus the first solution vector is

$$\mathbf{Y}^{(1)}(t) = \begin{pmatrix} 1 \\ -1 \end{pmatrix}e^{-3t}$$

To obtain a second solution vector, we choose

$$\mathbf{K}^{(2)} = \mathbf{K}^{(1)} = \begin{pmatrix} 1 \\ -1 \end{pmatrix}$$

and determine $\mathbf{L}^{(2)} = \begin{pmatrix} l_1 \\ l_2 \end{pmatrix}$ by solving [see Equation (89)]

$$\begin{pmatrix} -(4 + \lambda) & -1 \\ 1 & -(2 + \lambda) \end{pmatrix}\begin{pmatrix} l_1 \\ l_2 \end{pmatrix} = \begin{pmatrix} 1 \\ -1 \end{pmatrix}$$

Upon setting $\lambda = -3$, we obtain the equations

$$-l_1 - l_2 = 1$$
$$l_1 + l_2 = -1$$

a possible solution of which is $l_1 = 1$ and $l_2 = -2$. Thus, a second solution of the original system is

$$\mathbf{Y}^{(2)}(t) = \begin{pmatrix} 1 \\ -1 \end{pmatrix}te^{-3t} + \begin{pmatrix} 1 \\ -2 \end{pmatrix}e^{-3t}$$

$$= \begin{pmatrix} t + 1 \\ -(t + 2) \end{pmatrix}e^{-3t}$$

which, for arbitrary constants C_1 and C_2, leads to the general solution

$$\mathbf{Y} = C_1 \begin{pmatrix} 1 \\ -1 \end{pmatrix} e^{-3t} + C_2 \begin{pmatrix} t+1 \\ -(t+2) \end{pmatrix} e^{-3t}$$

In scalar form the general solution is

$$x(t) = C_1 e^{-3t} + C_2(t+1)e^{-3t}$$
$$y(t) = -C_1 e^{-3t} - C_2(t+2)e^{-3t}$$

$\blacksquare$

Case III. Complex Conjugate Roots

When the characteristic roots are complex, they will appear as complex conjugates, that is, $\lambda_1 = p + iq$ and $\lambda_2 = p - iq$. Following the approach used in Case I with real distinct roots, it follows that the resulting solution vectors

$$\mathbf{Y}^{(1)}(t) = \mathbf{K}^{(1)} e^{(p+iq)t}$$
$$\mathbf{Y}^{(2)}(t) = \mathbf{K}^{(2)} e^{(p-iq)t} \tag{91}$$

are linearly independent and thus can be used in the construction of a general solution. However, these are *complex* solution vectors and in most applications we desire *real* solutions.

To obtain real solutions from those in (91) we proceed in a fashion similar to that in Section 4.4.1. First, we recognize that both $\mathbf{K}^{(1)}$ and $\mathbf{K}^{(2)}$ will be complex eigenvectors in general. Moreover, if $\mathbf{K}^{(1)} = \begin{pmatrix} k_1 \\ k_2 \end{pmatrix} = \begin{pmatrix} a_1 \\ a_2 \end{pmatrix} + i \begin{pmatrix} b_1 \\ b_2 \end{pmatrix} = \mathbf{a} + i\mathbf{b}$ is an eigenvector belonging to the eigenvalue $\lambda_1 = p + iq$, it follows that $\mathbf{K}^{(2)} = \mathbf{a} - i\mathbf{b}$ is an eigenvector belonging to $\lambda_2 = p - iq$. Thus, (91) becomes

$$\mathbf{Y}^{(1)}(t) = (\mathbf{a} + i\mathbf{b})e^{(p+iq)t} = \mathbf{U}(t) + i\mathbf{V}(t)$$
$$\mathbf{Y}^{(2)}(t) = (\mathbf{a} - i\mathbf{b})e^{(p-iq)t} = \mathbf{U}(t) - i\mathbf{V}(t) \tag{92}$$

where

$$\mathbf{U}(t) = e^{pt}(\mathbf{a}\cos qt - \mathbf{b}\sin qt)$$
$$\mathbf{V}(t) = e^{pt}(\mathbf{a}\sin qt + \mathbf{b}\cos qt) \tag{93}$$

are the real and imaginary parts of the solutions. The functions

$$\mathbf{U}(t) = \frac{1}{2}\left[\mathbf{Y}^{(1)}(t) + \mathbf{Y}^{(2)}(t)\right]$$

and

$$\mathbf{V}(t) = \frac{1}{2i}\left[\mathbf{Y}^{(1)}(t) - \mathbf{Y}^{(2)}(t)\right]$$

are linear combinations of solutions of $\mathbf{Y}' = \mathbf{A}\mathbf{Y}$, and thus are also solutions (recall Theorem 8.5). Moreover, they form a fundamental solution set (why?), so the *general solution* in this case becomes

$$\mathbf{Y} = C_1\mathbf{U}(t) + C_2\mathbf{V}(t) \tag{94}$$

where C_1 and C_2 are arbitrary constants.

Remark. Although the general solution (94) appears to be quite complicated, this case will actually require fewer calculations in obtaining the general solution than either of the previous two cases.

EXAMPLE 24

Solve the homogeneous system

$$x' = 6x - y$$
$$y' = 5x + 4y$$

Solution

The characteristic equation is given by

$$\begin{vmatrix} 6 - \lambda & -1 \\ 5 & 4 - \lambda \end{vmatrix} = \lambda^2 - 10\lambda + 29 = 0$$

the roots of which are $\lambda_1 = 5 + 2i$ and $\lambda_2 = 5 - 2i$. To find an eigenvector of λ_1, we substitute $\lambda_1 = 5 + 2i$ into

$$\begin{pmatrix} 6 - \lambda & -1 \\ 5 & 4 - \lambda \end{pmatrix}\begin{pmatrix} a_1 + ib_1 \\ a_2 + ib_2 \end{pmatrix} = \begin{pmatrix} 0 \\ 0 \end{pmatrix}$$

from which we obtain (from the first equation)

$$(1 - 2i)(a_1 + ib_1) - (a_2 + ib_2) = 0$$

Therefore, if we choose $a_1 + ib_1 = 1$ and $a_2 + ib_2 = 1 - 2i$, it follows that $a_1 = 1$, $b_1 = 0$, $a_2 = 1$, and $b_2 = -2$. Hence, $\mathbf{a} = \begin{pmatrix} 1 \\ 1 \end{pmatrix}$ and $\mathbf{b} = \begin{pmatrix} 0 \\ -2 \end{pmatrix}$. Using (93) and (94), we can now immediately write down the general solution

$$\mathbf{Y} = C_1 e^{5t}\left\{\begin{pmatrix} 1 \\ 1 \end{pmatrix}\cos 2t - \begin{pmatrix} 0 \\ -2 \end{pmatrix}\sin 2t\right\}$$

$$+ C_2\left\{\begin{pmatrix} 1 \\ 1 \end{pmatrix}\sin 2t + \begin{pmatrix} 0 \\ -2 \end{pmatrix}\cos 2t\right\}$$

Equivalently, the scalar component form of this solution is

$$x(t) = e^{5t}(C_1 \cos 2t + C_2 \sin 2t)$$
$$y(t) = e^{5t}[(C_1 - 2C_2)\cos 2t + (2C_1 + C_2)\sin 2t]$$

8.8.2 Higher-Order Systems

The methods illustrated in Section 8.8.1 carry over to systems of equations involving more than two equations in a natural sort of way. For example, if $\lambda_1, \lambda_2, \ldots, \lambda_n$ are *real and distinct* eigenvalues of the matrix $\mathbf{A}$ in the equation $\mathbf{Y}' = \mathbf{AY}$, we can substitute them in turn into the equation

$$(\mathbf{A} - \lambda\mathbf{I})\mathbf{K} = 0$$

to solve for the corresponding eigenvectors $\mathbf{K}^{(1)}, \mathbf{K}^{(2)}, \ldots, \mathbf{K}^{(n)}$. In this case it can be shown that all particular solutions $\mathbf{Y}^{(j)}(t) = \mathbf{K}^{(j)}e^{\lambda_j t}$, $j = 1, 2, \ldots, n$ are linearly independent. The case of distinct sets of complex eigenvalues is similarly handled. Finally, if one of the eigenvalues λ is a *repeated root k* times, we say it has **multiplicity** k and we look for k linearly independent solutions of the system of equations. For example, if λ has multiplicity 3, we look for three solutions of the form

$$\mathbf{Y}^{(1)}(t) = \mathbf{K}^{(1)}e^{\lambda t}$$
$$\mathbf{Y}^{(2)}(t) = \left[\mathbf{K}^{(2)}t + \mathbf{L}^{(2)}\right]e^{\lambda t} \tag{95}$$
$$\mathbf{Y}^{(3)}(t) = \left[\mathbf{K}^{(3)}t^2 + \mathbf{L}^{(3)}t + \mathbf{M}^{(3)}\right]e^{\lambda t}$$

where

$$\left.\begin{array}{l} (\mathbf{A} - \lambda\mathbf{I})\mathbf{K}^{(1)} = 0 \\ (\mathbf{A} - \lambda\mathbf{I})\mathbf{K}^{(2)} = 0 \\ (\mathbf{A} - \lambda\mathbf{I})\mathbf{L}^{(2)} = \mathbf{K}^{(2)} \\ (\mathbf{A} - \lambda\mathbf{I})\mathbf{K}^{(3)} = 0 \\ (\mathbf{A} - \lambda\mathbf{I})\mathbf{L}^{(3)} = \mathbf{K}^{(3)} \\ (\mathbf{A} - \lambda\mathbf{I})\mathbf{M}^{(3)} = \mathbf{L}^{(3)} \end{array}\right\} \tag{96}$$

Analogous results hold for higher multiplicities. Even multiplicities of complex roots are handled in a similar manner.

EXAMPLE 25

Solve the homogeneous system

$$x' = x + z$$
$$y' = y + 2z$$
$$z' = x + 2y + 5z$$

Solution

In this case the characteristic equation is

$$\Delta(\lambda) = \begin{vmatrix} 1 - \lambda & 0 & 1 \\ 0 & 1 - \lambda & 2 \\ 1 & 2 & 5 - \lambda \end{vmatrix} = -(\lambda^3 - 7\lambda^2 + 6\lambda) = 0$$

Factoring this polynomial leads to

$$\lambda(\lambda - 1)(\lambda - 6) = 0$$

from which we deduce $\lambda_1 = 0$, $\lambda_2 = 1$, and $\lambda_3 = 6$. The substitution of $\lambda_1 = 0$ into

$$\begin{pmatrix} 1 - \lambda & 0 & 1 \\ 0 & 1 - \lambda & 2 \\ 1 & 2 & 5 - \lambda \end{pmatrix} \begin{pmatrix} k_1 \\ k_2 \\ k_3 \end{pmatrix} = 0$$

yields

$$k_1 + k_3 = 0$$
$$k_2 + 2k_3 = 0$$
$$k_1 + 2k_2 + 5k_3 = 0$$

Hence, one possible solution is $k_1 = 1$, $k_2 = 2$, $k_3 = -1$, or

$$\mathbf{K}^{(1)} = \begin{pmatrix} 1 \\ 2 \\ -1 \end{pmatrix}$$

Similarly, the substitution of $\lambda_2 = 1$ and $\lambda_3 = 6$, respectively, into the algebraic system gives us

$$\mathbf{K}^{(2)} = \begin{pmatrix} 2 \\ -1 \\ 0 \end{pmatrix}, \qquad \mathbf{K}^{(3)} = \begin{pmatrix} 1 \\ 2 \\ 5 \end{pmatrix}$$

In this case, the general solution is

$$\mathbf{Y} = C_1 \begin{pmatrix} 1 \\ 2 \\ -1 \end{pmatrix} + C_2 \begin{pmatrix} 2 \\ -1 \\ 0 \end{pmatrix} e^t + C_3 \begin{pmatrix} 1 \\ 2 \\ 5 \end{pmatrix} e^{6t}$$

or in scalar component form

$$x(t) = C_1 + 2C_2 e^t + C_3 e^{6t}$$
$$y(t) = 2C_1 - C_2 e^t + 2C_3 e^{6t}$$
$$z(t) = -C_1 + 5C_3 e^{6t}$$

∎

Exercises 8.8

In Problems 1 to 20, solve the homogeneous system of equations. If initial conditions are prescribed, find the particular solution that satisfies them.

1. $x' = 2x - y$
$y' = 3x - 2y$

2. $x' = 4x - 3y$
$y' = 5x - 4y$

3. $x' = 4x - 3y$
$y' = 8x - 6y$

4. $x' = 3x + 2y$
$y' = 6x - y$

5. $x' = x - 4y, x(0) = 1$
$y' = x + y, y(0) = 0$

■**6.** $x' = 3x - 2y, x(0) = 0$
$y' = 2x + 3y, y(0) = 2$

7. $x' = 6x - 5y$
$y' = x + 2y$

8. $x' = 3x + 2y$
$y' = -5x + y$

9. $x' = 3x - 18y$
$y' = 2x - 9y$

10. $x' = -2x - 3y$
$y' = 3x + 4y$

11. $\mathbf{Y}' = \begin{pmatrix} 1 & 3 \\ 1 & -1 \end{pmatrix} \mathbf{Y}, \mathbf{Y}(0) = \begin{pmatrix} 0 \\ 8 \end{pmatrix}$

12. $\mathbf{Y}' = \begin{pmatrix} 2 & -5 \\ 4 & -2 \end{pmatrix} \mathbf{Y}, \mathbf{Y}(0) = \begin{pmatrix} 2 \\ 3 \end{pmatrix}$

13. $x' = -x + y$
$y' = 2x$

14. $x' = x + y + z$
$y' = 2x + y - z$
$z' = -y + z$

15. $x' = x + z$
$y' = x + y$
$z' = -2x - z$

16. $x' = x + y + z$
$y' = 2x + y - z$
$z' = -3x + 2y + 4z$

17. $\mathbf{Y}' = \begin{pmatrix} 1 & 2 & 1 \\ 6 & -1 & 0 \\ -1 & -2 & -1 \end{pmatrix} \mathbf{Y}$

■**18.** $\mathbf{Y}' = \begin{pmatrix} 5 & 2 & -2 \\ 7 & 0 & -2 \\ 11 & 1 & -1 \end{pmatrix} \mathbf{Y}, \mathbf{Y}(0) = \begin{pmatrix} 1 \\ 0 \\ -1 \end{pmatrix}$

19. $\mathbf{Y}' = \begin{pmatrix} 0 & 2 & 2 \\ 2 & 0 & 2 \\ 2 & 2 & 0 \end{pmatrix} \mathbf{Y}$

*20. $\mathbf{Y}' = \begin{pmatrix} 3 & 1 & 0 \\ -1 & 0 & -1 \\ 1 & 2 & 3 \end{pmatrix} \mathbf{Y}$

*21. By analogy with the Maclaurin series

$$e^{ax} = 1 + ax + a^2 \frac{x^2}{2!} + a^3 \frac{x^3}{3!} + \cdots$$

we define the **matrix exponential function**

$$e^{\mathbf{A}t} = \mathbf{I} + \mathbf{A}t + \mathbf{A}^2 \frac{t^2}{2!} + \mathbf{A}^3 \frac{t^3}{3!} + \cdots$$

where $\mathbf{I}$ is the identity matrix and $\mathbf{A}^2 = \mathbf{AA}$, $\mathbf{A}^3 \overset{\perp}{=} \mathbf{A}(\mathbf{A}^2)$, and so on.
(a) Use the approximation

$$e^{\mathbf{A}t} \cong \mathbf{I} + \mathbf{A}t + \mathbf{A}^2 \frac{t^2}{2!}$$

to compute $e^{\mathbf{A}t}$ when $t = 1$ and

$$\mathbf{A} = \begin{pmatrix} 0 & 1 \\ 1 & 0 \end{pmatrix}$$

(b) Find a pattern and compute $e^{\mathbf{A}t}$ exactly for the matrix $\mathbf{A}$ given in (a) when $t = 1$.

*22. Use the result of Problem 21 to compute $e^{\mathbf{A}t}$ exactly for any t when

$$\mathbf{A} = \begin{pmatrix} 1 & 0 \\ 0 & 2 \end{pmatrix}$$

*23. Using the matrix exponential function defined in Problem 21, show that the matrix equation

$$\mathbf{Y}' = \mathbf{AY}$$

where $\mathbf{A}$ is an $n \times n$ matrix of constants, has the formal solution

$$\mathbf{Y} = e^{\mathbf{A}t}\mathbf{C}$$

where $\mathbf{C}$ is an arbitrary constant vector.

Hint. Assume that $\dfrac{d}{dt} e^{\mathbf{A}t} = \mathbf{A}e^{\mathbf{A}t}$.

*24. Use the results of Problems 21 to 23 to solve the matrix equations

(a) $\mathbf{Y}' = \begin{pmatrix} 0 & 1 \\ 1 & 0 \end{pmatrix} \mathbf{Y}$

(b) $\mathbf{Y}' = \begin{pmatrix} 1 & 0 \\ 0 & 2 \end{pmatrix} \mathbf{Y}$

8.9 Nonhomogeneous Systems

In this section we want to discuss a general method for solving the **nonhomogeneous** matrix equation

$$\mathbf{Y}' = \mathbf{A}(t)\mathbf{Y} + \mathbf{F}(t) \tag{97}$$

where $\mathbf{A}(t)$ is not restricted to constant elements. (We consider only examples featuring constant coefficients, however.) According to Theorem 8.7, the **general solution** of (97) has the form

$$\mathbf{Y} = \mathbf{Y}_H + \mathbf{Y}_P$$

$$= C_1\mathbf{Y}^{(1)}(t) + C_2\mathbf{Y}^{(2)}(t) + \mathbf{Y}_P(t) \tag{98}$$

where $\mathbf{Y}_P$ is any **particular solution** of (97) while $\mathbf{Y}_H$ is the general solution of the associated *homogeneous* system.

Once we have determined $\mathbf{Y}_H$, we can construct $\mathbf{Y}_P$ by the method of **variation of parameters.** To start, we assume that

$$\mathbf{Y}_P = \boldsymbol{\phi}(t)\mathbf{U}(t) \tag{99}$$

where $\boldsymbol{\phi}(t)$ is the fundamental matrix associated with the homogeneous system and $\mathbf{U}(t)$ is an unknown vector to be chosen in such a way that (99) satisfies (97).[†] The direct substitution of (99) into (97), using the product rule for the derivative, leads to

$$\boldsymbol{\phi}'(t)\mathbf{U}(t) + \boldsymbol{\phi}(t)\mathbf{U}'(t) = \mathbf{A}(t)\boldsymbol{\phi}(t)\mathbf{U}(t) + \mathbf{F}(t)$$

However, it can be shown that $\boldsymbol{\phi}'(t) = \mathbf{A}(t)\boldsymbol{\phi}(t)$ since each column vector of $\boldsymbol{\phi}(t)$ satisfies the homogeneous equation $\mathbf{Y}' = \mathbf{A}(t)\mathbf{Y}$. Thus the above expression reduces to

$$\boldsymbol{\phi}(t)\mathbf{U}'(t) = \mathbf{F}(t) \tag{100}$$

Since $\det[\boldsymbol{\phi}(t)] = W(t) \neq 0$ on the interval of interest, it follows that $\boldsymbol{\phi}^{-1}(t)$ exists. Multiplying (100) by $\boldsymbol{\phi}^{-1}(t)$ yields

$$\mathbf{U}'(t) = \boldsymbol{\phi}^{-1}(t)\mathbf{F}(t)$$

which is sufficient to choose $\mathbf{U}(t)$; that is, the integral of this last expression yields

$$\mathbf{U}(t) = \int \boldsymbol{\phi}^{-1}(t)\mathbf{F}(t)\,dt \tag{101}$$

where the constant vector of integration has been conveniently set to zero. Hence, our particular solution (99) becomes

$$\mathbf{Y}_P = \boldsymbol{\phi}(t)\int \boldsymbol{\phi}^{-1}(t)\mathbf{F}(t)\,dt \tag{102}$$

leading to the general solution form

$$\mathbf{Y} = \boldsymbol{\phi}(t)\mathbf{C} + \boldsymbol{\phi}(t)\int \boldsymbol{\phi}^{-1}(t)\mathbf{F}(t)\,dt \tag{103}$$

where $\mathbf{C}$ is any constant vector.

[†] The vector $\mathbf{U}(t)$ here is not the same as that in Section 8.8.1.

Lastly, if an initial condition of the form

$$\mathbf{Y}(0) = \mathbf{Y}_0 \tag{104}$$

is also prescribed, the solution of (97) satisfying (104) can be expressed as (see Problem 13 in Exercises 8.8)

$$\mathbf{Y} = \boldsymbol{\phi}(t)\boldsymbol{\phi}^{-1}(0)\mathbf{Y}_0 + \boldsymbol{\phi}(t) \int_0^t \boldsymbol{\phi}^{-1}(\tau)\mathbf{F}(\tau)\, d\tau \tag{105}$$

Remark. The reader should study the form of (103) in contrast to the solution of the first-order DE

$$y' + a(x)y = f(x)$$

which was given in Section 2.5.2 as

$$y = Cy_1(x) + y_1(x) \int \frac{f(x)}{y_1(x)}\, dx$$

where $y_1(x)$ is a fundamental solution of the associated homogeneous DE.

EXAMPLE 26

Use variation of parameters to solve

$$\mathbf{Y}' = \begin{pmatrix} -4 & 2 \\ 2 & -1 \end{pmatrix}\mathbf{Y} + \begin{pmatrix} \dfrac{1}{t} \\ 4 + \dfrac{2}{t} \end{pmatrix}$$

Solution

We first recognize that the coefficient matrix $\mathbf{A}$ has eigenvalues given by solutions of

$$\Delta(\lambda) = \begin{vmatrix} -(4 + \lambda) & 2 \\ 2 & -(\lambda + 1) \end{vmatrix} = \lambda^2 + 5\lambda = 0$$

which are $\lambda_1 = 0$ and $\lambda_2 = -5$. For $\lambda_1 = 0$, the eigenvector $\mathbf{K}^{(1)} = \begin{pmatrix} k_1 \\ k_2 \end{pmatrix}$ is a solution of the matrix equation

$$\begin{pmatrix} -4 & 2 \\ 2 & -1 \end{pmatrix}\begin{pmatrix} k_1 \\ k_2 \end{pmatrix} = \begin{pmatrix} 0 \\ 0 \end{pmatrix}$$

or

$$-4k_1 + 2k_2 = 0$$

By choosing $k_1 = 1$, it follows that $k_2 = 2$ and thus

$$\mathbf{K}^{(1)} = \begin{pmatrix} 1 \\ 2 \end{pmatrix}$$

In the same manner, we find that

$$\mathbf{K}^{(2)} = \begin{pmatrix} 2 \\ -1 \end{pmatrix}$$

The general solution of the homogeneous system is therefore

$$\mathbf{Y}_H = C_1 \begin{pmatrix} 1 \\ 2 \end{pmatrix} + C_2 \begin{pmatrix} 2 \\ -1 \end{pmatrix} e^{-5t}$$

from which we obtain the fundamental matrix

$$\boldsymbol{\phi}(t) = \begin{pmatrix} 1 & 2e^{-5t} \\ 2 & -e^{-5t} \end{pmatrix}$$

In order to construct $\mathbf{Y}_P$, we need the inverse $\boldsymbol{\phi}^{-1}(t)$. Here we see that $\det[\boldsymbol{\phi}(t)] = -5e^{-5t}$, which leads to

$$\boldsymbol{\phi}^{-1}(t) = -\frac{1}{5} e^{5t} \begin{pmatrix} -e^{-5t} & -2e^{-5t} \\ -2 & 1 \end{pmatrix} = \begin{pmatrix} \frac{1}{5} & \frac{2}{5} \\ \frac{2}{5}e^{5t} & -\frac{1}{5}e^{5t} \end{pmatrix}$$

Hence, from (103) we have

$$\mathbf{Y}_P = \boldsymbol{\phi}(t) \int \boldsymbol{\phi}^{-1}(t)\mathbf{F}(t)\, dt$$

$$= \begin{pmatrix} 1 & 2e^{-5t} \\ 2 & -e^{-5t} \end{pmatrix} \int \begin{pmatrix} \frac{1}{5} & \frac{2}{5} \\ \frac{2}{5}e^{5t} & -\frac{1}{5}e^{5t} \end{pmatrix} \begin{pmatrix} \frac{1}{t} \\ 4 + \frac{2}{t} \end{pmatrix} dt$$

$$= \begin{pmatrix} 1 & 2e^{-5t} \\ 2 & -e^{-5t} \end{pmatrix} \int \begin{pmatrix} \frac{8}{5} + \frac{1}{t} \\ -\frac{4}{5}e^{5t} \end{pmatrix} dt$$

$$= \begin{pmatrix} 1 & 2e^{-5t} \\ 2 & -e^{-5t} \end{pmatrix} \begin{pmatrix} \frac{8}{5}t + \log t \\ -\frac{4}{25}e^{5t} \end{pmatrix}$$

$$= \begin{pmatrix} \frac{8}{5}t + \log t - \frac{8}{25} \\ \frac{16}{5}t + 2\log t + \frac{4}{25} \end{pmatrix}$$

Combining the homogeneous and particular solutions, we finally obtain

$$\mathbf{Y} = \begin{pmatrix} 1 & 2e^{-5t} \\ 2 & -e^{-5t} \end{pmatrix} \begin{pmatrix} C_1 \\ C_2 \end{pmatrix} + \begin{pmatrix} \frac{8}{5}t + \log t - \frac{8}{25} \\ \frac{16}{5}t + 2\log t + \frac{4}{25} \end{pmatrix}$$

or, equivalently,

$$\mathbf{Y} = C_1 \begin{pmatrix} 1 \\ 2 \end{pmatrix} + C_2 \begin{pmatrix} 2 \\ -1 \end{pmatrix} e^{-5t} + \begin{pmatrix} \frac{8}{5} \\ \frac{16}{5} \end{pmatrix} t + \begin{pmatrix} 1 \\ 2 \end{pmatrix} \log t + \begin{pmatrix} -\frac{8}{25} \\ \frac{4}{25} \end{pmatrix}$$

Exercises 8.9

In Problems 1 to 10, use matrix methods to solve the given nonhomogeneous system of equations. If initial conditions are prescribed, find the particular solution that satisfies them.

1. $x' = 2x + y + e^t$
$y' = 4x - y - e^t$

2. $x' = 2x - y + t$
$y' = 3x - 2y + 2t$

3. $x' = 2x - 5y - \sin 2t$
$y' = x - 2y + t$

4. $x' = 4x - 2y + e^t$
$y' = 5x + 2y - t$

5. $\mathbf{Y}' = \begin{pmatrix} 1 & 2 \\ -\frac{1}{2} & 1 \end{pmatrix} \mathbf{Y} + \begin{pmatrix} 0 \\ e^t \tan t \end{pmatrix}$

6. $\mathbf{Y}' = \begin{pmatrix} 2 & -5 \\ 1 & -2 \end{pmatrix} \mathbf{Y} + \begin{pmatrix} \csc t \\ \sec t \end{pmatrix}$

7. $\mathbf{Y}' = \begin{pmatrix} -10 & 6 \\ -12 & 7 \end{pmatrix} \mathbf{Y} + \begin{pmatrix} 10e^{-3t} \\ 18e^{-3t} \end{pmatrix}$, $\mathbf{Y}(0) = \begin{pmatrix} -1 \\ 2 \end{pmatrix}$

8. $\mathbf{Y}' = \begin{pmatrix} 1 & -1 \\ 1 & 1 \end{pmatrix} \mathbf{Y} + \begin{pmatrix} 3e^t \\ 3e^t \end{pmatrix}$, $\mathbf{Y}(0) = \begin{pmatrix} 3 \\ 1 \end{pmatrix}$

9. $\mathbf{Y}' = \begin{pmatrix} 4 & 2 \\ 3 & -1 \end{pmatrix} \mathbf{Y} - \begin{pmatrix} 0 \\ 1 \end{pmatrix} e^{-2t}$, $\mathbf{Y}(0) = \begin{pmatrix} 0 \\ 0 \end{pmatrix}$

■10. $\mathbf{Y}' = \begin{pmatrix} 4 & 2 \\ 3 & -1 \end{pmatrix} \mathbf{Y} - \begin{pmatrix} 15 \\ 4 \end{pmatrix} te^{-2t}$, $\mathbf{Y}(0) = \begin{pmatrix} 1 \\ -1 \end{pmatrix}$

In Problems 11 and 12, find a particular solution vector $\mathbf{Y}_P$.

*11. $\mathbf{Y}' = \begin{pmatrix} 1 & -1 & 1 \\ 0 & 0 & 1 \\ 0 & -1 & 2 \end{pmatrix} \mathbf{Y} + \begin{pmatrix} 0 \\ 1 \\ 1 \end{pmatrix} e^t$

*12. $\mathbf{Y}' = \begin{pmatrix} 1 & 1 & 1 \\ 0 & -1 & 0 \\ -2 & -1 & -2 \end{pmatrix} \mathbf{Y} + \begin{pmatrix} 1 \\ 0 \\ 2e^{-t} \end{pmatrix}$

*13. Verify directly that

$$\mathbf{Y} = \phi(t)\phi^{-1}(0)\mathbf{Y}_0 + \phi(t) \int_0^t \phi^{-1}(\tau)\mathbf{F}(\tau)\, d\tau$$

is a solution of the IVP

$$\mathbf{Y}' = \mathbf{A}(t)\mathbf{Y} + \mathbf{F}(t), \quad \mathbf{Y}(0) = \mathbf{Y}_0$$

[0] 8.10 The Matrix Exponential Function

In this section we wish to further emphasize the similarity between solving first-order matrix equations and first-order scalar DEs. Recall from Chapter 2 that the first-order DE

$$y' = ay$$

where a is a constant, has the general solution

$$y = C_1 e^{at}$$

where C_1 is arbitrary. By analogy, we may expect that the first-order matrix equation

$$\mathbf{Y}' = \mathbf{A}\mathbf{Y} \tag{106}$$

where $\mathbf{A}$ is an $n \times n$ constant matrix, has a general solution of the form

$$\mathbf{Y} = e^{\mathbf{A}t}\mathbf{C} \tag{107}$$

where $\mathbf{C}$ is an arbitrary constant vector. It turns out that this is indeed the case, but to utilize this solution formula we need to define what we mean by the *exponential matrix* $e^{\mathbf{A}t}$.

In calculus we define the exponential function e^x by the Maclaurin series

$$e^x = 1 + x + \frac{x^2}{2!} + \cdots + \frac{x^n}{n!} + \cdots$$

In a similar fashion, if $\mathbf{A}$ is an $n \times n$ matrix, the **exponential matrix** $e^{\mathbf{A}t}$ is the $n \times n$ matrix defined by the series

$$e^{\mathbf{A}t} = \mathbf{I} + \mathbf{A}t + \mathbf{A}^2 \frac{t^2}{2!} + \cdots + \mathbf{A}^n \frac{t^n}{n!} + \cdots \tag{108}$$

where $\mathbf{I}$ is the identity matrix and where $\mathbf{A}^2 = \mathbf{AA}$, $\mathbf{A}^3 = \mathbf{AA}^2$, and so on. It can be shown that the series (108) converges for all t and has many of the same properties as the scalar exponential function e^{at}. In particular, it has the following useful properties:

$$e^{\mathbf{A}0} = e^{\mathbf{0}} = \mathbf{I} \tag{109}$$

$$e^{\mathbf{A}t} e^{-\mathbf{A}s} = e^{\mathbf{A}(t-s)} \tag{110}$$

$$(e^{\mathbf{A}t})^{-1} = e^{-\mathbf{A}t} \tag{111}$$

Another useful property involves the derivative of $e^{\mathbf{A}t}$, which leads to the same result as the derivative of the scalar function e^{at}. That is, by termwise differentiation of (108), we obtain

$$\frac{d}{dt}(e^{\mathbf{A}t}) = \frac{d}{dt}\left(\mathbf{I} + \mathbf{A}t + \mathbf{A}^2 \frac{t^2}{2!} + \cdots + \mathbf{A}^n \frac{t^n}{n!} + \cdots\right)$$

$$= \mathbf{A} + \mathbf{A}^2 t + \mathbf{A}^3 \frac{t^2}{2!} + \cdots + \mathbf{A}^n \frac{t^{n-1}}{(n-1)!} + \mathbf{A}^{n+1} \frac{t^n}{n!} + \cdots$$

$$= \mathbf{A}\left(\mathbf{I} + \mathbf{A}t + \mathbf{A}^2 \frac{t^2}{2!} + \cdots + \mathbf{A}^n \frac{t^n}{n!} + \cdots\right)$$

from which we deduce

$$\frac{d}{dt}(e^{\mathbf{A}t}) = \mathbf{A}e^{\mathbf{A}t} \tag{112}$$

Hence, it follows that $\mathbf{Y} = e^{\mathbf{A}t}$ is a solution of the matrix equation (106).

By comparing the solution (107) with the solution of (106) given by the fundamental matrix $\boldsymbol{\phi}(t)$, namely, $\mathbf{Y} = \boldsymbol{\phi}(t)\mathbf{C}$, we recognize that the exponential matrix $e^{\mathbf{A}t}$ is a *fundamental matrix* for the system given by (106). In this case, the exponential matrix $e^{\mathbf{A}t}$ can differ from $\boldsymbol{\phi}(t)$ by at most a multiplicative constant matrix.

If $\mathbf{A}$ is an $n \times n$ matrix, then we can use the exponential matrix to solve the IVP

$$\mathbf{Y}' = \mathbf{AY}, \qquad \mathbf{Y}(0) = \mathbf{Y}_0 \tag{113}$$

Specifically, the general solution of the matrix DE is that given by (107). The initial condition in (113) then requires that

$$\mathbf{Y}(0) = e^{\mathbf{0}}\mathbf{C} = \mathbf{Y}_0$$

Using property (109) above, we deduce that $\mathbf{C} = \mathbf{Y}_0$ and hence, the unique solution of (113) is given by

$$\mathbf{Y} = e^{\mathbf{A}t}\mathbf{Y}_0 \qquad (114)$$

From this result, we see that $e^{\mathbf{A}t} = \phi(t)\phi^{-1}(0)$, confirming that $e^{\mathbf{A}t}$ and $\phi(t)$ differ at most by a multiplicative constant matrix.

In order to use the matrix exponential to solve matrix equations, we need to learn how to form the fundamental matrix $e^{\mathbf{A}t}$. Of course, we could always rely on its series definition (108), but this may be difficult to use in general (recall Problems 21 to 24 in Exercises 8.8). However, if $\mathbf{A}$ is a diagonal matrix, then the formation of $e^{\mathbf{A}t}$ is straightforward. For example, if

$$\mathbf{A} = \begin{pmatrix} \lambda_1 & 0 \\ 0 & \lambda_2 \end{pmatrix}$$

then

$$\mathbf{A}^2 = \begin{pmatrix} \lambda_1^2 & 0 \\ 0 & \lambda_2^2 \end{pmatrix}, \qquad \mathbf{A}^3 = \begin{pmatrix} \lambda_1^3 & 0 \\ 0 & \lambda_2^3 \end{pmatrix}, \qquad \cdots$$

whereas in general

$$\mathbf{A}^n = \begin{pmatrix} \lambda_1^n & 0 \\ 0 & \lambda_2^n \end{pmatrix}$$

From these results, it follows that

$$e^{\mathbf{A}t} = \sum_{n=0}^{\infty} \mathbf{A}^n \frac{t^n}{n!} = \begin{pmatrix} e^{\lambda_1 t} & 0 \\ 0 & e^{\lambda_2 t} \end{pmatrix}$$

It is easy to generalize this result to the case of any $n \times n$ diagonal matrix. For example, if $\lambda_1, \lambda_2, \ldots, \lambda_n$ are the diagonal elements of the $n \times n$ matrix $\mathbf{A}$, and all other elements are zero, then $e^{\mathbf{A}t}$ is the $n \times n$ diagonal matrix with $e^{\lambda_1 t}, e^{\lambda_2 t}, \ldots, e^{\lambda_n t}$ down its main diagonal.

For general matrices $\mathbf{A}$, it is more difficult to use the series definition for $e^{\mathbf{A}t}$. However, it has been shown that $e^{\mathbf{A}t}$ can always be expressed as a *finite* sum of matrices involving powers of $\mathbf{A}$ and some unknown functions of t. That is, if $\mathbf{A}$ is an $n \times n$ matrix, then $e^{\mathbf{A}t}$ has a finite series representation of the form

$$e^{\mathbf{A}t} = \alpha_0(t)\mathbf{I} + \alpha_1(t)\mathbf{A}t + \cdots + \alpha_{n-1}(t)\mathbf{A}^{n-1}t^{n-1} \qquad (115)$$

where the α's are functions of t to be determined for each $\mathbf{A}$.

In the case of a diagonal matrix $\mathbf{A}$, it is easy to verify (115). For example, if $\mathbf{A} = \begin{pmatrix} \lambda_1 & 0 \\ 0 & \lambda_2 \end{pmatrix}$, then it follows for some $\alpha_0(t)$ and $\alpha_1(t)$ that

$$e^{\mathbf{A}t} = \begin{pmatrix} e^{\lambda_1 t} & 0 \\ 0 & e^{\lambda_2 t} \end{pmatrix} = \alpha_0(t)\begin{pmatrix} 1 & 0 \\ 0 & 1 \end{pmatrix} + \alpha_1(t)\begin{pmatrix} \lambda_1 & 0 \\ 0 & \lambda_2 \end{pmatrix}t$$

which leads to the simultaneous equations

$$e^{\lambda_1 t} = \alpha_0(t) + \alpha_1(t)\lambda_1 t$$
$$e^{\lambda_2 t} = \alpha_0(t) + \alpha_1(t)\lambda_2 t \tag{116}$$

From these equations, we now determine

$$\alpha_0(t) = \frac{1}{\lambda_2 - \lambda_1}(\lambda_2 e^{\lambda_1 t} - \lambda_1 e^{\lambda_2 t})$$

$$\alpha_1(t) = \frac{1}{(\lambda_2 - \lambda_1)t}(e^{\lambda_1 t} - e^{\lambda_2 t}) \tag{117}$$

It can be shown that if λ_1 and λ_2 are the *eigenvalues* of $\mathbf{A}$, then Equations (116) and (117) are still valid. More generally, if $\mathbf{A}$ is an $n \times n$ matrix, then each eigenvalue λ_i satisfies a relation like

$$e^{\lambda_i t} = R_n(\lambda_i t) \tag{118}$$

where

$$R_n(\lambda t) = \alpha_0(t) + \alpha_1(t)\lambda t + \cdots + \alpha_{n-1}(t)(\lambda t)^{n-1} \tag{119}$$

However, if λ_i has an eigenvalue of multiplicity k, for $k > 1$, then (118) gives us only one equation instead of k equations. In this case it can be shown that

$$e^{\lambda_i t} = R'_n(\lambda_i t) = R''_n(\lambda_i t) = \cdots = R_n^{(k-1)}(\lambda_i t) \tag{120}$$

which provides us with the remaining equations necessary for the complete determination of the α's. Hence, by finding the n eigenvalues (including multiplicities) of the matrix $\mathbf{A}$, and substituting these sequentially into (118), or possibly (120), we are led to n equations in the unknown α's. We can then solve these equations simultaneously for $\alpha_0(t), \alpha_1(t), \ldots, \alpha_{n-1}(t)$, from which we compute $e^{\mathbf{A}t}$ via Equation (115). Let us illustrate with some examples.

EXAMPLE 27

Find $e^{\mathbf{A}t}$, where

$$\mathbf{A} = \begin{pmatrix} 1 & 2 \\ 4 & 3 \end{pmatrix}$$

Solution

Because $n = 2$, we look for the representation

$$e^{\mathbf{A}t} = \alpha_0(t)\mathbf{I} + \alpha_1(t)\mathbf{A}t$$

The eigenvalues of $\mathbf{A}$ are solutions of $\det(\mathbf{A} - \lambda) = 0$, which leads to

$$\begin{vmatrix} 1 - \lambda & 2 \\ 4 & 3 - \lambda \end{vmatrix} = \lambda^2 - 4\lambda - 5 = 0$$

Factoring, we find $(\lambda + 1)(\lambda - 5) = 0$, from which we obtain

$$\lambda_1 = -1, \qquad \lambda_2 = 5$$

Thus, $\alpha_0(t)$ and $\alpha_1(t)$ satisfy

$$e^{-t} = R_2(-t) = \alpha_0(t) - t\alpha_1(t)$$

$$e^{5t} = R_2(5t) = \alpha_0(t) + 5t\alpha_1(t)$$

the simultaneous solution of which is

$$\alpha_0(t) = \frac{1}{6}(e^{5t} + 5e^{-t})$$

$$\alpha_1(t) = \frac{1}{6t}(e^{5t} - e^{-t})$$

Hence, it follows that

$$e^{\mathbf{A}t} = \frac{1}{6}(e^{5t} + 5e^{-t})\begin{pmatrix} 1 & 0 \\ 0 & 1 \end{pmatrix} + \frac{1}{6t}(e^{5t} - e^{-t})\begin{pmatrix} 1 & 2 \\ 4 & 3 \end{pmatrix}t$$

$$= \frac{1}{6}\begin{pmatrix} 2e^{5t} + 4e^{-t} & 2e^{5t} - 2e^{-t} \\ 4e^{5t} - 4e^{-t} & 4e^{5t} + 2e^{-t} \end{pmatrix}$$

EXAMPLE 28

Find $e^{\mathbf{A}t}$, where

$$\mathbf{A} = \begin{pmatrix} 0 & 1 \\ -4 & -4 \end{pmatrix}$$

Solution

The eigenvalues are solutions of

$$\begin{vmatrix} -\lambda & -4 \\ -4 & -(4 + \lambda) \end{vmatrix} = \lambda^2 + 4\lambda + 4 = 0$$

which yields $\lambda_1 = \lambda_2 = -2$. In this case, we write

$$e^{-2t} = R_2(-2t) = \alpha_0(t) - 2t\alpha_1(t)$$

$$e^{-2t} = R'_2(-2t) = \alpha_1(t)$$

the simultaneous solution of which yields

$$\alpha_0(t) = (1 + 2t)e^{-2t}$$

$$\alpha_1(t) = e^{-2t}$$

Thus, we obtain

$$e^{\mathbf{A}t} = \alpha_0(t)\mathbf{I} + \alpha_1(t)\mathbf{A}t$$

$$= (1 + 2t)e^{-2t}\begin{pmatrix} 1 & 0 \\ 0 & 1 \end{pmatrix} + te^{-2t}\begin{pmatrix} 0 & 1 \\ -4 & -4 \end{pmatrix}$$

or

$$e^{\mathbf{A}t} = \begin{pmatrix} (1 + 2t)e^{-2t} & te^{-2t} \\ -4te^{-2t} & (1 - 2t)e^{-2t} \end{pmatrix}$$

EXAMPLE 29   Use the exponential matrix to solve the IVP

$$\mathbf{Y}' = \begin{pmatrix} 0 & 1 \\ -16 & 0 \end{pmatrix}\mathbf{Y}, \qquad \mathbf{Y}_0 = \begin{pmatrix} 0 \\ 2 \end{pmatrix}$$

Solution

Here we see that

$$\begin{vmatrix} -\lambda & 1 \\ -16 & -\lambda \end{vmatrix} = \lambda^2 + 16 = 0$$

or $\lambda = \pm 4i$. Hence, it follows that

$$e^{4it} = \alpha_0(t) + 4it\alpha_1(t)$$

$$e^{-4it} = \alpha_0(t) - 4it\alpha_1(t)$$

the simultaneous solution of which leads to

$$\alpha_0(t) = \frac{1}{2}(e^{4it} + e^{-4it}) = \cos 4t$$

$$\alpha_1(t) = \frac{1}{8it}(e^{4it} - e^{-4it}) = \frac{1}{4t}\sin 4t$$

In this case the exponential matrix is

$$e^{\mathbf{A}t} = \begin{pmatrix} \cos 4t & \frac{1}{4}\sin 4t \\ -4\sin 4t & \cos 4t \end{pmatrix}$$

and the solution of the IVP is therefore given by

$$\mathbf{Y} = e^{\mathbf{A}t}\mathbf{Y}_0$$

$$= \begin{pmatrix} \cos 4t & \frac{1}{4}\sin 4t \\ -4\sin 4t & \cos 4t \end{pmatrix}\begin{pmatrix} 0 \\ 2 \end{pmatrix}$$

or

$$\mathbf{Y} = \begin{pmatrix} \frac{1}{2}\sin 4t \\ 2\cos 4t \end{pmatrix}$$

Finally, let us consider the nonhomogeneous matrix equation

$$\mathbf{Y}' = \mathbf{A}\mathbf{Y} + \mathbf{F}(t) \tag{121}$$

where $\mathbf{A}$ is an $n \times n$ matrix with constant elements. Since $e^{\mathbf{A}t}$ is a fundamental matrix, that is, $e^{\mathbf{A}t} = \boldsymbol{\phi}(t)$, it follows from (111) that $\boldsymbol{\phi}^{-1}(t) = e^{-\mathbf{A}t}$. Hence, by comparison with Equation (103) in Section 8.9, we deduce that the **general solution** of (121) can also be expressed in the form

$$Y = e^{At}C + e^{At} \int e^{-At}F(t)\, dt \tag{122}$$

Moreover, if the initial condition

$$Y(0) = Y_0 \tag{123}$$

is also prescribed along with (121) then, analogous to Equation (105) in Section 8.9, we have the unique solution

$$Y = e^{At}Y_0 + \int_0^t e^{A(t-\tau)}F(\tau)\, d\tau \tag{124}$$

where we are using the property $e^{At}e^{-A\tau} = e^{A(t-\tau)}$. Equation (124) is particularly simpler to use than Equation (105), or even (122), since (124) does not require the evaluation of an inverse matrix, merely the replacement of t by $t - \tau$ in the matrix exponential function.

EXAMPLE 30

Solve the IVP

$$Y' = \begin{pmatrix} 0 & 1 \\ 8 & -2 \end{pmatrix} Y + \begin{pmatrix} 0 \\ e^t \end{pmatrix}, \qquad Y(0) = \begin{pmatrix} 1 \\ -4 \end{pmatrix}$$

Solution

The eigenvalues of the coefficient matrix A are solutions of

$$\begin{vmatrix} -\lambda & 1 \\ 8 & -(2+\lambda) \end{vmatrix} = \lambda^2 + 2\lambda - 8 = 0$$

from which we deduce $\lambda_1 = 2$ and $\lambda_2 = -4$. Hence, from (117) we see that

$$\alpha_0(t) = \frac{1}{6}(4e^{2t} + 2e^{-4t})$$

$$\alpha_1(t) = \frac{1}{6t}(e^{2t} - e^{-4t})$$

which leads to

$$e^{At} = \frac{1}{6}\begin{pmatrix} 4e^{2t} + 2e^{-4t} & e^{2t} - e^{-4t} \\ 8e^{2t} - 8e^{-4t} & 2e^{2t} + 4e^{-4t} \end{pmatrix}$$

The solution of the IVP that we seek is given by

$$Y = e^{At}Y_0 + \int_0^t e^{A(t-\tau)}F(\tau)\, d\tau$$

where

$$e^{At}Y_0 = \frac{1}{6}\begin{pmatrix} 4e^{2t} + 2e^{-4t} & e^{2t} - e^{-4t} \\ 8e^{2t} - 8e^{-4t} & 2e^{2t} + 4e^{-4t} \end{pmatrix}\begin{pmatrix} 1 \\ -4 \end{pmatrix} = \begin{pmatrix} e^{-4t} \\ -4e^{-4t} \end{pmatrix}$$

and

$$\int_0^t e^{A(t-\tau)}F(\tau)\,d\tau$$

$$= \frac{1}{6}\int_0^t \begin{pmatrix} 4e^{2(t-\tau)} + 2e^{-4(t-\tau)} & e^{2(t-\tau)} - e^{-4(t-\tau)} \\ 8e^{2(t-\tau)} - 8e^{-4(t-\tau)} & 2e^{2(t-\tau)} + 4e^{-4(t-\tau)} \end{pmatrix}\begin{pmatrix} 0 \\ e^t \end{pmatrix}d\tau$$

$$= \frac{1}{6}\int_0^t \begin{pmatrix} e^{2t}e^{-\tau} - e^{-4t}e^{5\tau} \\ 2e^{2t}e^{-\tau} + 4e^{-4t}e^{5\tau} \end{pmatrix}d\tau$$

$$= \frac{1}{6}\begin{pmatrix} e^{2t} + \frac{1}{5}e^{-4t} & -\frac{6}{5}e^t \\ 2e^{2t} - \frac{4}{5}e^{-4t} & -\frac{6}{5}e^t \end{pmatrix}$$

Therefore, combining results, we finally obtain

$$\mathbf{Y} = \begin{pmatrix} e^{-4t} \\ -4e^{-4t} \end{pmatrix} + \frac{1}{6}\begin{pmatrix} e^{2t} + \frac{1}{5}e^{-4t} - \frac{6}{5}e^t \\ 2e^{2t} - \frac{4}{5}e^{-4t} - \frac{6}{5}e^t \end{pmatrix}$$

$$= \begin{pmatrix} \frac{1}{6}e^{2t} + \frac{31}{30}e^{-4t} - \frac{1}{5}e^t \\ \frac{1}{3}e^{2t} - \frac{62}{15}e^{-4t} - \frac{1}{5}e^t \end{pmatrix}$$ ∎

Exercises 8.10

In Problems 1 to 20, formulate the exponential matrix e^{At} and use it to solve the homogeneous system of equations. If initial conditions are prescribed, find the particular solution that satisfies them.

1. $x' = 2x - y$
 $y' = 3x - 2y$

2. $x' = 4x - 3y$
 $y' = 5x - 4y$

3. $x' = 4x - 3y$
 $y' = 8x - 6y$

4. $x' = 3x + 2y$
 $y' = 6x - y$

5. $x' = x - 4y,\ x(0) = 1$
 $y' = x + y,\ y(0) = 0$

6. $x' = 3x - 2y,\ x(0) = 0$
 $y' = 2x + 3y,\ y(0) = 2$

7. $x' = 6x - 5y$
 $y' = x + 2y$

∎8. $x' = 3x + 2y$
 $y' = -5x + y$

9. $x' = 3x - 18y$
 $y' = 2x - 9y$

10. $x' = -2x - 3y$
 $y' = 3x + 4y$

11. $\mathbf{Y}' = \begin{pmatrix} 1 & 3 \\ 1 & -1 \end{pmatrix}\mathbf{Y},\ \mathbf{Y}(0) = \begin{pmatrix} 0 \\ 8 \end{pmatrix}$

12. $\mathbf{Y}' = \begin{pmatrix} 2 & -5 \\ 4 & -2 \end{pmatrix}\mathbf{Y},\ \mathbf{Y}(0) = \begin{pmatrix} 2 \\ 3 \end{pmatrix}$

13. $x' = -x + y$
 $y' = 2x$

14. $x' = x + y + z$
 $y' = 2x + y - z$
 $z' = -y + z$

15. $x' = x + z$
 $y' = x + y$
 $z' = -2x - z$

∎16. $x' = x + y + z$
 $y' = 2x + y - z$
 $z' = -3x + 2y + 4z$

17. $\mathbf{Y}' = \begin{pmatrix} 1 & 2 & 1 \\ 6 & -1 & 0 \\ -1 & -2 & -1 \end{pmatrix} \mathbf{Y}$

18. $\mathbf{Y}' = \begin{pmatrix} 5 & 2 & -2 \\ 7 & 0 & -2 \\ 11 & 1 & -1 \end{pmatrix} \mathbf{Y}$

19. $\mathbf{Y}' = \begin{pmatrix} 0 & 2 & 2 \\ 2 & 0 & 2 \\ 2 & 2 & 0 \end{pmatrix} \mathbf{Y}$

*20. $\mathbf{Y}' = \begin{pmatrix} 3 & 1 & 0 \\ -1 & 0 & -1 \\ 1 & 2 & 3 \end{pmatrix} \mathbf{Y}$

*21. Let $\mathbf{P}$ denote a matrix whose columns are the eigenvectors $\mathbf{K}_1, \mathbf{K}_2, \ldots, \mathbf{K}_n$ associated with the *distinct* eigenvalues $\lambda_1, \lambda_2, \ldots, \lambda_n$ belonging to the $n \times n$ matrix $\mathbf{A}$. Assuming that

$$\mathbf{A} = \mathbf{P}\mathbf{D}\mathbf{P}^{-1}$$

where $\mathbf{D}$ is a diagonal matrix whose elements down the main diagonal are the eigenvalues $\lambda_1, \lambda_2, \ldots, \lambda_n$, show that

$$e^{\mathbf{A}t} = \mathbf{P}e^{\mathbf{D}t}\mathbf{P}^{-1}$$

*22. Use the result of Problem 21 to find $e^{\mathbf{A}t}$, where $\mathbf{A}$ is the 2×2 matrix in Example 27.

In Problems 23 to 32, evaluate the exponential matrix $e^{\mathbf{A}t}$ and use it to solve the nonhomogeneous system of equations. If an initial condition is prescribed, find the particular solution that satisfies it.

23. $x' = 2x - 5y - \sin 2t$
$y' = x - 2y + t$

24. $x' = 2x - y + t$
$y' = 3x - 2y + 2t$

25. $x' = 2x + y + e^t$
$y' = 4x - y - e^t$

26. $x' = 4x - 2y + e^t$
$y' = 5x + 2y - t$

27. $\mathbf{Y}' = \begin{pmatrix} 1 & 2 \\ -\frac{1}{2} & 1 \end{pmatrix} \mathbf{Y} + \begin{pmatrix} 0 \\ e^t \tan t \end{pmatrix}$

28. $\mathbf{Y}' = \begin{pmatrix} 2 & -5 \\ 1 & -2 \end{pmatrix} \mathbf{Y} + \begin{pmatrix} \csc t \\ \sec t \end{pmatrix}$

29. $\mathbf{Y}' = \begin{pmatrix} -10 & 6 \\ -12 & 7 \end{pmatrix} \mathbf{Y} + \begin{pmatrix} 10e^{-3t} \\ 18e^{-3t} \end{pmatrix}, \ \mathbf{Y}(0) = \begin{pmatrix} -1 \\ 2 \end{pmatrix}$

■30. $\mathbf{Y}' = \begin{pmatrix} 1 & -1 \\ 1 & 1 \end{pmatrix} \mathbf{Y} + \begin{pmatrix} 3e^t \\ 3e^t \end{pmatrix}, \ \mathbf{Y}(0) = \begin{pmatrix} 3 \\ 1 \end{pmatrix}$

31. $\mathbf{Y}' = \begin{pmatrix} 4 & 2 \\ 3 & -1 \end{pmatrix} \mathbf{Y} - \begin{pmatrix} 0 \\ 1 \end{pmatrix} e^{-2t}, \ \mathbf{Y}(0) = \begin{pmatrix} 0 \\ 0 \end{pmatrix}$

32. $\mathbf{Y}' = \begin{pmatrix} 4 & 2 \\ 3 & -1 \end{pmatrix} \mathbf{Y} - \begin{pmatrix} 15 \\ 4 \end{pmatrix} te^{-2t}, \ \mathbf{Y}(0) = \begin{pmatrix} 1 \\ -1 \end{pmatrix}$

8.11 Chapter Summary

In this chapter we have presented several techniques for solving **linear systems** of DEs. The most basic method for solving such equations is the **operator method,** which can be further systematized by the use of determinants. In theory it can be applied to any number of equations, each of arbitrary order in the unknown functions, but in practice it is cumbersome to use if the number of unknowns exceeds two or three.

Another technique that is fairly basic is the method of **Laplace transforms,** which may be easier than the operator method to use when initial conditions are prescribed. The Laplace transform applied to the system of DEs reduces it to a system of algebraic equations, which can be solved by elementary methods. The solution of the original system of DEs is then obtained by applying the inverse Laplace transform to the solution of the algebraic equations. The basic properties of the Laplace transform are discussed in Chapter 6.

The terminology associated with systems of DEs is the same as that for linear DEs in general. For instance, the first-order system

$$x' = a_{11}(t)x + a_{12}(t)y + f_1(t)$$
$$y' = a_{21}(t)x + a_{22}(t)y + f_2(t) \tag{125}$$

is called **homogeneous** if $f_1(t) \equiv f_2(t) \equiv 0$ on some interval I; otherwise it is called **nonhomogeneous.** Every second-order linear DE can be expressed in this form.

The use of **matrix methods** often facilitates the solution process, particularly when the number of equations is large. In matrix notation the system described by (125), for example, assumes the form

$$\mathbf{Y}' = \mathbf{A}(t)\mathbf{Y} + \mathbf{F}(t) \tag{126}$$

where

$$\mathbf{Y} = \begin{pmatrix} x(t) \\ y(t) \end{pmatrix}, \quad \mathbf{A}(t) = \begin{pmatrix} a_{11}(t) & a_{12}(t) \\ a_{21}(t) & a_{22}(t) \end{pmatrix}, \quad \text{and} \quad \mathbf{F}(t) = \begin{pmatrix} f_1(t) \\ f_2(t) \end{pmatrix}$$

If, on an interval I, $\mathbf{Y}^{(1)}(t)$ and $\mathbf{Y}^{(2)}(t)$ are linearly independent solution vectors of the associated **homogeneous system**

$$\mathbf{Y}' = \mathbf{A}(t)\mathbf{Y} \tag{127}$$

we say they form a **fundamental solution set** on I and write the **general solution** of (127) as

$$\mathbf{Y}_H = C_1 \mathbf{Y}^{(1)}(t) + C_2 \mathbf{Y}^{(2)}(t) \tag{128}$$

By using the solution vectors $\mathbf{Y}^{(1)}(t)$ and $\mathbf{Y}^{(2)}(t)$ as columns of a 2×2 matrix, we obtain the **fundamental matrix**

$$\phi(t) = \begin{pmatrix} x_1(t) & x_2(t) \\ y_1(t) & y_2(t) \end{pmatrix} \tag{129}$$

In terms of the fundamental matrix, we can express the general solution of (127) in the equivalent form

$$\mathbf{Y}_H = \phi(t)\mathbf{C} \tag{130}$$

where $\mathbf{C} = \begin{pmatrix} C_1 \\ C_2 \end{pmatrix}$ is an arbitrary constant matrix. To find the fundamental solution vectors in the special case when the coefficient matrix $\mathbf{A}$ is a **constant matrix,** we need to first find the **eigenvalues** of $\mathbf{A}$ and the corresponding **eigenvectors.**

In solving the **nonhomogeneous system** (126), we must find a **particular solution** vector $\mathbf{Y}_P$ and combine it with the homogeneous solution $\mathbf{Y}_H$ to obtain the **general solution**

$$\mathbf{Y} = \mathbf{Y}_H + \mathbf{Y}_P \tag{131}$$

The particular solution $\mathbf{Y}_P$ can always be constructed by the method of variation of parameters, which leads to

$$\mathbf{Y}_P = \phi(t) \int \phi^{-1}(t)\mathbf{F}(t) \, dt \tag{132}$$

where $\phi^{-1}(t)$ denotes a **matrix inverse.**

Review Exercises

In Problems 1 to 4, use the operator method to solve the given system of equations.

1. $x' = 2x - 3y$, $x(0) = 8$
$y' = -2x + y$, $y(0) = 3$

2. $x' = -y + t$, $x(0) = 1$
$y' = -4x$, $y(0) = -1$

3. $x' = 4x + 2y - 8t$, $x(0) = 2/25$
$y' = 3x - y + 2t + 3$, $y(0) = 1/25$

4. $x' = y$, $x(\pi) = 1$
$y' = -x + 3$, $y(\pi) = 2$

In Problems 5 to 8, use the method of Laplace transforms to solve the given problem.

5. Problem 1

6. Problem 2

7. Problem 3

8. Problem 4

In Problems 9 to 12, write the given IVP in matrix notation.

9. Problem 1

10. Problem 2

11. Problem 3

12. Problem 4

In Problems 13 to 16, find the eigenvalues of the given matrix **A**.

13. $\mathbf{A} = \begin{pmatrix} 0 & 1 \\ -64 & -20 \end{pmatrix}$

14. $\mathbf{A} = \begin{pmatrix} 0 & 1 \\ -1 & 0 \end{pmatrix}$

15. $\mathbf{A} = \begin{pmatrix} 0 & 1 \\ 8 & -2 \end{pmatrix}$

16. $\mathbf{A} = \begin{pmatrix} 0 & 1 \\ -25 & -8 \end{pmatrix}$

In Problems 17 to 20, use matrix methods to solve the given problem.

17. Problem 1

18. Problem 2

19. Problem 3

20. Problem 4

CHAPTER 9
Nonlinear Systems and Stability

In this chapter we provide a brief treatment of **autonomous** systems of nonlinear equations. Since exact solutions of nonlinear systems are often not obtainable, we concentrate on the **qualitative** aspects of these solutions. The general questions of concern are mainly associated with the idea of stability of equilibrium points of the system.

In Section 9.2 we introduce the notions of **critical point** (or **equilibrium point**) and **stability** by studying a single DE. These ideas are carried over to systems of two equations along with the notion of a **phase plane** in Section 9.3. Certain curves in the phase plane, known as **trajectories,** provide much information about the general behavior of the solutions, particularly in the neighborhood of a critical point. If the solution eventually reaches a critical point along some trajectory, we say that point is **asymptotically stable.** On the other hand, if it continues to move away from the critical point, it is called **unstable.**

The **stability of linear systems** with constant coefficients is discussed in detail in Section 9.4. These results are then related in Section 9.5 to the stability of certain nonlinear systems, called **perturbed linear systems,** whose stability in the vicinity of critical points is compared with that of the related linear system. The **nonlinear pendulum problem** and **predator-prey problem** are discussed in this context.

9.1 Introduction

A DE provides the basic model from which a mathematical analyst pursues investigations into the existence, characterization, and construction of solutions to physical problems. Once we have the solution of the DE it is usually rather easy to answer questions concerning basic characteristics of the system under study, such as its oscillatory nature and stability. But finding the solution of the DE may be either difficult or impossible. For this reason it is necessary to seek techniques, called **qualitative methods,** for describing the fundamental nature of the solution without explicitly solving the DE. For example, we used qualitative methods in Section 5.6 in studying the *oscillatory characteristics* of the solutions of certain second-order *linear* equations. We now wish to use qualitative methods once again in studying systems of *nonlinear* equations, but this time with regard primarily to the behavior of the solutions of such systems as time $t \to \infty$. Throughout our discussion we consider only **autonomous** systems, that is, those DEs for which the independent variable t does not explicitly appear. The treatment of **nonautonomous** systems is more difficult, and we will not pursue it.

9.2 Systems of One Equation

To provide a brief introduction to the idea of stability, let us first consider a system described by a *single* DE of first order having the general form

$$y' = F(y) \tag{1}$$

where $y' = dy/dt$. Such DEs play an important role in many applications, some of which were featured in Chapter 3.

Points for which $F(y) = 0$ are called **critical points** of (1) (also called **equilibrium points**). If y_0 is a critical point, it follows that $y = y_0$ is then a solution of the DE, called an **equilibrium solution** since, if the function y starts at the value y_0, it remains at this value. The equilibrium solution for a mechanical system, for example, may be one in which the velocity y' is zero. On the other hand, if y denotes a population size, then the critical point might represent that size for which the population is in "equilibrium" with its environment. If the solution function starts at some point other than y_0 and is "well-behaved," it may approach the critical point as $t \to \infty$. Let us illustrate these concepts by means of a specific example.

EXAMPLE 1

Find the critical points of

$$y' = 6y - 2y^2, \qquad y(0) = \beta$$

and discuss the nature of the solution when β is "near" a critical point.

Solution

By setting

$$6y - 2y^2 = 2y(3 - y) = 0$$

we see that the critical points are $y = 0$ and $y = 3$.

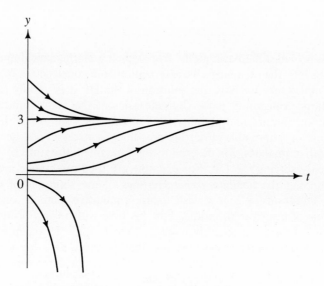

Figure 9.1 Solution curves of $y' = 6y - 2y^2$.

To discuss the nature of the critical points, we solve the DE by separating variables (see Section 2.2) to find

$$y(t) = \frac{3\beta}{\beta + (3 - \beta)e^{-6t}}$$

Observe that the equilibrium solutions $y = 0$ (if $\beta = 0$) and $y = 3$ (if $\beta = 3$) are both included in this solution formula.

The two critical points divide the y axis into three separate regions: $y < 0$, $0 < y < 3$, $y > 3$. First, let us suppose that β is in one of the two regions $0 < \beta < 3$ or $\beta > 3$. It is easy to see in this case that

$$\lim_{t \to \infty} y(t) = 3$$

Thus, if β is a positive value (including $\beta = 3$), the solution will eventually approach the equilibrium solution $y = 3$. On the other hand, if $\beta < 0$, there is a critical time $t_0 > 0$, given by

$$t_0 = \frac{1}{6} \log\left(1 - \frac{3}{\beta}\right)$$

for which

$$\lim_{t \to t_0} y(t) = -\infty, \qquad (\beta < 0)$$

the verification of which we leave to the exercises (see Problem 7 in Exercises 9.2). Typical solution curves illustrating each of these cases are shown in Figure 9.1. ∎

If y denotes a population size in thousands in Example 1, then we can interpret the results as saying that the population $y(t)$ will eventually approach an equilibrium size of 3000 so long as we start with at least one member of the population. This is probably not very realistic for most types of populations. There are other factors to consider which, under the right conditions, might lead to *extinction*. For example, a variation of the *logistic equation* $y' = ay - by^2$ [recall Equation (30) in Section 3.3], such as

$$y' = ay - by^2 - H \tag{2}$$

might describe population growth with constant *harvesting* rate H. An example of this is to consider a population of rabbits on an isolated island where exactly H of them are killed by hunters each year. If β rabbits are placed on the island at some time, say $t = 0$, we may be interested in whether enough rabbits will be available for the hunters each year. If $H < a^2/4b$ and $\beta > H$, it can be shown that there will always be at least H rabbits. Even if $\beta < H$, there may be enough rabbits for all the hunters. However, if $H > a^2/4b$, then the rabbits will eventually become *extinct* at some time. In Problems 9 to 12 in Exercises 9.2 we examine some of these situations.

What we are discussing here is the basic concept of *stability*. If y_0 is a critical point of $y' = F(y)$, we say that y_0 is **stable** if all solutions that start "near" y_0 remain "near" y_0 for all future time; otherwise, we say that y_0 is **unstable**. More precisely, we have the following definition.

Definition 9.1 If y_0 is a critical point of the autonomous first-order IVP

$$y' = F(y), \qquad y(0) = \beta$$

we say that y_0 is **stable** provided that, given $\epsilon > 0$, there exists a $\delta > 0$ such that

$$|y_0 - \beta| < \delta \quad \text{implies} \quad |y(t) - \beta| < \epsilon$$

for all $t > 0$. If y_0 is not stable, we say it is **unstable.**

An easy way to determine the stability of a critical point y_0 is to check the sign of $F(y)$ for $y < y_0$ and $y > y_0$. Points for which $y' = F(y) > 0$ indicate the solution is increasing while points for which $y' = F(y) < 0$ indicate the solution is decreasing. If $y_1 < y_0$ and $F(y_1) > 0$, and if $y_2 > y_0$ and $F(y_2) < 0$, then the solution is always moving toward y_0. In this case we conclude that y_0 is a *stable* critical point. However, if $F(y_1) < 0$ or $F(y_2) > 0$, the critical point is *unstable*. For instance, based on this sign analysis, we see in Example 1 that $F(2) = 4 > 0$ and $F(4) = -8 < 0$, showing that solutions near the critical point $y = 3$ move toward it. Therefore, $y = 3$ is a stable critical point. On the other hand, $F(-1) = -8 < 0$ and $F(1) = 4 > 0$, showing that solutions move away from $y = 0$. Hence, $y = 0$ is an unstable critical point.

Exercises 9.2

In Problems 1 to 6, solve explicitly for y assuming the initial condition $y(0) = \beta$. Determine all critical points and discuss their stability.

1. $y' = -2y$

2. $y' = 2y - 1$

3. $y' = y - y^2$

4. $y' = y^2 - y$

5. $y' = 4y - y^2 - 3$

6. $y' = y^2 - 5y + 6$

7. Show that when $\beta > 0$, the solution in Example 1 satisfies

$$\lim_{t \to t_0} y(t) = -\infty$$

where $t_0 = \frac{1}{6}\log(1 - 3/\beta)$.

8. Given the DE

$$y' = ay - y^3$$

(a) Show that $y = 0$ is a stable critical point if $a \le 0$.

(b) If $a > 0$, show that $y = 0$ is unstable.

(c) If $a = k^2 > 0$, show that $y = \pm k$ are both stable critical points.

In Problems 9 to 12, consider the IVP

$$y' = ay - by^2 - H, \qquad y(0) = \beta$$

where $a \ge 0$, $b \ge 0$, and $H \ge 0$.

9. Determine $\lim_{t \to \infty} y(t)$ if

(a) $H = 0$

(b) $H = a^2/4b$

10. When $H < a^2/4b$, show that

$$\lim_{t \to \infty} y(t) = \frac{a + \sqrt{a^2 - 4bH}}{2b}$$

11. When $H > a^2/4b$, show that the harvesting rate H is too large and extinction occurs.

■**12.** Let $a = 4$, $b = 1$, $\beta = 2$, and $H = 5$.

(a) Determine at what time extinction occurs.

(b) What is the largest value of H for which extinction will not occur?

9.3 Systems of Two Equations and the Phase Plane

The general nonlinear systems that concern us are characterized by *autonomous* systems of two equations of the form

$$x' = P(x, y)$$
$$y' = Q(x, y) \tag{3}$$

where $P(x, y)$ and $Q(x, y)$ are continuous functions and have continuous first partial derivatives in some domain of the xy plane. Under such conditions, it can be shown that for a given set of values (α, β) within this domain, there exists a unique solution set $\{x(t), y(t)\}$ on some interval $a \le t \le b$ satisfying the initial conditions

$$x(t_0) = \alpha, \qquad y(t_0) = \beta \tag{4}$$

where $a < t_0 < b$. We can think of the solution functions $x = x(t)$ and $y = y(t)$ as **parametric equations** of an arc in the xy plane that passes through the point (α, β). From this viewpoint, it is customary to refer to the xy plane as the **phase plane.** Any arc described parametrically by a solution of (3) is called a **trajectory,** or sometimes a **path** or an **orbit,** with the positive direction along the trajectory defined by the direction of increasing t.

In principle the trajectories of a system can always be found by eliminating the parameter t between the solution functions $x = x(t)$ and $y = y(t)$. Even if the solution set $\{x(t), y(t)\}$ is not known, which is the usual situation in practice, the trajectories may still be found! In fact, it turns out that determining the trajectories of a nonlinear system is generally much easier to do than finding the solution of the system.

To do so, we start by forming the ratio

$$\frac{y'}{x'} = \frac{Q(x, y)}{P(x, y)}$$

which simplifies to the first-order DE

$$\frac{dy}{dx} = \frac{Q(x, y)}{P(x, y)} \qquad (5)$$

The trajectories are then simply the solutions of (5). Let us illustrate both techniques of producing the trajectories.

EXAMPLE 2

Determine the trajectories associated with the second-order DE

$$mx'' + kx = 0, \qquad m, k > 0$$

which governs the free oscillations of a spring-mass system (see Section 5.2).

Solution

Although the DE is linear, we can use it to illustrate the general technique. By setting $x' = y$, we can rewrite the DE as the first-order system of equations.

$$x' = y$$

$$y' = -\frac{kx}{m}$$

Next, using techniques from Chapter 8, it can be shown that the solution is

$$x(t) = C_1 \cos(\sqrt{k/m}t) + C_2 \sin(\sqrt{k/m}t)$$

$$= A \cos(\sqrt{k/m}t - \phi)$$

$$y(t) = -\sqrt{k/m} \, A \sin(\sqrt{k/m}t - \phi)$$

where $A = \sqrt{C_1^2 + C_2^2}$ and $\phi = \tan^{-1}(C_2/C_1)$. To eliminate the parameter t, we note that

$$\frac{k}{m} x^2(t) + y^2(t) = \frac{A^2 k}{m} \left[\cos^2(\sqrt{k/m}t) + \sin^2(\sqrt{k/m}t) \right]$$

$$= \frac{A^2 k}{m}$$

which can be expressed in the form

$$kx^2 + my^2 = C, \qquad C > 0$$

where C is an arbitrary constant.[†] By allowing C to assume various numerical values, we generate the one-parameter family of ellipses shown in Figure 9.2. Each trajectory

---

[†] The equation $kx^2 + my^2 = C$ is actually a statement of the *conservation of energy principle*, namely, the sum of potential and kinetic energies is constant.

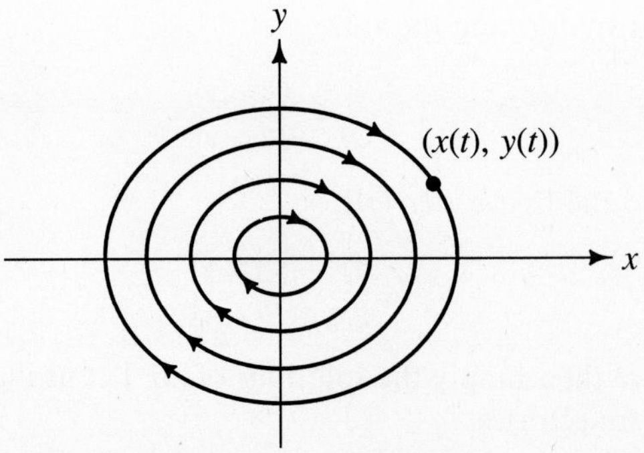

Figure 9.2 Trajectories of a harmonic oscillator.

represents a possible motion of the system (depending upon the prescribed initial conditions), and each point on a given trajectory represents an instantaneous state of the system. The direction of the arrows in Figure 9.2 suggests that the representative point moves clockwise along a given trajectory. This is so since $y = x'$, and $y > 0$ implies that $x(t)$ is increasing and $y < 0$ implies that $x(t)$ is decreasing.

By using Equation (5) to determine the trajectories, we form the ratio y'/x', which leads to

$$\frac{dy}{dx} = -\frac{kx}{my}$$

The general solution of this DE can be found through separation of variables. Thus,

$$kx\,dx + my\,dy = 0$$

the integration of which yields the same equation as above, namely,

$$kx^2 + my^2 = C \qquad \blacksquare$$

■────────────
EXAMPLE 3

Find the trajectories associated with the system

$$x' = x + 2y + x\cos y$$

$$y' = -y - \sin y$$

Solution

This time the system of equations is nonlinear. Dividing the second equation by the first, we obtain the first-order DE

$$\frac{dy}{dx} = \frac{-y - \sin y}{x + 2y + x\cos y}$$

which can also be written as

$$(y + \sin y)\,dx + (x + 2y + x\cos y)\,dy = 0$$

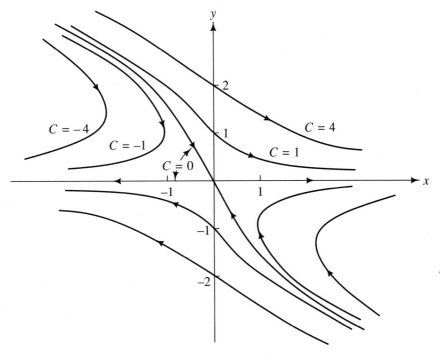

Figure 9.3 Trajectories described by $xy + y^2 + x \sin y = C$.

Checking, we see that $\partial M/\partial y = 1 + \cos y = \partial N/\partial x$, which proves that the DE is *exact*. Hence, we can use the method of Section 2.3 to find the family of solutions

$$xy + y^2 + x \sin y = C$$

where C is an arbitrary constant. Typical configurations of the trajectories defined by this family of solutions are shown in Figure 9.3 for various values of C. ∎

> **Remark.** Since the parameterization of a curve is not unique, the terms "solution" and "trajectory" are not synonymous. For example, $x = e^t$, $y = e^{2t}$ constitutes a solution set of $x' = \sqrt{y}$, $y' = 2x^2$ and forms the trajectory $y = x^2$, $x > 0$. However, $x = t$, $y = t^2$ is another parameterization of the trajectory $y = x^2$, $x > 0$, but clearly does not represent a solution of the differential system.

Knowledge of the trajectories gives us valuable information about the system under study. For example, let $x(t)$ and $y(t)$ denote the respective populations of two species competing for the same limited resources and suppose that a trajectory corresponding to a specific set of initial conditions is that shown in Figure 9.4. By studying this trajectory, we see that when $x(t) = x_1$, an increase in the population $x(t)$ causes a decrease in population $y(t)$ until it reaches y_1. At this time both populations increase

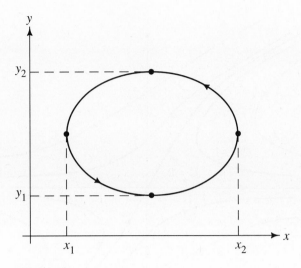

Figure 9.4

until $x(t)$ reaches its maximum value x_2. As time continues further, the second population $y(t)$ will continue to increase, while $x(t)$ decreases, until $y(t)$ reaches its maximum population y_2. Then both populations begin to decrease and the process repeats itself; that is, it is *periodic*. Notice that all this information is available directly from the trajectory without any explicit solution.

9.3.1 Critical Points and Stability

Points in the xy plane at which

$$P(x, y) = 0, \qquad Q(x, y) = 0 \tag{6}$$

are called **critical points** of the system (3). If (x_0, y_0) is a critical point of (3), then $x = x_0$, $y = y_0$ is a solution of the system, called an **equilibrium solution.** Thus, a critical point may also be called an **equilibrium point.** Note, however, that the trajectory of an equilibrium solution is simply a *single point*.

In problems in mechanics the functions $x(t)$ and $y(t)$ usually denote the position and velocity, respectively, of a particle in motion. A critical point (x_0, y_0), or equilibrium solution, represents a point where both velocity and acceleration vanish, and hence the motion *stops*. Alternatively, the two functions $x(t)$ and $y(t)$ could represent the populations of two species of animals living in the same environment and competing for the same food supply (or one may prey on the other). In the latter case a critical point (x_0, y_0) specifies the constant populations of these two competing species that can coexist in the same environment. Other interpretations of $x(t)$ and $y(t)$, and their relation to critical points, are also possible.

EXAMPLE 4

Determine the critical points of

$$x' = 84x - 3x^2 - 2xy$$
$$y' = 120y - 3y^2 - 4xy$$

Solution

By setting

$$84x - 3x^2 - 2xy = x(84 - 3x - 2y) = 0$$

$$120y - 3y^2 - 4xy = y(120 - 3y - 4x) = 0$$

and solving simultaneously, we see that either

$$x = 0 \quad \text{or} \quad 84 - 3x - 2y = 0$$

and either

$$y = 0 \quad \text{or} \quad 120 - 3y - 4x = 0$$

Clearly $x = y = 0$ is one solution. Now suppose that $x = 0$ and $y \neq 0$. In this case we find that the last equation reduces to $120 - 3y = 0$, or $y = 40$. Next, if $y = 0$ and $x \neq 0$, the second equation yields $84 - 3x = 0$, or $x = 28$. Finally, if $x \neq 0$ and $y \neq 0$, we must solve the simultaneous equations

$$3x + 2y = 84$$

$$4x + 3y = 120$$

leading to the solution $x = 12$, $y = 24$. Hence, this system has four critical points given by

$$(0, 0), \quad (0, 40), \quad (28, 0), \quad \text{and} \quad (12, 24)$$

If we interpret $x(t)$ as representing a population of "large" fish in a lake that feed upon the "small" fish in the lake, and $y(t)$ denotes a population of fishermen who fish for the small fish, then the first three critical points above all involve either the large fish or the fishermen missing from the lake. That is, $(0, 0)$ corresponds to zero large fish and no fishermen, $(0, 40)$ corresponds to zero large fish and 40 fishermen, and $(28, 0)$ corresponds to 28 large fish and no fishermen. The only possibility for coexistence of constant *nonzero* populations of both large fish and fishermen is represented by the critical point $(12, 24)$, which corresponds to 12 large fish and 24 fishermen. ■

The next example illustrates that critical points need not be isolated.

EXAMPLE 5

Determine the critical points of

$$x' = 3y(y + 2x)$$

$$y' = (1 - x)(y + 2x)$$

Solution

By considering the simultaneous equations

$$3y(y + 2x) = 0$$

$$(1 - x)(y + 2x) = 0$$

we see that $(1, 0)$ is a critical point, as is *every* point on the line $y + 2x = 0$. ■

If the solution set $\{x(t), y(t)\}$ has a finite limit as t tends to infinity, that limit point must be a critical point. More formally, we have the following theorem.

Theorem 9.1 *If* $\{x(t), y(t)\}$ *is a solution set of the system of equations*

$$x' = P(x, y)$$

$$y' = Q(x, y)$$

and

$$\lim_{t \to \infty} x(t) = x_0$$

$$\lim_{t \to \infty} y(t) = y_0$$

where x_0 *and* y_0 *are finite, then* (x_0, y_0) *is a critical point of the system.*

Once the critical points have been determined and the trajectories sketched, it is easy to determine the general behavior of the solutions of the system under study. For example, we can see if certain solutions remain bounded, and further, if those solutions continue to move in a certain direction as $t \to \infty$.

Because critical points in problems in mechanics (or in population growth) correspond to points where the motion stops (or sizes where the population is in equilibrium with its environment), it is natural to ask what happens to a particle (or population) that is slightly displaced from a critical point. Specifically, we want to know if the particle (or population)

1. Returns to the critical point
2. Moves away from the critical point
3. Moves about the critical point but does not approach it

If a particle slightly displaced from its critical point returns to the critical point, we say the critical point is **asymptotically stable.** For example, a pendulum bob that hangs vertically downward will return to this position after a small displacement if there is a damping force such as air resistance (see Figure 9.5a). On the other hand, if the pendulum bob is in a vertical upward position, then any small movement will cause it to move away from this position (see Figure 9.5b); in this case we say the equilibrium point is **unstable.** Finally, if there is no resistive force, then the pendulum bob slightly displaced from a vertically downward position will move back and forth forever, never again coming to rest (see Figure 9.5c). In this last situation the critical point is said to be **stable,** but not asymptotically stable.

As we have seen, critical points play a significant role in the general analysis of nonlinear systems. In particular, it can be shown that the global behavior of trajectories in the phase plane is closely tied to the location of the critical points and the local behavior of the trajectories near them. Without proof, we offer the following statements about trajectories.

1. There exists at most one trajectory passing through any point of the phase plane that is not a critical point.

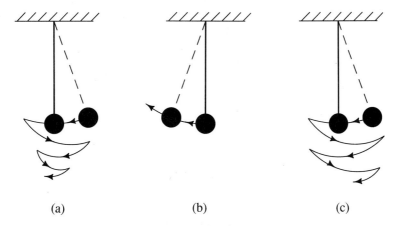

(a)   (b)   (c)

Figure 9.5 Stability of a pendulum bob: (a) with damping forces, (b) with or without damping forces, and (c) no damping forces.

2. A particle starting at a point other than a critical point cannot reach a critical point in a finite amount of time (if indeed it reaches one at all). If a critical point is unstable, the particle will move away from it, no matter how close to the critical point the particle starts.

3. If a trajectory crosses itself at a point of the phase plane that is not a critical point, that trajectory is a closed path and corresponds to a periodic solution of the system (see Examples 2 and 6).

EXAMPLE 6

In Example 2 we discussed the trajectories associated with the free oscillations of a spring-mass system whose governing equation is the linear DE $mx'' + kx = 0$. In some cases, however, the restoring force $-kx$ in the spring may not be linear. Assuming this force is proportional to the cube of the distance from equilibrium, the equation of motion becomes

$$mx'' + kx^3 = 0, \qquad k > 0$$

Determine the trajectories and discuss the possible motions of the mass.

Solution

Setting $y = x'$ we obtain the system of equations

$$x' = y$$

$$y' = -\frac{kx^3}{m}$$

Clearly, the origin $(0, 0)$ is the only critical point and the trajectories are found by solving

$$\frac{dy}{dx} = -\frac{kx^3}{my}$$

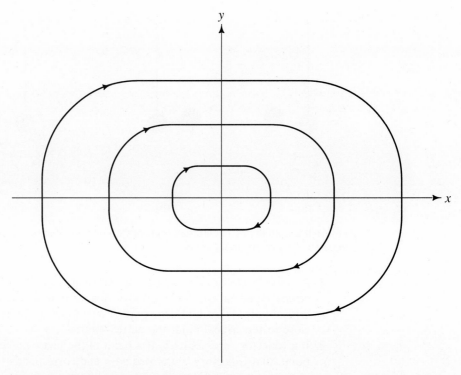

Figure 9.6 Trajectories of $mx'' + kx^3 = 0$.

Through separation of variables, it follows that

$$kx^4 + 2my^2 = C, \qquad C > 0$$

Some representative trajectories are sketched in Figure 9.6, all of which are closed paths about the critical point $(0, 0)$. Hence, all motions of this particular nonlinear spring-mass system are *periodic* and the critical point $(0, 0)$ is stable, but not asymptotically stable. ∎

Exercises 9.3

In Problems 1 to 10, find the one-parameter family of trajectories and determine all critical points.

1. $x' = y$
$y' = x + 1$

2. $x' = 2x - 2y$
$y' = 4x - 2y + 5$

3. $x' = 3x + y$
$y' = x + 3y$

4. $x' = -2x^3y$
$y' = x^4 + y^4$

5. $x' = x - y^2$
$y' = x^2 - y$

■**6.** $x' = 2xy - \sin y$
$y' = 1 - y^2$

7. $x' = y - xy$
$\quad y' = x^2 - x$

8. $x' = 3(y + 2)(y - x^2)$
$\quad y' = -x(y - x^2)$

9. $x' = -x(y + 4)$
$\quad y' = 2x^3(y + 4)$

*10. $x' = 3x^2y^2 - 3x + 4y$
$\quad y' = y(3 - 2xy^2)$

In Problems 11 to 14, sketch the trajectory corresponding to the solution satisfying the prescribed initial conditions, and indicate the positive direction of motion.

11. $x' = -y$, $x(0) = 3$
$\quad y' = x$, $y(0) = 0$

12. $x' = y$, $x(0) = 1$
$\quad y' = x$, $y(0) = 0$

13. $x' = y$, $x(0) = -1$
$\quad y' = x$, $y(0) = 0$

■14. $x' = 2x + 4y$, $x(0) = 4$
$\quad y' = -2x + 6y$, $y(0) = 0$

15. Consider the linear autonomous system

$$x' = x$$

$$y' = x + y$$

(a) Find the particular solution satisfying the initial conditions $x(0) = 1$, $y(0) = 3$.
(b) Repeat (a) for $x(4) = e$, $y(4) = 4e$.
(c) Show that both (a) and (b) have the same trajectory.

16. Sketch a phase plot for the IVP

$$x' = y, \qquad x(0) = \alpha$$

$$y' = 2xy, \qquad y(0) = \beta$$

(a) Determine $\lim_{t \to \infty} x(t)$ and $\lim_{t \to \infty} y(t)$ when $\alpha = 2$ and $\beta = -1$.
(b) Repeat (a) when $\alpha = -2$ and $\beta = 3$.
(c) Determine all values of α and β such that

$$\lim_{t \to \infty} x(t) = \lim_{t \to \infty} y(t) = 0$$

*17. Find the trajectories for the **nonautonomous** system

$$x' = tx, \qquad x(t_0) = 1$$

$$y' = -y, \qquad y(t_0) = 1$$

and graph them for

(a) $t_0 = 0$
(b) $t_0 = 1$
(c) $t_0 = 2$

9.4 Stability of Linear Systems

To determine if a critical point of a nonlinear system is stable or not, it is sometimes useful to examine a similar *linear* system for which the exact solution can be found. By using the solution of the linear system, the stability of the critical point of the linear system can be established and, in many cases, the stability of the same critical point of the related nonlinear system is similar. To recall the notation and language that is used below in our discussion of linear systems, the reader is advised to review Section 8.8.

The general first-order linear system with *constant coefficients* and two unknowns is given by

$$x' = ax + by + h_1, \qquad ad - bc \neq 0 \tag{7}$$

$$y' = cx + dy + h_2,$$

where a, b, c, d, h_1, and h_2 are constants.[†] This system has only one critical point, which we designate by (x_0, y_0). When $h_1 = h_2 = 0$, we obtain the special system

$$x' = ax + by$$
$$y' = cx + dy$$

(8)

for which the critical point is $(0, 0)$. Note, however, that if we make the simple transformation $X = x - x_0$, $Y = y - y_0$, then (7) is reduced to a linear system of the form (8). Hence, there is no real loss of generality by considering only systems of the form (8).

> **Remark.** The requirement that $ad - bc \neq 0$ is to ensure that only isolated critical points occur. For example, if $ad - bc \neq 0$, all points on the line $ax + by = 0$ (or the line $cx + dy = 0$) are critical points of (8).

The *characteristic equation* associated with (8) is (see Section 8.8.1)

$$\lambda^2 - (a + d)\lambda + (ad - bc) = 0$$

(9)

with roots λ_1 and λ_2. Observe that $\lambda = 0$ cannot be a root since $ad - bc \neq 0$. Also, based on our discussion in Section 8.8.1, we know there are three possible cases of solution of (8), depending on the nature of the roots λ_1 and λ_2.

Case I. Real and Distinct Roots

When $\lambda_1 \neq \lambda_2$, and both are real, the solution vectors of (8) take the form $\mathbf{Y}^{(1)} = \mathbf{K}^{(1)}e^{\lambda_1 t}$ and $\mathbf{Y}^{(2)} = \mathbf{K}^{(2)}e^{\lambda_2 t}$. Notice that if either λ_1 or λ_2 is positive, then one or both of these solution vectors becomes infinite as $t \to \infty$. Hence, in this case we say the critical point $(0, 0)$ is *unstable*. On the other hand, if both λ_1 and λ_2 are negative, then the critical point $(0, 0)$ is *asymptotically stable* since both solution vectors approach this point as $t \to \infty$.

EXAMPLE 7

Discuss the stability of the linear system

$$x' = -x$$
$$y' = -2y$$

Solution

The characteristic equation is given by

$$\begin{vmatrix} -(1 + \lambda) & 0 \\ 0 & -(2 + \lambda) \end{vmatrix} = (1 + \lambda)(2 + \lambda) = 0$$

[†] Recall that the matrix formulation of (7) takes the form

$$\mathbf{Y}' = \begin{pmatrix} a & b \\ c & d \end{pmatrix}\mathbf{Y} + \begin{pmatrix} h_1 \\ h_2 \end{pmatrix}, \quad \text{where } \mathbf{Y} = \begin{pmatrix} x \\ y \end{pmatrix}$$

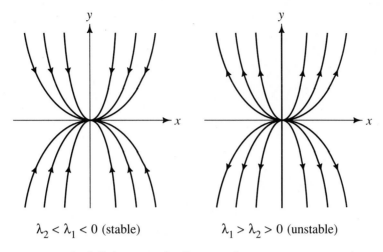

$$\lambda_2 < \lambda_1 < 0 \text{ (stable)} \qquad \lambda_1 > \lambda_2 > 0 \text{ (unstable)}$$

Figure 9.7 Real distinct roots leading to nodes.

which has the roots $\lambda_1 = -1$ and $\lambda_2 = -2$. Since both roots are negative, the origin is *asymptotically stable*. This type of critical point, called a (stable) *node,* has trajectories as shown in Figure 9.7a. Notice the direction of the arrows in this figure and also in Figure 9.7b. ∎

EXAMPLE 8

Discuss the stability of the system

$$x' = 4x - y$$
$$y' = 6x - 3y$$

Solution

Here we find

$$\begin{vmatrix} 4 - \lambda & -1 \\ 6 & -(3 + \lambda) \end{vmatrix} = \lambda^2 - \lambda - 6 = 0$$

with roots $\lambda_1 = 3$ and $\lambda_2 = -2$. In this case the origin is unstable since λ_1 is positive. Such a critical point is called a *saddle point,* an example of which is illustrated in Figure 9.8. ∎

Case II. Equal Roots

When $\lambda_1 = \lambda_2 = \lambda$, the solution vectors are $\mathbf{Y}^{(1)} = \mathbf{K}^{(1)}e^{\lambda t}$ and $\mathbf{Y}^{(2)} = [\mathbf{K}^{(2)}t + \mathbf{L}^{(2)}]e^{\lambda t}$. If $\lambda > 0$, the solution vectors move away from the critical point $(0, 0)$ as $t \to \infty$. Hence, we say that the origin is *unstable*. However, if $\lambda < 0$, then both solution vectors approach the origin as $t \to \infty$ and we say the critical point is *asymptotically stable*.

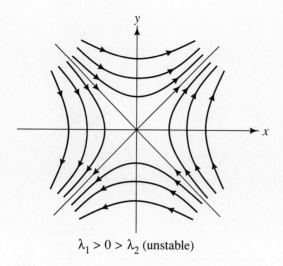

$$\lambda_1 > 0 > \lambda_2 \text{ (unstable)}$$

Figure 9.8 Real distinct roots leading to a saddle point.

EXAMPLE 9

Discuss the stability of the system

$$x' = -4x - y$$
$$y' = x - 2y$$

Solution

This time the auxiliary equation

$$\begin{vmatrix} -(4 + \lambda) & -1 \\ 1 & -(2 + \lambda) \end{vmatrix} = \lambda^2 + 6\lambda + 9 = 0$$

has the double root $\lambda = -3$. Hence, the origin is asymptotically stable. In this case the critical point is again called a (stable) *node,* which in general has the form of the trajectories shown in Figure 9.9a. ∎

Case III. Complex Roots

If $\lambda_1 = p + iq$ and $\lambda_2 = p - iq$, then the solution sets are given by $\mathbf{Y}^{(1)} = e^{pt}(\mathbf{a} \cos qt - \mathbf{b} \sin qt)$ and $\mathbf{Y}^{(2)} = e^{pt}(\mathbf{a} \sin qt + \mathbf{b} \cos qt)$. Here we see that the critical point $(0, 0)$ is *unstable* if $p > 0$ and *asymptotically stable* if $p < 0$. On the other hand, if $p = 0$, then $(0, 0)$ is *stable* but not asymptotically stable.

EXAMPLE 10

Discuss the stability of the system

$$x' = x - 3y$$
$$y' = x - y$$

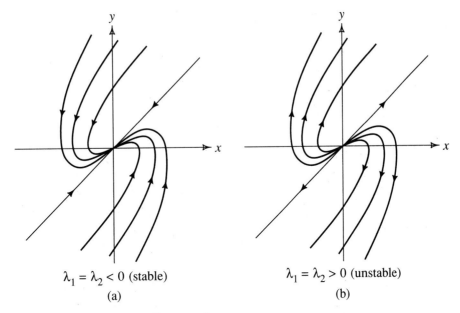

$\lambda_1 = \lambda_2 < 0$ (stable) $\lambda_1 = \lambda_2 > 0$ (unstable)

(a)   (b)

Figure 9.9 Equal roots leading to nodes.

Solution

The auxiliary equation is

$$\begin{vmatrix} 1 - \lambda & -3 \\ 1 & -(1 + \lambda) \end{vmatrix} = \lambda^2 + 2 = 0$$

with pure imaginary roots $\lambda = \pm i\sqrt{2}$. Thus, the origin $(0, 0)$ is stable, but not asymptotically stable. In this case we say the origin is a *center*, illustrated in Figure 9.10. ∎

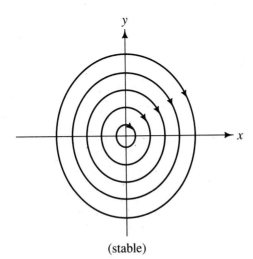

(stable)

Figure 9.10 Pure imaginary roots $\lambda = \pm iq$ leading to a center.

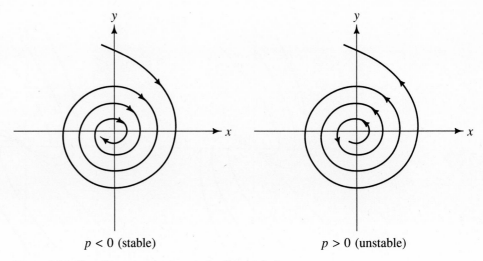

$p < 0$ (stable) $p > 0$ (unstable)

Figure 9.11 Complex roots $\lambda = p \pm iq$ leading to foci.

TABLE 9.1   STABILITY OF LINEAR SYSTEMS

Eigenvalues	Type of critical point
Real: $\lambda_1 \le \lambda_2 < 0$	Stable node
Real: $\lambda_1 < 0 < \lambda_2$	Saddle point (unstable)
Real: $0 < \lambda_1 \le \lambda_2$	Unstable node
Complex: $p \pm iq$, $p < 0$	Stable focus
Complex: $p \pm iq$, $p > 0$	Unstable focus
Complex: $p \pm iq$, $p = 0$	Center (stable)

In general, complex roots for which $p \ne 0$ lead to critical points that are called *foci*. Typical examples are illustrated in Figure 9.11.

Since periodic solutions are often of special interest, it should be noted that they arise only when λ_1 and λ_2 are pure imaginary (Case III). The trajectories in this situation are closed curves (ellipses), and the critical point is a center (recall Examples 2 and 10). In other words, *closed trajectories imply periodic motion.*

In Table 9.1 we have summarized the above results for linear systems.

Exercises 9.4

In Problems 1 to 12, determine whether the critical point $(0, 0)$ is *stable, asymptotically stable,* or *unstable.*

1. $x' = -x + y$
    $y' = 4x - y$

2. $x' = y$
    $y' = -x + 2y$

3. $x' = x - y$
    $y' = 5x - y$

4. $x' = -y$
    $y' = 5x - 2y$

5. $x' = 2x - 7y$
$y' = 3x - 8y$

6. $x' = 2x + 4y$
$y' = -2x + 6y$

7. $x' = x + 3y$
$y' = 3x + y$

■**8.** $x' = 2x + 5y$
$y' = x - 2y$

9. $x' = x - 2y$
$y' = 4x + 5y$

10. $x' = x - y$
$y' = x + 5y$

11. $x' = x + 7y$
$y' = 3x + 5y$

12. $x' = -x + y$
$y' = -4x - y$

In Problems 13 to 15, describe the nature of the critical point $(0, 0)$ for real values of the constant a.

13. $x' = -x + ay$
$y' = x - y$

■**14.** $x' = -ax + y$
$y' = -x - y$

15. $x' = y$
$y' = -x + ay$

*****16.** Given the linear system

$$x' = ax - y$$
$$y' = x + ay$$

(a) Show that the trajectories are spirals when $a \neq 0$.

(b) Discuss the stability of the critical point $(0, 0)$ for both cases $a > 0$ and $a < 0$.

(c) What are the trajectories when $a = 0$?

Hint. Transform to polar coordinates to show spiral trajectories.

17. The damped free oscillations of a spring-mass system are described by solutions of the second-order DE

$$mx'' + cx' + kx = 0, \qquad m, c, k > 0$$

Write the DE as a system of first-order equations and discuss the nature of the origin under the conditions

(a) $c^2 > 4mk$

(b) $c^2 = 4mk$

(c) $c^2 < 4mk$

9.5 Stability of Nonlinear Systems

Based on the results of Section 9.4, we now wish to examine nonlinear systems of the form

$$x' = P(x, y)$$
$$y' = Q(x, y)$$
(10)

for which $P(0, 0) = Q(0, 0) = 0$. Thus, the origin $(0, 0)$ is a critical point that we assume is isolated, but there may be others. In addition, we assume that $P(x, y)$ and $Q(x, y)$ are continuous functions and possess at least continuous second-order partial derivatives in a neighborhood of $(0, 0)$. In this case, we can express $P(x, y)$ and $Q(x, y)$ in Taylor expansions of the form

$$P(x, y) = P(0, 0) + P_x(0, 0)x + P_y(0, 0)y + P_{xx}(0, 0)\frac{x^2}{2!}$$

$$+ P_{xy}(0, 0)xy + P_{yy}(0, 0)\frac{y^2}{2!} + \cdots$$

and

$$Q(x, y) = Q(0, 0) + Q_x(0, 0)x + Q_y(0, 0)y + Q_{xx}(0, 0)\frac{x^2}{2!}$$

$$+ Q_{xy}(0, 0)xy + Q_{yy}(0, 0)\frac{y^2}{2!} + \cdots$$

Now, using the fact that $P(0, 0) = Q(0, 0) = 0$ and setting $a = P_x(0, 0)$, $b = P_y(0, 0)$, $c = Q_x(0, 0)$, and $d = Q_y(0, 0)$, we see that $P(x, y)$ and $Q(x, y)$ can be written as

$$P(x, y) = ax + by + p(x, y)$$
$$Q(x, y) = cx + dy + q(x, y)$$

(11)

where $p(x, y)$ and $q(x, y)$ denote all the remaining terms. In this case we find that (10) can be written in the form

$$x' = ax + by + p(x, y)$$
$$y' = cx + dy + q(x, y)$$

(12)

If $p(x, y)$ and $q(x, y)$ are "small enough" in the vicinity of the origin in the sense that

$$\lim_{x, y \to 0} \frac{p(x, y)}{\sqrt{x^2 + y^2}} = \lim_{x, y \to 0} \frac{q(x, y)}{\sqrt{x^2 + y^2}} = 0$$

(13)

then (12) is called a **perturbed linear system** or **almost linear system**.

> *Remark.* The term "perturbed linear system" means that the nonlinear system (12) behaves like a small perturbation of the linear system
>
> $$x' = ax + by$$
> $$y' = cx + dy$$
>
> *near the origin;* that is, it has characteristics similar to the linear system near the origin.

EXAMPLE 11

Show that the following system is a perturbed linear system:

$$x' = 3x - 2y + x^3$$
$$y' = -x + y - 5y^{3/2}$$

Solution

Here we see that $p(x, y) = x^3$ and $q(x, y) = -5y^{3/2}$. Thus, for $p(x, y)$ we obtain

$$\lim_{x, y \to 0} \frac{x^3}{\sqrt{x^2 + y^2}} = \lim_{x \to 0} \frac{x^3}{\sqrt{x^2}} = 0$$

where we have taken the first limit as $y \to 0$. In the second case, we see that

$$\lim_{x,\,y \to 0} \frac{-5y^{3/2}}{\sqrt{x^2 + y^2}} = \lim_{y \to 0} \frac{-5y^{3/2}}{\sqrt{y^2}} = 0$$

where this time we first let $x \to 0$. Hence, this system is a perturbed linear system. ∎

In most cases, the general behavior of the perturbed linear system (12) in the neighborhood of the critical point $(0, 0)$ is closely related to the general behavior of the corresponding linear system arising when $p(x, y) \equiv q(x, y) \equiv 0$. In particular, we have the following theorem, which we state without proof.

Theorem 9.2 *Let $(0, 0)$ be an isolated critical point of the perturbed linear system*

$$x' = ax + by + p(x, y)$$

$$y' = cx + dy + q(x, y)$$

where $ac - bd \neq 0$. If the critical point $(0, 0)$ of the associated linear system

$$x' = ax + by$$

$$y' = cx + dy$$

is

(a) Asymptotically stable, then $(0, 0)$ is an asymptotically stable point of the nonlinear system.

(b) Unstable, then $(0, 0)$ is an unstable point of the nonlinear system.

(c) Stable, but not asymptotically stable, then $(0, 0)$ of the nonlinear system may be asymptotically stable, stable, or unstable (i.e., no conclusion can be reached).

Although Theorem 9.2 addresses only the critical point $(0, 0)$, it applies equally well to other critical points. If the original nonlinear system has a critical point at (x_0, y_0), it is usually best to make the transformation $X = x - x_0$, $Y = y - y_0$, which transforms the critical point to $(0, 0)$ in the XY plane.

EXAMPLE 12

Determine the nature of the critical point $(0, 0)$ of the perturbed linear system

$$x' = 3x - 2y + x^3$$

$$y' = -x + y - 5y^{3/2}$$

by analyzing the associated linear system.

Solution

By inspection, the associated linear system is

$$x' = 3x - 2y$$

$$y' = -x + y$$

which leads to the auxiliary equation

$$\begin{vmatrix} 3 - \lambda & -2 \\ -1 & 1 - \lambda \end{vmatrix} = \lambda^2 - 4\lambda + 1 = 0$$

Hence, the eigenvalues are $\lambda = 2 \pm i\sqrt{3}$, and since the real part is positive, we deduce that $(0, 0)$ is an unstable critical point. Based on Theorem 9.2, the point $(0, 0)$ is also an unstable critical point of the given nonlinear system. ∎

Even though the stability of a critical point of a perturbed linear system may be the same as that of the associated linear system, the trajectories of the nonlinear system may differ greatly from those of the linear system. For example, if we move sufficiently far from a stable critical point of a nonlinear system, the solution may then move away from the critical point as $t \to \infty$. In the case of linear systems, however, the solution always approaches a stable critical point or stays close to the point no matter how far away we start from the point. For this reason, an important consideration in solving nonlinear systems is to determine the possible initial conditions under which a trajectory will approach or stay sufficiently close to a stable critical point.

Finally, we mention that, although we won't further discuss it, there exists a more direct method for determining the stability of a nonlinear system that rests upon the construction of a suitable auxiliary function called a *Liapunov function*. This more powerful method, due to A. M. Liapunov (1857–1918), is a generalization of the physical principles associated with a conservative system.[†]

9.5.1 Nonlinear Pendulum Problem

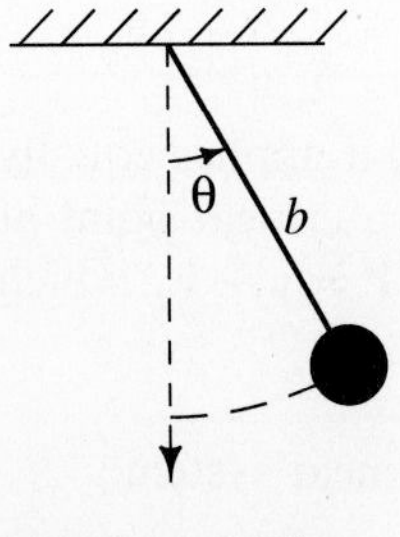

Figure 9.12 The simple pendulum.

In Section 5.2.2 we showed that the *undamped* oscillations of a simple pendulum are governed by the nonlinear DE

$$b\theta'' + g \sin \theta = 0 \tag{14}$$

where g is the gravitational constant and b is the length of the rod (see Figure 9.12). By introducing $\omega = \sqrt{g/b}$, we can rewrite (14) in the more convenient form

$$\theta'' + \omega^2 \sin \theta = 0 \tag{15}$$

If we wish to consider the possibility of resistance proportional to velocity, we must then analyze the more general *damped* oscillations model

$$\theta'' + c\theta' + \omega^2 \sin \theta = 0, \qquad c > 0 \tag{16}$$

To begin, let us consider the undamped case for which we may assume $c = 0$. Setting $x = \theta$ and $y = \theta'$, we can rewrite the DE (15) as the system of nonlinear

[†] A discussion of the Liapunov method is given in G. Birkhoff and G.-C. Rota, *Ordinary Differential Equations,* 3d ed., Wiley, New York, 1978.

first-order equations

$$x' = y$$
$$y' = -\omega^2 \sin x \qquad (17)$$

Clearly, the critical points occur where $y = 0$ and $\sin x = 0$. There are therefore an infinite number of them located at $(0, 0)$ and $(n\pi, 0)$, $n = \pm 1, \pm 2, \pm 3, \ldots$.

Let us first check the stability of the critical point $(0, 0)$. To do so we start with the series expansion about $x = 0$

$$\sin x = x - \frac{x^3}{3!} + \cdots$$

from which we obtain the *linear* approximation $\sin x \cong x$. Thus, (17) reduces to the *linear system*

$$x' = y$$
$$y' = -\omega^2 x \qquad (18)$$

The auxiliary equation associated with this linear system is

$$\begin{vmatrix} -\lambda & 1 \\ -\omega^2 & -\lambda \end{vmatrix} = \lambda^2 + \omega^2 = 0 \qquad (19)$$

which has pure imaginary roots $\lambda = \pm i\omega$. Hence, $(0, 0)$ is a center of the linear system but, based on Theorem 9.2, we cannot conclude anything about the stability of the nonlinear system (17) near $(0, 0)$.

The critical point $(0, 0)$ corresponds physically to the pendulum hanging vertically downward. Experience tells us that small displacements from this vertical position always lead to some form of small oscillation about it. On the basis of physical arguments, therefore, we conclude that this is a *stable* critical point of the nonlinear system. The critical point is actually a center because the near trajectories are simple closed curves about it (see Figure 9.13).

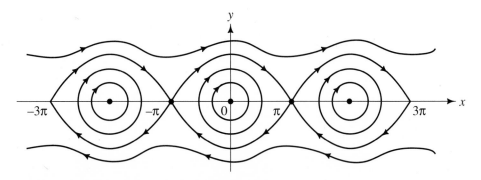

Figure 9.13 Trajectories of an undamped pendulum.

Next, let us discuss the critical point $(\pi, 0)$. This time we expand $\sin x$ about the point $x = \pi$, which yields

$$\sin x = (\pi - x) - \frac{1}{3!}(\pi - x)^3 + \cdots$$

The corresponding *linear* system is therefore

$$\begin{aligned} x' &= y \\ y' &= \omega^2(x - \pi) \end{aligned} \tag{20}$$

Now we let $X = x - \pi$, which reduces this linear system to

$$\begin{aligned} X' &= y \\ y' &= \omega^2 X \end{aligned} \tag{21}$$

the characteristic equation of which is $\lambda^2 - \omega^2 = 0$. Thus, the roots $\lambda_1 = \omega$ and $\lambda_2 = -\omega$ lead to the conclusion that $(\pi, 0)$ is an *unstable* critical point of the linear system, which is of the variety called a saddle point. Based on Theorem 9.2, $(\pi, 0)$ is also an unstable critical point of the related nonlinear system.

At the point where $x = \pi$, the pendulum is pointing vertically upward. Such a position is clearly unstable based on physical arguments alone without the above analysis. In fact, it can be shown that all critical points $(n\pi, 0)$, $n = \pm 1, \pm 3, \pm 5, \ldots,$ are also unstable (why?). Moreover, the remaining critical points $(n\pi, 0)$, $n = \pm 2, \pm 4, \pm 6, \ldots,$ are stable. That is, the critical points are alternately centers (stable, but not asymptotically stable) and saddle points (unstable).

The trajectories of this system are shown in Figure 9.13. Close to the stable critical points are the closed paths suggesting periodic motion. The trajectories that cross at the unstable critical points are called **separatrices** since they separate the possible oscillatory motions from other types of motion. Finally, the wavy paths outside the separatrices correspond to whirling motions of the pendulum.

Let us now consider the damped pendulum case described by (16). Again we set $\theta = x$, $\theta' = y$, and replace (16) by the system of equations

$$\begin{aligned} x' &= y \\ y' &= -\omega^2 \sin x - cy \end{aligned} \tag{22}$$

Observe that the origin $(0, 0)$ is still a critical point, but now it is *asymptotically stable*. To see this, let us examine the associated linear system

$$\begin{aligned} x' &= y \\ y' &= -\omega^2 x - cy \end{aligned} \tag{23}$$

whose auxiliary equation is

$$\begin{vmatrix} -\lambda & 1 \\ -\omega^2 & -(\lambda + c) \end{vmatrix} = \lambda^2 + c\lambda + \omega^2 = 0$$

Using the quadratic formula, we find that the roots are

$$\lambda_1, \lambda_2 = \frac{-c \pm \sqrt{c^2 - 4\omega^2}}{2}, \qquad c > 0 \tag{24}$$

and hence, the origin is a *stable* node if $c \geq 2\omega$, or a *stable* focus if $c < 2\omega$. According to Theorem 9.2, the original nonlinear system is therefore also asymptotically stable at $(0, 0)$. Clearly, the critical points such as $(\pi, 0)$ are still unstable in this case since these correspond to the pendulum pointing vertically upward.

9.5.2 Predator-Prey Problem

There are numerous instances in nature where one species of animal feeds on another species of animal for survival. The first species is called a **predator** and the second species is called the **prey.** A classic example of this is to suppose that $x(t)$ denotes the population of rabbits at any time on an isolated island, and $y(t)$ the number of foxes at any time on this same island. The foxes eat the rabbits, and the rabbits eat clover, which we assume is in ample supply. When the rabbits are plentiful, so too are the foxes and their population grows. When the foxes become too numerous and eat too many rabbits, they enter a period of famine and their population begins to diminish. The rabbits eventually become relatively safe again and their population increases, which also initiates new increases in the number of foxes. Under the proper ecological balance, these cycles of population increases and decreases can be repeated incessantly. On the other hand, if the balance of nature is disturbed in the right way, one or both species of animals could die out.

If the predators didn't exist, we might expect that the population of prey would *increase* at a rate that is roughly proportional to their number at any instant of time, so that

$$x' = ax, \qquad a > 0 \tag{25}$$

Likewise, if there were no prey, the predators might *decrease* at a similar rate so that

$$y' = -cy, \qquad c > 0 \tag{26}$$

Under these conditions the population of prey $x(t) \sim e^{at}$ would increase indefinitely for $t \to \infty$, while the predator population $y(t) \sim e^{-ct}$ would eventually die out. Of course, the actual situation is somewhere between these two extremes. To obtain a more realistic mathematical model of the predator-prey problem, we must include additional terms in (25) and (26) that take into account the interaction between the two species. One of the simplest models that takes interaction into account is given by the **Lotka-Volterra equations**

$$\begin{aligned} x' &= ax - bxy \\ y' &= -cy + dxy \end{aligned} \tag{27}$$

where both b and d are positive constants. Thus, we assume that the interaction is actually an *encounter* between predator and prey, and that the frequency of such encounters is proportional to the product xy.

Exact solutions to the Lotka-Volterra equations are unknown, but nevertheless we can learn a lot about the solutions from a qualitative analysis. First, by solving the equations

$$ax - bxy = x(a - by) = 0$$

$$-cy + dxy = y(-c + dx) = 0$$

we see that there are two critical points $(0, 0)$ and $(c/d, a/b)$. The associated linear system in the vicinity of the point $(0, 0)$ is that given by (25) and (26), that is, the product terms involving xy are omitted since they are much smaller than the terms involving only x or y. Hence, since one solution becomes unbounded and the other approaches zero, we see that $(0, 0)$ is an *unstable* critical point. Based on Theorem 9.2, the point $(0, 0)$ is also an unstable critical point of the nonlinear system (27).

The critical point $(c/d, a/b)$ is of greater interest since it corresponds to an equilibrium state in which the predator and prey can coexist without extinction. To analyze this critical point we start with the transformation

$$X = x - c/d$$

$$Y = y - a/b$$

which, when substituted into (27), yields

$$X' = -\frac{bc}{d} Y - bXY$$

$$Y' = \frac{ad}{b} X - dXY \tag{28}$$

This system of equations has $(0, 0)$ as its critical point, corresponding to the critical point $(c/d, a/b)$ of the original system (27). The linear system related to (28) is obtained by neglecting the product terms XY, that is,

$$X' = -\frac{bc}{d} Y$$

$$Y' = \frac{ad}{b} X \tag{29}$$

with auxiliary equation $\lambda^2 + ac = 0$. Hence, we obtain the pure imaginary roots $\lambda_1, \lambda_2 = \pm i\sqrt{ac}$, corresponding to a *stable* center of (29) at $(0, 0)$. We deduce therefore that $(c/d, a/b)$ is a stable center of the linear system associated with (27) in the neighborhood of $(c/d, a/b)$.

Unfortunately, based on the above results and Theorem 9.2, we cannot conclude anything about the stability of the critical point $(c/d, a/b)$ of the original *nonlinear* system (27). It can be shown, however, that in the vicinity of the critical point the trajectories are simple closed paths enclosing $(c/d, a/b)$, which leads us to conclude that $(c/d, a/b)$ is also a stable center of the nonlinear system (for example, see Problem 24 in Exercises 9.5). For these trajectories the populations $x(t)$ and $y(t)$ describe *periodic* behavior, that is, periodic fluctuations in the sizes of each population about the equilibrium solution.

Exercises 9.5

In Problems 1 to 10, determine all critical points of the system. If the system is a perturbed linear system about any of these critical points, determine the related linear system.

1. $x' = y$
$y' = -x$

2. $x' = x$
$y' = x + y$

3. $x' = 3x + y$
$y' = x + 3y$

4. $x' = x + 2y - 3$
$y' = 3x + 2y + 1$

5. $x' = 2y - 4$
$y' = x + 3$

6. $x' = x^2 - 2x$
$y' = x^2 - y$

7. $x' = -2x + y - 1$
$y' = -3x + xy$

8. $x' = y$
$y' = -x + x^3 - y$

9. $x' = x - xy$
$y' = y + 2xy$

■10. $x' = x - x^2 - xy$
$y' = \frac{1}{2}y - \frac{1}{4}y^2 - \frac{3}{4}xy$

In Problems 11 to 20, determine the nature of the critical point $(0, 0)$ by analyzing the related linear system.

11. $x' = 2 \sin x - 1 + e^y$
$y' = xy - y$

12. $x' = 1 - y - e^x$
$y' = y - \sin x$

13. $x' = x^2 - y - y^3 - 3x$
$y' = -\frac{1}{2} \sin 2y + x^2$

14. $x' = x(1 - y^2) - y$
$y' = x$

15. $x' = x + 2y + x \cos y$
$y' = -y - \sin y$

16. $x' = y$
$y' = -x - y^2$

17. $x' = y$
$y' = -x - y^3$

■18. $x' = 3x + 4y + x^2$
$y' = 4x - 3y - 2xy$

19. $x' = x + 2y + 2 \sin y$
$y' = -3y - xe^x$

20. $x' = e^{-x+y} - \cos x$
$y' = \sin(x - 3y)$

In Problems 21 to 23, determine all critical points and discuss the nature of their stability.

21. $x' = 8x - y^2$
$y' = -6y + 6x^2$

22. $x' = 2y + x^2$
$y' = -2x - 4y$

23. $x' = 1 - xy$
$y' = x - y^3$

*24. The equations for the predator-prey problem are given by

$$x' = ax - bxy$$

$$y' = -cy + dxy$$

(a) Show that the change of variables $x = cX/d$, $y = aY/b$ leads to the system

$$X' = a(X - XY)$$

$$Y' = -c(Y - XY)$$

with trajectories given by

$$(e^X/X)^c = K(Y/e^Y)^a, \qquad K \text{ constant}$$

(b) The point $(0, 0)$ is obviously a critical point of the system in part (a). Find a second critical point, and discuss the stability of the system at each critical point when $a = 2$ and $c = 1$.

*25. The population sizes of two species of animals competing for the same food sources in an environment can be modeled by the system of equations

$$x' = a_1 x - b_1 x^2 - c_1 xy$$

$$y' = a_2 y - b_2 y^2 - c_2 xy$$

where the coefficients are all positive numbers.

(a) Show that $(0, 0)$, $(0, a_2/b_2)$, and $(a_1/b_1, 0)$ are all critical points.

(b) Find a fourth critical point in the special case when $a_1 = 60$, $b_1 = 4$, $c_1 = 3$, $a_2 = 42$, $b_2 = 2$, and $c_2 = 3$, and show that it is an unstable saddle point.

(c) If $a_1 = 60$, $a_2 = 42$, $b_1 = b_2 = 3$, $c_1 = 4$, and $c_2 = 2$, find the fourth critical point and show that it is now an asymptotically stable node.

9.6 Chapter Summary

In this chapter we have briefly introduced the notion of **stability** in connection with the study of **nonlinear systems of equations.** Specifically, we have studied **autonomous** systems of two equations whose general form is

$$x' = P(x, y)$$
$$y' = Q(x, y) \tag{30}$$

Since exact solutions of such equations are often difficult or impossible to obtain, we settle for obtaining qualitative information about the behavior of the solutions. In particular, we may examine the **trajectories** of the system in the xy plane, or **phase plane,** which are found by solving the first-order DE

$$\frac{dy}{dx} = \frac{Q(x, y)}{P(x, y)} \tag{31}$$

Points in the phase plane for which $P(x, y) = Q(x, y) = 0$ are called **critical points** or **equilibrium points.** Questions about the stability of the nonlinear system (30) are usually concerned with the behavior of the solutions of the system "near" the critical points. That is, if the solution moves toward the critical point for increasing time t, we say the critical point is **stable.** On the other hand, it is called **unstable** if it moves away from the critical point.

To study the stability of a nonlinear system in the vicinity of one of its critical points, we often examine a similar *linear* system that has the same critical point. Nonlinear systems that permit this kind of analysis are called **perturbed linear systems.** Generally speaking, the stability behavior of the perturbed linear system in the neighborhood of a critical point is closely related to the stability behavior of the corresponding linear system.

CHAPTER *10*
Numerical Methods

Numerical methods are becoming more and more important in the solution of differential equations, partly because of the difficulties encountered in obtaining exact analytical solutions but also, more recently, because of the ease with which numerical techniques can be used in conjunction with today's high-speed computers.

In Section 10.2 we introduce two simple numerical procedures for solving first-order DEs. The first is the **Euler method,** which is the easiest of all numerical methods to apply. It is generally not very accurate and hence is seldom used in practice. The second technique, called the **improved Euler method,** involves an additional computation at each step of the procedure but is usually far more accurate. A brief discussion on the types of **errors** that commonly arise in numerical schemes is presented in Section 10.2.3.

The **Runge-Kutta method,** which is perhaps the most widely used numerical procedure, is introduced in Section 10.3. Although the number of computations at each step of the process is more than required by either of the Euler methods, it is a very accurate method. Finally, in Section 10.4 we look briefly at **systems of first-order equations** and **higher-order equations.**

10.1 Introduction

When possible, exact analytical solutions of DEs are generally desired, mostly because they provide valuable information about the qualitative nature of the solution, including their dependence on certain parameters in the DE or initial conditions. However, if the analytical solution is a complicated function, it may be difficult to use in obtaining numerical values. Hence, because of this or the fact that exact solutions of DEs are often difficult or impossible to find, particularly when solving nonlinear equations, it becomes either convenient or necessary to employ some method that yields accurate numerical estimates of the true behavior of the system under study.

In this chapter we briefly consider several numerical methods that are fairly easy to develop but yield the "solution" of a given IVP only at discrete points over a specified interval. For instance, a numerical solution of the IVP

$$2y' + y = 0, \qquad y(0) = 3 \tag{1}$$

where t is the independent variable, would consist of a sequence of approximations $y_1, y_2, \ldots, y_n$ to the true solution values $y(t_1), y(t_2), \ldots, y(t_n)$, where $y = 3e^{-t/2}$ is the true solution. Normally, the approximations are equally spaced over some interval $0 \le t \le T$, where T is a fixed value (e.g., see Figure 10.1).

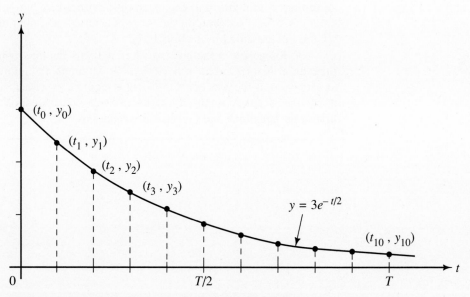

Figure 10.1 Numerical approximation to the function $3e^{-t/2}$.

10.2 Simple Numerical Methods for First-Order Equations

Let us consider the IVP

$$y' = F(t, y), \qquad y(t_0) = y_0 \tag{2}$$

where we assume that F and $\partial F/\partial y$ are continuous functions in some rectangular domain in the ty plane containing the point (t_0, y_0). Under these conditions, according to Theorem 2.1, there is a unique solution of (2).

The most common numerical methods are called **one-step** or **step-by-step procedures,** wherein each calculated value for the solution makes use of the previously calculated value. We start at the initial point (t_0, y_0) on the solution curve and increase t_0 by a fixed **step size** $h > 0$ to get $t_1 = t_0 + h$, and then compute a value y_1 that approximates the true solution value $y(t_1)$. In the second step of the procedure, we find an approximate value y_2 for $y(t_2)$, where $t_2 = t_1 + h = t_0 + 2h$, and continue in this fashion. A method that makes use of more than one previous value is called a **multistep** or **continuing method.** All multistep methods, however, make use of a one-step method at the first step to determine y_1. If the multistep method requires more than two previous values, then the additional values must also be determined by a one-step procedure. In this chapter we discuss only one-step methods.

The calculations for a one-step method are all done by the same formula at each step of the process. These formulas are suggested by the Taylor series

$$y(t + h) = y(t) + hy'(t) + \tfrac{1}{2}h^2 y''(t) + \cdots \tag{3}$$

By using (2), we see that $y'(t) = F(t, y)$ and consequently, $y''(t) = (dF/dt)(t, y)$, and so forth. Now substituting these results into (3), we obtain

$$y(t + h) = y(t) + hF(t, y) + \frac{1}{2} h^2 \frac{dF}{dt}(t, y) + \cdots \tag{4}$$

In the first step of the procedure, we set $t = t_0$ and $y(t_0) = y_0$, and then calculate

$$y_1 = y_0 + hF(t_0, y_0) + \frac{1}{2} h^2 \frac{dF}{dt}(t_0, y_0) + \cdots$$

For the second step, we set $t = t_0 + h = t_1$ and calculate

$$y_2 = y_1 + hF(t_1, y_1) + \frac{1}{2} h^2 \frac{dF}{dt}(t_1, y_1) + \cdots$$

whereas in general

$$y_{n+1} = y_n + hF(t_n, y_n) + \frac{1}{2} h^2 \frac{dF}{dt}(t_n, y_n) + \cdots, \qquad n = 0, 1, 2, \ldots \tag{5}$$

For practical reasons, the series in (5) must be truncated after a finite number of terms, leading to what is called a **truncation error** or **formula error.** If we neglect all terms in (5) after $hF(t_n, y_n)$, we obtain a **first-order approximation,** and in this case the truncation error *per step* is then of the order h^2 (see Section 10.2.3). Another source of error is **round-off error,** due to the limited number of digits that are used in any computation. Round-off errors, therefore, depend very much upon the machine on which the calculations are made. We briefly discuss both types of error in more detail in Section 10.2.3.

Finally, when additional terms in (5) are required for greater accuracy, leading to **higher-order approximations,** the derivatives of F must be computed. Hence, using (5) directly is often not feasible since the derivatives must be calculated by hand. In such cases it is usually preferable to replace the derivatives by certain numerical approximations so that the formulas lend themselves to computer calculations for any number of steps m (see Section 10.3).

10.2.1 The Euler Method

The simplest numerical method for solving IVPs is the **Euler method** in which calculations are made using the formula

$$y_{n+1} = y_n + hF(t_n, y_n), \qquad n = 0, 1, 2, \ldots \tag{6}$$

obtained from (5) by truncating all terms involving $h^2, h^3, \ldots$. Hence, (6) is an example of a *first-order approximation.*

Geometrically, the Euler method approximates the true solution curve with a polygon whose first side is tangent to the solution curve at (t_0, y_0) (see Figure 10.2). For this reason, it is also known as the **method of tangent lines.** Let us illustrate Euler's method with a simple example.

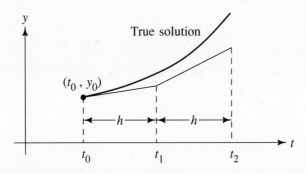

Figure 10.2

EXAMPLE 1

Use Euler's method to approximate the solution of the IVP

$$y' + 2ty^2 = 0, \qquad y(0) = 1$$

on the interval $0 \le t \le 1.0$, first using step size $h = 0.1$ and then step size $h = 0.05$.

Solution

By rearranging the equation as $y' = -2ty^2$, we recognize that $F(t, y) = -2ty^2$. Hence, $F(t_n, y_n) = -2t_n y_n^2$, and so Equation (6) becomes

$$y_{n+1} = y_n - 2ht_n y_n^2, \qquad n = 0, 1, 2, \ldots$$

For step size $h = 0.1$, the above formula reduces to

$$y_{n+1} = y_n - (0.2)t_n y_n^2, \qquad n = 0, 1, 2, \ldots$$

Setting $n = 0$, we first calculate

$n = 0$:
$$y_1 = y_0 - (0.2)t_0 y_0^2$$
$$= 1 - (0.2)(0)(1)^2$$
$$= 1$$

which is our estimate to the value $y(0.1)$. Continuing, we find

$n = 1$:
$$y_2 = y_1 - (0.2)t_1 y_1^2$$
$$= 1 - (0.2)(0.1)(1)^2$$
$$= 0.9800$$

$n = 2$:
$$y_3 = y_2 - (0.2)t_2 y_2^2$$
$$= 0.98 - (0.2)(0.2)(0.98)^2$$
$$= 0.9416$$

and so on. These, plus a few additional calculations, along with the exact values, are listed in Table 10.1. All values are rounded off to four decimal places. [By separating the variables, the exact solution is found to be $y = 1/(t^2 + 1)$.]

Now using the value $h = 0.05$ and repeating the above calculations, we see that

$n = 0$:
$$y_1 = y_0 - (0.1)t_0 y_0^2$$
$$= 1 - (0.1)(0)(1)^2$$
$$= 1$$

$n = 1$:
$$y_2 = y_1 - (0.1)t_1 y_1^2$$
$$= 1 - (0.1)(0.05)(1)^2$$
$$= 0.9950$$

$n = 2$:
$$y_3 = y_2 - (0.1)t_2 y_2^2$$
$$= 0.995 - (0.1)(0.1)(0.995)^2$$
$$= 0.9851$$

and so on. To compare with the results of Table 10.1, only the values $y_0, y_2, y_4, \ldots, y_{20}$, along with the exact values, are listed in Table 10.2. For still further comparison, we have also included results for $h = 0.01$.

Observe that in the case $h = 0.05$, because of the smaller step size h, it takes twice as many calculations as when $h = 0.1$ to reach the point where $t = 1.0$. For $h = 0.01$, it takes 10 times as many calculations to reach $t = 1.0$. However, the error incurred at each step in these other cases is smaller than that in the case $h = 0.1$. ∎

TABLE 10.1 EULER'S METHOD FOR $y' = -2ty^2$, $y(0) = 1$, WITH $h = 0.1$

t	y_n	Exact y	Error	Percentage error
0.0	1.0000	1.0000	0.0000	0.00
0.1	1.0000	0.9901	0.0099	1.00
0.2	0.9800	0.9615	0.0185	1.92
0.3	0.9416	0.9174	0.0242	2.64
0.4	0.8884	0.8621	0.0263	3.05
0.5	0.8253	0.8000	0.0253	3.16
0.6	0.7571	0.7353	0.0218	2.96
0.7	0.6884	0.6711	0.0173	2.58
0.8	0.6220	0.6098	0.0122	2.00
0.9	0.5601	0.5525	0.0076	1.38
1.0	0.5036	0.5000	0.0036	0.72

TABLE 10.2 EULER'S METHOD FOR $y' = -2ty^2$, $y(0) = 1$, WITH $h = 0.05$ AND $h = 0.01$

t	$h = 0.05$ y_n	$h = 0.01$ y_n	Exact y
0.0	1.0000	1.0000	1.0000
0.1	0.9950	0.9911	0.9901
0.2	0.9705	0.9633	0.9615
0.3	0.9291	0.9197	0.9174
0.4	0.8746	0.8645	0.8621
0.5	0.8120	0.8023	0.8000
0.6	0.7456	0.7373	0.7353
0.7	0.6793	0.6727	0.6711
0.8	0.6156	0.6109	0.6098
0.9	0.5562	0.5532	0.5525
1.0	0.5018	0.5004	0.5000

Since the exact solution was available for comparison in Example 1, we calculated and listed in Table 10.1 the *cumulative error* defined by $|y(t_n) - y_n|$ that was incurred in the method at each step of the procedure. The *percentage error,* defined by the ratio of error to exact value, also appears in Table 10.1. Of special interest is the **maximum percentage error** over the interval of computation $0 \le t \le 1$. In Example 1 this occurs at $t = 0.5$ and is found to be

$$\frac{|\text{Error}|}{\text{Exact value}} \times 100 = \frac{0.0253}{0.8000} \times 100$$

$$= 3.16\%$$

For step size $h = 0.05$, the maximum percentage error is

$$\frac{0.0120}{0.8000} \times 100 = 1.50\%$$

while for $h = 0.01$ the maximum percentage error is only 0.29 percent.

Based on Example 1, we see that our approximations y_n to the exact values $y(t_n)$ improve in accuracy as the step size h is decreased. In general, of course, this is the case. Also, we might conclude from Example 1 that the Euler method works quite well in spite of its simplicity, since the above errors are acceptable in many applications. This kind of accuracy in Euler's method, however, is quite rare and should not be expected in general (for instance, see Example 2 below).

Although we have performed the calculations for the first few values in Example 1 by hand, one rarely tries to produce all the results like those in Tables 10.1 and 10.2 through hand calculations. Almost everyone today has access to a programmable pocket calculator or microcomputer in which to make numerical computations. A simple BASIC program (which should run on most computers that accept BASIC) that was used to compute the data given in Tables 10.1 and 10.2 is listed in Figure

```
100  DEF FNF(T,Y) = -2*T*Y*Y

110  INPUT "Initial Values of t and y";T,Y
120  INPUT "Step size h";H
130  INPUT "Total Number of Steps";M
140  INPUT "Print Step k ≤ m";K

150  FOR N = 1 TO M

160      Y = Y + H*FNF(T,Y)
170      T = T + H
180      IF INT(N/K) = N/K THEN PRINT T,Y

190  NEXT N
200  END
```

Figure 10.3 BASIC program for Euler method.

10.3.[†] Line 100 of the program defines the function $F(t, y) = -2ty^2$ which appears in the DE in Example 1. It is the only line of the program that needs to be changed when solving another IVP. The INPUT "Print step number k" is used to select which steps of the calculations are to be printed. For example, if each step size h is to be printed, we simply set $k = 1$; if every other step size h is to be printed, we set $k = 2$; and so on. To produce the data occurring in the column where $h = 0.01$ in Table 10.2, for example, we entered the values $x = 0$, $y = 1$; $h = 0.01$; $m = 100$; and $k = 10$.

EXAMPLE 2

Use Euler's method with step size $h = 0.1$ to approximate the solution of the IVP

$$y' = 5y + t - 1, \qquad y(0) = 2$$

on the interval $0 \leq t \leq 1.0$.

Solution

Here we find that $F(t_n, y_n) = 5y_n + t_n - 1$, and thus with $h = 0.1$ our general formula becomes

$$y_{n+1} = y_n + h(5y_n + t_n - 1)$$
$$= y_n + (0.1)(5y_n + t_n - 1), \qquad n = 0, 1, 2, \ldots$$

The first two calculations yield

$n = 0$:
$$y_1 = y_0 + (0.1)(5y_0 + t_0 - 1)$$
$$= 2 + (0.1)(10 + 0 - 1)$$
$$= 2.9$$

$n = 1$:
$$y_2 = y_1 + (0.1)(5y_1 + t_1 - 1)$$
$$= 2.9 + (0.1)(14.5 + 0.1 - 1)$$
$$= 4.26$$

[†] All programs listed in this chapter were run on an IBM PC-AT to produce the data in the tables.

TABLE 10.3   EULER'S METHOD FOR $y' = 5y + t - 1$,
$y(0) = 2$, WITH $h = 0.1$, $h = 0.05$, AND
$h = 0.01$

t	$h = 0.1$ y_n	$h = 0.05$ y_n	$h = 0.01$ y_n	Exact y
0.0	2.0000	2.0000	2.0000	2.0000
0.1	2.9000	3.0150	3.1372	3.1736
0.2	4.2600	4.6122	5.0021	5.1216
0.3	6.3100	7.1190	8.0524	8.3463
0.4	9.3950	11.047	13.034	13.676
0.5	14.033	17.196	21.160	22.476
0.6	20.999	26.816	34.410	36.997
0.7	31.458	41.857	56.005	60.952
0.8	47.157	65.370	91.193	100.46
0.9	70.716	102.12	148.52	165.61
1.0	106.06	159.55	241.92	273.04

Additional values are listed in Table 10.3 along with those using step sizes $h = 0.05$ and $h = 0.01$. Exact values computed from the true solution $y = \frac{46}{25}e^{5t} + \frac{4}{25} - \frac{1}{5}t$ are also included for comparison.   ∎

The maximum percentage error in Example 2 for step size $h = 0.1$ is 61.2 percent, significantly greater than that in Example 1 for $h = 0.1$. By reducing the step size in Example 2 to $h = 0.01$, we can reduce the maximum percentage error to 11.4 percent; reducing the step size to $h = 0.001$ reduces the maximum percentage error to 1.24 percent. Further reductions in step size will lead to still more accurate results, but the number of calculations might have to increase drastically to achieve a desired accuracy. Moreover, the error normally builds up as we move farther away from the initial point on the solution curve so that it eventually exceeds what is deemed acceptable, no matter how small we choose the value of h.[†] That is, we obtain an inherent formula error at each step due to the linear approximation in the general formula (6). But since y_n is only an approximation to the true value $y(t_n)$ for each n, the error at each step is actually a **cumulative error,** that is, an accumulation of all formula errors introduced at the previous steps (see Figure 10.1).

Euler's method is seldom used in practice, even with today's high-speed digital computers, because there are more accurate methods at our disposal. Euler's method is introduced here primarily because it is very simple to develop and use. It may also provide insight into the general understanding and use of other more sophisticated numerical methods.

[†] Example 1 is an exception to this rule wherein the maximum percentage error does not occur at the endpoint of the interval.

10.2.2 The Improved Euler Method

Euler's method can be made more accurate for a fixed value of h by first computing the auxiliary value

$$y_{n+1}^* = y_n + hF(t_n, y_n) \tag{7}$$

and then the new value

$$y_{n+1} = y_n + \tfrac{1}{2}h[F(t_n, y_n) + F(t_{n+1}, y_{n+1}^*)] \tag{8}$$

This technique, called the **improved Euler method** (or **Heun's method**), is an example of what is called a **predictor-corrector method.** That is, at each step of the procedure we "predict" a value by (7) and then "correct" it by (8).

The geometric interpretation of this new method is that we approximate the true solution y in the interval from t_n to $t_n + \tfrac{1}{2}h$ by the straight line through (t_n, y_n) with slope $F(t_n, y_n)$, and then along a new line with slope $F(t_{n+1}, y_{n+1}^*)$ up to t_{n+1} (see Figure 10.4). Therefore, we might interpret the sum

$$\tfrac{1}{2}[F(t_n, y_n) + F(t_{n+1}, y_{n+1}^*)]$$

as some average slope over the interval $t_n \leq t \leq t_{n+1}$. It can be shown that (8) is equivalent to a Taylor series like (4) through the term containing h^2, but it has the advantage that the derivative dF/dt does not have to be explicitly calculated.

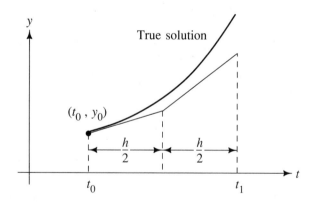

Figure 10.4

EXAMPLE 3

Use the improved Euler's formula with step size $h = 0.1$ to obtain an approximate solution of the IVP

$$y' = -2ty^2, \qquad y(0) = 1$$

on the interval $0 \leq t \leq 1.0$.

Solution

This is the same IVP as that in Example 1, so we can make a comparison of methods.

TABLE 10.4   APPROXIMATE SOLUTIONS OF
$y' = -2ty^2$, $y(0) = 1$, USING
THE EULER METHOD WITH
STEP SIZE $h = 0.05$ AND
THE IMPROVED EULER
METHOD WITH $h = 0.1$

t	Euler $h = 0.05$ y_n	Improved Euler $h = 0.1$ y_n	Exact y
0.0	1.0000	1.0000	1.0000
0.1	0.9950	0.9900	0.9901
0.2	0.9705	0.9614	0.9615
0.3	0.9291	0.9172	0.9174
0.4	0.8746	0.8620	0.8621
0.5	0.8120	0.8000	0.8000
0.6	0.7456	0.7355	0.7353
0.7	0.6793	0.6716	0.6711
0.8	0.6156	0.6104	0.6098
0.9	0.5562	0.5533	0.5525
1.0	0.5018	0.5009	0.5000

Using (7) followed by (8), we obtain at the first step

$n = 0$:
$$y_1^* = y_0 - (0.2)t_0 y_0^2$$
$$= 1 - (0.2)(0)(1)^2$$
$$= 1$$

and

$$y_1 = y_0 + (0.05)[-2t_0 y_0^2 - 2t_1(y_1^*)^2]$$
$$= 1 - (0.05)[(2)(0)(1)^2 + (2)(0.1)(1)^2]$$
$$= 0.9900$$

The remaining values are provided in Table 10.4 along with those obtained by Euler's method using step size $h = 0.05$ and the exact values. Again all entries are rounded off to four decimal places.   ■

Examples 1 and 3 illustrate that the results using the improved Euler method with step size $h = 0.1$ are better than those obtained by the Euler method, even for Euler's method with step size $h = 0.05$. In general this is the case, and although a few more calculations are required at each step using the improved Euler's method, the greater accuracy is worth the extra effort.

EXAMPLE 4

Use the improved Euler's formula with step size $h = 0.1$ to obtain an approximate solution of the IVP

$$y' = 5y + t - 1, \qquad y(0) = 2$$

on the interval $0 \le t \le 1.0$.

Solution

This problem is the same as that solved in Example 2 by Euler's method. In this case we find at the first step

$n = 0$:
$$y_1^* = y_0 + (0.1)(5y_0 + t_0 - 1)$$
$$= 2 + (0.1)(10 + 0 - 1)$$
$$= 2.9$$

and

$$y_1 = y_0 + (0.05)[(5y_0 + t_0 - 1) + (5y_1^* + t_1 - 1)]$$
$$= 2 + (0.05)[(10 + 0 - 1) + (14.5 + 0.1 - 1)]$$
$$= 3.13$$

The remaining values are listed in Table 10.5. Once again we have included those values from Euler's method with step size $h = 0.05$ and the exact solution for comparison.

TABLE 10.5 APPROXIMATE SOLUTIONS OF $y' = 5y + t - 1$, $y(0) = 2$, USING THE EULER METHOD WITH STEP SIZE $h = 0.05$ AND THE IMPROVED EULER METHOD WITH $h = 0.1$

t	Euler $h = 0.05$ y_n	Improved Euler $h = 0.1$ y_n	Exact y
0.0	2.0000	2.0000	2.0000
0.1	3.0150	3.1300	3.1736
0.2	4.6122	4.9788	5.1216
0.3	7.1190	7.9955	8.3463
0.4	11.047	12.910	13.676
0.5	17.196	20.909	22.476
0.6	26.816	33.920	36.997
0.7	41.857	55.074	60.952
0.8	65.370	89.463	100.46
0.9	102.12	145.36	165.61
1.0	159.55	236.20	273.04

TABLE 10.6 APPROXIMATE SOLUTIONS OF
$y' = 5y + t - 1$, $y(0) = 2$, USING
THE IMPROVED EULER
METHOD WITH STEP SIZES
$h = 0.05$ AND $h = 0.01$

	$h = 0.05$	$h = 0.01$	Exact
t	y_n	y_n	y
0.0	2.0000	2.0000	2.0000
0.1	3.1605	3.1730	3.1736
0.2	5.0785	5.1196	5.1216
0.3	8.2399	8.3413	8.3463
0.4	13.443	13.665	13.676
0.5	21.996	22.453	22.476
0.6	36.050	36.953	36.997
0.7	59.134	60.867	60.952
0.8	97.042	100.30	100.46
0.9	159.28	165.31	165.61
1.0	261.47	272.49	273.04

In Table 10.6 we present further numerical results using the improved Euler method with step sizes $h = 0.05$ and $h = 0.01$. ■

In Example 4 the difference in accuracy between Euler's method and the improved Euler's method is even more pronounced than it was in Example 3. That is, the maximum percentage error in this last example using the improved Euler formula is only 0.2 percent compared with 11.4 percent by Euler's method, both with step size $h = 0.01$.

The BASIC computer program used to generate the data in Tables 10.4 to 10.6 is listed in Figure 10.5. Observe that it is very similar to that in Figure 10.3 for the Euler method.

```
100   DEF FNF(T,Y) = -2*T*Y*Y

110   INPUT "Initial Values of t and y";T,Y
120   INPUT "Step Size h";H
130   INPUT "Total Number of Steps";M
140   INPUT "Print Step k ≤ m";K

150   FOR N = 1 TO M

160       FO = FNF(T,Y)
170       Z = Y + H*FO
180       X = X + H
190       F1 = FNF(T,Z)
200       Y = Y + 0.5*H*(FO + F1)
210       IF INT(N/K) = N/K THEN PRINT T,Y

220   NEXT N
230   END
```

Figure 10.5 BASIC program for improved Euler method.

10.2.3 Errors

In practice, one should pay close attention to the various types of errors that may arise in solving IVPs by any numerical scheme. For example, if the accumulation of errors is too great for a particular method, an alternate procedure should be considered. The analysis of errors is therefore an important aspect of the implementation of any numerical method. In this brief chapter a detailed study of errors would take us too far afield for our purposes, so we simply make some general comments about them in addition to our previous discussion.

As we have previously mentioned, there are primarily two sources of error—*formula* (or *truncation*) *error* and *round-off error*. Recall that the formula error is due to terminating the Taylor series expansion (5) after a finite number of terms. If we assume that our computer can retain an *infinite* number of significant digits, the formula error, being the only error, is simply the difference between the exact solution and the approximate solution at a given point. For example, from Equations (5) and (6), it follows that the formula error at the first step of the Euler method is

$$E_1 = y(t_1) - y_1$$

$$= \frac{1}{2} h^2 \frac{dF}{dt}(t_0, y_0) + \cdots$$

Relying on *Taylor's theorem with remainder* from calculus, we can then show that this error has the upper bound

$$|E_1| \le Kh^2 \tag{9}$$

where K is a positive constant (independent of h) proportional to the maximum value of dF/dt on the interval $t_0 \le t \le t_1$.

At the nth step the formula error is

$$E_n = y(t_n) - y_n$$

but, as previously pointed out, this is actually an *accumulated* error since the error incurred at each step is carried into the calculations at the next step. The maximum formula error at each step satisfies an inequality such as (9); thus, the maximum cumulative error at the nth step is roughly n times that given by (9). We therefore have at the point $t_n = t_0 + nh$ the cumulative error bound

$$|E_n| \le nK_1 h^2 \tag{10}$$

where K_1 is another constant that may differ from K. However, since $n = (t_n - t_0)/h$, we find that the *cumulative error* after n steps satisfies the inequality

$$|E_n| = |y(t_n) - y_n| \le C_1 h \tag{11}$$

where $C_1 = K_1(t_n - t_0)$. A similar analysis applied to the improved Euler method leads to the cumulative error bound

$$|E_n| = |y(t_n) - y_n| \le C_2 h^2 \tag{12}$$

where C_2 is also a constant (independent of h).

Because of limitations that exist in all computers, the number of significant digits carried throughout the calculations cannot exceed a certain finite value. Thus, we always obtain some round-off error in the calculations in addition to the accumulated formula errors. Round-off errors are difficult to analyze, however, since they are tied to the type of computer used in the computations, the sequence in which the computations take place, the number of calculations, and so on. Just as in the case of formula errors, we actually obtain an *accumulated* error at any given step due to the round-off that must take place at all previous steps of the procedure. If Y_n denotes the value actually determined by the numerical procedure at the nth step and y_n is our rounded-off value at this point, the accumulated round-off error is then defined by

$$R_n = Y_n - y_n$$

The combined error due to both round-off and formula errors is actually bounded by the sum of the absolute values of each type of error. To see this, we simply form the difference

$$|y(t_n) - Y_n| = |y(t_n) - y_n + y_n - Y_n|$$

and by making use of the triangle inequality,[†] we obtain

$$|y(t_n) - Y_n| \le |y(t_n) - y_n| + |y_n - Y_n| \le |E_n| + |R_n| \tag{13}$$

For further discussion regarding errors, the reader should refer to a text on numerical analysis.

Exercises 10.2

In Problems 1 to 10, use the Euler formula with $h = 0.1$ to obtain an approximation, to *four* places after the decimal, to the indicated value of y.

1. $y' = 2ty$, $y(1) = 1$; $y(1.5) = ?$

2. $y' = 1 + y^2$, $y(0) = 0$; $y(0.5) = ?$

3. $y' = (t + y - 1)^2$, $y(0) = 2$; $y(0.5) = ?$

4. $y' = t + y^2$, $y(0) = 1$; $y(0.5) = ?$

5. $y' = t^2 + y^2$, $y(0) = 1$; $y(0.5) = ?$

6. $y' = ty + \sqrt{y}$, $y(0) = 1$; $y(0.5) = ?$

7. $y' = ty^2 - y/t$, $y(1) = 1$; $y(1.5) = ?$

8. $y' = \sin(t + y)$, $y(0) = 0$; $y(0.5) = ?$

9. $y' = e^{-y}$, $y(0) = 0$; $y(0.5) = ?$

10. $y' = y - y^2$, $y(0) = \frac{1}{2}$; $y(0.5) = ?$

In Problems 11 to 20, use the improved Euler formula with $h = 0.1$ to obtain an approximation, to *four* places after the decimal, to the indicated value of y.

11. $y' = 2ty$, $y(1) = 1$; $y(1.5) = ?$

12. $y' = 1 + y^2$, $y(0) = 0$; $y(0.5) = ?$

13. $y' = (t + y - 1)^2$, $y(0) = 2$; $y(0.5) = ?$

14. $y' = t + y^2$, $y(0) = 1$; $y(0.5) = ?$

15. $y' = t^2 + y^2$, $y(0) = 1$; $y(0.5) = ?$

16. $y' = ty + \sqrt{y}$, $y(0) = 1$; $y(0.5) = ?$

17. $y' = ty^2 - y/t$, $y(1) = 1$; $y(1.5) = ?$

18. $y' = \sin(t + y)$, $y(0) = 0$; $y(0.5) = ?$

19. $y' = e^{-y}$, $y(0) = 0$; $y(0.5) = ?$

20. $y' = y - y^2$, $y(0) = \frac{1}{2}$; $y(0.5) = ?$

[†] The triangle inequality is $|a + b| \le |a| + |b|$.

21. Using Euler's formula with $h = 0.2$, find an approximate value for $y(1)$, where y is the solution of

$$y' = y, \qquad y(0) = 1$$

(Your answer approximates the value of e. Can you explain why?)

22. Solve Problem 21 using the improved Euler's formula.

23. Using Euler's formula with $h = 0.2$,

   (a) Find an approximate value for $y(1)$, where y is the solution of

$$y' = \frac{1}{t^2 + 1}, \qquad y(0) = 0$$

   (b) Use your answer in (a) to approximate the value of π.

   Hint. The exact solution in (a) is $y = \tan^{-1} t$.

24. Solve Problem 23 using the improved Euler's formula.

*25. Use both Euler methods to approximate the value $y(1.4)$, where y is the solution of

$$y' = t^2 + y^3, \qquad y(1) = 1$$

   (a) Use step size $h = 0.2$.
   (b) Use step size $h = 0.1$.
   (c) Use step size $h = 0.05$.

*26. Derive Euler's formula (6) by integrating both sides of the DE $y' = F(t, y)$ from t_n to t_{n+1} and then approximating the integral by

$$\int_a^b f(t)\, dt \cong (b - a)f(a)$$

Give some justification for using the above approximation for the integral.

10.3 The Runge-Kutta Method

Perhaps the most commonly used and most accurate technique for numerically solving IVPs is the **Runge-Kutta method.**[†] It requires considerably more calculations at each step, however, than either of the Euler methods discussed in Section 10.2. For example, at each step we must first compute the four auxiliary quantities

$$\begin{aligned}
k_1 &= F(t_n, y_n) \\
k_2 &= F(t_n + h/2, y_n + hk_1/2) \\
k_3 &= F(t_n + h/2, y_n + hk_2/2) \\
k_4 &= F(t_{n+1}, y_n + hk_3)
\end{aligned} \qquad (14)$$

and then calculate the new value

$$y_{n+1} = y_n + \frac{h}{6}(k_1 + 2k_2 + 2k_3 + k_4), \qquad n = 0, 1, 2, \dots \qquad (15)$$

The purpose of the Runge-Kutta method is to achieve the accuracy of a Taylor series expansion without having to calculate higher-order derivatives. For instance, the algorithm given by (14) and (15) was derived by obtaining appropriate constants A, B, C, and D so that

$$y_{n+1} = y_n + Ak_1 + Bk_2 + Ck_3 + Dk_4$$

[†] Named after the German mathematicians Carl Runge (1856–1927) and Wilhelm Kutta (1867–1944).

agrees with that by the Taylor expansion (5) out to h^4, or the fifth term of the series (see Problem 18 in Exercises 10.3). Thus, the technique is often called the *fourth-order Runge-Kutta method*. The term $\frac{1}{6}(k_1 + 2k_2 + 2k_3 + k_4)$ appearing in (15) can be interpreted as an average slope over the interval $t_n \leq t \leq t_{n+1}$.

Although the Runge-Kutta formula is more complex than either of the Euler formulas, it yields results that are much more accurate than those by either of the Euler formulas. And it generally achieves this accuracy with larger comparative increments h. With the use of a modern computer, the calculations are rather routine and can be performed in a few seconds or less for many problems; thus, the additional complexity is of little concern.

Finally, we note that if the function F appearing in the IVP

$$y' = F(t, y), \qquad y(t_0) = y_0$$

does not explicitly depend upon y, that is, if $F = F(t)$, then (14) reduces to

$$\begin{aligned} k_1 &= F(t_n) \\ k_2 &= k_3 = F(t_n + h/2) \\ k_4 &= F(t_n + h) \end{aligned} \tag{16}$$

and (15) takes on the simpler form

$$y_{n+1} = y_n + \frac{h}{6}\left[F(t_n) + 4F(t_n + h/2) + F(t_n + h)\right] \tag{17}$$

which is actually equivalent to Simpson's[†] rule of integration on the interval $t_n \leq t \leq t_{n+1}$.

EXAMPLE 5

Use the Runge-Kutta method with step size $h = 0.2$ to find an approximate solution of

$$y' = -2ty^2, \qquad y(0) = 1$$

on the interval $0 \leq t \leq 1$.

Solution

We recognize this problem as the same one that appeared in Examples 1 and 3. With $F(t, y) = -2ty^2$, we first set up the expressions

$$\begin{aligned} k_1 &= -2t_n y_n^2 \\ k_2 &= -2(t_n + 0.1)(y_n + 0.1k_1)^2 \\ k_3 &= -2(t_n + 0.1)(y_n + 0.1k_2)^2 \\ k_4 &= -2(t_{n+1})(y_n + 0.2k_3)^2 \end{aligned}$$

[†] Named after Thomas Simpson (1710–1761).

TABLE 10.7 APPROXIMATE SOLUTIONS OF $y' = -2ty^2$,
$y(0) = 1$, USING THE EULER METHODS AND
THE RUNGE-KUTTA METHOD WITH $h = 0.2$

t	Euler y_n	Improved Euler y_n	Runge-Kutta y_n	Improved Euler $y_n(h = 0.1)$	Exact y
0.00	1.0000	1.0000	1.0000	1.0000	1.0000
0.20	1.0000	0.9600	0.9615	0.9614	0.9615
0.40	0.9200	0.8603	0.8621	0.8620	0.8621
0.60	0.7846	0.7350	0.7353	0.7356	0.7353
0.80	0.6369	0.6115	0.6098	0.6104	0.6098
1.00	0.5071	0.5033	0.5000	0.5009	0.5000

At the first step, therefore, our calculations yield

$n = 0$: $k_1 = 0$, $k_2 = -0.2$, $k_3 = -0.192$, $k_4 = -0.37$

and

$$y_1 = 1 - \frac{(0.2)}{6} \left[2(0.2) + 2(0.192) + 0.37 \right]$$

$$= 0.9615$$

The remaining calculations follow in a similar manner and are given in Table 10.7 along with the exact values. For comparison purposes we have also provided numerical results from both Euler methods for $h = 0.2$ and, in addition, the improved Euler method with $h = 0.1$. ■

EXAMPLE 6

Use the Runge-Kutta method with step size $h = 0.1$ to find an approximate solution of

$$y' = 5y + t - 1, \qquad y(0) = 2$$

on the interval $0 \le t \le 1$.

Solution

We previously solved this problem in Examples 2 and 4 using the Euler methods. Here $F(t, y) = 5y + t - 1$, which leads to

$n = 0$:

$$k_1 = F(0, 2) = 9$$

$$k_2 = F(0 + 0.05, 2 + 0.45) = 11.3$$

$$k_3 = F(0 + 0.05, 2 + 0.565) = 11.875$$

$$k_4 = F(0.1, 2 + 1.1875) = 15.0375$$

TABLE 10.8 APPROXIMATE SOLUTIONS OF $y' = 5y + t - 1$, $y(0) = 2$, USING THE EULER METHODS WITH STEP SIZE $h = 0.05$ AND THE RUNGE-KUTTA METHOD WITH $h = 0.1$

t	Euler $h = 0.05$ y_n	Improved Euler $h = 0.05$ y_n	Runge-Kutta $h = 0.1$ y_n	Exact y
0.0	2.0000	2.0000	2.0000	2.0000
0.1	3.0150	3.1605	3.1731	3.1736
0.2	4.6122	5.0785	5.1199	5.1216
0.3	7.1190	8.2399	8.3421	8.3463
0.4	11.047	13.443	13.667	13.676
0.5	17.196	21.996	22.457	22.476
0.6	26.816	36.050	36.959	36.997
0.7	41.857	59.134	60.879	60.952
0.8	65.370	97.042	100.32	100.46
0.9	102.12	159.28	165.36	165.61
1.0	159.55	261.47	272.57	273.04

TABLE 10.9 APPROXIMATE SOLUTIONS OF $y' = 5y + t - 1$, $y(0) = 2$, USING THE RUNGE-KUTTA METHOD WITH STEP SIZES $h = 0.05$ AND $h = 0.01$

t	$h = 0.05$ y_n	$h = 0.01$ y_n	Exact y
0.0	2.0000	2.0000	2.0000
0.1	3.1736	3.1736	3.1736
0.2	5.1215	5.1216	5.1216
0.3	8.3460	8.3463	8.3463
0.4	13.675	13.676	13.676
0.5	22.474	22.476	22.476
0.6	36.994	36.997	36.997
0.7	60.947	60.952	60.952
0.8	100.45	100.46	100.46
0.9	165.59	165.61	165.61
1.0	273.00	273.04	273.04

and consequently,

$$y_1 = 1 + \frac{(0.1)}{6} [9 + 2(11.3) + 2(11.875) + 15.0375]$$

$$= 3.1731$$

A tabulation of the remaining values is given in Table 10.8 as well as results obtained from the Euler methods. The superiority of the Runge-Kutta method is clearly demonstrated by this example.

Further numerical results using the Runge-Kutta method with step sizes $h = 0.05$ and $h = 0.01$ are given in Table 10.9. ∎

The BASIC computer program used to calculate the data in Tables 10.7 to 10.9 is given in Figure 10.6.

It can be shown that the *cumulative error* in using the Runge-Kutta method satisfies

$$|E_n| = |y(t_n) - y_n| \le Ch^4 \tag{18}$$

where C is a constant independent of h. Although this method provides far greater accuracy than either Euler method, in some cases the calculations required of the Runge-Kutta method at each step may be too plentiful if a large number of steps is required. Such calculations can be greatly reduced by use of certain **predictor-corrector methods,** such as the **Adams-Moulton method.** However, such methods create some inconveniences of their own in many cases by requiring a change in step size h as the calculations proceed. The proper numerical procedure to choose in a

```
100  DEF FNF(T,Y) = -2*T*Y*Y

110  INPUT "Initial Values of t and y";T,Y
120  INPUT "Step Size h";H
130  INPUT "Total Number of Steps m";M
140  INPUT "Print Step p ≤ m";P

150  FOR N = 1 TO M

160      K1 = FNF(T,Y)
170      K2 = FNF(T+H/2,Y+H*K1/2)
180      K3 = FNF(T+H/2,Y+H*K2/2)
190      K4 = FNF(T+H,Y+H*K3)
200      Y = Y + (H/6)*(K1 + 2*K2 + 2*K3 + K4)

210      X = X + H
220      IF INT(N/P) = N/P THEN PRINT T,Y

230  NEXT N
240  END
```

Figure 10.6 BASIC program for Runge-Kutta method.

particular situation will greatly depend upon the application and also upon the computer time available for making the necessary computer calculations.

Exercises 10.3

In Problems 1 to 10, use the Runge-Kutta method with $h = 0.1$ to obtain an approximation, to *four* places after the decimal, to the indicated value of y.

1. $y' = 2ty$, $y(1) = 1$; $y(1.5) = ?$

2. $y' = 1 + y^2$, $y(0) = 0$; $y(0.5) = ?$

3. $y' = (t + y - 1)^2$, $y(0) = 2$; $y(0.5) = ?$

4. $y' = t + y^2$, $y(0) = 1$; $y(0.5) = ?$

5. $y' = t^2 + y^2$, $y(0) = 1$; $y(0.5) = ?$

6. $y' = ty + \sqrt{y}$, $y(0) = 1$; $y(0.5) = ?$

7. $y' = ty^2 - y/t$, $y(1) = 1$; $y(1.5) = ?$

8. $y' = \sin(t + y)$, $y(0) = 0$; $y(0.5) = ?$

9. $y' = e^{-y}$, $y(0) = 0$; $y(0.5) = ?$

10. $y' = y - y^2$, $y(0) = \frac{1}{2}$; $y(0.5) = ?$

11. Using the Runge-Kutta method with $h = 0.2$, find an approximate value for $y(1)$, where y is the solution of

$$y' = y, \qquad y(0) = 1$$

(Your answer should approximate the value of e. Why?)

12. Using the Runge-Kutta method with $h = 0.2$,
   (a) Find an approximate value for $y(1)$, where y is the solution of

$$y' = \frac{1}{t^2 + 1}, \qquad y(0) = 0$$

   (b) Use your answer in (a) to approximate the value of π.
   Hint. The exact solution in (a) is $y = \tan^{-1} t$.

13. Using the step size $h = 0.2$, approximate the value $y(1.4)$, where y is the solution of

$$y' = t^2 + y^3, \qquad y(1) = 1$$

   (a) By Euler's formula
   (b) By the improved Euler's formula
   (c) By the Runge-Kutta formula

*14. Show that the Runge-Kutta method reduces to Simpson's rule of integration on the interval $t_n \le t \le t_{n+1}$ when the DE is of the form $y' = F(t)$. [See (15).]

*15. (*Three-term Taylor formula*) Using the first three terms of the Taylor series (5), show that it leads to the approximation formula

$$y_{n+1} = y_n + hy_n' + \tfrac{1}{2}h^2 y_n'', \qquad n = 0, 1, 2, \ldots$$

where $y' = F(t, y)$ and $y'' = (dF/dt)(t, y)$

*16. Use the three-term Taylor formula in Problem 15 to approximate the value $y(0.5)$ for the solution of

$$y' = -2ty^2, \qquad y(0) = 1$$

using $h = 0.1$. Compare your answer with those listed in Table 10.5.

*17. Show that when $F(t, y)$ is linear in t and y, the three-term Taylor formula given in Problem 15 reduces to the improved Euler's formula.

*18. Prove that the fourth-order Runge-Kutta formula (15) agrees with the Taylor expansion (5) out to terms involving h^4; that is, show that (15) and (5) are equivalent through the fifth term of (5).

10.4 Systems of Equations

The methods discussed in the previous sections for solving initial value problems featuring first-order DEs can be extended to higher-order DEs or, equivalently, systems of first-order equations. For instance, let us consider the system of two equations

$$\begin{aligned} x' &= F(t, x, y), & x(t_0) &= x_0 \\ y' &= G(t, x, y), & y(t_0) &= y_0 \end{aligned} \tag{19}$$

in which both F and G are suitably well behaved so that a unique solution exists.

For illustrative purposes we generalize the Euler method of Section 10.2.1, since it is the simplest of all to apply. Generalizations of the improved Euler method and Runge-Kutta method follow in a similar manner. In developing the Euler method for (19), we start by calculating

$$\begin{aligned} x_1 &= x_0 + hF(t_0, x_0, y_0) \\ y_1 &= y_0 + hG(t_0, x_0, y_0) \end{aligned}$$

for some prechosen positive increment h. In general, after n steps, our calculations involve

$$\begin{aligned} x_{n+1} &= x_n + hF(t_n, x_n, y_n) \\ y_{n+1} &= y_n + hG(t_n, x_n, y_n) \end{aligned} \tag{20}$$

where $t_n = t_0 + nh$, $n = 1, 2, 3, \ldots$.

EXAMPLE 7

Use Euler's method to approximate the solution of

$$x' = x - y + e^t, \qquad x(0) = 1$$
$$y' = 2x + 3y + e^{-t}, \qquad y(0) = 0$$

at the points $t = 0.1$ and $t = 0.2$ with $h = 0.1$.

Solution

Our first calculation leads to

$n = 0$:
$$x_1 = x_0 + h(x_0 - y_0 + e^{t_0})$$
$$= 1 + (0.1)(1 - 0 + 1)$$
$$= 1.2$$

and

$$y_1 = y_0 + h(2x_0 + 3y_0 + e^{-t_0})$$
$$= 0 + (0.1)[(2)(1) + (3)(0) + 1]$$
$$= 0.3$$

Similarly,

$n = 1$:
$$x_2 = x_1 + h(x_1 - y_1 + e^{t_1})$$
$$= 1.2 + (0.1)(1.2 - 0.3 + e^{0.1})$$
$$= 1.4005$$

and

$$y_2 = y_1 + h(2x_1 + 3y_1 + e^{-t_1})$$
$$= 0.3 + (0.1)[(2)(1.2) + (3)(0.3) + e^{-0.1}]$$
$$= 0.7205$$

For comparison purposes, the exact solution is

$$x(t) = \tfrac{1}{10}e^{2t}(21 \cos t - 13 \sin t) - e^t - \tfrac{1}{10}e^{-t}$$
$$y(t) = \tfrac{1}{5}e^{2t}(-4 \cos t + 17 \sin t) + e^t - \tfrac{1}{5}e^{-t}$$

from which we calculate

$$x(0.1) = 1.1980, \qquad y(0.1) = 0.3665; \qquad [x_1 = 1.2, y_1 = 0.3]$$
$$x(0.2) = 1.3818, \qquad y(0.2) = 0.8957; \qquad [x_2 = 1.4005, y_2 = 0.7205] \qquad \blacksquare$$

10.4.1 Higher-Order Equations

In Chapter 8 we found that higher-order DEs can always be reduced to a system of first-order DEs. For example, the IVP

$$x'' = G(t, x, x'), \qquad x(0) = x_0, \qquad x'(0) = y_0 \qquad (21)$$

is equivalent to the system of first-order equations

$$x' = y, \qquad x(0) = x_0$$
$$y' = G(t, x, y), \qquad y(0) = y_0 \tag{22}$$

and thus can be solved by methods just illustrated.[†]

■ **EXAMPLE 8**

Use Euler's method to approximate the solution of

$$x'' + t^2x' + 3x = t, \qquad x(0) = 1, \qquad x'(0) = 2$$

at the points $t = 0.1$ and $t = 0.2$ using $h = 0.1$.

Solution

The equivalent system of equations is

$$x' = y, \qquad x(0) = 1$$
$$y' = t - 3x - t^2y, \qquad y(0) = 2$$

Recalling (20), we obtain

$n = 0$:

$$x_1 = x_0 + hy_0$$
$$= 1 + (0.1)(2)$$
$$= 1.2$$

$$y_1 = y_0 + h(t_0 - 3x_0 - t_0^2 y_0)$$
$$= 2 + (0.1)[0 - (3)(1) - 0]$$
$$= 1.7$$

and

$n = 1$:

$$x_2 = x_1 + hy_1$$
$$= 1.2 + (0.1)(1.7)$$
$$= 1.37$$

$$y_2 = y_1 + h(t_1 - 3x_1 - t_1^2 y_1)$$
$$= 1.7 + (0.1)[0.1 - (3)(1.2) - (0.1)^2(1.7)]$$
$$= 1.3483$$

Hence, the solutions we seek are $x(0.1) \cong 1.2$ and $x(0.2) \cong 1.37$. Notice that y_2 is not needed for these values, but would be required at the next increment.   ■

[†] Some numerical analysts believe that greater accuracy is achieved by applying a numerical procedure directly to the higher-order DE rather than reducing the DE to a system of first-order equations and then applying a numerical procedure. For example, see Peter Henrici, *Discrete Variable Methods in Ordinary Differential Equations*, Wiley, New York, 1962.

Exercises 10.4

In Problems 1 to 5, use Euler's method with $h = 0.1$ to determine approximate values of the solution at $t = 0.1$ and $t = 0.2$.

1. $x' = x - 4y$, $x(0) = 1$
  $y' = -x + y$, $y(0) = 0$

2. $x' = x + y$, $x(0) = 1$
  $y' = x - y$, $y(0) = 2$

3. $x' = 2x + ty$, $x(0) = 1$
  $y' = xy$, $y(0) = 1$

4. $x' = x + y + t$, $x(0) = 1$
  $y' = 4x - 2y$, $y(0) = 0$

5. $x' = -tx - y - 1$, $x(0) = 1$
  $y' = x$, $y(0) = 1$

6. The generalizations of the Runge-Kutta method to the system of equations

$$x' = F(t, x, y), \qquad x(t_0) = x_0$$

$$y' = G(t, x, y), \qquad y(t_0) = y_0$$

leads to the formulas

$$x_{n+1} = x_n + \frac{h}{6}(K_1 + 2K_2 + 2K_3 + K_4)$$

$$y_{n+1} = y_n + \frac{h}{6}(L_1 + 2L_2 + 2L_3 + L_4)$$

where

$$K_1 = F(t_n, x_n, y_n)$$

$$L_1 = G(t_n, x_n, y_n)$$

$$K_2 = F(t_n + h/2, x_n + hK_1/2, y_n + hL_1/2)$$

$$L_2 = G(t_n + h/2, x_n + hK_1/2, y_n + hL_1/2)$$

$$K_3 = F(t_n + h/2, x_n + hK_2/2, y_n + hL_2/2)$$

$$L_3 = G(t_n + h/2, x_n + hK_2/2, y_n + hL_2/2)$$

$$K_4 = F(t_n + h, x_n + hK_3, y_n + hL_3)$$

$$L_4 = G(t_n + h, x_n + hK_3, y_n + hL_3)$$

Use these formulas to solve Problem 1 at $t = 0.1$.

7. Use the Runge-Kutta formulas in Problem 6 to solve Problem 2 at $t = 0.1$.

8. Use the Runge-Kutta formulas in Problem 6 to solve Problem 3 at $t = 0.1$.

9. Use the Runge-Kutta formulas in Problem 6 to solve Problem 4 at $t = 0.1$.

10. Use the Runge-Kutta formulas in Problem 6 to solve Problem 5 at $t = 0.1$.

In Problems 11 and 12, convert to a system of first-order DEs and use the Euler method with $h = 0.1$ to determine approximate values of the solution at $t = 0.1$ and $t = 0.2$.

11. $x'' + tx' + x = 0$, $x(0) = 1$, $x'(0) = 2$

12. $x'' + t^2x' + 3x = t$, $x(0) = 1$, $x'(0) = 1$

13. Use the Runge-Kutta formulas in Problem 6 to solve Problem 11 at $t = 0.1$.

14. Use the Runge-Kutta formulas in Problem 6 to solve Problem 12 at $t = 0.1$.

10.5 Chapter Summary

In this chapter we have briefly discussed three methods of obtaining numerical approximations to the true solution of IVPs of the form

$$y' = F(t, y), \qquad y(t_0) = y_0 \tag{23}$$

The first method, called **Euler's method,** makes use of the recurrence formula

$$y_{n+1} = y_n + hF(t_n, y_n), \qquad n = 0, 1, 2, \ldots \tag{24}$$

whereas the **improved Euler's method** utilizes the recurrence formula

$$y_{n+1} = \tfrac{1}{2}h[F(t_n, y_n) + F(t_{n+1}, y^*_{n+1})], \qquad n = 0, 1, 2, \ldots \tag{25a}$$

where

$$y^*_{n+1} = y_n + hF(t_n, y_n) \tag{25b}$$

The third and most important numerical technique, called the **Runge-Kutta method,** is based on the formula

$$y_{n+1} = y_n + \frac{h}{6}(k_1 + 2k_2 + 2k_3 + k_4), \qquad n = 0, 1, 2, \ldots \tag{26}$$

where

$$k_1 = F(t_n, y_n)$$
$$k_2 = F(t_n + h/2, y_n + hk_1/2)$$
$$k_3 = F(t_n + h/2, y_n + hk_2/2) \tag{27}$$
$$k_4 = F(t_{n+1}, y_n + hk_3)$$

In each case the number h is the step size or distance between t_{n+1} and t_n, that is, $t_{n+1} = t_n + h, n = 0, 1, 2, \ldots$. Generally speaking, greater accuracy can be achieved in all methods by choosing smaller values of h. Of course, the number of computations necessary to reach a particular t value will increase with decreasing h.

CHAPTER 11

Boundary Value Problems and Fourier Series

If the auxiliary conditions of a DE are specified at two points on the interval of interest (usually the endpoints), the resulting problem is called a (two-point) **boundary value problem.** Unless the boundary conditions are carefully chosen, however, a given boundary value problem may not lead to a unique solution.

In Section 11.2 we look briefly at a physical application involving the **static displacements of an elastic string** supporting a load and then introduce the **general theory of boundary value problems.** Homogeneous DEs with homogeneous boundary conditions are separately examined in Section 11.3 in connection with **eigenvalue problems,** a class of problems for which the solution is dependent upon a parameter λ. By proper choices of λ (called **eigenvalues**), an infinite collection of solutions (called **eigenfunctions**) can often be found.

We take up the topic of **Fourier series** in Sections 11.4 to 11.6. In Section 11.4 we motivate the subject by forming a connection with eigenvalue problems, while in Section 11.5 a more traditional approach to the subject is presented in terms of representations of **periodic functions. Sturm-Liouville problems** and **generalized** Fourier series are briefly discussed in Section 11.6, where we also introduce the notion of **orthogonality.**

The last section contains some **historical comments.**

11.1 Introduction

Boundary value problems (BVPs) are closely associated with problems of *static equilibrium configurations* and *steady-state phenomena*. Most of the DEs occurring in these areas of application are second order, the general form of which is given by

$$A_2(x)y'' + A_1(x)y' + A_0(x)y = F(x), \qquad a < x < b \qquad (1)$$

where a and b denote the endpoints of the interval of interest. For discussion purposes we generally assume the coefficients $A_0(x)$, $A_1(x)$, and $A_2(x)$ are continuous functions on $a < x < b$ and that $A_2(x) \neq 0$ on $a \leq x \leq b$. Recall that in the case of IVPs the auxiliary conditions are all specified at a single point, say at $x = x_0$, and we seek solutions on the semi-infinite interval $x > x_0$. In the case of BVPs, however, we normally specify auxiliary conditions at two points, say at $x = a$ and $x = b$, and then seek solutions on the interval $a < x < b$. Typical auxiliary conditions, called **boundary conditions,** are various specializations of

$$\begin{aligned} h_1 y(a) + h_2 y'(a) = \alpha \\ h_3 y(b) + h_4 y'(b) = \beta \end{aligned} \qquad (2)$$

where h_1, h_2, h_3, h_4, α, and β are constants. Unless the boundary conditions for a particular problem are carefully chosen, a given BVP may have *no solution* or *more than one solution,* rather than a *unique solution* as is generally the case for IVPs (recall Theorem 5.1).

11.2 Boundary Value Problems

Because the existence of a unique solution is so closely tied to the prescription of boundary conditions, the theory of BVPs is inherently more complicated than that for IVPs. Before discussing the general theory, however, it may be instructive to consider a simple physical example leading to a BVP.

11.2.1 Deflections of an Elastic String

Let us consider an elastic string stretched tightly with constant tension T between supports located at $x = 0$ and $x = b$ (see Figure 11.1). In addition, we assume the string is supporting a distributed vertical load of intensity (force per unit length) $q(x)$. The problem is to determine the equilibrium displacement y of the string. Such a problem might physically correspond to a telephone cable stretched between two poles where the distributed load is simply the force of gravity acting on the cable. If the cable has uniform mass density along its length, the distributed load $q(x)$ is constant, but is a function of x in case the cable has variable mass density.

To obtain a simple model from which to determine the displacements, it is necessary to make certain simplifying assumptions. Primarily, we assume the displacement from the x axis is a "small" movement in the vertical plane and that the string is perfectly elastic with constant mass density.

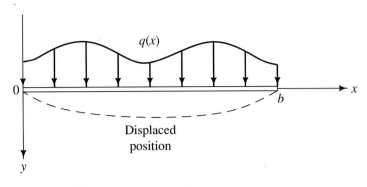

Figure 11.1 Elastic string under a load.

Let us now focus our attention on a small element of the string from x to $x + \Delta x$, as shown in Figure 11.2. The total force acting on the elemental length of string Δx is simply $q(x) \, \Delta x$ (i.e., the force is essentially constant over the short length Δx). For small displacements we only need to consider forces acting in the vertical plane, which include the external force $q(x) \, \Delta x$ and the y components of the tensile force T at x and at $x + \Delta x$. From Figure 11.2, the y component of the force at x is given by

$$-T \sin \theta_1 \cong -T \tan \theta_1$$

$$\cong -Ty'(x) \tag{3}$$

where we are using the small angle approximation $\sin \theta_1 \cong \tan \theta_1$ and the fact that $\tan \theta_1$ is simply the slope $y'(x)$ of the string at x. Likewise, at $x + \Delta x$ the y component of the tensile force is

$$T \sin \theta_2 \cong T \tan \theta_2$$

$$\cong Ty'(x + \Delta x) \tag{4}$$

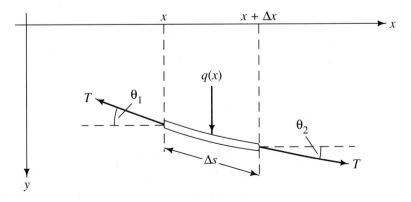

Figure 11.2 Small element of the elastic string.

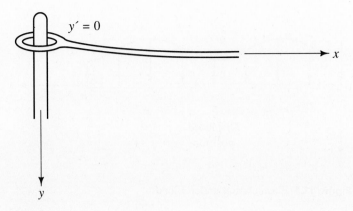

Figure 11.3 Free end support.

The sum of vertical forces must be zero, and this leads to

$$-Ty'(x) + Ty'(x + \Delta x) + q(x)\,\Delta x = 0$$

or, upon division by Δx and T,

$$\frac{y'(x + \Delta x) - y'(x)}{\Delta x} + \frac{1}{T}\,q(x) = 0 \tag{5}$$

Finally, by taking the limit as $\Delta x \to 0$, we obtain the *equation of equilibrium*

$$y'' = -\frac{q(x)}{T}, \qquad 0 < x < b \tag{6}$$

The string may be attached to various types of support at the ends $x = 0$ and $x = b$. Most common are fixed supports, which describe the displacements at the endpoints. These are mathematically described by

$$y(0) = \alpha, \qquad y(b) = \beta \tag{7}$$

where α and β are constants. Other boundary conditions are also possible for the string, although they are physically difficult to achieve in practice. For instance, we might prescribe the condition $y' = 0$ at one endpoint; physically, this corresponds to a "free end." This condition could be realized, for example, by considering the free end looped over a frictionless peg that maintains a zero slope on the string at this end while allowing small vertical movements (see Figure 11.3).

EXAMPLE 1

Find the shape of a telephone cable stretched tightly between poles of the same height located at $x = 0$ and $x = b$ and hanging under its own weight.

Solution

We assume the mass is uniformly distributed along the cable so that we may set $q(x) = mg$, where m is the (constant) mass per unit length and g is the gravitational constant. Also, because the poles have the same height, we assume the displacement at the endpoints is zero, that is, $y(0) = 0$, $y(b) = 0$. Hence, the governing BVP becomes

$$y'' = -\frac{mg}{T}, \qquad 0 < x < b, \qquad y(0) = 0, \qquad y(b) = 0$$

Integrating the above DE twice, we obtain the general solution

$$y = -\frac{mg}{2T}x^2 + C_1 x + C_2$$

The boundary conditions require that $C_1 = mgb/2T$ and $C_2 = 0$, and thus the solution is

$$y = -\frac{mg}{2T}x^2 + \frac{mg}{2T}bx$$

$$= \frac{mg}{2T}x(b - x)$$

which says the telephone cable hangs in a parabolic arc.[†]   ■

If, in Example 1, we had imposed the boundary conditions $y'(0) = y'(b) = 0$, which physically correspond to "free ends" at the endpoints, the problem would have *no solution*. Physically, we may be able to understand why a solution doesn't exist, but the mathematical reasoning why this should be so is less clear.

11.2.2 General Theory

To further illustrate some of the subtleties associated with solving BVPs, let us consider the following examples.

EXAMPLE 2

Solve

$$y'' + y = 0, \qquad 0 < x < \frac{\pi}{2}, \qquad y(0) = 0, \qquad y(\pi/2) = 3$$

Solution

The general solution of the DE is

$$y = C_1 \cos x + C_2 \sin x$$

Applying the first boundary condition, we see that

$$y(0) = C_1 = 0$$

[†] Actually, a more accurate model of the displacement of a telephone line leads to a nonlinear DE, the solution of which is a *catenary* (see Problem 35 in Exercises 11.2).

The second condition applied to the remaining solution yields

$$y(\pi/2) = C_2 = 3$$

and thus we obtain the unique solution

$$y = 3 \sin x$$ ∎

EXAMPLE 3

Solve

$$y'' + y = 0, \qquad 0 < x < \pi, \qquad y(0) = 0, \qquad y(\pi) = 3$$

Solution

Again the general solution is

$$y = C_1 \cos x + C_2 \sin x$$

and the first boundary condition requires that $C_1 = 0$. By imposing the second boundary condition, we are led to

$$y(\pi) = C_2 \cdot 0 = 3$$

However, there is no value of C_2 for which this last relation is satisfied. We conclude, therefore, that this BVP has *no solution*. ∎

Notice that only the interval of interest is different in Examples 2 and 3. Yet, in Example 2 we found a unique solution while no solution exists for Example 3. By slightly changing the second boundary condition in Example 3, we can arrive at still another conclusion as illustrated below in Example 4.

EXAMPLE 4

Solve

$$y'' + y = 0, \qquad 0 < x < \pi, \qquad y(0) = 0, \qquad y(\pi) = 0$$

Solution

Based on Examples 2 and 3, the solution satisfying $y(0) = 0$ is given by

$$y = C_2 \sin x$$

If we impose the remaining boundary condition on this solution, we see that

$$y(\pi) = C_2 \cdot 0 = 0$$

which is satisfied by any constant C_2. Hence, in this case we obtain the *infinite family of solutions*

$$y = C_2 \sin x$$ ∎

where C_2 is any constant.

These examples illustrate the sharp contrast between the general behavior of BVPs and that of IVPs, namely, that the boundary conditions are crucial in determining whether the BVP has a unique solution. In the following development of the general theory of BVPs we find it helpful to simplify notation by introducing the **differential operator** M and **boundary operators** B_1 and B_2 defined, respectively, by

$$M[y] = A_2(x)y'' + A_1(x)y' + A_0(x)y \qquad (8)$$

and

$$B_1[y] = h_1 y(a) + h_2 y'(a)$$
$$B_2[y] = h_3 y(b) + h_4 y'(b) \qquad (9)$$

In terms of these operators, we can express the general BVP of interest more compactly by

$$M[y] = F(x), \qquad a < x < b, \qquad B_1[y] = \alpha, \qquad B_2[y] = \beta \qquad (10)$$

For discussion purposes let us consider separately the cases of *homogeneous* and *nonhomogeneous* DEs.

Case I. Homogeneous DEs

When $F(x) \equiv 0$ on $a \le x \le b$, the problem described by (10) reduces to

$$M[y] = 0, \qquad a < x < b, \qquad B_1[y] = \alpha, \qquad B_2[y] = \beta \qquad (11)$$

Let us assume the general solution of $M[y] = 0$ is given by

$$y = C_1 y_1(x) + C_2 y_2(x) \qquad (12)$$

where C_1 and C_2 are arbitrary constants. The boundary conditions in (11) imposed upon this general solution lead to the *system of linear equations*

$$C_1 B_1[y_1] + C_2 B_1[y_2] = \alpha$$
$$C_1 B_2[y_1] + C_2 B_2[y_2] = \beta \qquad (13)$$

To solve (13) for C_1 and C_2, we begin by multiplying the first equation by $B_2[y_2]$ and the second equation by $B_1[y_2]$. Thus, we have

$$C_1 B_1[y_1]B_2[y_2] + C_2 B_1[y_2]B_2[y_2] = \alpha B_2[y_2]$$
$$C_1 B_1[y_2]B_2[y_1] + C_2 B_1[y_2]B_2[y_2] = \beta B_1[y_2]$$

and by subtracting the second equation from the first, we obtain the solution

$$C_1 = \frac{1}{\Delta}(\alpha B_2[y_2] - \beta B_1[y_2]) \qquad (14)$$

where

$$\Delta = B_1[y_1]B_2[y_2] - B_1[y_2]B_2[y_1] \qquad (15)$$

In a similar fashion we find that

$$C_2 = \frac{1}{\Delta}(\beta B_1[y_1] - \alpha B_2[y_1]) \qquad (16)$$

The formal solution of the BVP described by (11) is therefore that given by (12), where C_1 and C_2 are defined by (14) to (16). If $\Delta \neq 0$, the solution of (11) will be *unique* (see Example 2), but if $\Delta = 0$, either *no solution* exists (see Example 3) or there are *infinitely many solutions* (see Example 4). If $\Delta \neq 0$ and $\alpha = \beta = 0$, there is only the **trivial solution** $y = 0$ since $C_1 = C_2 = 0$ in this case. When $\Delta = 0$, the case $\alpha = \beta = 0$ is special since, unlike IVPs with homogeneous DE and initial conditions, *nontrivial solutions* are possible. In particular, we have the following theorem.

Theorem 11.1 *If y_1 and y_2 are linearly independent solutions of the homogeneous equation $M[y] = 0$, then the BVP*

$$M[y] = 0, \quad a < x < b, \quad B_1[y] = 0, \quad B_2[y] = 0$$

has only the trivial solution if $\Delta \neq 0$, where

$$\Delta = B_1[y_1]B_2[y_2] - B_1[y_2]B_2[y_1]$$

Nontrivial solutions exist if and only if $\Delta = 0$.

Case II. Nonhomogeneous DEs

For the nonhomogeneous DE and boundary conditions described by

$$M[y] = F(x), \quad a < x < b, \quad B_1[y] = \alpha, \quad B_2[y] = \beta \qquad (17)$$

we assume the general solution is given by

$$y = C_1 y_1(x) + C_2 y_2(x) + y_P(x) \qquad (18)$$

where C_1 and C_2 are arbitrary constants and y_P is any *particular solution*. Imposing the boundary conditions on the solution, we obtain the system of linear equations

$$\begin{aligned} C_1 B_1[y_1] + C_2 B_1[y_2] &= \alpha - B_1[y_P] \\ C_1 B_2[y_1] + C_2 B_2[y_2] &= \beta - B_2[y_P] \end{aligned} \qquad (19)$$

which, by analogy with (13), has the solution

$$C_1 = \frac{1}{\Delta}(\gamma_1 B_2[y_2] - \gamma_2 B_1[y_2])$$

$$C_2 = \frac{1}{\Delta}(\gamma_2 B_1[y_1] - \gamma_1 B_2[y_1]) \qquad (20)$$

where Δ is defined by (15) and

$$\begin{aligned} \gamma_1 &= \alpha - B_1[y_P] \\ \gamma_2 &= \beta - B_2[y_P] \end{aligned} \qquad (21)$$

Once again we see that (17) has a *unique solution* if and only if $\Delta \neq 0$, but if $\Delta = 0$ then (17) has either *no solution* or *infinitely many solutions*. Let us consider some examples.

EXAMPLE 5

Solve

$$y'' + y = x - 1, \qquad 0 < x < 1, \qquad y(0) = -1, \qquad y(1) = 1$$

Solution

By inspection, we see that $y_P = x - 1$ is a particular solution. Thus, the general solution is

$$y = C_1 \cos x + C_2 \sin x + x - 1$$

The first boundary condition requires that

$$y(0) = C_1 - 1 = -1$$

or $C_1 = 0$, while the second condition leads to

$$y(1) = C_2 \sin 1 = 1$$

from which we find $C_2 = 1/(\sin 1)$. In this case we have the *unique solution*

$$y = \frac{\sin x}{\sin 1} + x - 1 \qquad \blacksquare$$

EXAMPLE 6

Solve

$$y'' + y = x - 1, \qquad 0 < x < \pi, \qquad y(0) = -1, \qquad y(\pi) = 1$$

Solution

From Example 5, the solution satisfying the first boundary condition is

$$y = C_2 \sin x + x - 1$$

The second boundary condition requires that

$$y(\pi) = \pi - 1 = 1$$

which is impossible. We conclude, therefore, that this problem has *no solution*. $\blacksquare$

EXAMPLE 7

Solve

$$y'' + y = x, \qquad 0 < x < \pi, \qquad y(0) = \pi, \qquad y(\pi) = 0$$

Solution

A particular solution is $y_P = x$, and therefore the general solution is

$$y = C_1 \cos x + C_2 \sin x + x$$

The boundary conditions lead to the system of equations

$$y(0) = C_1 = \pi$$
$$y(\pi) = -C_1 + \pi = 0$$

which are both satisfied by $C_1 = \pi$ and arbitrary C_2. Hence, we obtain the *infinite collection of solutions*

$$y = \pi \cos x + C_2 \sin x + x$$

where C_2 is arbitrary.   ■

Summarizing the results of our observations here, we have the following theorem.

Theorem 11.2 *If y_1 and y_2 are linearly independent solutions of the homogeneous equation $M[y] = 0$, then the BVP*

$$M[y] = F(x), \qquad a < x < b, \qquad B_1[y] = \alpha, \qquad B_2[y] = \beta$$

has a unique solution if and only if $\Delta \neq 0$, where

$$\Delta = B_1[y_1]B_2[y_2] - B_1[y_2]B_2[y_1]$$

If $\Delta = 0$, the problem has either no solution or infinitely many solutions.

Exercises 11.2

In Problems 1 to 20, find nontrivial solutions (if possible). State whether the solution is unique or not.

1. $y'' = 0$, $y(0) = 0$, $y(1) = 0$

2. $y'' + y = 0$, $y(0) = 0$, $y'(\pi) = 0$

3. $y'' + 9y = 0$, $y'(0) = 0$, $y'(\pi) = 0$

4. $y'' + y = 0$, $y(0) = 0$, $y(\pi) = 0$

5. $y'' + \pi^2 y = 0$, $y(0) + y(1) = 0$, $y'(0) + y'(1) = 0$

6. $y'' - y = 0$, $y(0) = 0$, $y(1) = 0$

7. $y'' - 3y' + 2y = 0$, $y(0) = 0$, $y(1) = 0$

8. $y'' + y = 0$, $y(-1) = 0$, $y(1) = 1$

9. $y'' + y = 0$, $y(0) = 3$, $y(\pi) = -2$

■**10.** $y'' + y = 0$, $y(0) = 3$, $y(\pi) = -3$

11. $y'' + k^2 y = 0$, $y(0) = 0$, $y(1) = 2$, $(k > 0)$

12. $y'' - y = 0$, $y(0) = 1$, $y(1) = -1$

13. $y'' - y = 0$, $y'(0) + 3y(0) = 0$, $y'(1) + y(1) = 1$

14. $y'' - 3y' + 2y = 0$, $y(0) = 1$, $y(1) = 0$

15. $y'' - 6y' + 9y = 0$, $y(0) = 0$, $y(1) = 3$

16. $y'' - y' - 6y = 0$, $y(0) = 0$, $y(1) = e^3$

17. $x^2 y'' - 5xy' + 25y = 0$, $y(1) = 0$, $y(e^\pi) = 0$

*****18.** $x^2 y'' - 5xy' + 5y = 0$, $y(0) = 0$, $y(1) = 0$

19. $x^2 y'' - xy' + y = 0$, $y(1) = 1$, $y(e^2) = 0$

■**20.** $x^2 y'' + xy' - y = 0$, $y(1) = 0$, $y(e) = \frac{1}{2}e$

In Problems 21 to 32, find the solution (if possible). State whether the solution is unique or not.

21. $y'' = x$, $y(0) = 1$, $y(1) - y'(1) = 0$

22. $y'' = \sin \pi x$, $y(0) = 0$, $y'(1) = 0$

23. $y'' = 3 \sin \pi x$, $y(0) = -2$, $y'(1) = 7/\pi$

24. $y'' = 1$, $y'(0) = 0$, $y'(1) = 1$

25. $y'' + 4y = e^{-x}$, $y(0) = -1$, $y'(\pi/2) = -\frac{1}{5}e^{-\pi/2}$

26. $y'' - y = e^x$, $y(0) = 0$, $y'(1) = 1$

27. $y'' + 2y' + y = x$, $y(0) = -3$, $y(1) = -1$

■**28.** $y'' - 6y' + 9y = 9$, $y(0) = 1$, $y(1) = 3$

29. $y'' - y' - 6y = e^x$, $y(0) = 0$, $y(1) = 1$

30. $x^2 y'' + xy' - y = x$, $y(1) = 0$, $y(e) = \frac{1}{2}e$

*****31.** $2x^2 y'' + 3xy' - y = 4x\sqrt{x}$, $y(0) = 0$, $y(1) = 0$

32. $xy'' + y' = x$, $y(1) = 0$, $y(e) = \frac{1}{4}e^2$

*****33.** When the tension T in a taut string is not constant, the linear equation of equilibrium is given by

$$\frac{d}{dx}[T(x)y'] = -q(x)$$

Solve this DE when $T(x) = 1/(1 + x)$, $q(x) = 1$, and the boundary conditions are $y(0) = 0$, $y(1) = 0$.

34. Solve Example 1 when the boundary conditions are $y(0) = \alpha$, $y(b) = \beta$, where α and β are constants. Is the solution still a parabola?

*35. A telephone cable between supports at $x = 0$ and $x = 1$ hangs at rest under its own weight. For small displacements the equation of equilibrium is approximately that given by (6), while for larger displacements the governing DE is the *nonlinear* DE

$$y'' = -\frac{mg}{T}\sqrt{1 + (y')^2},$$

$$y(0) = 0, \qquad y(1) = 0$$

By letting $v = y'$, solve the resulting DE by separation of variables and show that

$$y = \frac{T}{mg}\left\{\cosh\frac{mg}{2T} - \cosh\left[\frac{mg}{T}\left(x - \frac{1}{2}\right)\right]\right\}$$

11.3 Eigenvalue Problems

In some cases the DE contains a parameter λ that may assume various values. Usually the DE is then of the form

$$M[y] + \lambda y = 0, \qquad a < x < b \tag{22}$$

where $M = A_2(x)D^2 + A_1(x)D + A_0(x)$. The solutions in such instances must depend upon both x and λ. Thus, if y_1 and y_2 constitute linearly independent solutions of (22), we write the general solution as

$$y = C_1 y_1(x, \lambda) + C_2 y_2(x, \lambda)$$

Determining which nontrivial solutions of this general family (if any) satisfy the **homogeneous boundary conditions**

$$B_1[y] = 0, \qquad B_2[y] = 0 \tag{23}$$

where B_1 and B_2 are defined by (9), is what we call an **eigenvalue problem.** That is, for certain values of λ there may exist nontrivial solutions of the boundary value problem (22) and (23). These special values of λ are called **eigenvalues,** and the corresponding nontrivial solutions are called **eigenfunctions.**

> *Remark.* Eigenvalues and eigenfunctions are also called **characteristic values** and **characteristic functions.**

In applications the eigenvalues and eigenfunctions have important physical interpretations. For example, in vibration problems the eigenvalues are proportional to the squares of the natural frequencies of vibration, while the eigenfunctions provide the natural configuration modes of the system. The eigenvalues denote the possible energy states of a system in quantum mechanics problems and the eigenfunctions are the wave functions. In buckling problems the eigenvalues are related to the critical compressive loads that a column can withstand without buckling and the eigenfunctions provide us with the possible buckling configurations of the column. Thus, the interpretations vary widely, as do the areas of application.

■───────
EXAMPLE 8

Find the eigenvalues and eigenfunctions of

$$y'' + \lambda y = 0, \qquad 0 < x < p, \qquad y(0) = 0, \qquad y(p) = 0$$

Solution

We first attempt to find the eigenvalues, which we tacitly assume are real. Because the solution of the DE may assume different functional forms, depending upon the value of λ, we must consider three separate cases corresponding to these different functional forms.[†]

Case I. Let us first assume that $\lambda = 0$, which leads to the general solution

$$y = C_1 + C_2 x$$

The first boundary condition $y(0) = 0$ requires that $C_1 = 0$, and the second condition likewise leads to $C_2 = 0$. Thus, $\lambda = 0$ is not an eigenvalue since the only possible solution is the trivial solution $y = 0$.

Case II. If λ is negative, it is helpful to set $\lambda = -k^2 < 0$, and then the general solution can be expressed in either the form

$$y = C_1 e^{-kx} + C_2 e^{kx}$$

or

$$y = C_1 \cosh kx + C_2 \sinh kx$$

In solving eigenvalue problems on *finite domains,* it is usually preferred to express the general solution in terms of hyperbolic functions rather than exponential functions (although either solution form is acceptable). By applying the boundary conditions to the general solution written in terms of hyperbolic functions, we find

$$y(0) = C_1 = 0$$

$$y(p) = C_2 \sinh kp = 0$$

but since $\sinh kp \neq 0$ for $kp \neq 0$, we deduce that $C_1 = C_2 = 0$, which leads again to the trivial solution $y = 0$. Therefore, there are no negative eigenvalues.

Case III. Next we set $\lambda = k^2 > 0$, and the general solution is

$$y = C_1 \cos kx + C_2 \sin kx$$

Imposing the boundary conditions leads to $C_1 = 0$ and

$$C_2 \sin kp = 0$$

───────
[†] The auxiliary equation is $m^2 + \lambda = 0$, with roots $m = \pm\sqrt{-\lambda}$. Thus, the three cases correspond to $\lambda = 0$, $\lambda < 0$, and $\lambda > 0$.

If we restrict $C_2 \neq 0$, this last condition can be satisfied only if kp is an integral multiple of π, that is, only if $k = n\pi/p$ $(n = 1, 2, 3, \ldots)$.[†] Hence, the eigenvalues are given by

$$\lambda_n = k_n^2 = \frac{n^2\pi^2}{p^2}, \qquad n = 1, 2, 3, \ldots$$

and the corresponding eigenfunctions are any (nonzero) multiples of

$$y = \phi_n(x) = \sin \frac{n\pi x}{p}, \qquad n = 1, 2, 3, \ldots$$

where we set $C_2 = 1$ for convenience.   ■

EXAMPLE 9

Find the eigenvalues and eigenfunctions belonging to

$$y'' + \lambda y = 0, \qquad 0 < x < 1, \qquad y(0) = 0, \qquad y(1) + y'(1) = 0$$

Solution

Again it is necessary to consider the same three cases as in Example 8.

Case I.   For $\lambda = 0$, the general solution is

$$y = C_1 + C_2 x$$

The two boundary conditions demand that $C_1 = C_2 = 0$, forcing the trivial solution. Therefore, $\lambda = 0$ is not an eigenvalue.

Case II.   For negative λ, we again set $\lambda = -k^2 < 0$, leading to the general solution

$$y = C_1 \cosh kx + C_2 \sinh kx$$

The first boundary condition requires that $C_1 = 0$ while the second condition yields

$$C_2(\sinh k + k \cosh k) = 0$$

For $C_2 \neq 0$, this last relation is the same as

$$k = -\tanh k$$

To see if there are any solutions of this *transcendental equation,* we plot the graphs of $u = k$ and $u = -\tanh k$ and look for intersections. From Figure 11.4 it is clear that no intersections of these curves occur for $k > 0$, and hence we conclude that there are no negative eigenvalues.

Case III.   If $\lambda = k^2 > 0$, the general solution is

$$y = C_1 \cos kx + C_2 \sin kx$$

[†] Since $\lambda = k^2$, we can take k positive without loss of generality.

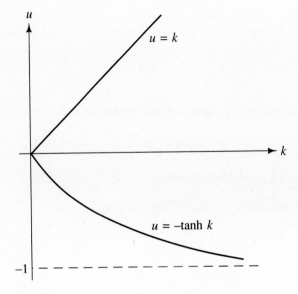

Figure 11.4

The first boundary condition requires that $C_1 = 0$ and the second condition leads to the relation

$$C_2(\sin k + k \cos k) = 0$$

or, for $C_2 \neq 0$,

$$k = -\tan k$$

Once again we plot graphs of $u = k$ and $u = -\tan k$ to see if any intersections occur. Looking at Figure 11.5 we see that there are an infinite number of values k_n ($n = 1, 2, 3, \ldots$) where intersections occur, but their exact values must be determined numerically. Nonetheless, we claim the eigenvalues are given by

$$\lambda_n = k_n^2, \qquad n = 1, 2, 3, \ldots$$

with corresponding eigenfunctions

$$\phi_n(x) = \sin k_n x, \qquad n = 1, 2, 3, \ldots$$

Although we did not determine the exact numerical values of the eigenvalues, we do note that the intersections of $u = k$ with $u = -\tan k$ are close to the vertical asymptotes of $-\tan k$ for large values of k. Hence, the larger eigenvalues satisfy the approximate relation

$$\lambda_n \cong \tfrac{1}{4}(2n - 1)^2\pi^2, \qquad n \gg 1 \qquad \blacksquare$$

The three cases identified in the hunt for eigenvalues in Examples 8 and 9 are directly related to the form of the equation $y'' + \lambda y = 0$. For more general DEs the cases are usually different. For instance, in solving the second-order, constant-

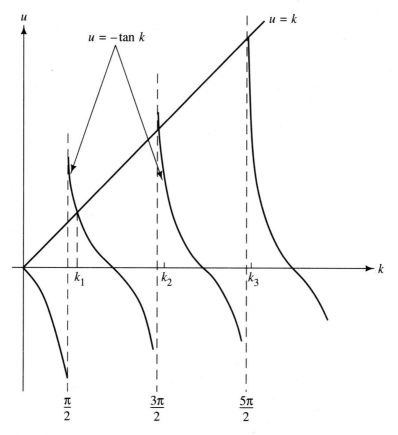

Figure 11.5

coefficient DE

$$ay'' + by' + \lambda y = 0 \qquad (24)$$

we are led to the auxiliary equation

$$am^2 + bm + \lambda = 0$$

The roots of this auxiliary equation are given by

$$m = \frac{-b \pm \sqrt{b^2 - 4a\lambda}}{2a}$$

and thus the three cases involved in the search for eigenvalues correspond to

$$b^2 - 4a\lambda = 0$$
$$b^2 - 4a\lambda > 0 \qquad (25)$$
$$b^2 - 4a\lambda < 0$$

EXAMPLE 10   Find the eigenvalues and eigenfunctions of

$$y'' + y' + \lambda y = 0, \qquad 0 < x < 3, \qquad y(0) = 0, \qquad y(3) = 0$$

Solution

The given DE has the auxiliary equation

$$m^2 + m + \lambda = 0$$

with roots

$$m = -\tfrac{1}{2} \pm \tfrac{1}{2}\sqrt{1 - 4\lambda}$$

Thus, the three cases to consider are:

$$1 - 4\lambda = 0$$
$$1 - 4\lambda > 0$$
$$1 - 4\lambda < 0$$

We leave it to the reader this time to show that the first two cases lead to only the trivial solution $y = 0$. For the last case we set $1 - 4\lambda = -k^2 < 0$, which yields

$$m = -\frac{1}{2} \pm i\,\frac{k}{2}$$

Hence, the general solution is

$$y = e^{-x/2}\left(C_1 \cos \frac{kx}{2} + C_2 \sin \frac{kx}{2}\right)$$

and applying the boundary conditions, we find that $C_1 = 0$ and

$$y(3) = C_2 e^{-3/2} \sin \frac{3k}{2} = 0$$

For $C_2 \neq 0$, it follows that $3k/2 = n\pi$, or $k = 2n\pi/3$ $(n = 1, 2, 3, \ldots)$. The eigenvalues are therefore given by

$$\lambda_n = \frac{1}{4}(1 + k_n^2) = \frac{1}{4} + \frac{n^2 \pi^2}{9}, \qquad n = 1, 2, 3, \ldots$$

and the eigenfunctions by

$$\phi_n(x) = e^{-x/2} \sin \frac{n\pi x}{3}, \qquad n = 1, 2, 3, \ldots \qquad \blacksquare$$

In the examples considered here we found that there were an infinite number of eigenvalues $\lambda_1, \lambda_2, \ldots, \lambda_n, \ldots$ and corresponding eigenfunctions $\phi_1(x), \phi_2(x), \ldots, \phi_n(x), \ldots$. In fact, under rather general conditions, it can be shown that the eigenvalues always form an infinite sequence of real numbers of ever-increasing size, namely,

$$\lambda_1 < \lambda_2 < \lambda_3 < \cdots < \lambda_n < \cdots$$

each with a corresponding eigenfunction. An important special class of eigenvalue problems, known as *Sturm-Liouville problems,* that have such properties will be further investigated in Section 11.6.

11.3.1 Buckling of a Long Column

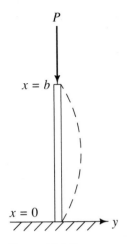

Figure 11.6 Long column under axial load P.

One of the oldest eigenvalue problems concerns the buckling of a long, slender rod or column supporting a compressive load. Vertical columns have been used extensively in Greek and Roman structures throughout the centuries. Prior to the time of Euler in the eighteenth century, columns were designed according to empirical formulas developed by the architects of the early civilizations. Needless to say, by today's standards, such empirical formulas give only a crude estimate of the weight a particular column design can support without collapsing. As a measure of safety such columns were often made much larger than necessary to ensure support of the structure above them. Euler developed the first truly mathematical model that can rather accurately predict the "critical compressive load" that a column can withstand before deformation, or "buckling," takes place.

Consider a long column or rod of length b that is supporting an axial compressive force P applied at the top as shown in Figure 11.6. By "long" we mean that the length of the column is much greater than its diameter. It is the magnitude of the compressive force P that can be applied to the column without the occurrence of buckling that we want to determine, and if buckling does occur, we then want to know the possible modes of lateral deflection of the column from the equilibrium position ($y = 0$).

Using the principles of elementary beam theory, it can be shown that the governing equation for this problem is the fourth-order DE[†]

$$\frac{d^2}{dx^2}\left[EI\, y''\right] + Py'' = 0, \qquad 0 < x < b \tag{26}$$

where E is *Young's modulus* of the beam material and I is the *moment of inertia* of the beam's cross section. This same DE occurs in studying the deflection of beams in various structures that are likewise subject to compressive forces. In many cases of interest the cross section of the beam is uniform throughout the beam. Hence, in this case both E and I are constant and then the above DE is usually expressed in the form

$$y^{(4)} + \lambda y'' = 0, \qquad 0 < x < b \tag{27}$$

where $\lambda = P/EI$.

Equation (27), along with prescribed homogeneous boundary conditions, constitutes an eigenvalue problem. The allowed values of λ correspond to *critical buckling loads*

$$P_n = \lambda_n EI, \qquad n = 1, 2, 3, \ldots \tag{28}$$

[†] For example, see Chapter 1 in L. C. Andrews, *Elementary Partial Differential Equations with Boundary Value Problems,* Academic Press, Orlando, 1986.

The first critical load P_1 is also known as the **Euler load** and is generally the one of greatest interest. If the axial load P is less than the Euler load, the column is stable and will remain in its undeformed equilibrium position. However, if the axial load P reaches the value P_1, the theory predicts that the column (under a small disturbance) will assume the shape predicted by the first eigenfunction $\phi_1(x)$, called the **fundamental buckling mode.**

The types of end conditions that may be prescribed for buckling beams or columns are more varied than in other kinds of problems. Partly this is because the governing DE is fourth-order and therefore requires *four* separate boundary conditions involving combinations of y, y', y'', and y'''. Usually we prescribe two conditions at each end, corresponding to the actual physical constraint that holds the beam in place. Some of these end conditions are described below:

> Fixed end: $y = 0$, $y' = 0$
> Free end: $y'' = 0$, $y''' = 0$
> Simple support: $y = 0$, $y'' = 0$
> Sliding clamped end: $y' = 0$, $y''' = 0$

EXAMPLE 11

Determine the Euler load and the corresponding fundamental buckling mode of a simply supported beam of length b under an axial compressive force P (see Figure 11.7).

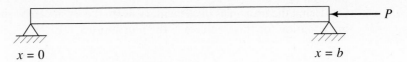

$x = 0$ $\qquad\qquad\qquad\qquad\qquad\qquad\qquad\qquad\qquad$ $x = b$

Figure 11.7 Simply supported beam.

Solution

A simply supported beam is one for which there is no displacement or bending moment at the endpoint. Mathematically, we describe such conditions by specifying $y = y'' = 0$ at the endpoint. Thus, the eigenvalue problem we wish to solve is given by

$$y^{(4)} + \lambda y'' = 0, \qquad 0 < x < b^\dagger$$

$$y(0) = y''(0) = 0, \qquad y(b) = y''(b) = 0$$

Physical considerations suggest that λ is positive and thus we consider only the case $\lambda = k^2 > 0$. In this situation the auxiliary equation is $m^2(m^2 + k^2) = 0$, with roots

---

† When the beam is simply supported, it can be shown that the eigenequation $y^{(4)} + \lambda y'' = 0$ can be replaced by the simpler second-order DE $y'' + \lambda y = 0$ without any real loss of generality.

$m = 0, 0, \pm ik$, and the general solution is therefore given by

$$y = C_1 + C_2 x + C_3 \cos kx + C_4 \sin kx$$

The boundary conditions at the endpoint $x = 0$ require that

$$y(0) = C_1 + C_3 = 0$$
$$y''(0) = -k^2 C_3 = 0$$

from which we deduce that $C_1 = C_3 = 0$. The remaining boundary conditions at the endpoint $x = b$ then lead to the pair of equations

$$C_2 b + C_4 \sin kb = 0$$
$$k^2 C_4 \sin kb = 0$$

For nontrivial solutions to exist, we must select $k = n\pi/b$ and set $C_2 = 0$. Hence, the eigenvalues and associated eigenfunctions are

$$\lambda_n = \frac{n^2 \pi^2}{b^2}, \qquad \phi_n(x) = \sin \frac{n\pi x}{b}, \qquad n = 1, 2, 3, \ldots$$

For these "allowed" values of λ, we find that the critical buckling loads become

$$P_n = \frac{n^2 \pi^2 EI}{b^2}, \qquad n = 1, 2, 3, \ldots$$

The largest load the beam can withstand before possible buckling is the *Euler load*

$$P_1 = \frac{\pi^2 EI}{b^2}$$

corresponding to the *fundamental buckling mode*

$$\phi_1(x) = \sin \frac{\pi x}{b} \qquad\qquad \blacksquare$$

Notice that the critical load P_1 in Example 11 can be *increased* if the moment of inertia I of the beam is increased or if the length of the beam is decreased. In general, the moment of inertia is large for cross sections that have their masses "far" from the centroid of the beam, leading to stronger beams. It is for this reason that we often find cross-sectional shapes in beams that resemble the letters I and H.

Although the mathematical theory predicts an infinite number of possible critical loads and deflection modes, only the Euler load and corresponding fundamental buckling mode are actually realized by the beam or column in most situations (if buckling does indeed occur). Notice also that each buckling mode (because of the arbitrary constant) only provides us with the possible "shape" of the buckled beam, but not its maximum displacement from equilibrium. This is a direct consequence of the fact that we have used a *linear* model to solve what is essentially a *nonlinear problem*. Fortunately, there are other techniques for finding the maximum deflection of the beam under various loadings.

Exercises 11.3

In Problems 1 to 14, find all real eigenvalues and corresponding eigenfunctions of the given boundary value problem.

1. $y'' + \lambda y = 0$, $y(0) = 0$, $y'(1) = 0$

2. $y'' + \lambda y = 0$, $y'(0) = 0$, $y(2) = 0$

3. $y'' + \lambda y = 0$, $y'(0) = 0$, $y'(p) = 0$, $p > 0$

4. $y'' + \lambda y = 0$, $y(0) = 0$, $y(1) - y'(1) = 0$

5. $y'' + 2y' + (1 - \lambda)y = 0$, $y(0) = 0$, $y(1) = 0$

6. $y'' + 4y' + (4 + 9\lambda)y = 0$, $y(0) = 0$, $y(2) = 0$

7. $y'' + y' + \lambda y = 0$, $y(0) = 0$, $y(1) = 0$

■**8.** $y'' + 2y' + \lambda y = 0$, $y(0) = 0$, $y'(1) = 0$

9. $y'' - 3y' + 2\lambda y = 0$, $y(0) = 0$, $y(1) = 0$

***10.** $y'' + \lambda y = 0$, $y(0) - y(\pi) = 0$, $y'(0) - y'(\pi) = 0$

11. $x^2 y'' + xy' + \lambda y = 0$, $y(1) = 0$, $y(e) = 0$

12. $x^2 y'' - xy' + \lambda y = 0$, $y(1) = 0$, $y(e) = 0$

13. $(d/dx)(xy') + \dfrac{\lambda}{x} y = 0$, $y'(1) = 0$, $y(2) = 0$

***14.** $(d/dx)(x^3 y') + \lambda xy = 0$, $y(1) = 0$, $y'(e^\pi) = 0$

15. The governing DE for an undamped spring-mass system is (see Chapter 5)

$$my'' + ky = 0$$

Suppose the mass m is initially ($t = 0$) at its equilibrium position ($y = 0$) with an unknown positive initial velocity. Find a relationship between the mass and spring constant k so that after 1 s the mass has returned to its equilibrium position.

16. Given the eigenvalue problem

$$y'' + \lambda y = 0, \qquad y(-\pi) = y(\pi),$$

$$y'(-\pi) = y'(\pi)$$

(a) Show that $\lambda = 0$ is an eigenvalue and find its eigenfunction.

(b) For $\lambda > 0$, show that the nth eigenvalue is $\lambda_n = n^2$ and that $\cos nx$ and $\sin nx$ are both associated eigenfunctions.

***17.** Given the eigenvalue problem

$$y'' + \lambda y = 0, \qquad hy(0) + y'(0) = 0, \qquad y(1) = 0$$

(a) Show that if $\lambda = 0$ is an eigenvalue, then $h = 1$.

(b) If $h > 1$, show that there is exactly one negative eigenvalue.

(c) If $h < 0$, show that there are no negative eigenvalues.

■**18.** A string of linear mass density ρ is tightly stretched and connected at $x = 0$ and $x = b$ to the centers of two synchronized shafts whose axes of rotation coincide with the axis of the string (see figure). If the tension in the string is T and the synchronized shafts are rotating with uniform angular speed ω, the deflection modes of the taut string are governed by the eigenequation

$$y'' + \lambda y = 0, \qquad 0 < x < b$$

where $\lambda = \rho \omega^2 / T$. If the physical constraints are such that the prescribed boundary conditions are $y(0) = 0$, $y'(b) = 0$,

(a) Determine the eigenvalues and eigenfunctions.

(b) From (a), determine the first *critical speed* ω_1 at which the string will bow out of its equilibrium position.

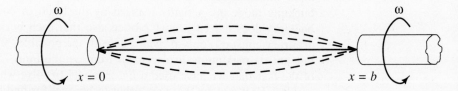

Problem 18 Rotating string.

19. Repeat Problem 18 when the boundary conditions are

$$y'(0) = 0, \qquad y(b) = 0$$

*20. The deflection modes of a 4-ft straight shaft of constant density ρ, rotating with uniform angular speed ω about its equilibrium position along the x axis, are solutions of the eigenvalue problem

$$y^{(4)} - \lambda y = 0, \qquad y(0) = y''(0) = 0,$$
$$y(4) = y''(4) = 0$$

where $\lambda = \rho \omega^2 / EI$. Find the lowest critical speed ω_1 for which the shaft will bow out of its equilibrium position.

11.4 Fourier Series—Part I

Power series representations (see Chapter 7) date back to Maclaurin and Taylor in the early eighteenth century. By the middle of that century, mathematicians like D. Bernoulli began to examine other types of infinite series in solving problems involving vibrating strings. Later, in the early nineteenth century, J. Fourier investigated series similar to those of Bernoulli in his classic work on heat conduction.

The infinite series investigated by both Bernoulli and Fourier involved sinusoidal functions that were actually the eigenfunctions of some eigenvalue problem. In general, the problem they faced was to find a series expansion of the form

$$f(x) = \sum_{n=1}^{\infty} c_n \phi_n(x) \tag{29}$$

where f is a given function, $\{\phi_n(x)\}$ is a set of eigenfunctions belonging to a particular eigenvalue problem, and the c's are constants to be determined. Like Bernoulli and Fourier, our major interest is focused on series where the eigenfunctions are *sines* and/or *cosines*. These special series, now called **Fourier series,** are important in a variety of applications.

11.4.1 Sine Series

An important type of Fourier series involves the eigenfunctions of the BVP

$$y'' + \lambda y = 0, \qquad 0 < x < p, \qquad y(0) = 0, \qquad y(p) = 0 \tag{30}$$

which we showed in Example 8 to be given by

$$\phi_n(x) = \sin \frac{n\pi x}{p}, \qquad n = 1, 2, 3, \ldots \tag{31}$$

Hence, given a suitable function f defined on $0 \le x \le p$, we may be interested in series expansions of the form

$$f(x) = \sum_{n=1}^{\infty} b_n \sin \frac{n\pi x}{p}, \qquad 0 < x < p \tag{32}$$

called a **Fourier sine series** or, more simply, a **sine series.** To determine the constants b_n, we assume that (32) is a valid series representation and proceed in a purely formal fashion.

First, by using the trigonometric identity

$$\sin A \sin B = \tfrac{1}{2}[\cos(A - B) - \cos(A + B)]$$

we see that

$$\int_0^p \sin\frac{n\pi x}{p} \sin\frac{m\pi x}{p}\, dx = \frac{1}{2}\int_0^p \left\{\cos\left[\frac{(n - m)\pi x}{p}\right] - \cos\left[\frac{(n + m)\pi x}{p}\right]\right\} dx$$

$$= \frac{p}{2\pi}\left\{\frac{\sin[(n - m)\pi x/p]}{n - m} - \frac{\sin[(n + m)\pi x/p]}{n + m}\right\}\Big|_0^p$$

$$= 0, \qquad n \neq m \tag{33}$$

Next, if we multiply both sides of the series (32) by the function $\sin(m\pi x/p)$ and integrate the result from 0 to p, assuming that termwise integration is justified, we obtain

$$\int_0^p f(x) \sin\frac{m\pi x}{p}\, dx = \sum_{n=1}^{\infty} b_n \int_0^p \sin\frac{n\pi x}{p} \sin\frac{m\pi x}{p}\, dx$$

But, because all integrals on the right-hand side vanish except for the one when $n = m$, this expression reduces to

$$\int_0^p f(x) \sin\frac{m\pi x}{p}\, dx = b_m \int_0^p \left(\sin\frac{m\pi x}{p}\right)^2 dx$$

$$= \frac{p}{2} b_m$$

where the evaluation of the integral on the right is left to the exercises (see Problem 21 in Exercises 11.4). Now, replacing the dummy index m with n in this last expression, we have formally shown that the constants b_n are defined by

$$b_n = \frac{2}{p}\int_0^p f(x) \sin\frac{n\pi x}{p}\, dx, \qquad n = 1, 2, 3, \ldots \tag{34}$$

The constants b_n are called the **Fourier coefficients** of the sine series (32). What we have shown is that if the series (32) converges to $f(x)$ and termwise integration of the series is permitted, then the Fourier coefficients are defined by (34). However, if the function f is integrable on the interval $[0, p]$, then we can simply define the Fourier coefficients by (34) and study the resulting series (32) on its own merits. That is the point of view we wish to take here and delay any general discussions about convergence until later.

EXAMPLE 12

Find the Fourier sine series of

$$f(x) = x, \qquad 0 \leq x \leq \pi$$

Solution

Because the interval of interest is $0 \leq x \leq \pi$, we see that $p = \pi$, and so the series in this case assumes the form

$$x = \sum_{n=1}^{\infty} b_n \sin nx, \qquad 0 < x < \pi$$

where

$$b_n = \frac{2}{\pi} \int_0^\pi x \sin nx \, dx$$

$$= \frac{2}{\pi} \left(-\frac{x}{n} \cos nx + \frac{\sin nx}{n^2} \right) \Big|_0^\pi$$

the last step obtained through integration by parts. Upon substituting the limits of integration, we find

$$b_n = -\frac{2}{n} \cos n\pi, \qquad n = 1, 2, 3, \ldots$$

However, $\cos n\pi = (-1)^n$, $n = 1, 2, 3, \ldots$, and therefore we can write the Fourier coefficient b_n more conveniently in the form

$$b_n = \frac{2}{\pi}(-1)^{n-1}, \qquad n = 1, 2, 3, \ldots$$

The series we seek is

$$x = 2 \sum_{n=1}^\infty \frac{(-1)^{n-1}}{n} \sin nx, \qquad 0 < x < \pi$$

or, upon writing out the first few terms,

$$x = 2(\sin x - \tfrac{1}{2} \sin 2x + \tfrac{1}{3} \sin 3x - \cdots), \qquad 0 < x < \pi \qquad ■$$

Observe that the sine series in Example 12 sums to zero at $x = \pi$, since each term of the series is zero at this point. Hence, it clearly does not converge to $f(\pi) = \pi$ at $x = \pi$. Yet, it can be shown that for each point on the interval $0 \leq x < \pi$ the series indeed converges to the proper function value $f(x) = x$.

EXAMPLE 13

Find the Fourier sine series of the function

$$f(x) = 10 + 2x, \qquad 0 < x < 10$$

Solution

With $p = 10$, the series this time takes the form

$$10 + 2x = \sum_{n=1}^\infty b_n \sin \frac{n\pi x}{10}, \qquad 0 < x < 10$$

where, using integration by parts, we obtain

$$b_n = \frac{1}{5} \int_0^{10} (10 + 2x) \sin \frac{n\pi x}{10} \, dx$$

$$= \frac{20}{n\pi} [1 - (-1)^n 3], \qquad n = 1, 2, 3, \ldots$$

Hence, the desired series expansion is

$$10 + 2x = \frac{20}{\pi} \sum_{n=1}^{\infty} \frac{[1 - (-1)^n 3]}{n} \sin \frac{n\pi x}{10}, \qquad 0 < x < 10 \qquad \blacksquare$$

11.4.2 Cosine Series

Another important set of functions arises as the set of eigenfunctions of the BVP

$$y'' + \lambda y = 0, \qquad 0 < x < p, \qquad y'(0) = 0, \qquad y'(p) = 0 \qquad (35)$$

It can be shown that (see Problem 3 in Exercises 11.3) the eigenvalues are

$$\lambda_n = \begin{cases} 1, & n = 0 \\ \dfrac{n^2 \pi^2}{p^2}, & n = 1, 2, 3, \ldots \end{cases} \qquad (36a)$$

with corresponding eigenfunctions

$$\phi_n(x) = \begin{cases} 1, & n = 0 \\ \cos \dfrac{n\pi x}{p}, & n = 1, 2, 3, \ldots \end{cases} \qquad (36b)$$

Thus, for a suitable function f, we may consider a Fourier series expansion of the form

$$f(x) = A_0 + \sum_{n=1}^{\infty} a_n \cos \frac{n\pi x}{p}, \qquad 0 < x < p \qquad (37)$$

called a **Fourier cosine series** (or **cosine series**). We can determine expressions for the Fourier coefficients A_0 and a_n in a manner similar to that illustrated in Section 11.4.1 for the sine series, except that here we make use of the integral relations

$$\int_0^p \cos \frac{n\pi x}{p} \, dx = 0, \qquad n = 1, 2, 3, \ldots \qquad (38)$$

and

$$\int_0^p \cos \frac{m\pi x}{p} \cos \frac{n\pi x}{p} \, dx = 0, \qquad m \neq n \qquad (39)$$

In this case the Fourier coefficients are defined by (see Problem 22 in Exercises 11.4)

$$A_0 = \frac{1}{p} \int_0^p f(x) \, dx \qquad (40)$$

and

$$a_n = \frac{2}{p} \int_0^p f(x) \cos \frac{n\pi x}{p} \, dx, \qquad n = 1, 2, 3, \ldots \qquad (41)$$

EXAMPLE 14

Find the Fourier cosine series of

$$f(x) = x, \qquad 0 < x < \pi$$

Solution

The substitution of $f(x) = x$ into (40) and (41) leads to

$$A_0 = \frac{1}{\pi} \int_0^\pi x \, dx = \frac{\pi}{2}$$

and

$$a_n = \frac{2}{\pi} \int_0^\pi x \cos nx \, dx = \frac{2}{\pi n^2} (\cos n\pi - 1)$$

or

$$a_n = \frac{2}{\pi n^2} [(-1)^n - 1] = \begin{cases} -\dfrac{4}{\pi n^2}, & n = 1, 3, 5, \ldots \\ 0, & n = 2, 4, 6, \ldots \end{cases}$$

Thus, the cosine series becomes

$$x = \frac{\pi}{2} - \frac{4}{\pi} \sum_{\substack{n=1 \\ (\text{odd})}}^{\infty} \frac{\cos nx}{n^2}, \qquad 0 < x < \pi$$

the first few terms of which are

$$x = \frac{\pi}{2} - \frac{4}{\pi} \left(\cos x + \frac{\cos 3x}{9} + \frac{\cos 5x}{25} + \cdots \right), \qquad 0 < x < \pi \qquad \blacksquare$$

Exercises 11.4

In Problems 1 to 10, find the Fourier sine series of the given function.

1. $f(x) = 1, \qquad 0 < x < 1$

2. $f(x) = \begin{cases} 1, & 0 < x < 1 \\ 0, & 1 < x < 2 \end{cases}$

3. $f(x) = x, \qquad 0 < x < 2$

4. $f(x) = x^2, \qquad 0 < x < \pi$

5. $f(x) = x - x^2, \qquad 0 < x < 1$

■**6.** $f(x) = \begin{cases} \pi - x, & 0 < x < \pi/2 \\ x, & \pi/2 < x < \pi \end{cases}$

7. $f(x) = \sin x - 4 \sin 3x, \qquad 0 < x < \pi$

8. $f(x) = 2 \sin^2 x, \qquad 0 < x < \pi$
   Hint. $\sin^2 x = \frac{1}{2}(1 - \cos 2x)$

9. $f(x) = e^x, \qquad 0 < x < 1$

10. $f(x) = \cosh x, \qquad 0 < x < 1$

In Problems 11 to 20, find the Fourier cosine series of the given function.

11. See Problem 1.

12. See Problem 2.

13. See Problem 3.

14. See Problem 4.

15. See Problem 5.

16. See Problem 6.

17. See Problem 7.

■**18.** See Problem 8.

19. See Problem 9.

20. See Problem 10.

21. For $m, n = 1, 2, 3, \ldots$, show that

(a) $\int_0^p \left(\sin \dfrac{m\pi x}{p} \right)^2 dx$

$$= \int_0^p \left(\cos \dfrac{m\pi x}{p} \right)^2 dx = \dfrac{p}{2}$$

(b) $\int_0^p \cos \dfrac{n\pi x}{p} \, dx = 0$

(c) $\int_0^p \cos \dfrac{m\pi x}{p} \cos \dfrac{n\pi x}{p} \, dx = 0, \qquad m \neq n$

22. (a) By integrating (37) termwise, deduce that A_0 is given by (40).

(b) Multiply (37) by $\cos(m\pi x/p)$ and integrate the result to deduce that a_n is given by (41).

11.5 Fourier Series—Part II

A function f is called **periodic** if there exists a constant $T > 0$ for which $f(x + T) = f(x)$ for all x. The smallest value of T for which the property holds is called the **period.** It follows that if $f(x + T) = f(x)$, then also

$$f(x \pm T) = f(x \pm 2T) = f(x \pm 3T) = \cdots = f(x)$$

An example of a periodic function is shown in Figure 11.8.

Periodic functions appear in a variety of physical problems, such as those concerning vibrating springs and membranes, planetary motion, a swinging pendulum, and musical sounds, to name a few. In some of these problems the periodic function may be quite complicated, so in order to better understand its basic nature it may be convenient to represent it in a series of simple periodic functions. Because trigonometric functions are the simplest examples of periodic functions, we usually look for series representations in terms of sines and cosines.

Observe that all members of the set $\{1, \cos(n\pi x/p), \sin(n\pi x/p)\}$, $n = 1, 2, 3, \ldots$, have the same period $T = 2p$.[†] Therefore, if f is any periodic function with period $T = 2p$, we may look for a series representation such as

$$f(x) = \dfrac{1}{2} a_0 + \sum_{n=1}^{\infty} \left(a_n \cos \dfrac{n\pi x}{p} + b_n \sin \dfrac{n\pi x}{p} \right) \qquad (42)$$

where $a_0, a_1, a_2, \ldots, b_1, b_2, \ldots$, are constants. Series of this form are called **Fourier trigonometric series** or, more simply, **Fourier series.**

In order to obtain expressions for the constants in (42), we find that we need the following integral formulas

$$\int_{-p}^p \cos \dfrac{n\pi x}{p} \, dx = \int_{-p}^p \sin \dfrac{n\pi x}{p} \, dx = \int_{-p}^p \sin \dfrac{n\pi x}{p} \cos \dfrac{m\pi x}{p} \, dx = 0 \qquad (43)$$

[†] Any constant is a periodic function (see Problem 1 in Exercises 11.5).

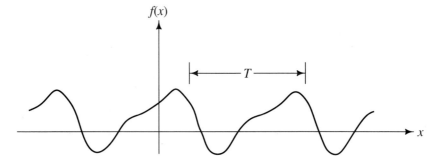

Figure 11.8 A periodic function.

and

$$\int_{-p}^{p} \cos \frac{n\pi x}{p} \cos \frac{m\pi x}{p}\, dx = \int_{-p}^{p} \sin \frac{n\pi x}{p} \sin \frac{m\pi x}{p}\, dx = \begin{cases} 0, & n \neq m \\ p, & n = m \end{cases} \qquad (44)$$

where m and n are any nonzero integers. We can now use a device similar to that in Section 11.4.1 to find the constants a_0, a_n, and b_n.

To begin, let us integrate the series (42) termwise from $x = -p$ to $x = p$, assuming that this is permissible, to obtain

$$\int_{-p}^{p} f(x)\, dx = \frac{1}{2} a_0 \int_{-p}^{p} dx + \sum_{n=1}^{\infty} \left(a_n \int_{-p}^{p} \cos \frac{n\pi x}{p}\, dx + b_n \int_{-p}^{p} \sin \frac{n\pi x}{p}\, dx \right)$$

In view of the integral relations (43), we see that each term of the series integrates to zero, and from the remaining nonzero integrals, it follows that

$$a_0 = \frac{1}{p} \int_{-p}^{p} f(x)\, dx \qquad (45)$$

Next, we multiply (42) by $\cos(m\pi x/p)$ and integrate termwise again to find

$$\int_{-p}^{p} f(x) \cos \frac{m\pi x}{p}\, dx$$
$$= \frac{1}{2} a_0 \int_{-p}^{p} \cos \frac{m\pi x}{p}\, dx + \sum_{n=1}^{\infty} \left(a_n \int_{-p}^{p} \cos \frac{n\pi x}{p} \cos \frac{m\pi x}{p}\, dx + b_n \int_{-p}^{p} \sin \frac{n\pi x}{p} \sin \frac{m\pi x}{p}\, dx \right)$$

Because of (43) and (44), all terms integrate to zero except for the coefficient of a_n corresponding to $n = m$, and here we get

$$\int_{-p}^{p} f(x) \cos \frac{m\pi x}{p}\, dx = a_m \int_{-p}^{p} \left(\cos \frac{m\pi x}{p} \right)^2 dx = p a_m$$

from which we deduce

$$a_m = \frac{1}{p} \int_{-p}^{p} f(x) \cos \frac{m\pi x}{p}\, dx \qquad (46)$$

In the same fashion, if we multiply the series (42) by $\sin(m\pi x/p)$ and integrate the result termwise, we obtain the final formula

$$b_m = \frac{1}{p} \int_{-p}^{p} f(x) \sin(m\pi x/p) \, dx \qquad (47)$$

The constants defined by (45) to (47) are known as **Fourier coefficients** (also called **Euler's formulas**). It is customary to change the dummy index m back to n in (46) and (47), and combine (45) and (46) into a single formula. In this case we write

$$a_n = \frac{1}{p} \int_{-p}^{p} f(x) \cos \frac{n\pi x}{p} \, dx, \qquad n = 0, 1, 2, \ldots \qquad (48)$$

and

$$b_n = \frac{1}{p} \int_{-p}^{p} f(x) \sin \frac{n\pi x}{p} \, dx, \qquad n = 1, 2, 3, \ldots \cdot \qquad (49)$$

Remark. Writing the constant term in (42) as $\frac{1}{2}a_0$ is customary, but not necessary. It does, however, enable us to combine the formulas for a_0 and a_n, $n = 1, 2, 3, \ldots$ into a single formula as given by (48).

EXAMPLE 15

Find the Fourier series of the periodic function

$$f(x) = \begin{cases} 0, & -\pi < x < 0 \\ x, & 0 < x < \pi \end{cases} \qquad f(x + 2\pi) = f(x)$$

whose graph is shown in Figure 11.9.

Solution

Based on (48) and (49) with $p = \pi$, the Fourier coefficients of f are

$$a_0 = \frac{1}{\pi} \int_{-\pi}^{\pi} f(x) \, dx = \frac{1}{\pi} \int_{0}^{\pi} x \, dx = \frac{\pi}{2}$$

$$a_n = \frac{1}{\pi} \int_{0}^{\pi} x \cos nx \, dx = \begin{cases} -\dfrac{2}{\pi n^2}, & n = 1, 3, 5, \ldots \\ 0, & n = 2, 4, 6, \ldots \end{cases}$$

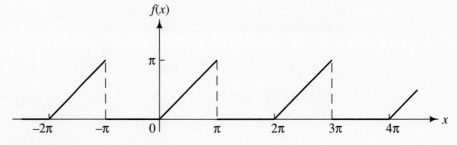

Figure 11.9

and

$$b_n = \frac{1}{\pi} \int_0^\pi x \sin nx \, dx = \frac{(-1)^{n-1}}{n}, \qquad n = 1, 2, 3, \ldots$$

Substituting these results into the series (42), we obtain

$$f(x) = \frac{\pi}{4} - \frac{2}{\pi} \sum_{\substack{n=1 \\ (\text{odd})}}^\infty \frac{\cos nx}{n^2} - \sum_{n=1}^\infty \frac{(-1)^n}{n} \sin nx$$

or, changing the index $n \to 2n - 1$ in the first summation,

$$f(x) = \frac{\pi}{4} - \sum_{n=1}^\infty \left[\frac{2}{\pi} \frac{\cos(2n-1)x}{(2n-1)^2} + \frac{(-1)^n}{n} \sin nx \right] \qquad \blacksquare$$

11.5.1 Convergence of the Series

As in Section 11.4, our treatment of Fourier series thus far has been purely formal. Nonetheless, we can take the point of view that (42) *defines* the Fourier series of a periodic function f, provided the integrals in (48) and (49) exist, regardless of whether the series converges to $f(x)$ (or indeed converges at all). And these integrals always exist if the periodic function is at least *piecewise continuous* on the interval $[-p, p]$. Recall from Chapter 6 (Definition 6.1) that a piecewise continuous function on a finite interval is one which is continuous everywhere on the interval except possibly at isolated points, and at such points the discontinuities are *finite*. The function is called piecewise continuous for all x if it is piecewise continuous on every finite interval. If f and its derivative f' are both piecewise continuous, then we are led to the following important theorem.

Theorem 11.3 Convergence.   *If $f(x + 2p) = f(x)$ for all x, and f and f' are piecewise continuous functions, then the Fourier series*

$$f(x) = \frac{1}{2} a_0 + \sum_{n=1}^\infty \left(a_n \cos \frac{n\pi x}{p} + b_n \sin \frac{n\pi x}{p} \right)$$

where

$$a_n = \frac{1}{p} \int_{-p}^p f(x) \cos \frac{n\pi x}{p} \, dx, \qquad n = 0, 1, 2, \ldots$$

and

$$b_n = \frac{1}{p} \int_{-p}^p f(x) \sin \frac{n\pi x}{p} \, dx, \qquad n = 1, 2, 3, \ldots$$

converges for all values of x. The sum of the series equals $f(x)$ at all points of continuity of f and equals the average value $\frac{1}{2}[f(x^+) + f(x^-)]$ at every point of discontinuity of f.[†]

[†] A proof of Theorem 11.3 is given in L. C. Andrews, *Elementary Partial Differential Equations with Boundary Value Problems,* Academic Press, Orlando, 1986.

The conditions listed in Theorem 11.3 are only *sufficient conditions* for convergence, not necessary conditions. For example, the function $f(x) = |x|^{1/2}$ does not have a piecewise continuous derivative on any interval containing $x = 0$, and yet it has a convergent Fourier series. Also, it should come as no surprise that the series does not converge to the function value $f(x)$ at a point of discontinuity. That is, we can always change the function value at a finite number of points without changing the Fourier series.

In order to illustrate the convergence of a Fourier series at a point of discontinuity, let us consider the periodic function (see Figure 11.10)

$$f(x) = \begin{cases} 0, & -\pi < x < 0 \\ K, & 0 \le x \le \pi \end{cases} \qquad f(x + 2\pi) = f(x) \qquad (50)$$

Because of its shape, this function is called a *square wave* in the engineering literature. Some simple calculations reveal that

$$a_0 = \frac{1}{\pi} \int_{-\pi}^{\pi} f(x)\, dx = \frac{K}{\pi} \int_0^{\pi} dx = K$$

$$a_n = \frac{1}{\pi} \int_{-\pi}^{\pi} f(x) \cos nx\, dx = \frac{K}{\pi} \int_0^{\pi} \cos nx\, dx = 0, \qquad n \ne 0$$

and

$$b_n = \frac{K}{\pi} \int_0^{\pi} \sin nx\, dx = \frac{K}{\pi n}[1 - (-1)^n] = \begin{cases} \dfrac{2K}{\pi n}, & n = 1, 3, 5, \ldots \\ 0, & n = 2, 4, 6, \ldots \end{cases}$$

Combining results, the series takes the form

$$f(x) = \frac{K}{2} + \frac{2K}{\pi} \sum_{n=1}^{\infty} \frac{\sin(2n-1)x}{2n-1} \qquad (51)$$

Because of discontinuities in the function f at $x = 0$ and multiples of π, the series (51) converges to the average value $K/2$ of the left-hand and right-hand limits of f at these points (marked by $\times$ in Figure 11.10). Elsewhere, both f and its derivative

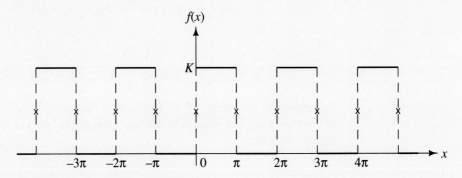

Figure 11.10 A square wave function.

f' are continuous and thus the series converges to $f(x)$. For instance, setting $x = \pi/2$ in (51) leads to

$$f(\pi/2) = K = \frac{K}{2} + \frac{2K}{\pi}\left(1 - \frac{1}{3} + \frac{1}{5} - \frac{1}{7} + \cdots\right)$$

from which we deduce the interesting result

$$1 - \frac{1}{3} + \frac{1}{5} - \frac{1}{7} + \cdots = \sum_{n=1}^{\infty} \frac{1}{2n-1} = \frac{\pi}{4} \tag{52}$$

Leibniz (around 1673) was the first to discover this relation between π and the reciprocal of the odd integers, but he obtained it from geometrical considerations alone. There are numerous other relations of this type that can be derived through the use of Fourier series, some of which are taken up in the exercises.

By taking only a finite number of terms of the series (51), we get the **partial sums**

$$S_N(x) = \frac{K}{2} + \frac{2K}{\pi} \sum_{n=1}^{N} \frac{\sin(2n-1)x}{2n-1}, \qquad N = 1, 2, 3, \ldots \tag{53}$$

which are plotted over one period in Figure 11.11 for $N = 1, 2, 3$, and 12. Convergence of the partial sums $S_N(x)$ to $f(x)$ for increasing N is fairly clear. However, at the discontinuities of f at $x = 0$ and $x = \pm\pi$ the partial sums tend to overshoot their mark at first and then approach the value K or 0. This feature is typical of Fourier series in the vicinity of a discontinuity of the function, and is known as the **Gibbs' phenomenon.**[†]

In the use of Fourier series it is often necessary to perform various operations on the series, such as differentiation. When required in our analysis, we usually make the assumption that *termwise differentiation* of the series is permitted. However, unless certain conditions are satisfied, this may lead to some difficulties. For example, the square-wave function defined by (50) has the convergent series

$$f(x) = \frac{K}{2} + \frac{2K}{\pi} \sum_{n=1}^{\infty} \frac{\sin(2n-1)x}{2n-1}$$

Since f is either K or 0, its derivative at every point of continuity is clearly zero. Termwise differentiation of the above series, however, leads to the relation

$$0 = 0 + \frac{2K}{\pi} \sum_{n=1}^{\infty} \cos(2n-1)x$$

which clearly has something seriously wrong with it! That is, for $x = 0$, for example, we obtain the absurd result

$$0 = 0 + \frac{2K}{\pi}(1 + 1 + 1 + \cdots)$$

[†] Named after Josiah Willard Gibbs (1839–1903), who is more famous for his work on vector analysis and statistical mechanics.

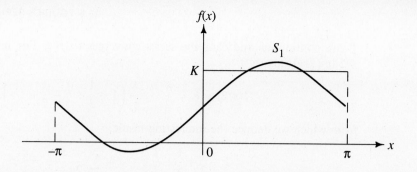

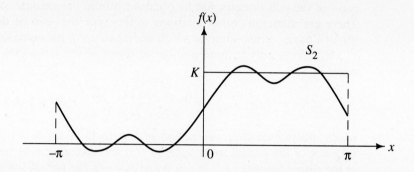

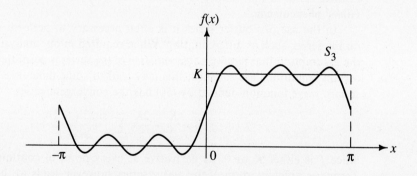

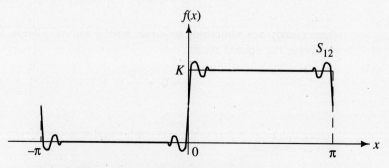

Figure 11.11 Partial sums and Gibbs' phenomenon.

Based on this example, it is clear that convergence alone is not sufficient to justify termwise differentiation of the series. It turns out that we need f to be *continuous*, and both f' and f'' to be piecewise continuous.

Theorem 11.4 Termwise Differentiation. *If f is a continuous function and $f(x + 2p) = f(x)$ for all x, and f' and f'' are piecewise continuous functions, then the Fourier series*

$$f(x) = \frac{1}{2} a_0 + \sum_{n=1}^{\infty} \left(a_n \cos \frac{n\pi x}{p} + b_n \sin \frac{n\pi x}{p} \right)$$

where

$$a_n = \frac{1}{p} \int_{-p}^{p} f(x) \cos \frac{n\pi x}{p} \, dx, \qquad n = 0, 1, 2, \dots$$

and

$$b_n = \frac{1}{p} \int_{-p}^{p} f(x) \sin \frac{n\pi x}{p} \, dx, \qquad n = 1, 2, 3, \dots$$

can be differentiated termwise to produce the series

$$f'(x) = \frac{\pi}{p} \sum_{n=1}^{\infty} n \left(-a_n \sin \frac{n\pi x}{p} + b_n \cos \frac{n\pi x}{p} \right)$$

which converges to $f'(x)$ wherever $f''(x)$ exists.

11.5.2 Even and Odd Functions: Cosine and Sine Series

Many of the functions that arise in Fourier series representations are either *even* or *odd* functions. If g is defined on either the entire x axis or some finite interval so that $x = 0$ is the midpoint of the interval, and if $g(-x) = g(x)$ for all x, we say that g is an **even function.** On the other hand, we say that g is an **odd function** if $g(-x) = -g(x)$ for all x. See Figure 11.12 for an illustration of an even and an odd function.

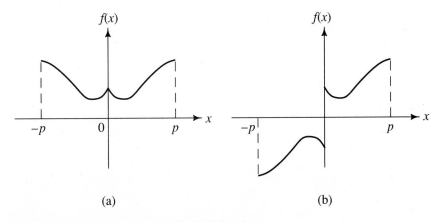

(a)   (b)

Figure 11.12 (a) Even function. (b) Odd function.

Determine whether the following functions are even, odd, or neither: **(a)** $\cos(n\pi x/p)$, **(b)** $\sin(n\pi x/p)$, **(c)** 1, **(d)** x^4, **(e)** $1 - x$.

Solution

Applying the definition, we find that replacing x by $-x$ yields

(a) $\cos\left(-\dfrac{n\pi x}{p}\right) = \cos\dfrac{n\pi x}{p}$

(b) $\sin\left(-\dfrac{n\pi x}{p}\right) = -\sin\dfrac{n\pi x}{p}$

(c) $1 = 1$

(d) $(-x)^4 = x^4$

(e) $1 - (-x) = 1 + x$

and thus conclude that **(a)**, **(c)**, and **(d)** are *even* functions, **(b)** is an *odd* function, and **(e)** is *neither*.   ■

Provided the integrals exist, it follows from the geometric interpretation of integrals that if g is an even function, then

$$\int_{-p}^{p} g(x)\, dx = 2 \int_{0}^{p} g(x)\, dx \tag{54}$$

and if g is an odd function,

$$\int_{-p}^{p} g(x)\, dx = 0 \tag{55}$$

for any p (see Problem 25 in Exercises 11.5).

If f is an even function, the product $f(x)\cos(n\pi x/p)$ is an even function and the product $f(x)\sin(n\pi x/p)$ is an odd function (see Problem 26 in Exercises 11.5). Hence, from (54) and (55), we see that the Fourier coefficients of f satisfy

$$
\begin{aligned}
a_n &= \frac{1}{p}\int_{-p}^{p} f(x)\cos\frac{n\pi x}{p}\, dx \\[4pt]
&= \frac{2}{p}\int_{0}^{p} f(x)\cos\frac{n\pi x}{p}\, dx, \qquad n = 0, 1, 2, \ldots
\end{aligned}
\tag{56}
$$

and

$$b_n = \frac{1}{p}\int_{-p}^{p} f(x)\sin\frac{n\pi x}{p}\, dx = 0, \qquad n = 1, 2, 3, \ldots \tag{57}$$

The Fourier series of an *even function*, therefore, reduces to a *cosine series*; that is, (42) becomes

$$f(x) = \frac{1}{2}a_0 + \sum_{n=1}^{\infty} a_n \cos\frac{n\pi x}{p} \tag{58}$$

where the Fourier coefficients are defined by (56). Similarly, if f is an *odd function*, its Fourier series (42) reduces to the *sine series*

$$f(x) = \sum_{n=1}^{\infty} b_n \sin \frac{n\pi x}{p} \qquad (59)$$

where $a_n = 0$, $n = 0, 1, 2, \ldots$, and

$$b_n = \frac{2}{p} \int_0^p f(x) \sin \frac{n\pi x}{p} \, dx, \qquad n = 1, 2, 3, \ldots \qquad (60)$$

EXAMPLE 17

Find the Fourier series of the periodic function (see Figure 11.13)

$$f(x) = x, \qquad -\pi < x < \pi, \qquad f(x + 2\pi) = f(x)$$

Solution

Because the function is an odd function, we seek a sine series representation for which $p = \pi$. Thus, from (60) we obtain

$$b_n = \frac{2}{\pi} \int_0^\pi x \sin nx \, dx$$

$$= -\frac{2}{n} \cos n\pi, \qquad n = 1, 2, 3, \ldots$$

or, since $\cos n\pi = (-1)^n$,

$$b_n = \frac{2}{n}(-1)^{n-1}, \qquad n = 1, 2, 3, \ldots$$

Hence, the sine series we seek takes the form

$$f(x) = 2 \sum_{n=1}^{\infty} \frac{(-1)^{n-1}}{n} \sin nx \qquad \blacksquare$$

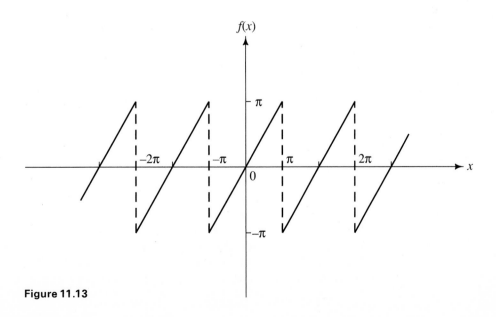

Figure 11.13

EXAMPLE 18

Find the Fourier series of the triangular wave function defined by (see Figure 11.14)

$$f(x) = |x|, \quad -\pi < x < \pi, \quad f(x + 2\pi) = f(x)$$

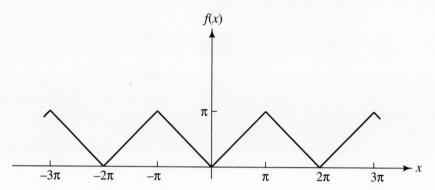

Figure 11.14 Triangular wave.

Solution

This time the function is an even function and so we look for a cosine series. The substitution of $f(x) = |x|$ into (59) leads to

$$a_0 = \frac{2}{\pi} \int_0^\pi x \, dx = \pi$$

and

$$a_n = \frac{2}{\pi} \int_0^\pi x \cos nx \, dx = \frac{2}{n^2 \pi} (\cos n\pi - 1), \quad n = 1, 2, 3, \ldots$$

Again, we write $\cos n\pi = (-1)^n$, $n = 1, 2, 3, \ldots$, and thus we have

$$a_n = \frac{2}{n^2 \pi} [(-1)^n - 1] = \begin{cases} -4/n^2\pi, & n = 1, 3, 5, \ldots \\ 0, & n = 2, 4, 6, \ldots \end{cases}$$

The desired cosine series now takes the form

$$f(x) = \frac{\pi}{2} - \frac{4}{\pi} \sum_{\substack{n=1 \\ (\text{odd})}}^{\infty} \frac{\cos nx}{n^2}$$

11.5.3 Periodic Extensions

If a Fourier series converges, it must necessarily converge to a periodic function having the same period as the sines and/or cosines appearing in the series. This would at first seem to suggest that only periodic functions have Fourier series representations, but this is not the case. For example, recall that in Section 11.4 we considered functions that were defined only on the finite interval $[0, p]$. The sine series (or cosine series) for such a function is the same as that for the *odd* (or *even*) *periodic extension* of this function over the entire x axis. If f is defined only on $[0, p]$, we define the **odd periodic extension** of f to be the function f_0, where

$$f_0(x) = \begin{cases} f(x), & 0 < x < p \\ -f(-x), & -p < x < 0 \end{cases} \qquad (61)$$

and $f_0(x + 2p) = f_0(x)$ for all x. Similarly, the **even periodic extension** f_e is defined by

$$f_e(x) = f(|x|), \qquad -p < x < p \qquad (62)$$

and $f_e(x + 2p) = f_e(x)$ for all x.

To better understand the concept of periodic extensions, observe that the periodic functions in Examples 17 and 18 led to the identical series found in Examples 12 and 14 in Section 11.4, but in Examples 12 and 14 the functions were defined only over the finite interval $[0, \pi]$. We recognize, of course, that over the finite interval $[0, \pi]$ the functions defined in Examples 12, 14, 17, and 18 all take on the same function values. The odd periodic extension of the function given in Examples 12 and 14 leads to the function defined in Example 17. The even periodic extension of this same function gives us the function appearing in Example 18.

Series obtained by either even or odd periodic extensions of a function f defined only over the finite range $[0, p]$ are sometimes called **half-range expansions.** That is, the full period covers the interval $[-p, p]$ but the series are determined completely over the "half-range" $[0, p]$. In some cases the function f may be defined on the full interval $[-p, p]$ but is not even or odd. In this case the periodic extension will lead to a *full* Fourier series, that is, both sines and cosines. For instance, the periodic extension of the function f shown in Figure 11.15 is the function designated by f^* in that figure. As in the case of half-range expansions, the full Fourier series (42) for both f and f^* will be the same.

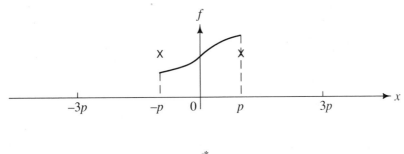

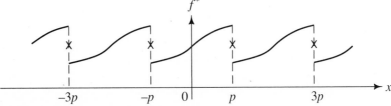

Figure 11.15 Periodic extension of $f(x)$.

Finally, we remark that if a function f is defined on the finite interval $[0, p]$, we can form a half-range expansion as a sine series or a cosine series by considering the period to be $T = 2p$, or we might form a full-range expansion (involving both sines and cosines) by considering the period to be $T = p$. See the following example.

> **Remark.** If f is a periodic function with period $T = 2p$ and defined on the interval $[c, c + 2p]$ rather than $[-p, p]$, Its Fourier coefficients are then given by
>
> $$a_n = \frac{1}{p} \int_c^{c + 2p} f(x) \cos \frac{n\pi x}{p} \, dx, \qquad n = 0, 1, 2, \ldots$$
>
> and
>
> $$b_n = \frac{1}{p} \int_c^{c + 2p} f(x) \sin \frac{n\pi x}{p} \, dx, \qquad n = 1, 2, 3, \ldots$$

■ EXAMPLE 19

Find a Fourier series representation of

$$f(x) = x^2, \qquad 0 < x < 1$$

(a) As a sine series   **(b)** As a cosine series   **(c)** As a full Fourier series

See Figure 11.16 for the periodic extension used in each of these series.

Solution

(a) Using the half-range formula for the Fourier coefficients given by (60) with $p = 1$, we obtain

$$b_n = 2 \int_0^1 x^2 \sin n\pi x \, dx$$

$$= \frac{2}{n\pi} (-1)^{n-1} + \frac{4}{n^3 \pi^3} [(-1)^n - 1], \qquad n = 1, 2, 3, \ldots$$

where we have performed integration by parts two times. Therefore,

$$f(x) = \frac{2}{\pi} \sum_{n=1}^{\infty} \left\{ \frac{(-1)^{n-1}}{n} + \frac{2}{n^3 \pi^2} [(-1)^n - 1] \right\} \sin n\pi x$$

(b) For the cosine series, we first calculate

$$a_0 = 2 \int_0^1 x^2 \, dx = \frac{2}{3}$$

Next, integrating by parts, we find that

$$a_n = 2 \int_0^1 x^2 \cos n\pi x \, dx = \frac{4(-1)^n}{n^2 \pi^2}, \qquad n = 1, 2, 3, \ldots$$

and thus the series we seek becomes

$$f(x) = \frac{1}{3} + \frac{4}{\pi^2} \sum_{n=1}^{\infty} \frac{(-1)^n}{n^2} \cos n\pi x$$

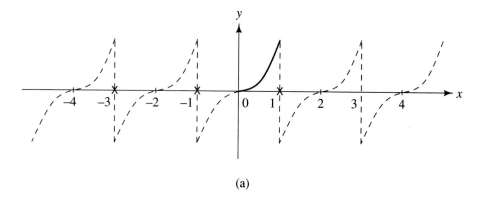

(a)

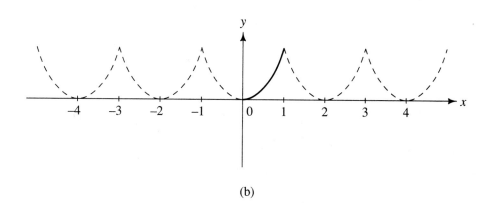

(b)

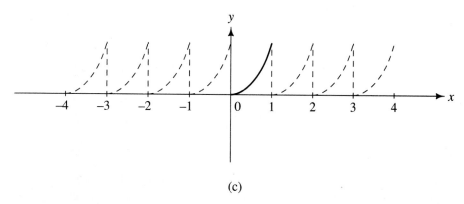

(c)

Figure 11.16 (a) Sine series. (b) Cosine series. (c) Full Fourier series.

(c) For the full Fourier series we first recognize that the period is $2p = 1$, or $p = 1/2$. Therefore,

$$a_0 = 2 \int_0^1 x^2 \, dx = \frac{2}{3}$$

$$a_n = 2 \int_0^1 x^2 \cos 2n\pi x \, dx = \frac{1}{n^2 \pi^2}, \qquad n = 1, 2, 3, \ldots$$

and

$$b_n = 2 \int_0^1 x^2 \sin 2n\pi x \, dx = -\frac{1}{n\pi}, \qquad n = 1, 2, 3, \ldots$$

from which we deduce

$$f(x) = \frac{1}{3} + \frac{1}{\pi} \sum_{n=1}^{\infty} \left(\frac{1}{n^2 \pi} \cos 2n\pi x - \frac{1}{n} \sin 2n\pi x \right)$$

Note that this result is not simply the sum of series in (a) and (b). ∎

Exercises 11.5

1. Verify that the constant function $f(x) = 1$ is periodic. What is its period?

2. Verify the integral relations (43) and (44).
   Hint. It may helpful to recall the trigonometric identities

$$\sin A \sin B = \tfrac{1}{2}[\cos(A - B) - \cos(A + B)]$$

$$\cos A \cos B = \tfrac{1}{2}[\cos(A - B) + \cos(A + B)]$$

$$\sin A \cos B = \tfrac{1}{2}[\sin(A - B) + \sin(A + B)]$$

3. Sketch two or more periods of the following functions:

   (a) $f(x) = x$, $-\pi < x < \pi$, $f(x + 2\pi) = f(x)$
   (b) $f(x) = x^2$, $-\pi < x < \pi$, $f(x + 2\pi) = f(x)$
   (c) $f(x) = e^x$, $-1 < x < 1$, $f(x + 2) = f(x)$

4. If $f(x)$ and $g(x)$ are periodic functions with common period T, show that their sum $[f(x) + g(x)]$ and product $f(x)g(x)$ are also periodic functions with the same period T.

In Problems 5 to 14, find the Fourier series of the periodic function $f(x)$, whose definition over one period is given.

5. $f(x) = \begin{cases} 0, & -1 < x < 0 \\ -1, & 0 < x < 1 \end{cases}$

6. $f(x) = \begin{cases} -1, & -\pi < x < 0 \\ 2, & 0 < x < \pi \end{cases}$

7. $f(x) = x$, $\quad -\pi < x < \pi$

8. $f(x) = x^2$, $\quad -\pi < x < \pi$

9. $f(x) = \begin{cases} x, & -2 < x < 0 \\ 0, & 0 < x < 2 \end{cases}$

■**10.** $f(x) = \begin{cases} 0, & -1 < x < 0 \\ x^2, & 0 < x < 1 \end{cases}$

11. $f(x) = e^{-|x|}$, $\quad -\pi < x < \pi$

12. $f(x) = \begin{cases} -e^x, & -\pi < x < 0 \\ e^{-x}, & 0 < x < \pi \end{cases}$

13. $f(x) = x(1 - x)$, $\quad -p < x < p$

14. $f(x) = \begin{cases} x, & -\pi < x < 0 \\ \pi - x, & 0 < x < \pi \end{cases}$

15. If $f(x + 2\pi) = f(x)$ and

$$f(x) = \begin{cases} 0, & -\pi < x < 0 \\ 1, & 0 < x < \pi/2 \\ 4, & \pi/2 < x < \pi \end{cases}$$

to what numerical value will the Fourier series converge at

(a) $x = 0$?   **(b)** $x = \pi/2$?
(c) $x = \pi$?   **(d)** $x = 2\pi$?
(e) $x = \pi/3$?

16. Use the Fourier series in Example 18 to deduce that

$$1 + \frac{1}{3^2} + \frac{1}{5^2} + \cdots = \sum_{n=1}^{\infty} \frac{1}{(2n-1)^2} = \frac{\pi^2}{8}$$

17. Given that the function $f(x) = x^2$, $-\pi < x < \pi$, has the Fourier cosine series

$$x^2 = \frac{\pi^2}{3} + 4 \sum_{n=1}^{\infty} \frac{(-1)^n}{n^2} \cos nx$$

use this result to show that

(a) $1 + \frac{1}{2^2} + \frac{1}{3^2} + \cdots = \sum_{n=1}^{\infty} \frac{1}{n^2} = \frac{\pi^2}{6}$

(b) $1 - \frac{1}{2^2} + \frac{1}{3^2} - \cdots = \sum_{n=1}^{\infty} \frac{(-1)^n}{n^2} = \frac{\pi^2}{12}$

(c) From (a) and (b), deduce that

$$1 + \frac{1}{3^2} + \frac{1}{5^2} + \cdots = \sum_{n=1}^{\infty} \frac{1}{(2n-1)^2} = \frac{\pi^2}{8}$$

■**18.** Use Equation (51) to derive the result

$$\frac{\sqrt{2}}{2}\left(1 + \frac{1}{3} - \frac{1}{5} - \frac{1}{7} + \cdots\right) = \frac{\pi}{4}$$

In Problems 19 to 24, identify the function as *even, odd,* or *neither.*

19. $f(x) = x$
20. $f(x) = \sin^5 x$
21. $f(x) = \tan x$

22. $f(x) = x^3 \sin 2x$
23. $f(x) = \cos(1 - x)$
24. $f(x) = e^{-x^2}$

25. If
(a) g is an even function, show that

$$\int_{-p}^{p} g(x)\, dx = 2 \int_{0}^{p} g(x)\, dx, \qquad p > 0$$

(b) g is an odd function, show that

$$\int_{-p}^{p} g(x)\, dx = 0, \qquad p > 0$$

26. Prove that
(a) The product of two odd functions is an even function.
(b) The product of two even functions is an even function.
(c) The product of an even and an odd function is an odd function.

27. Find a full Fourier series of the given function defined on a finite interval.

(a) $f(x) = x, 0 < x < 1$
(b) $f(x) = e^{-x}, 0 < x < \pi$

■**28.** Given $f(x) = x(1 - x)$, $0 < x < 1$, find
(a) A sine series (with period 2)
(b) A cosine series (with period 2)
(c) A full Fourier series (with period 1)

29. Given $f(x) = \pi - x$, $0 < x < \pi$, find
(a) A sine series (with period 2π)
(b) A cosine series (with period 2π)
(c) A full Fourier series (with period π)

*****30.** If f is a periodic function with period $T > 0$ and g is likewise a periodic function with period $S > 0$,
(a) Under what condition on T and S is the sum $f + g$ a periodic function?
(b) If $f(x) = \cos \pi x$ and $g(x) = \cos \sqrt{2}\pi x$, is the sum $f + g$ periodic?

*31. Suppose we wish to approximate a given function f on the interval $-\pi \le x \le \pi$ by the trigonometric polynomial

$$S_N(x) = \frac{1}{2}\alpha_0 + \sum_{n=1}^{N} (\alpha_n \cos nx + \beta_n \sin nx),$$

$$-\pi \le x \le \pi$$

where $\alpha_0, \alpha_1, \alpha_2, \ldots, \alpha_N, \beta_1, \beta_2, \ldots, \beta_N$ are not necessarily the Fourier coefficients of f. In this case we define the **mean square error** of the approximation by the expression

$$E_N = \int_{-\pi}^{\pi} [f(x) - S_N(x)]^2 \, dx$$

Show that E_N is minimized by choosing the α's and β's to be the Fourier coefficients a_0, $a_1, a_2, \ldots, a_N$ and $b_1, b_2, \ldots, b_N$.
Hint. By differentiating under the integral sign, set

$$\frac{\partial E_N}{\partial \alpha_k} = 0 \quad (k = 0, 1, 2, \ldots, N)$$

$$\frac{\partial E_N}{\partial \beta_k} = 0 \quad (k = 1, 2, \ldots, N)$$

*32. If f and f' are piecewise continuous, show that

(a) $\displaystyle\int_{-\pi}^{\pi} f(x)S_N(x)\, dx = \frac{\pi}{2}a_0^2 + \pi \sum_{n=1}^{N} (a_n^2 + b_n^2)$

(b) $\displaystyle\int_{-\pi}^{\pi} [S_N(x)]^2\, dx = \frac{\pi}{2}a_0^2 + \pi \sum_{n=1}^{N} (a_n^2 + b_n^2)$

*33. Using the results of Problem 32, show that

(a) $\displaystyle\int_{-\pi}^{\pi} [f(x) - S_N(x)]^2\, dx$

$$= \int_{-\pi}^{\pi} [f(x)]^2\, dx - \frac{\pi}{2}a_0^2 - \pi \sum_{n=1}^{N} (a_n^2 + b_n^2)$$

(b) From (a), deduce **Bessel's inequality** (for all N)

$$\frac{1}{2}a_0^2 + \sum_{n=1}^{N} (a_n^2 + b_n^2) \le \frac{1}{\pi} \int_{-\pi}^{\pi} [f(x)]^2\, dx$$

*34. A Fourier series is said to **converge in the mean** if and only if

$$\lim_{N \to \infty} \int_{-\pi}^{\pi} [f(x) - S_N(x)]^2\, dx = 0$$

Under this assumption and the results of Problem 33, derive **Parseval's equality**

$$\frac{1}{2}a_0^2 + \sum_{n=1}^{\infty} (a_n^2 + b_n^2) = \frac{1}{\pi} \int_{-\pi}^{\pi} [f(x)]^2\, dx$$

*35. Use the result of Problem 34 to obtain

(a) The result of Problem 16

(b) $\displaystyle\sum_{n=1}^{\infty} \frac{1}{n^4} = \frac{\pi^4}{90}$

Hint. See Problem 17.

[0] 11.6 Sturm-Liouville Problems and Eigenfunction Expansions

In this section we wish to extend the notion of eigenfunction expansions beyond that involving either *sine* or *cosine* series.

Suppose we start with the general eigenvalue problem

$$A_2(x)y'' + A_1(x)y' + [A_0(x) + \lambda]y = 0, \quad a < x < b$$

$$h_1 y(a) + h_2 y'(a) = 0, \quad h_3 y(b) + h_4 y'(b) = 0 \tag{63}$$

where $A_2(x) \ne 0$ on $[a, b]$. Multiplying the DE by $p(x)/A_2(x)$ leads to the **self-adjoint form** (see Problem 1 in Exercises 11.6), and then we can rewrite (63) in the more conventional form

$$\frac{d}{dx}[p(x)y''] + [q(x) + \lambda r(x)]y = 0$$

$$h_1 y(a) + h_2 y'(a) = 0, \qquad h_3 y(b) + h_4 y'(b) = 0$$

(64)

where

$$p(x) = \exp\left[\int \frac{A_1(x)}{A_2(x)}\, dx\right]$$

(65)

and where $q(x) = p(x)A_0(x)/A_2(x)$ and $r(x) = p(x)/A_2(x)$. For example, multiplying the DE

$$y'' - 2xy' + \lambda xy = 0$$

by e^{-x^2}, and recognizing that $e^{-x^2}y'' - 2xe^{-x^2}y' = (d/dx)(e^{-x^2}y')$, we obtain the self-adjoint form

$$\frac{d}{dx}(e^{-x^2}y') + \lambda xe^{-x^2}y = 0$$

Among other reasons, putting the DE in self-adjoint form identifies the function $r(x)$, called a **weighting function.** In the above example, the weighting function is $r(x) = xe^{-x^2}$.

Boundary conditions of the kind featured in (64) are called **unmixed** or **separated,** and the general eigenvalue problem belongs to a larger class of problems known as **Sturm-Liouville problems,** named in honor of the two French mathematicians who investigated problems of this nature in the 1830s.[†] Since we have assumed $A_2(x) \neq 0$ on $[a, b]$, it follows that $p(x) > 0$ on $[a, b]$, and the resulting Sturm-Liouville problem is then called **regular.** Such problems are further characterized by the fact that the eigenvalues are all *real* and form an infinite sequence ordered in increasing magnitude so that

$$\lambda_1 < \lambda_2 < \cdots < \lambda_n < \cdots$$

and where $\lambda_n \to \infty$ as $n \to \infty$. The eigenfunctions of regular Sturm-Liouville problems are also real, but their most important property is that they are mutually *orthogonal* (see Theorem 11.5 below).

Definition 11.1 A set of real-valued functions $\{\phi_n(x)\}$ is said to be **orthogonal** on the interval $a \leq x \leq b$ if

$$\int_a^b \phi_n(x)\phi_m(x)\, dx = 0, \qquad m \neq n$$

[†] Jacques C. F. Sturm (1803–1855) and Joseph Liouville (1809–1882). See Section 11.8 for a short historical account of these two mathematicians.

When $m = n$ in the integral in Definition 11.1, we are led to the real number

$$\|\phi_n(x)\| = \sqrt{\int_a^b [\phi_n(x)]^2 \, dx}, \qquad n = 1, 2, 3, \ldots \qquad (66)$$

called the **norm** of the function $\phi_n(x)$. An orthogonal set $\{\phi_n(x)\}$ for which $\|\phi_n(x)\| = 1$ for all n is called an **orthonormal set.**

In some cases the orthogonality condition involves a *weighting function* $r(x)$, where $r(x) > 0$ on $a < x < b$. That is,

$$\int_a^b r(x)\phi_n(x)\phi_m(x) \, dx = 0, \qquad m \neq n \qquad (67)$$

and we say that $\{\phi_n(x)\}$ is an **orthogonal set with respect to the weighting function** $r(x)$. The related *norm* is then defined by

$$\|\phi_n(x)\| = \sqrt{\int_a^b r(x)[\phi_n(x)]^2 \, dx}, \qquad n = 1, 2, 3, \ldots \qquad (68)$$

The following theorem, which we state without proof, is a central result in the general theory of Sturm-Liouville problems.

Theorem 11.5 Orthogonality. *If $\phi_m(x)$ and $\phi_n(x)$ are eigenfunctions of the regular Sturm-Liouville problem (64) belonging to distinct eigenvalues λ_m and λ_m, respectively, then*

$$\int_a^b r(x)\phi_n(x)\phi_m(x) \, dx = 0, \qquad m \neq n$$

EXAMPLE 20

The eigenfunctions of

$$y'' + \lambda y = 0, \qquad y(0) = 0, \qquad y(p) = 0$$

are $\phi_n(x) = \sin(n\pi x/p)$, $n = 1, 2, 3, \ldots$. Verify that they are mutually orthogonal and find the norm $\|\phi_n(x)\|$.

Solution

From Equation (33), we have

$$\int_0^p \sin \frac{m\pi x}{p} \sin \frac{n\pi x}{p} \, dx = 0, \qquad m \neq n$$

and thus the eigenfunctions are indeed mutually orthogonal.

To find the norm $\|\phi_n(x)\|$, we use the trigonometric identity $\sin^2 x = \frac{1}{2}(1 - \cos 2x)$ and calculate

$$\|\phi_n(x)\|^2 = \int_0^p \sin^2 \frac{n\pi x}{p} \, dx = \frac{1}{2} \int_0^p \left(1 - \cos \frac{2n\pi x}{p} \right) dx$$

which, upon integration and taking the square root, reduces to

$$\|\phi_n(x)\| = \sqrt{\frac{p}{2}}, \qquad n = 1, 2, 3, \ldots$$

11.6.1 Generalized Fourier Series

If $\phi_1(x), \phi_2(x), \phi_3(x), \ldots$ denotes an orthogonal set of eigenfunctions on the interval $[a, b]$, with weight function $r(x)$, and f is a suitable function on this interval, then we may consider series expansions of the form

$$f(x) = \sum_{n=1}^{\infty} c_n \phi_n(x), \qquad a < x < b \tag{69}$$

where the c's are constants to be determined. An expression for the constants can be obtained formally by multiplying both sides of (69) by $r(x)\phi_m(x)$ and integrating the result over the interval $[a, b]$. This action leads to

$$\int_a^b r(x)f(x)\phi_m(x)\, dx = \int_a^b \left[\sum_{n=1}^{\infty} c_n \phi_n(x) \right] r(x)\phi_m(x)\, dx$$
$$= \sum_{n=1}^{\infty} c_n \int_a^b r(x)\phi_n(x)\phi_m(x)\, dx \tag{70}$$

where we have assumed that it is permissible to interchange the order of summation and integration. Because of the orthogonality property (67), the integrals on the right are zero for every n except for $n = m$; hence, Equation (70) reduces to

$$\int_a^b r(x)f(x)\phi_m(x)\, dx = c_m \int_a^b r(x)[\phi_m(x)]^2\, dx$$
$$= c_m \|\phi_m(x)\|^2 \tag{71}$$

If we now solve (71) for c_m and change the dummy index from m to n, we see that the coefficients c_n in (69) are given by

$$c_n = \|\phi_n(x)\|^{-2} \int_a^b r(x)f(x)\phi_n(x)\, dx, \qquad n = 1, 2, 3, \ldots \tag{72}$$

Infinite series expansions like (69) are called **generalized Fourier series,** or **eigenfunction expansions,** and the coefficients c_n are called the **Fourier coefficients** of the function f. The following theorem, stated without proof, generalizes Theorem 11.3.

Theorem 11.6 *If $\{\phi_n(x)\}$ is the set of eigenfunctions of a regular Sturm-Liouville problem on $[a, b]$, and if f and f' are piecewise continuous functions on $[a, b]$, then the eigenfunction expansion*

$$f(x) = \sum_{n=1}^{\infty} c_n \phi_n(x), \qquad a < x < b$$

where

$$c_n = \|\phi_n(x)\|^{-2} \int_a^b r(x)f(x)\phi_n(x)\, dx, \qquad n = 1, 2, 3, \ldots$$

converges on the open interval $a < x < b$ to $f(x)$ at points where f is continuous and to $\frac{1}{2}[f(x^+) + f(x^-)]$ at points where f is discontinuous.

EXAMPLE 21   Find an expression for the Fourier coefficients associated with the generalized Fourier series arising from the eigenfunctions of

$$y'' + y' + \lambda y = 0, \qquad 0 < x < 3, \qquad y(0) = 0, \qquad y(3) = 0$$

Solution

We must first solve the eigenvalue problem to identify the eigenfunctions $\{\phi_n(x)\}$. This was done in Example 10, which led to the result

$$\phi_n(x) = e^{-x/2} \sin \frac{n\pi x}{3}, \qquad n = 1, 2, 3, \ldots$$

Next, we multiply the DE by e^x from which we obtain the self-adjoint form

$$\frac{d}{dx}(e^x y') + \lambda e^x y = 0$$

Hence, we see that the weighting function is $r(x) = e^x$, and the normalization factor is

$$\|\phi_n(x)\|^2 = \int_0^3 e^x \left[e^{-x/2} \sin \frac{n\pi x}{3} \right]^2 dx$$

$$= \int_0^3 \sin^2 \frac{n\pi x}{3} \, dx$$

$$= \frac{3}{2}, \qquad n = 1, 2, 3, \ldots$$

Based on this result, the associated generalized Fourier series for a function f satisfying the conditions of Theorem 11.6 takes the form

$$f(x) = \sum_{n=1}^{\infty} c_n e^{-x/2} \sin \frac{n\pi x}{3}$$

$$= e^{-x/2} \sum_{n=1}^{\infty} c_n \sin \frac{n\pi x}{3}$$

where the Fourier coefficients are given by

$$c_n = \frac{2}{3} \int_0^3 e^x f(x) e^{-x/2} \sin \frac{n\pi x}{3} \, dx$$

$$= \frac{2}{3} \int_0^3 e^{x/2} f(x) \sin \frac{n\pi x}{3} \, dx, \qquad n = 1, 2, 3, \ldots \qquad \blacksquare$$

EXAMPLE 22   Find a generalized Fourier series expansion of the function $f(x) = 1, 0 < x < 1$, in terms of the eigenfunctions of

$$y'' + \lambda y = 0, \qquad 0 < x < 1, \qquad y(0) = 0, \qquad y(1) + y'(1) = 0$$

Solution

In Example 9 we found that the eigenfunctions are given by

$$\phi_n(x) = \sin k_n x, \qquad n = 1, 2, 3, \ldots$$

where $k_1, k_2, k_3, \ldots$ are the positive roots of $k = -\tan k$.
The corresponding generalized Fourier series we seek has the form

$$f(x) = \sum_{n=1}^{\infty} c_n \sin k_n x, \qquad 0 < x < 1$$

To calculate the Fourier coefficients, we first observe that

$$\|\sin k_n x\|^2 = \int_0^1 \sin^2 k_n x \, dx = \frac{1}{2}\left(1 + \frac{\sin 2k_n}{2k_n}\right)$$

or, since $\sin 2k_n = 2 \sin k_n \cos k_n$ and $\sin k_n = -k_n \cos k_n$,

$$\|\sin k_n x\|^2 = \tfrac{1}{2}(1 + 2\cos^2 k_n)$$

Also,

$$\int_0^1 \sin k_n x \, dx = \frac{1 - \cos k_n}{k_n}$$

from which we deduce

$$c_n = \frac{2(1 - \cos k_n)}{k_n(1 + 2\cos^2 k_n)}, \qquad n = 1, 2, 3, \ldots$$

Consequently, the above series becomes

$$f(x) = 2 \sum_{n=1}^{\infty} \frac{1 - \cos k_n}{k_n(1 + 2\cos^2 k_n)} \sin k_n x, \qquad 0 < x < 1 \qquad \blacksquare$$

Exercises 11.6

1. Show that multiplying the DE

$$A_2(x)y'' + A_1(x)y' + [A_0(x) + \lambda]y = 0$$

by $p(x)/A_2(x)$ leads to the *self-adjoint form* given by

$$p(x)y'' + p'(x)y' + [q(x) + \lambda r(x)]y = 0$$

or, equivalently by

$$\frac{d}{dx}[p(x)y'] + [q(x) + \lambda r(x)]y = 0$$

where $p(x) = \exp\left[\int \dfrac{A_1(x)}{A_2(x)} \, dx\right]$,

$q(x) = p(x)A_0(x)/A_2(x)$, and $r(x) = p(x)/A_2(x)$.

In Problems 2 to 6, put the given equation in self-adjoint form.

2. $xy'' + \lambda y = 0, \ x > 0$

3. $y'' - y' + \lambda y = 0$

4. $xy'' + (1 - x)y' + \lambda y = 0, \ x > 0$

5. $x^2 y'' + xy' + (\lambda x^2 - n^2)y = 0, \ x > 0$

■**6.** $x^2(x^2 + 1)y'' + 2x^3 y' + \lambda y = 0, \ x > 0$

7. Show that $\{1, \cos(n\pi x/p)\}$, $n = 1, 2, 3, \ldots$, forms an orthogonal set of functions on the interval $0 \leq x \leq p$, that is, show that

$$\int_0^p \cos\frac{n\pi x}{p}\, dx = 0, \qquad n = 1, 2, 3, \ldots$$

and

$$\int_0^p \cos\frac{n\pi x}{p} \cos\frac{m\pi x}{p}\, dx = 0, \qquad n = 1, 2, 3, \ldots$$

Also, find the *norm* of each function in the set.

8. Show that $\{e^{-x/2} \sin n\pi x\}$, $n = 1, 2, 3, \ldots$, forms an orthogonal set of functions on the interval $0 \leq x \leq 1$ with respect to the weighting function $r(x) = e^x$. What is the norm of each function in the set?

In Problems 9 to 12, show that each set of functions is orthogonal on the given interval and find the norm of each function in the set.

9. $\left\{\sin\dfrac{(2n-1)\pi x}{2}\right\}$, $n = 1, 2, 3, \ldots, 0 \leq x \leq 1$

10. $\left\{\cos\dfrac{(2n-1)\pi x}{2}\right\}$, $n = 1, 2, 3, \ldots, 0 \leq x \leq 1$

11. $\{1, \cos nx\}$, $n = 1, 2, 3, \ldots, -\pi \leq x \leq \pi$

12. $\{1, \cos nx, \sin nx\}$, $n = 1, 2, 3, \ldots,$ $-\pi \leq x \leq \pi$

In Problems 13 to 16, find a generalized Fourier series for the function f in terms of the eigenfunctions of the given Sturm-Liouville problem.

13. $f(x) = 1, 0 < x < 1$; $y'' + \lambda y = 0$, $y(0) = 0, y'(0) = 0$

14. $f(x) = x, 0 < x < \pi$; $y'' + \lambda y = 0$, $y'(0) = 0, y'(\pi) = 0$

15. $f(x) = 1 + x, -\pi < x < \pi$; $y'' + \lambda y = 0$, $y(-\pi) = y(\pi), y'(-\pi) = y'(\pi)$

■**16.** $f(x) = 1, 0 < x < 2$; $y'' + 4y' + (4 + 9\lambda)y = 0$, $y(0) = 0, y(2) = 0$

*****17.** Consider the eigenvalue problem

$$y'' + \lambda y = 0, \qquad y(0) = 0, \qquad y'(1) = 0$$

Without explicitly solving for them, show that if $\phi_m(x)$ and $\phi_n(x)$ are eigenfunctions corre-

sponding to the eigenvalues λ_m and λ_n, respectively, with $\lambda_m \neq \lambda_n$, then

$$\int_0^1 \phi_m(x)\phi_n(x)\, dx = 0$$

Hint. Observe that

$$\phi_m'' + \lambda_m\phi_m = 0, \qquad \phi_n'' + \lambda_n\phi_n = 0$$

Multiply the first of these equations by ϕ_n, the second by ϕ_m, subtract the two equations, and then integrate the result from 0 to 1 using integration by parts and taking into account the boundary conditions.

11.7 Chapter Summary

In this chapter we have examined the general theory concerning BVPs of the form

$$A_2(x)y'' + A_1(x)y' + A_0(x)y = F(x), \qquad a < x < b$$
$$h_1y(a) + h_2y'(a) = \alpha, \qquad h_3y(b) + h_4y'(b) = \beta \tag{73}$$

A major distinction that occurs in solving such problems, that doesn't occur in solving IVPs, is that unique solutions do not always exist. In fact, the problem may not even have a solution! The reason is that the prescription of boundary conditions is very critical in solving BVPs.

In some cases the DE and boundary conditions are all **homogeneous,** and further, the DE contains a parameter λ that may assume various values. In such problems we try to determine the values of λ that permit nontrivial solutions. These values of λ are called **eigenvalues** and the corresponding nontrivial solutions are called **eigenfunctions.**

Another important topic discussed in this chapter is the representation of certain **periodic functions** f in a **Fourier series** of the form

$$f(x) = \frac{1}{2} a_0 + \sum_{n=1}^{\infty} \left(a_n \cos \frac{n\pi x}{p} + b_n \sin \frac{n\pi x}{p} \right) \tag{74}$$

where the **Fourier coefficients** are defined by

$$a_n = \frac{1}{p} \int_{-p}^{p} f(x) \cos \frac{n\pi x}{p} \, dx, \qquad n = 0, 1, 2, \ldots \tag{75}$$

and

$$b_n = \frac{1}{p} \int_{-p}^{p} f(x) \sin \frac{n\pi x}{p} \, dx, \qquad n = 1, 2, 3, \ldots \tag{76}$$

If the function f is even, then $b_n = 0$, $n = 1, 2, 3, \ldots$, and (75) can be written in the form

$$a_n = \frac{2}{p} \int_{0}^{p} f(x) \cos \frac{n\pi x}{p} \, dx, \qquad n = 0, 1, 2, \ldots \tag{77}$$

On the other hand, if f is an odd function, then $a_n = 0$, $n = 0, 1, 2, \ldots$, and

$$b_n = \frac{2}{p} \int_{0}^{p} f(x) \sin \frac{n\pi x}{p} \, dx, \qquad n = 1, 2, 3, \ldots \tag{78}$$

When the function f is defined on the finite interval $0 < x < p$, then we can consider either an even or an odd **periodic extension** of the function over the entire x axis. Thus, we may represent f in such cases by either a **cosine series** or a **sine series,** depending on whether we use (77) or (78) to determine the Fourier coefficients. We refer to such series representations as **half-range expansions.**

11.8 Historical Comments

Leonhard Euler solved the first **eigenvalue problem** when he developed a model for describing the "buckling" modes of a vertical column. However, the general theory of eigenvalue problems for second-order DEs, now known as the **Sturm-Liouville theory,** originated in the work of J. Sturm, a professor of mechanics at the Sorbonne, and J. Liouville, a professor of mathematics at the College de France. Sturm is also recognized for making the first accurate determination of the velocity of sound in water and his essay on compressible fluids. Liouville, in addition to his work on differential equations, provided the first proofs of the existence of transcendental numbers and he also made contributions in the fields of number theory and differential geometry.

Joseph Fourier is one of several famous French mathematicians who flourished during the time of Napoleon. In 1794, Napoleon offered Fourier the chair of mathematics at the Ecole Normale in Paris. He left this position in 1798 to accompany Napoleon and a group of other scientists to Egypt, where he remained for four years, establishing the scientific institute of Cairo. In 1802, Fourier returned to France to become prefect of the department of Isere at Grenoble, in the French Alps.

The theory concerned with the representation of functions by **trigonometric series,** now widely known as **Fourier series,** is credited to Fourier, who came across such representations in his classic studies in the theory of heat conduction. His basic papers, presented to the Academy of Sciences in Paris in 1807 and 1811, were criticized by the referees (most strongly by Lagrange) for a lack of rigor and consequently were not then published.

Fourier was called to Paris by the Academy of Sciences in 1816, whereupon he succeeded Laplace as president of the board of the Ecole Polytechnique. When publishing the classic *Théorie analytique de la Chaleur* in 1822, he also incorporated his earlier work that was previously rejected, almost without change. Fourier died in Paris on May 16, 1830, without finishing his final publication which was subsequently completed by Navier in 1831.

CHAPTER 12

Applications Involving Partial Differential Equations

In general, partial differential equations are prominent in those physical and geometrical problems involving functions that depend on more than one independent variable. Our treatment of such equations is intentionally brief, however, focusing on only the basic equations of heat conduction, wave phenomena, and potential theory. The primary solution technique in all cases is **separation of variables.**

In Section 12.1 we introduce the notions of **order, linearity,** and **homogeneity** in connection with partial differential equations. We solve the **heat equation** in Section 12.2 to determine the temperature distribution in a rod of finite length whose endpoints are subject to various types of boundary conditions. In Section 12.3 a similar analysis concerning the **wave equation** is presented, but in connection with the vibration modes of a stretched string set in motion. **Laplace's equation,** which governs steady-state phenomena, is solved in Section 12.4 for both rectangular and circular domains.

In Section 12.5 we briefly examine some applications involving **convective heat transfer** boundary conditions and **heat sources,** both of which lead to **generalized Fourier series.** Also featured here is a short section on applications involving **Bessel functions.** The last section contains some final **historical comments.**

12.1 Introduction

When partial derivatives are required in the mathematical formulation of some physical phenomenon, the resulting equation is called a **partial differential equation** (PDE). The variables involved may be time and/or one or more spatial coordinates. It is convenient to indicate partial derivatives by writing independent variables as subscripts. Thus, we write

$$u_x \text{ for } \frac{\partial u}{\partial x}, \qquad u_{xx} \text{ for } \frac{\partial^2 u}{\partial x^2}, \qquad u_{xy} \text{ for } \frac{\partial^2 u}{\partial y\, \partial x}$$

and so on. It is generally assumed that u satisfies conditions so that $u_{xy} = u_{yx}$.

PDEs are classified as to *order* and *linearity* in much the same way as ordinary differential equations (ODEs). For example, the **order** of a PDE is the order of the partial derivative of highest order appearing in the equation. The equation is said to be **linear** if the unknown function and all its derivatives are of the first degree algebraically. For example, the general linear PDE of second order in two variables has the form

$$Au_{xx} + Bu_{xy} + Cu_{yy} + Du_x + Eu_y + Gu = F \tag{1}$$

where $A, B, C, \ldots, F$ are functions of x and y, but not u. If $F(x, y) \equiv 0$ on some interval I, we say that (1) is **homogeneous;** otherwise it is **nonhomogeneous.** Only linear PDEs are discussed in this chapter.

By **solution** of a PDE we mean a function u that has all partial derivatives occurring in the PDE that, when substituted into the equation, reduce it to an identity for all independent variables. Like ODEs, PDEs generally have many solutions. For example, each of the functions

$$u(x, y) = x + 2y$$

$$u(x, y) = \cos(x + 2y)$$

$$u(x, y) = 3e^{x + 2y}$$

is a solution of the first-order linear PDE

$$2u_x - u_y = 0$$

Of course, there are numerous other solutions (can you think of some?)

A **general solution** of a PDE is a collection of all the solutions of the equation. For instance, it can be shown that

$$u(x, y) = G(x + 2y)$$

where G is any differentiable function of $x + 2y$, is a general solution of $2u_x - u_y = 0$. To show that it is a solution, observe that

$$u_x = G'(x + 2y)$$

$$u_y = 2G'(x + 2y)$$

and thus

$$2u_x - u_y = 2G'(x + 2y) - 2G'(x + 2y) = 0$$

Here we find one of the most fundamental differences between the general solution of an ODE and that of a PDE—the general solution of an ODE contains *arbitrary constants,* whereas that of a PDE involves *arbitrary functions.*

Because it is usually very difficult or impossible to use a general solution to find the particular solution of a PDE satisfying certain *auxiliary conditions,* we ordinarily don't seek general solutions. Instead, we often use a technique called **separation of variables,** which is the only general technique that we discuss in this chapter. Also, rather than give a general treatment of PDEs, our interest focuses on three equations

$$u_{xx} = a^{-2}u_t$$
$$u_{xx} = c^{-2}u_{tt} \tag{2}$$
$$u_{xx} + u_{yy} = 0$$

known, respectively, as the **heat equation, wave equation,** and **Laplace's equation.** Each of these PDEs has several areas of application and each is representative of a larger class of PDEs.

12.2 The Heat Equation

In this section we develop solutions of the *one-dimensional* **heat equation**

$$u_{xx} = a^{-2}u_t \tag{3}$$

Among other areas of application, Equation (3) governs the temperature distribution in a long rod or wire whose lateral surface is impervious to heat (i.e., insulated). In such problems the constant a^2 in (3) is called the **thermal diffusivity** of the material. Some values of a^2 for different materials are provided in Table 12.1. For modeling purposes we assume the rod coincides with a portion of the x axis, is made of uniform material, and has uniform cross section. Further, we assume that the temperature $u(x, t)$ is the same at any point of a particular cross section of the rod but may change from cross section to cross section. That is, heat flows only in the x direction from warmer to cooler parts of the rod.[†]

TABLE 12.1   SOME VALUES OF THERMAL DIFFUSIVITY

Material	a^2 (cm^2/s)
Silver	1.71
Copper	1.14
Aluminum	0.86
Cast iron	0.12

[†] See Section 12.2.3 for a derivation of the heat equation.

To solve for the temperature distribution $u(x, t)$ in a rod of finite length p, extending from $x = 0$ to $x = p$, it is necessary to prescribe certain *auxiliary conditions* on the solutions of the heat equation (3). For example, we usually specify the temperature function $f(x)$ at time $t = 0$, which gives the *initial condition*

$$u(x, 0) = f(x) \tag{4}$$

Also, we need to prescribe conditions at the exposed endpoints of the rod. Generally, this means that we prescribe a fixed temperature at each endpoint, which gives the *boundary conditions*

$$u(0, t) = T_1, \qquad u(p, t) = T_2. \tag{5}$$

If $T_1 = T_2 = 0$, we say the boundary conditions are **homogeneous,** and **nonhomogeneous** otherwise. Other boundary conditions are also possible, such as

$$u_x(0, t) = 0, \qquad u_x(p, t) = 0 \tag{6}$$

which correspond to *insulated* endpoints. In some cases, the rod may be insulated at one end and have a fixed temperature prescribed at the other end. Finding the solution of (3) subject to initial and boundary conditions is called an **initial boundary value problem.**

12.2.1 Homogeneous Boundary Conditions

We first consider some examples featuring *homogeneous* boundary conditions. Let us assume a thin rod having length p is initially heated to some temperature $f(x)$ and the ends of the rod are held in ice packs at temperature $0°C$. The temperature in the rod at any later time is a solution of (see Figure 12.1)

$$u_{xx} = a^{-2}u_t, \qquad 0 < x < p, \qquad t > 0$$

BC: $\qquad u(0, t) = 0, \qquad u(p, t) = 0, \qquad t > 0 \tag{7}$

IC: $\qquad\qquad u(x, 0) = f(x), \qquad 0 < x < p$

Remark. We have marked the boundary conditions in (7) by BC and the initial conditions by IC to aid the reader.

To solve (7) by the method of **separation of variables,** we seek a solution that is a product of a function of x alone by a function of t alone; that is, we seek a solution

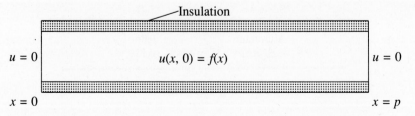

$$u = 0 \qquad\qquad u(x, 0) = f(x) \qquad\qquad u = 0$$

Insulation

$x = 0 \qquad\qquad\qquad\qquad\qquad\qquad\qquad\qquad\qquad x = p$

Figure 12.1

of the form

$$u(x, t) = X(x)T(t) \tag{8}$$

Because each factor depends on only one variable, we find

$$u_{xx}(x, t) = X''(x)T(t), \qquad u_t(x, t) = X(x)T'(t)$$

(For notational ease, we use primes throughout to indicate differentiation with respect to whatever variable is in parentheses.) The substitution of these expressions for u_{xx} and u_t into the heat equation in (7) leads to

$$X''(x)T(t) = a^{-2}X(x)T'(t)$$

and by dividing by the product $X(x)T(t)$, we obtain

$$\frac{X''(x)}{X(x)} = \frac{T'(t)}{a^2 T(t)} \tag{9}$$

In (9) we have "separated the variables" since the left-hand side contains only functions of x and the right-hand side only functions of t. Because the left-hand side is independent of t, the same is true of the right-hand side. Similarly, the right-hand side does not depend on x and thus the left-hand side cannot. If (9) is independent of x and t, both sides can be equated to a common constant, $-\lambda$; that is,

$$\frac{X''(x)}{X(x)} = -\lambda, \qquad \frac{T'(t)}{a^2 T(t)} = -\lambda$$

The negative sign of the constant is not required but merely conventional.[†] These last two relations lead to separate ODEs for the unknown factors $X(x)$ and $T(t)$, which are

$$X''(x) + \lambda X(x) = 0, \qquad 0 < x < p \tag{10}$$

and

$$T'(t) + \lambda a^2 T(t) = 0, \qquad t > 0 \tag{11}$$

Under the assumption (8), the boundary conditions in (7) take the form

$$u(0, t) = X(0)T(t) = 0$$

$$u(p, t) = X(p)T(t) = 0$$

For $T(t) \neq 0$, these relations are satisfied by setting

$$X(0) = 0, \qquad X(p) = 0 \tag{12}$$

Equation (10) subject to the boundary conditions (12) is an eigenvalue problem of the type studied in Section 11.3. Depending upon the value of λ, the solution of

---

[†] No loss of generality occurs by writing $-\lambda$, since any real constant written this way can still be positive, negative, or zero.

(10) assumes one of the forms

$$X(x) = \begin{cases} C_1 \cosh \sqrt{-\lambda}x + C_2 \sinh \sqrt{-\lambda}x, & \lambda < 0 \\ C_1 + C_2 x & \lambda = 0 \\ C_1 \cos \sqrt{\lambda}x + C_2 \sin \sqrt{\lambda}x & \lambda > 0 \end{cases} \tag{13}$$

Based on our previous analysis of this problem, we recall that nontrivial solutions exist only if $\lambda > 0$. In particular, we found that (see Example 8 in Section 11.3)

$$\lambda = \lambda_n = \frac{n^2 \pi^2}{p^2}, \qquad n = 1, 2, 3, \ldots \tag{14}$$

These "allowed" values of λ are called **eigenvalues.** The corresponding nontrivial solutions are called **eigenfunctions,** which are found to be

$$X(x) \equiv \phi_n(x) \equiv \sin \frac{n\pi x}{p}, \qquad n = 1, 2, 3, \ldots \tag{15}$$

where we have set $C_2 = 1$ for mathematical convenience. With λ restricted to the values (14), Equation (11) yields the collection of solutions

$$T_n(t) = c_n e^{-a^2 n^2 \pi^2 t/p^2}, \qquad n = 1, 2, 3, \ldots \tag{16}$$

where the c's are arbitrary constants.

Combining (15) and (16) leads to the family of solutions

$$u_n(x, t) = \phi_n(x) T_n(t)$$

$$= c_n \sin\left(\frac{n\pi x}{p}\right) e^{-a^2 n^2 \pi^2 t/p^2}, \qquad n = 1, 2, 3, \ldots \tag{17}$$

where each $u_n(x, t)$, $n = 1, 2, 3, \ldots$, is itself a solution of the heat equation and boundary conditions in (7). Unfortunately, individual solutions like (17) will not usually satisfy the specified initial condition in (7) since this requires

$$u_n(x, 0) = f(x) = c_n \sin \frac{n\pi x}{p}$$

Hence, unless $f(x)$ is a multiple of an eigenfunction we cannot satisfy this relation. However, new solution forms can be derived from (17) through use of the *superposition principle.*

Theorem 12.1 Superposition Principle. *If u_1 and u_2 are two solutions of a homogeneous, linear PDE in some domain R, then*

$$u = C_1 u_1 + C_2 u_2$$

is also a solution in R for any constants C_1 and C_2. More generally, if u_1, $u_2, \ldots, u_n, \ldots$ are all solutions in R, then

$$u = \sum_{n=1}^{\infty} C_n u_n$$

is also a solution in R for any set of constants $C_1, C_2, \ldots, C_n, \ldots$ for which the series converges.

By invoking the superposition principle on the set of solutions (17), we obtain the series

$$u(x, t) = \sum_{n=1}^{\infty} c_n \sin\left(\frac{n\pi x}{p}\right) e^{-a^2 n^2 \pi^2 t / p^2} \tag{18}$$

Since each term of the series (18) satisfies the heat equation and boundary conditions in (7), we assume the same is true of the complete series. Moreover, by letting $t = 0$, we now have that

$$u(x, 0) = f(x) = \sum_{n=1}^{\infty} c_n \sin\frac{n\pi x}{p}, \qquad 0 < x < p \tag{19}$$

We recognize this last expression as a **Fourier sine series** of the function $f(x)$, for which (see Section 11.4)

$$c_n = \frac{2}{p} \int_0^p f(x) \sin\frac{n\pi x}{p}\, dx, \qquad n = 1, 2, 3, \ldots \tag{20}$$

Thus, we say that (18) is a formal solution of the initial boundary value problem (7), where the constants c_n, $n = 1, 2, 3, \ldots$ are defined by (20).

EXAMPLE 1

Solve

$$u_{xx} = a^{-2} u_t, \qquad 0 < x < \pi, \qquad t > 0$$

BC: $\qquad\qquad u_x(0, t) = 0, \qquad u_x(\pi, t) = 0, \qquad t > 0$

IC: $\qquad\qquad u(x, 0) = x, \qquad 0 < x < \pi$

Solution

This problem is a variation of that described above. Specifically, we can physically interpret this problem as that corresponding to the situation where the rod is initially heated to the temperature $f(x) = x$, while the ends at $x = 0$ and $x = \pi$ are insulated (i.e., there is no heat flow into the surrounding medium at the ends).

The separation of variables technique applied this time leads to the ODEs

$$X'' + \lambda X = 0, \qquad X'(0) = 0, \qquad X'(\pi) = 0$$

and

$$T' + \lambda a^2 T = 0$$

The problem for X is a standard eigenvalue problem for which

$$\lambda_0 = 0, \ \phi_0(x) = 1, \qquad \lambda_n = n^2, \ \phi_n(x) = \cos nx, \qquad n = 1, 2, 3, \ldots$$

With λ restricted to these values, the equation in T yields

$$T_0(t) = c_0$$

$$T_n(t) = c_n e^{-a^2 n^2 t}, \qquad n = 1, 2, 3, \ldots$$

and hence, by use of the superposition principle, we obtain

$$u(x, t) = c_0 + \sum_{n=1}^{\infty} c_n \cos(nx) e^{-a^2 n^2 t}$$

Lastly, imposing the prescribed initial condition leads to

$$u(x, 0) = x = c_0 + \sum_{n=1}^{\infty} c_n \cos nx$$

from which we deduce (recall Example 14 in Section 11.4)

$$c_0 = \frac{\pi}{2}$$

$$c_n = \begin{cases} -\dfrac{4}{\pi n^2}, & n = 1, 3, 5, \ldots \\ 0, & n = 2, 4, 6, \ldots \end{cases}$$

The solution we seek is therefore

$$u(x, t) = \frac{\pi}{2} - \frac{4}{\pi} \sum_{\substack{n=1 \\ (\text{odd})}}^{\infty} \frac{\cos nx}{n^2} e^{-a^2 n^2 t} \qquad\blacksquare$$

It sometimes happens that the prescribed boundary conditions at $x = 0$ and/or $x = p$ do not agree with the values at these points prescribed by the initial condition. Of course, in practice discontinuities of this nature cannot actually occur. They are part of the math model simply because the resulting problem is easier to solve than the actual problem. Fortunately, such discontinuities generally cause only small errors in the solution near the endpoints of the rod, and only during the initial stages of heat flow. That is, diffusion processes in general tend to "smooth out" discontinuities over a period of time.

12.2.2 Nonhomogeneous Boundary Conditions

Suppose we now consider the problem described by (see Figure 12.2)

$$u_{xx} = a^{-2} u_t, \qquad 0 < x < p, \qquad t > 0$$

BC: $\qquad u(0, t) = T_1, \qquad u(p, t) = T_2, \qquad t > 0 \qquad\qquad (21)$

IC: $\qquad\qquad u(x, 0) = f(x), \qquad 0 < x < 0$

where T_1 and T_2 are constants. When $T_1 = T_2 = 0$, this problem reduces to that in Section 12.2.1. Thus, here we assume that at least one endpoint condition is *non-homogeneous*.

The separation of variables technique will not usually work in a direct manner when one or both boundary conditions are nonhomogeneous. For this reason we need to modify the approach used in the previous section. Based on physical observations, we expect the temperature $u(x, t)$ to approach a **steady-state temperature** $S(x)$ in the limit as $t \to \infty$. Hence, $S(x)$ must satisfy the equation $u_{xx} = S'' = 0$, which the

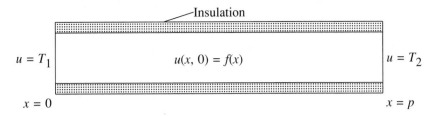

Figure 12.2

heat equation reduces to when $u_t = 0$. Therefore, let us assume the solution of (21) can be expressed in the form

$$u(x, t) = S(x) + v(x, t) \tag{22}$$

where $v(x, t)$, called the **transient solution,** must vanish in the limit as $t \to \infty$. The substitution of (22) into (21) leads to

$$S'' + v_{xx} = a^{-2}v_t, \qquad 0 < x < p, \qquad t > 0$$

BC: $\qquad\qquad S(0) + v(0, t) = T_1, \qquad S(p) + v(p, t) = T_2 \tag{23}$

IC: $\qquad\qquad\qquad S(x) + v(x, 0) = f(x)$

By setting $S'' = 0$ and requiring $S(x)$ to satisfy the nonhomogeneous boundary conditions, we obtain the *steady-state problem*

$$S'' = 0, \qquad S(0) = T_1, \qquad S(p) = T_2 \tag{24}$$

Two successive integrations of the DE yield

$$S(x) = C_1 + C_2 x$$

and by imposing the boundary conditions, we find that

$$S(0) = C_1 = T_1$$
$$S(p) = C_1 + C_2 p = T_2$$

from which we deduce the *steady-state solution*

$$S(x) = T_1 + (T_2 - T_1)\frac{x}{p} \tag{25}$$

Having selected $S(x)$ to be that given by (25), the problem described by (23) now reduces to

$$v_{xx} = a^{-2}v_t, \qquad 0 < x < p, \qquad t > 0$$

BC: $\qquad\qquad v(0, t) = 0, \qquad v(p, t) = 0$

IC: $\qquad\qquad v(x, 0) = f(x) - S(x)$

$$= f(x) - T_1 - (T_2 - T_1)\frac{x}{p} \tag{26}$$

Here we see that the transient solution $v(x, t)$ satisfies a heat conduction problem of the type for which the boundary conditions are homogeneous. Hence, the separation of variables method applied to (26) leads to the formal solution

$$v(x, t) = \sum_{n=1}^{\infty} c_n \sin\left(\frac{n\pi x}{p}\right) e^{-a^2 n^2 \pi^2 t/p^2} \qquad (27)$$

where the c's are defined by

$$c_n = \frac{2}{p} \int_0^p [f(x) - S(x)] \sin\frac{n\pi x}{p} dx$$

$$= \frac{2}{p} \int_0^p f(x) \sin\frac{n\pi x}{p} dx - \frac{2}{p} \int_0^p \left[T_1 + (T_2 - T_1)\frac{x}{p} \right] \sin\frac{n\pi x}{p} dx$$

or

$$c_n = \frac{2}{p} \int_0^p f(x) \sin\frac{n\pi x}{p} dx + \frac{2}{\pi n} [(-1)^n T_2 - T_1], \qquad n = 1, 2, 3, \ldots \quad (28)$$

Combining results, our formal solution of (22) becomes

$$u(x, t) = S(x) + v(x, t)$$

$$= T_1 + (T_2 - T_1)\frac{x}{p} + \sum_{n=1}^{\infty} c_n \sin\left(\frac{n\pi x}{p}\right) e^{-a^2 n^2 \pi^2 t/p^2} \qquad (29)$$

where the c's are given by (28).

EXAMPLE 2   Solve

$$u_{xx} = a^{-2} u_t, \qquad 0 < x < 10, \qquad t > 0$$

BC: $u(0, t) = 10, \qquad u(10, t) = 30, \qquad t > 0$

IC: $u(x, 0) = 0, \qquad 0 < x < 10$

Solution

Because of the nonhomogeneous boundary conditions, we write the solution as

$$u(x, t) = S(x) + v(x, t)$$

where $S(x)$, the steady-state solution, satisfies

$$S'' = 0, \qquad S(0) = 10, \qquad S(10) = 30$$

and the transient solution $v(x, t)$ satisfies

$$v_{xx} = a^{-2} v_t, \qquad 0 < x < 10, \qquad t > 0$$

BC: $v(0, t) = 0, \qquad v(10, t) = 0$

IC: $v(x, 0) = -S(x)$

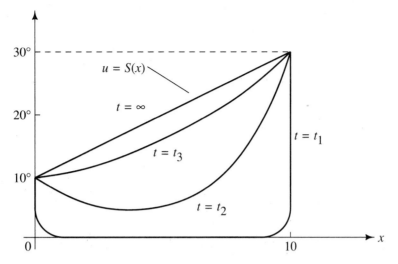

Figure 12.3

The solution of the steady-state problem is readily found to be

$$S(x) = 10 + 2x$$

while the transient solution leads to

$$v(x, t) = \sum_{n=1}^{\infty} c_n \sin\left(\frac{n\pi x}{10}\right) e^{-a^2 n^2 \pi^2 t/100}$$

Lastly, by imposing the initial condition, we have

$$v(x, 0) = -(10 + 2x) = \sum_{n=1}^{\infty} c_n \sin \frac{n\pi x}{10}$$

which leads to (see Example 13 in Section 11.4)

$$c_n = \frac{20}{\pi n}[(-1)^n 3 - 1], \qquad n = 1, 2, 3, \ldots.$$

Combining results, we obtain

$$u(x, t) = S(x) + v(x, t)$$

$$= 10 + 2x + \frac{20}{\pi} \sum_{n=1}^{\infty} \frac{[(-1)^n 3 - 1]}{n} \sin\left(\frac{n\pi x}{10}\right) e^{-a^2 n^2 \pi^2 t/100}$$

In Figure 12.3 we show how this solution converges from a 0°C initial temperature distribution to the steady-state solution $S(x) = 10 + 2x$ for increasing time. Because of the exponential function in the solution, $u(x, t)$ will normally converge quite rapidly to the steady-state solution, depending somewhat of course on the constant a^2 (i.e., the material forming the rod). ∎

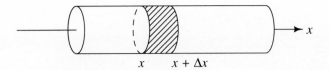

Figure 12.4 Rod of finite length.

[0] 12.2.3 Derivation of the Governing Equation

To derive the governing equation of heat flow in a uniform rod from first principles, we apply the law of conservation of thermal energy to a slice of the rod between x and a nearby point $x + \Delta x$ as shown in Figure 12.4. The **law of conservation of thermal energy** states that

> the rate of heat entering a region plus that which is generated inside the region equals the rate of heat leaving the region plus that which is stored.

Suppose we let $Q(x, t)$ be the rate of heat flow at the point x at time t, also known as the **heat flux**. If A denotes the cross-sectional area of the uniform rod, then $AQ(x, t)$ represents the rate at which heat enters the slice of the rod through the flat surface at x. Accordingly, the rate at which heat vacates the slice at $x + \Delta x$ is $AQ(x + \Delta x, t)$. If a heat source is present within the region, the rate at which heat is generated in the slice is $A \, \Delta x \, F(x, t)$, where $F(x, t)$ is the rate of heat generation per unit volume. Lastly, the rate of heat energy storage in the slice is proportional to the time rate change of temperature $u_t(x, t)$. To find the constant of proportionality necessitates the introduction of the **specific heat** constant c, defined as the heat energy that must be supplied to a unit mass of the rod to raise it one unit of temperature. Thus, if ρ is the (constant) **mass density** of the rod (i.e., mass per unit volume), the rate of change of heat energy is approximately $A \, \Delta x \, \rho c u_t(x, t)$.

If we now invoke the law of conservation of energy, it follows that

$$AQ(x, t) + A \, \Delta x \, F(x, t) = AQ(x + \Delta x, t) + A \, \Delta x \, \rho c u_t$$

which can be rearranged as

$$-\left(\frac{Q(x + \Delta x, t) - Q(x, t)}{\Delta x} \right) = \rho c u_t - F(x, t) \tag{30}$$

By allowing $\Delta x \to 0$, we find that (30) becomes

$$-Q_x = \rho c u_t - F(x, t) \tag{31}$$

We can relate $Q(x, t)$ and the temperature $u(x, t)$ by use of **Fourier's law of heat conduction,** which for the present problem reads

$$Q = -k u_x \tag{32}$$

where k is a positive constant.[†] In words, Fourier's law says that the heat flux $Q(x, t)$ at any point x is proportional to the temperature gradient $u_x(x, t)$ at that point. The

---

[†] The negative sign in (32) reflects the fact that heat flows from hotter to cooler regions.

proportionality constant k is characteristic of the material of the rod, called the **heat conductivity.** Fourier's law (32) combined with (31) yields the governing *nonhomogeneous* equation

$$u_{xx} = a^{-2}u_t - \frac{1}{k}F(x, t) \tag{33}$$

where $a^2 = k/\rho c$ is the **thermal diffusivity** of the material forming the rod. When the heat source $F(x, t)$ is not present, this PDE reduces to the *homogeneous* equation

$$u_{xx} = a^{-2}u_t \tag{34}$$

called the *one-dimensional* **heat equation.**

Exercises 12.2

In Problems 1 to 3, solve the heat conduction problem with zero temperatures on the boundary.

1. $u_{xx} = a^{-2}u_t,\ 0 < x < 1,\ t > 0$
BC: $u(0, t) = 0,\ u(1, t) = 0$
IC: $u(x, 0) = 3\sin \pi x - 5\sin 4\pi x$

2. $u_{xx} = a^{-2}u_t,\ 0 < x < \pi,\ t > 0$
BC: $u(0, t) = 0,\ u(\pi, t) = 0$
IC: $u(x, 0) = \begin{cases} x, & 0 < x < \pi/2 \\ \pi - x, & \pi/2 < x < \pi \end{cases}$

3. $u_{xx} = a^{-2}u_t,\ 0 < x < p,\ t > 0$
BC: $u(0, t) = 0,\ u(p, t) = 0$
IC: $u(x, 0) = x(p - x)$

In Problems 4 to 6, find the steady-state solution.

4. $u_{xx} = a^{-2}u_t,\ 0 < x < p,\ t > 0$
BC: $u(0, t) - u_x(0, t) = T_1,\ u(p, t) = 0$

5. $u_{xx} = a^{-2}u_t,\ 0 < x < p,\ t > 0$
BC: $u(0, t) = T_1,\ u(p, t) + u_x(p, t) = T_2$

*6. $\dfrac{\partial}{\partial x}\left[(1 + x)\dfrac{\partial u}{\partial x}\right] = \dfrac{\partial u}{\partial t},\ 0 < x < p,\ t > 0$
BC: $u(0, t) = T_1,\ u(p, t) = T_2$

In Problems 7 to 15, solve the given heat conduction problem.

7. $u_{xx} = a^{-2}u_t,\ 0 < x < 1,\ t > 0$
BC: $u(0, 1) = 1,\ u(1, t) = 0$

8. $u_{xx} = a^{-2}u_t,\ 0 < x < 2,\ t > 0$
BC: $u(0, t) = T_1,\ u(2, t) = T_2$
IC: $u(x, 0) = T_0$

9. $u_{xx} = a^{-2}u_t,\ 0 < x < p,\ t > 0$
BC: $u_x(0, t) = 0,\ u_x(p, t) = 0$
IC: $u(x, 0) = T_0\sin^2(\pi x/p)$

■10. $u_{xx} = a^{-2}u_t,\ 0 < x < p,\ t > 0$
BC: $u_x(0, t) = 0,\ u(p, t) = T_0$
IC: $u(x, 0) = T_0$

11. $u_{xx} = a^{-2}u_t,\ 0 < x < 1,\ t > 0$
BC: $u(0, t) = T_0,\ u(1, t) = T_0$
IC: $u(x, 0) = T_0 + x(1 - x)$

*12. $u_{xx} = a^{-2}u_t,\ 0 < x < 1,\ t > 0$
BC: $u(0, t) = T_1,\ u_x(1, t) = 0$
IC: $u(x, 0) = x$

*13. $u_{xx} = a^{-2}u_t,\ -\pi < x < \pi,\ t > 0$
BC: $u(-\pi, t) = u(\pi, t),\ u_x(-\pi, t) = u_x(\pi, t)$
IC: $u(x, 0) = |x|$

*14. $u_{xx} = a^{-2}u_t,\ 0 < x < 1,\ t > 0$
BC: $u(0, t) = 0,\ 2u(1, t) + u_x(1, t) = 0$
IC: $u(x, 0) = T_0$

*15. $u_{xx} = a^{-2}u_t,\ 0 < x < \pi,\ t > 0$
BC: $u(0, t) - u_x(0, t) = 100,\ u_x(\pi, t) = 0$
IC: $u(x, 0) = 200$

■**16.** The ends $x = 0$ and $x = 100$ of a rod 100 cm in length, with insulated lateral surfaces, are held at temperatures $0°$ and $100°C$, respectively, until steady-state conditions prevail. Then, at time $t = 0$, the temperatures of the two ends are interchanged. Find the subsequent temperature distribution throughout the rod.

17. A rod 2 m long is given the initial temperature distribution

$$u(x, 0) = \begin{cases} 50x & 0 < x < 1 \\ 100 - 50x, & 1 < x < 2 \end{cases}$$

If both ends of the rod are maintained at $0°C$, approximately how long will it take for the center of the rod to reach a temperature of $30°C$? Assume the lateral surface is insulated. **Hint.** Use only the first nonzero term of the series.

18. Repeat Problem 17 if the ends of the rod are insulated. What is the temperature in the rod after a long time?

*19. If the lateral surface of the rod is not insulated, there is a heat exchange by convection into the surrounding medium. If the surrounding medium has constant temperature T_0, the rate at which heat is lost from the rod is proportional to the difference $u - T_0$. The governing PDE in this situation is

$$a^2 u_{xx} = u_t + b(u - T_0), \qquad b > 0$$

Show that the change of variables $u(x, t) = T_0 + z(x, t)e^{-bt}$ leads to the heat equation in z.

*20. Use the technique of Problem 19 to solve

$$u_{xx} = u_t + 4u - 20, \qquad 0 < x < \pi, \qquad t > 0$$

BC: $\qquad u(0, t) = 5, \qquad u(\pi, t) = 5$

IC: $\qquad\qquad u(x, 0) = 5 + 2x$

12.3 The Wave Equation

Imagine a violin string under a large tension τ that is "plucked" and then allowed to vibrate freely. If the string has mass density (per unit length) ρ and initially lies along the x axis, then the transverse movement $u(x, t)$ after it is plucked is a solution of the *one-dimensional* **wave equation**

$$u_{xx} = c^{-2} u_{tt} \tag{35}$$

where $c = \sqrt{\tau/\rho}$ is a physical constant having the dimension of velocity. Equation (35) is essentially equivalent to Equation (6) in Section 11.2.1 which governs the deflections of an elastic string supporting an external load $q(x)$, where $q(x)$ is now the **inertial force** $-\rho u_{tt}$.

12.3.1 Free Motions of a Vibrating String

The study of free oscillations—those with no external forces other than tension—of a tightly stretched string between two supports is one of the most rudimentary problems in the theory of wave motion. Let us consider the free oscillations of a taut string of length p that is fixed at both ends (see Figure 12.5). The governing PDE and boundary conditions are prescribed by

$$u_{xx} = c^{-2} u_{tt}, \qquad 0 < x < p, \qquad t > 0$$

BC: $\qquad u(0,t) = 0, \qquad u(p, t) = 0, \qquad t > 0$ $\tag{36}$

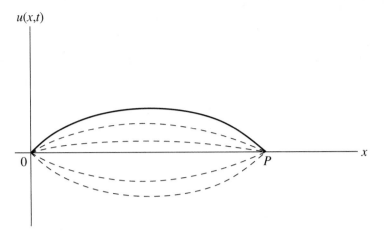

Figure 12.5 Freely vibrating string.

Of course, the motion of the string will also depend upon its deflection and velocity (speed) at the time of release. Denoting the initial deflection by $f(x)$ and the initial velocity by $g(x)$, the initial conditions are prescribed by

$$\text{IC:} \qquad u(x, 0) = f(x), \qquad u_t(x, 0) = g(x), \qquad 0 < x < p \qquad (37)$$

The problem described by (36) and (37) is once again called an **initial boundary value problem.**

Because the boundary conditions in (36) are homogeneous, we can resort immediately to the method of *separation of variables* used in Section 12.2 to solve the heat equation. Thus, we assume that

$$u(x, t) = X(x)T(t) \qquad (38)$$

and substitute this expression into the wave equation in (36). This action leads to

$$X''(x)T(t) = c^{-2}X(x)T''(t)$$

which, through "separating the variables," yields the two ODEs

$$X'' + \lambda X = 0, \qquad 0 < x < p \qquad (39)$$

$$T'' + \lambda c^2 T = 0, \qquad t > 0 \qquad (40)$$

where λ is the separation constant. Under the assumption (38), the boundary conditions in (36) take the form

$$X(0) = 0, \qquad X(p) = 0 \qquad (41)$$

The eigenvalue problem composed of (39) and (41) is one we have previously solved, and leads to

$$\lambda_n = \frac{n^2\pi^2}{p^2}, \qquad \phi_n(x) = \sin\frac{n\pi x}{p}, \qquad n = 1, 2, 3, \ldots \qquad (42)$$

For these values of λ, the solutions of (40) are

$$T_n(t) = a_n \cos \frac{nc\pi t}{p} + b_n \sin \frac{nc\pi t}{p}, \qquad n = 1, 2, 3, \ldots \tag{43}$$

where the a's and b's are arbitrary constants.

By combining (42) and (43), we obtain the set of solutions

$$u_n(x, t) = \left(a_n \cos \frac{nc\pi t}{p} + b_n \sin \frac{nc\pi t}{p} \right) \sin \frac{n\pi x}{p}, \qquad n = 1, 2, 3, \ldots \tag{44}$$

each of which satisfies (36). These solutions are called **standing waves** since, for a fixed value of t, each can be viewed as having the shape $\sin(n\pi x/p)$ with amplitude $T_n(t)$. The points where $\sin(n\pi x/p) = 0$ are called **nodes** and physically correspond to zero displacement of the string. The number of nodes depends on the value of n. For example, when $n = 1$ there is no node on the interval $0 < x < p$. When $n = 2$ there is one node, when $n = 3$ there are two nodes, and so forth (see Figure 12.6).

We can find another interpretation of (44) if we now think of x as fixed. In this case $u_n(x, t)$ represents the motion of a point on the string with abscissa x. Moreover, by writing $T_n(t)$ in the form,

$$T_n(t) = \sqrt{a_n^2 + b_n^2} \cos\left(\frac{nc\pi t}{p} - \alpha_n \right) \tag{45}$$

where $\alpha_n = \tan^{-1}(b_n/a_n)$, we see that for a fixed value of x, $u_n(x, t)$ represents **simple harmonic motion** of (angular) **frequency** $\omega_n = nc\pi/p$ and **amplitude** $\sqrt{a_n^2 + b_n^2} \sin(n\pi x/p)$ (recall our discussion in Section 5.2 with respect to a vibrating spring-mass system). The frequency ω_n is called the nth natural frequency (harmonic) of the system, whereas $\omega_1 = c\pi/p$ denotes the **fundamental frequency.**

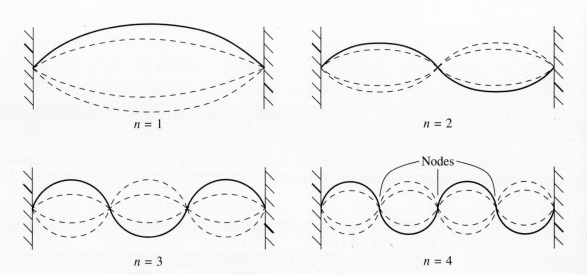

$n = 1$ $n = 2$

$n = 3$ $n = 4$

Figure 12.6 Harmonic motion and nodal points.

In order to satisfy the prescribed initial conditions (37), we must first impose the superposition principle to obtain

$$u(x, t) = \sum_{n=1}^{\infty} \left(a_n \cos \frac{nc\pi t}{p} + b_n \sin \frac{nc\pi t}{p} \right) \sin \frac{n\pi x}{p} \qquad (46)$$

The initial condition $u(x, 0) = f(x)$ requires that

$$f(x) = \sum_{n=1}^{\infty} a_n \sin \frac{n\pi x}{p}, \qquad 0 < x < p \qquad (47)$$

which we recognize as a **Fourier sine series.** Hence, we see that

$$a_n = \frac{2}{p} \int_0^p f(x) \sin \frac{n\pi x}{p} \, dx, \qquad n = 1, 2, 3, \dots \qquad (48)$$

By differentiating the series (46) termwise and imposing the remaining boundary condition $u_t(x, 0) = g(x)$, we have

$$g(x) = \sum_{n=1}^{\infty} \frac{n\pi c}{p} b_n \sin \frac{n\pi x}{p}, \qquad 0 < x < p \qquad (49)$$

from which we deduce

$$\frac{n\pi c}{p} b_n = \frac{2}{p} \int_0^p g(x) \sin \frac{n\pi x}{p} \, dx, \qquad n = 1, 2, 3, \dots \qquad (50)$$

EXAMPLE 3

Solve (see Figure 12.7)

$$u_{xx} = c^{-2} u_{tt}, \qquad 0 < x < 1, \qquad t > 0$$

BC: $\qquad u(0, t) = 0, \qquad u(1, t) = 0, \qquad t > 0$

IC:
$$\begin{cases} u(x, 0) = f(x) = \begin{cases} Ax, & 0 \le x \le 1/2 \\ A(1 - x), & 1/2 < x \le 1 \end{cases} \\ u_t(x, 0) = 0, \qquad 0 < x < 1 \end{cases}$$

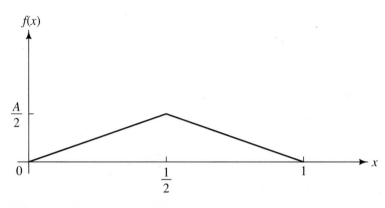

Figure 12.7 Initial shape of string.

Solution

By separating variables, we get the eigenvalue problem

$$X'' + \lambda X = 0, \qquad X(0) = 0, \qquad X(1) = 0$$

with eigenvalues and eigenfunctions given by

$$\lambda_n = n^2\pi^2, \qquad \phi_n(x) = \sin n\pi x, \qquad n = 1, 2, 3, \ldots$$

For these values of λ, the time-dependent equation becomes

$$T'' + n^2\pi^2c^2T = 0$$

with solutions

$$T_n(t) = a_n \cos n\pi ct + b_n \sin n\pi ct, \qquad n = 1, 2, 3, \ldots$$

By use of the superposition principle, we then obtain

$$u(x, t) = \sum_{n=1}^{\infty} (a_n \cos n\pi ct + b_n \sin n\pi ct) \sin n\pi x$$

Because the initial velocity $u_t(x, 0) = 0$, it follows that $b_n = 0, n = 1, 2, 3, \ldots$. Thus, setting $t = 0$ in the remaining solution yields

$$u(x, 0) = f(x) = \sum_{n=1}^{\infty} a_n \sin n\pi x$$

where

$$a_n = 2 \int_0^1 f(x) \sin n\pi x \, dx$$

$$= 2A \int_0^{1/2} x \sin n\pi x \, dx + 2A \int_{1/2}^1 (1 - x) \sin n\pi x \, dx$$

or

$$a_n = \frac{4A}{n^2\pi^2} \sin(n\pi/2), \qquad n = 1, 2, 3, \ldots$$

The solution then becomes

$$u(x, t) = \frac{4A}{\pi^2} \sum_{\substack{n=1 \\ (\text{odd})}}^{\infty} \frac{\sin(n\pi/2)}{n^2} \cos(n\pi ct) \sin(n\pi x)$$

or, by changing the index,

$$u(x, t) = \frac{4A}{\pi^2} \sum_{n=1}^{\infty} \frac{(-1)^{n-1}}{(2n-1)^2} \cos[(2n-1)\pi ct] \sin[(2n-1)\pi x]$$

∎

[0] 12.3.2 d'Alembert's Solution

Suppose we now imagine a string that is so long we consider it to be of "infinite extent." In this case the actual mathematical form of the boundary conditions is no longer important but only the initial deflection and velocity. Thus, we now consider the problem described by

$$u_{xx} = c^{-2}u_{tt}, \qquad -\infty < x < \infty$$

IC: $\qquad\qquad u(x, 0) = f(x), \qquad u_t(x, 0) = 0$ $\qquad\qquad$ (51)

The solution technique we are about to illustrate is attributed to Jean Le Rond d'Alembert (1717–1783), who discovered it six years before D. Bernoulli introduced the separation of variables method. At first d'Alembert thought he had developed a technique that could be applied to a large class of equations, but he soon realized that this was not the case. In fact, the wave equation is one of the rare examples for which d'Alembert's technique proves fruitful. However, in addition to being of historical value, the solution form developed by d'Alembert provides us with a more comprehensive understanding of the solutions of the wave equation in general.

We begin by introducing the change of variables

$$r = x + ct$$
$$s = x - ct$$
$$(52)$$

from which, through use of the chain rule, we get

$$u_x = u_r + u_s$$
$$u_t = cu_r - cu_s$$

One more application of the chain rule yields

$$u_{xx} = u_{rr} + 2u_{rs} + u_{ss}$$
$$u_{tt} = c^2(u_{rr} - 2u_{rs} + u_{ss})$$

and when these last expressions are substituted into the wave equation, we obtain[†]

$$u_{rs} = 0 \qquad\qquad (53)$$

By integrating (53) twice, once with respect to s and then with respect to r, we find that

$$u(r, s) = F(r) + G(s) \qquad\qquad (54)$$

where F and G are *arbitrary* differentiable functions. In terms of the original variables x and t, we then obtain the *general solution*

$$u(x, t) = F(x + ct) + G(x - ct) \qquad\qquad (55)$$

Imposing the initial conditions in (51) on the general solution (55), we see that

$$u(x, 0) = F(x) + G(x) = f(x)$$
$$u_t(x, 0) = cF'(x) - cG'(x) = g(x)$$

which is equivalent to

$$F(x) + G(x) = f(x)$$
$$F(x) - G(x) = \frac{1}{c}\int_0^x g(z)\, dz + C_1$$

[†] The equation $u_{rs} = 0$ is sometimes called the *canonical form* of the wave equation.

where C_1 is an arbitrary constant. The simultaneous solution of these equations leads to

$$F(x) = \tfrac{1}{2}f(x) + \frac{1}{2c} \int_0^x g(z) \, dz + \tfrac{1}{2}C_1$$

$$G(x) = \tfrac{1}{2}f(x) - \frac{1}{2c} \int_0^x g(z) \, dz - \tfrac{1}{2}C_1$$

(56)

Because we permit the arguments in F and G to extend over all numbers, the results (56) can be generalized to

$$\cdot F(x + ct) = \tfrac{1}{2}f(x + ct) + \frac{1}{2c} \int_0^{x+ct} g(z) \, dz + \tfrac{1}{2}C_1$$

$$G(x - ct) = \tfrac{1}{2}f(x - ct) - \frac{1}{2c} \int_0^{x-ct} g(z) \, dz - \tfrac{1}{2}C_1$$

Combining results according to (55) then yields

$$u(x, t) = \tfrac{1}{2}\big[f(x + ct) + f(x - ct)\big]$$

$$+ \frac{1}{2c} \int_0^{x+ct} g(z) \, dz - \frac{1}{2c} \int_0^{x-ct} g(z) \, dz$$

and by writing the sum of integrals as a single integral, we obtain **d'Alembert's solution**

$$u(x, t) = \tfrac{1}{2}\big[f(x + ct) + f(x - ct)\big] + \frac{1}{2c} \int_{x-ct}^{x+ct} g(z) \, dz \qquad (57)$$

The interesting observation here is that once we know the initial displacement $f(x)$ and the initial velocity $g(x)$, the solution (57) is immediately determined. This makes d'Alembert's solution easy to apply as compared with the infinite series solution obtained from the separation of variables method. For instance, if the initial deflection shape were described by the "Gaussian" function

$$u(x, 0) = f(x) = e^{-x^2}, \qquad -\infty < x < \infty$$

and $g(x) \equiv 0$, then (57) leads us at once to

$$u(x, t) = \tfrac{1}{2}\big[e^{-(x+ct)^2} + e^{-(x-ct)^2}\big]$$

The functions $F(x + ct)$ and $G(x - ct)$ are each separate solutions of the wave equation that have interesting physical interpretations as wave phenomena. The function $F(x + ct)$, plotted as a function of x alone, is exactly the same in shape as the function $F(x)$, but with every point on it displaced a distance ct to the left of the corresponding point in $F(x)$. Thus, the function $F(x + ct)$ represents a wave of displacement traveling to the left with velocity c. In the same manner we can interpret $G(x - ct)$ as a displacement wave traveling with velocity c to the right (see Figure 12.8). The solution (57) is then simply the superposition of these two **traveling waves.**

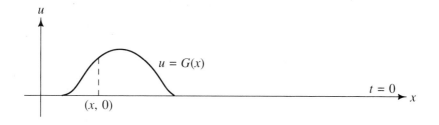

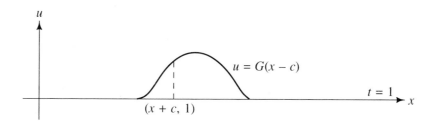

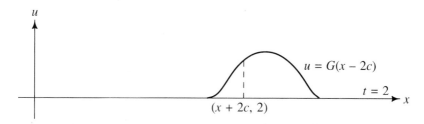

Figure 12.8 Traveling wave moving to the right.

Exercises 12.3

In Problems 1 to 4, solve the wave equation subject to the prescribed boundary and initial conditions.

1. $u_{xx} = c^{-2}u_{tt}$, $0 < x < 1$, $t > 0$
   BC: $u(0, t) = 0$, $u(1, t) = 0$
   IC: $u(x, 0) = 0$, $u_t(x, 0) = v_0$ (v_0 constant)

2. $u_{xx} = c^{-2}u_{tt}$, $0 < x < \pi$, $t > 0$
   BC: $u_x(0, t) = 0$, $u_x(\pi, t) = 0$
   IC: $u(x, 0) = 1 - 2 \cos 3x$, $u_t(x, 0) = 5 \cos 2x$

3. $u_{xx} = c^{-2}u_{tt}$, $0 < x < 1$, $t > 0$
   BC: $u(0, t) = 0$, $u(1, t) = 0$
   IC: $u(x, 0) = x(1 - x)$, $u_t(x, 0) = 0$

4. $u_{xx} = c^{-2}u_{tt}$, $0 < x < p$, $t > 0$
   BC: $u(0, t) = 0$, $u(p, t) = 0$
   IC: $U(x, 0) = p \sinh x - x \sinh p$, $u_t(x, 0) = 0$

5. Determine the relationship between the fundamental angular frequency of a vibrating string to its length p, the tension τ in the string, and the mass m per unit length.

■6. The initial displacement of a string of length 2 m is $0.1 \sin \pi x$. Assuming the string is released from rest, determine the maximum velocity in the string and state its location.

7. A string π meters long is stretched between fixed supports until the wave speed $c = 40$ m/s. If the string is given an initial velocity of $4 \sin x$ from its equilibrium position, calculate the maximum displacement and state its location.

8. A stretched string of unit length lies along the x axis with ends fixed at $(0, 0)$ and $(1, 0)$. If the string is initially displaced into the curve $u(x, 0) = A \sin^3 \pi x$, where A is a small constant, and then let go from rest, show that subsequent displacements are given by

$$u(x, t) = \frac{3A}{4} \sin \pi x \cos c\pi t$$

$$- \frac{A}{4} \sin 3\pi x \cos 3c\pi t$$

*9. The air resistance encountered by a vibrating string is proportional to the velocity of the string. Such conditions lead to the boundary value problem

$$u_{tt} = c^2 u_{xx} - 2ku_t, \qquad 0 < x < 1,$$
$$t > 0 \qquad (0 < k < \pi c)$$

BC: $\qquad u(0, t) = 0, \qquad u(1, t) = 0$

IC: $\qquad u(x, 0) = f(x), \qquad u_t(x, 0) = g(x)$

Find a solution of the problem for the special case when $f(x) = A \sin \pi x$ (A constant) and $g(x) \equiv 0$.

*10. Find a formal solution of

$$u_{xx} + 2u_x + u = u_{tt}, \qquad 0 < x < \pi, t > 0$$

BC: $\qquad u(0, t) = 0, \qquad u(\pi, t) = 0$

IC: $\qquad u(x, 0) = e^{-x}, \qquad u_t(x, 0) = 0$

11. A uniform infinite string is given the initial displacement

$$u(x, 0) = f(x) = \frac{1}{1 + 2x^2}$$

and released from rest, that is, $g(x) \equiv 0$.
(a) Determine its subsequent displacements.
(b) Plot the displacement curves corresponding to $ct = 0, 1/2, 1$.

12. A uniform infinite string is given the initial displacement

$$u(x, 0) = f(x) = \begin{cases} 1 - |x|, & |x| \le 1 \\ 0, & |x| > 1 \end{cases}$$

and released from rest. Determine its subsequent displacement.

12.4 Laplace's Equation

Perhaps the most important PDE in mathematical physics is **Laplace's equation,** also known as the **potential equation.** In *two dimensions,* Laplace's equation has the rectangular coordinate representation

$$u_{xx} + u_{yy} = 0 \qquad\qquad (58)$$

Laplace's equation arises in *steady-state heat conduction* problems involving homogeneous solids. This same equation is satisfied by the *gravitational potential* in free space, the *electrostatic potential* in a uniform dielectric, the *magnetic potential* in free space, the *electric potential* in the theory of steady flow of currents in solid conductors, and the *velocity potential* of inviscid, irrotational fluids.

Because the fields of application of Laplace's equation do not involve time, initial conditions are not prescribed for the solutions of (58). Rather, we find that it is proper to simply prescribe a *single* boundary condition (at each boundary point of the region of interest). Such problems are then called simply **boundary value problems** (BVPs).

The most common boundary conditions that might be prescribed along with Laplace's equation fall mainly into two categories, giving us two primary types of BVPs. If R denotes a region in the xy plane and C its boundary curve, then one type of problem is characterized by

$$u_{xx} + u_{yy} = 0 \quad \text{in } R$$
$$u = f \quad \text{on } C \tag{59}$$

which is called a **Dirichlet problem** or **boundary value problem of the first kind.** In this problem we specify the value of u at each point of the boundary by the function f. The second most common problem is characterized by

$$u_{xx} + u_{yy} = 0 \quad \text{in } R$$
$$\frac{\partial u}{\partial n} = f \quad \text{on } C \tag{60}$$

called a **Neumann problem** or **boundary value problem of the second kind.** The derivative $\partial u/\partial n$ denotes the **normal derivative** of u and is positive in the direction of the outward normal to the boundary curve C. (By definition, $\partial u/\partial n = \nabla u \cdot \mathbf{n}$, where $\mathbf{n}$ is the outward unit normal to C.)

12.4.1 Rectangular Domains

To begin, we wish to consider solutions of Laplace's equation in rectangular domains. A simple example of this is formulated by the **Dirichlet problem** (see Figure 12.9)

$$u_{xx} + u_{yy} = 0, \quad 0 < x < a, \quad 0 < y < b$$

BC: $\begin{cases} u(0, y) = 0, & u(a, y) = 0 \\ u(x, 0) = 0, & u(x, b) = f(x) \end{cases} \tag{61}$

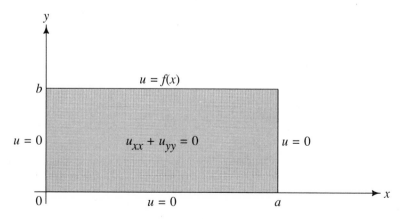

Figure 12.9 Dirichlet problem for a rectangle.

A problem like (61) would physically arise if three edges of a thin rectangular isotropic[†] plate with insulated flat surfaces were maintained at 0°C and the fourth edge was maintained at temperature f until steady-state conditions prevailed throughout the plate. In general, two-dimensional heat flow is governed by the *two-dimensional* heat equation

$$u_{xx} + u_{yy} = a^{-2}u_t \tag{62}$$

where a^2 is the thermal diffusivity (see Section 12.2). When steady-state conditions prevail, we have $u_t = 0$ throughout the region of interest and (62) reduces to Laplace's equation. Thus we can say the solution of (61) is the **steady-state temperature distribution** interior to the plate in this case. The same mathematical problem describes the steady-state temperature distribution in a long rectangular bar bounded by the planes $x = 0$, $x = a$, $y = 0$, $y = b$, whose temperature variation in the z direction may be neglected. Other physical situations also lead to the same mathematical problem.

Our solution technique is basically the same as that for the heat and wave equations. That is, we seek solutions of (61) in the product form $u(x, y) = X(x)Y(y)$ which, when substituted into the PDE, leads to

$$X''(x)Y(y) + X(x)Y''(y) = 0$$

Separating the variables, we obtain the ODEs

$$X'' + \lambda X = 0 \tag{63}$$

$$Y'' - \lambda Y = 0 \tag{64}$$

where λ is the separation constant. Since the first three boundary conditions in (61) are homogeneous, they reduce immediately to

$$X(0) = 0, \qquad X(a) = 0, \qquad Y(0) = 0 \tag{65}$$

The fourth condition, which is *nonhomogeneous,* must be handled separately.

The solution of (63) when subject to the first two boundary conditions in (65) is

$$\lambda_n = \frac{n^2\pi^2}{a^2}, \qquad X_n(x) = \sin\frac{n\pi x}{a}, \qquad n = 1, 2, 3, \dots \tag{66}$$

Corresponding to these values of λ, the solutions of (64) satisfying the remaining condition in (65) become

$$Y_n(y) = \sinh\frac{n\pi y}{a}, \qquad n = 1, 2, 3, \dots \tag{67}$$

Hence, for any choice of constants c_n ($n = 1, 2, 3, \dots$), the function

$$u(x, y) = \sum_{n=1}^{\infty} c_n \sin\frac{n\pi x}{a} \sinh\frac{n\pi y}{a} \tag{68}$$

[†] By *isotropic,* we mean the thermal conductivity at each point in the plate is independent of the direction of heat flow through the point.

satisfies Laplace's equation and the three homogeneous boundary conditions prescribed in (61). The remaining task, therefore, is to determine the constants c_n in such a way that the fourth boundary condition is satisfied. We do this by setting $y = b$ in (68), which yields

$$u(x, b) = f(x) = \sum_{n=1}^{\infty} c_n \sinh \frac{n\pi b}{a} \sin \frac{n\pi x}{a}, \qquad 0 < x < a \qquad (69)$$

From the theory of **Fourier series** it follows that

$$c_n \sinh \frac{n\pi b}{a} = \frac{2}{a} \int_0^a f(x) \sin \frac{n\pi x}{a} \, dx, \qquad n = 1, 2, 3, \ldots \qquad (70)$$

and the problem is formally solved.

A more realistic situation occurs when the temperature is prescribed by nonzero values along all four edges of the plate, rather than along just one edge as in (61). To solve this more general problem we simply superimpose two solutions, each of which corresponds to a problem in which temperatures of 0°C are prescribed along parallel edges of the plate (see Problems 5 to 8 in Exercises 12.4).

The Neumann problem for the rectangle is similarly solved, as are problems featuring a mix of Dirichlet and Neumann conditions. The following example features a mix of boundary conditions.

EXAMPLE 4

Solve

$$u_{xx} + u_{yy} = 0, \qquad 0 < x < \pi, \qquad 0 < y < 1$$

BC:
$$\begin{cases} u_x(0, y) = 0, & u_x(\pi, y) = 0 \\ u(x, 0) = T_0 \cos x, & u(x, 1) = T_0 \cos^2 x \end{cases}$$

Solution

Physically, this problem corresponds to a steady-state temperature distribution problem for a rectangular plate with temperatures prescribed along its edges $y = 0$ and $y = 1$, but whose edges $x = 0$ and $x = \pi$ are insulated.

Proceeding with separation of variables leads to

$$X'' + \lambda X = 0, \qquad X'(0) = 0, \qquad X'(\pi) = 0$$
$$Y'' - \lambda Y = 0$$

The problem for X is a standard eigenvalue problem for which

$$\lambda_0 = 0, \qquad \phi_0(x) = 1$$
$$\lambda_n = n^2, \qquad \phi_n(x) = \cos nx, \qquad n = 1, 2, 3, \ldots$$

Thus, the equation in Y yields the general solution

$$Y_n(y) = \begin{cases} a_0 + b_0 y, & n = 0 \\ a_n \cosh ny + b_n \sinh ny, & n = 1, 2, 3, \ldots \end{cases}$$

Combining the above solutions by the superposition principle, we then obtain

$$u(x, y) = a_0 + b_0 y + \sum_{n=1}^{\infty} (a_n \cosh ny + b_n \sinh ny) \cos nx$$

The boundary condition at $y = 0$ requires that

$$u(x, 0) = T_0 \cos x = a_0 + \sum_{n=1}^{\infty} a_n \cos nx$$

Because the boundary condition contains an eigenfunction, we can solve for the unknown constants by simply matching like terms on each side of the equation. This leads to $a_0 = 0$, $a_1 = T_0$, and $a_n = 0$, $n = 2, 3, 4, \ldots$. Therefore, our solution reduces to

$$u(x, y) = b_0 y + T_0 \cosh y \cos x + \sum_{n=1}^{\infty} b_n \sinh ny \cos nx$$

By imposing the remaining boundary condition at $y = 1$, we find

$$T_0 \cos^2 x = b_0 + T_0 \cosh 1 \cos x + \sum_{n=1}^{\infty} b_n \sinh n \cos nx$$

or, upon rearranging terms and writing $\cos^2 x = \frac{1}{2}(1 + \cos 2x)$,

$$\frac{1}{2} T_0(1 + \cos 2x) - T_0 \cosh 1 \cos x = b_0 + \sum_{n=1}^{\infty} b_n \sinh n \cos nx$$

Again matching like coefficients, we see that

$$b_0 = \frac{1}{2} T_0, \qquad b_1 = -\frac{T_0 \cosh 1}{\sinh 1}, \qquad b_2 = \frac{T_0}{2 \sinh 2}$$

and $b_n = 0$, $n = 3, 4, 5, \ldots$; thus,

$$u(x, y) = \frac{1}{2} T_0 y + T_0 \left(\cosh y - \frac{\cosh 1 \sinh y}{\sinh 1} \right) \cos x$$

$$+ T_0 \frac{\sinh 2y}{2 \sinh 2} \cos 2x$$

Finally, with the aid of the identity

$$\sinh(A - B) = \sinh A \cosh B - \cosh A \sinh B$$

our solution takes the more compact form

$$u(x, y) = T_0 \left(\frac{1}{2} y + \frac{\sinh(1 - y)}{\sinh 1} \cos x + \frac{\sinh 2y}{2 \sinh 2} \cos 2x \right) \qquad \blacksquare$$

Notice in Example 4 that the final solution is not in the form of an infinite series. This makes it particularly easy to calculate $u(x, y)$ at any point throughout the rectangle. The reason for the finite series is that the prescribed functions along $y = 0$

and $y = 1$ are actually composed of linear combinations of *eigenfunctions* belonging to the set $\phi_n(x)$, $n = 0, 1, 2, \ldots$. Thus, only a finite number of the eigenfunctions are required in the series solution.

12.4.2 Circular Domains

In solving potential problems by the separation of variables method in circular domains, it is necessary to express the problem in polar coordinates. By setting $x = r \cos \theta$, $y = r \sin \theta$, it can be shown that (see Problem 16 in Exercises 12.4)

$$u_{xx} + u_{yy} = u_{rr} + \frac{1}{r} u_r + \frac{1}{r^2} u_{\theta\theta} \tag{71}$$

Let us consider the steady-state heat conduction problem for a flat plate in the shape of a circular disk with boundary curve $x^2 + y^2 = \rho^2$. We assume that the plate is isotropic, that the flat surfaces are insulated, and that the temperature is known everywhere on the circular boundary. The temperature inside the disk is then a solution of the **Dirichlet problem** (see Figure 12.10)

$$u_{rr} + \frac{1}{r} u_r + \frac{1}{r^2} u_{\theta\theta} = 0, \quad 0 < r < \rho, \quad -\pi < \theta < \pi$$

BC: $\tag{72}$

$$u(\rho, \theta) = f(\theta), \quad -\pi < \theta < \pi$$

Before we attempt to solve (72), some observations about the solution are in order. First, we need to recognize that $r = 0$ is a "mathematical boundary" of the problem, although it is clearly not a physical boundary. Moreover, to obtain physically

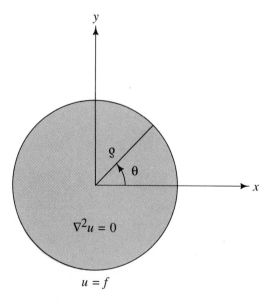

Figure 12.10 Dirichlet problem for a circle.

meaningful solutions we need to impose at $r = 0$ the *implicit* boundary condition

BC: $$|u(0, \theta)| < \infty \qquad (73)$$

which requires the solution to be bounded at the origin. Second, in order to allow θ to assume any value rather than be restricted to the interval $-\pi < \theta \leq \pi$, we specify that $f(\theta)$, and consequently $u(r, \theta)$, be *periodic* with period 2π. Therefore, we also require that

BC: $$u(r, -\pi) = u(r, \pi), \qquad u_\theta(r, -\pi) = u_\theta(r, \pi) \qquad (74)$$

which are actually continuity requirements along the slit $\theta = \pi$.

By writing $u(r, \theta) = R(r)H(\theta)$, we find that the potential equation in (72) separates according to

$$\frac{r^2 R'' + r R'}{R} = -\frac{H''}{H} = \lambda$$

Hence, we are led to the ODEs

$$H'' + \lambda H = 0, \qquad H(-\pi) = H(\pi), \qquad H'(-\pi) = H'(\pi) \qquad (75)$$
$$r^2 R'' + r R' - \lambda R = 0, \qquad |R(0)| < \infty \qquad (76)$$

where we have incorporated conditions (73) and (74).

For $\lambda < 0$, it can be shown that (75) has no nontrivial solution. However, for $\lambda = 0$ we find that

$$\lambda_0 = 0, \qquad \phi_0(\theta) = 1 \qquad (77a)$$

form an eigenvalue-eigenfunction pair. In the case of positive λ, we set $\lambda = k^2 > 0$, and the general solution of (75) is

$$H(\theta) = C_1 \cos k\theta + C_2 \sin k\theta$$

Clearly, in order for this solution to be periodic with period 2π we must restrict $k = n$, $n = 1, 2, 3, \ldots$. In this case, both C_1 and C_2 are arbitrary and thus we have the additional eigenvalues and eigenfunctions

$$\lambda_n = n^2, \qquad \phi_n(\theta) = a_n \cos n\theta + b_n \sin n\theta, \qquad n = 1, 2, 3, \ldots \qquad (77b)$$

where the a's and b's are arbitrary constants.

Equation (76) is a **Cauchy–Euler equation** (recall Section 4.9) whose general solutions for $\lambda = n^2$ are

$$R_n(r) = \begin{cases} C_1 + C_2 \log r, & n = 0 \\ C_3 r^n + C_4 r^{-n}, & n = 1, 2, 3, \ldots \end{cases} \qquad (78)$$

However, continuity requirements at $r = 0$ prescribed by (73) force us to set $C_2 = C_4 = 0$, since the terms $\log r$ and r^{-n} would lead to infinite discontinuities (and therefore infinite temperatures) at the center of the disk.

Collecting the solutions (77a,b) and (78), with $C_1 = 1$ and $C_3 = 1/\rho^n$ for mathematical convenience, we obtain the family of solutions

$$u_n(r, \theta) = \left(\frac{r}{\rho}\right)^n (a_n \cos n\theta + b_n \sin n\theta), \qquad n = 0, 1, 2, \ldots$$

each of which satisfies Laplace's equation and the implicit boundary conditions prescribed by (73) and (74). Thus, forming a linear combination of these solutions by use of the superposition principle yields the formal solution

$$u(r, \theta) = \frac{1}{2} a_0 + \sum_{n=1}^{\infty} \left(\frac{r}{\rho} \right)^n (a_n \cos n\theta + b_n \sin n\theta) \tag{79}$$

where we have written the constant term as $a_0/2$ for convenience. At $r = \rho$, the boundary condition in (72) leads to

$$f(\theta) = \frac{1}{2} a_0 + \sum_{n=1}^{\infty} (a_n \cos n\theta + b_n \sin n\theta) \tag{80}$$

which is a *full* **Fourier series.** Hence, we deduce that

$$a_n = \frac{1}{\pi} \int_{-\pi}^{\pi} f(\theta) \cos n\theta \, d\theta, \qquad n = 0, 1, 2, \ldots \tag{81}$$

and

$$b_n = \frac{1}{\pi} \int_{-\pi}^{\pi} f(\theta) \sin n\theta \, d\theta, \qquad n = 1, 2, 3, \ldots \tag{82}$$

Exercises 12.4

In Problems 1 to 4, solve the given potential problem on a rectangle.

1. $u_{xx} + u_{yy} = 0, 0 < x < \pi, 0 < y < 1$
BC: $u(0, y) = 0, u(\pi, y) = 0,$
$\quad u(x, 0) = 0, u(x, 1) = \sin x$

2. $u_{xx} + u_{yy} = 0, 0 < x < \pi, 0 < y < \pi$
BC: $u(0, y) = 0, u(\pi, y) = 1,$
$\quad u(x, 0) = 0, u(x, \pi) = 0$

3. $u_{xx} + u_{yy} = 0, 0 < x < 1, 0 < y < 1$
BC: $u_x(0, y) = 0, u_x(1, y) = 0,$
$\quad u(x, 0) = x^2, u_y(x, 1) = 0$

4. $u_{xx} + u_{yy} = 0, 0 < x < 1, 0 < y < 1$
BC: $u(0, y) = 0, u(1, y) = 0,$
$\quad u_y(x, 0) = 0, u_y(x, 1) = x(1 - x)$

5. Show that $u(x, y) = v(x, y) + w(x, y)$ is a solution of

$$u_{xx} + u_{yy} = 0, \qquad 0 < x < a, \qquad 0 < y < b$$

BC: $\begin{cases} u(x, 0) = f_1(x), & u(x, b) = f_2(x) \\ u(0, y) = g_1(y), & u(a, y) = g_2(y) \end{cases}$

given that $v(x, y)$ and $w(x, y)$ are solutions, respectively, of

$$v_{xx} + v_{yy} = 0, \qquad 0 < x < a, \qquad 0 < y < b$$

BC: $\begin{cases} v(x, 0) = f_1(x), & v(x, b) = f_2(x) \\ v(0, y) = 0, & v(a, y) = 0 \end{cases}$

and

$$w_{xx} + w_{yy} = 0, \qquad 0 < x < a, \qquad 0 < y < b$$

BC: $\begin{cases} w(x, 0) = 0, & w(x, b) = 0 \\ w(0, y) = g_1(y), & w(a, y) = g_2(y) \end{cases}$

6. Show that solutions of the potential problem for $v(x, y)$ in Problem 5 are of the form

$$v(x, y) = \sum_{n=1}^{\infty} \left(A_n \cosh \frac{n\pi y}{a} + B_n \sinh \frac{n\pi y}{a} \right) \sin \frac{n\pi x}{a}$$

Find formal expressions for the constants A_n and B_n.

7. Show that the solution in Problem 6 can also be expressed as

$$v(x, y) = \sum_{n=1}^{\infty} \left[\alpha_n \sinh \frac{n\pi y}{a} \right.$$

$$\left. + \beta_n \sinh \frac{n\pi(b - y)}{a} \right] \sin \frac{n\pi x}{a}$$

Find formal expressions for the constants α_n and β_n.

*8. Using the results of Problems 5 to 7, solve

$$u_{xx} + u_{yy} = 0, \qquad 0 < x < \pi, \qquad 0 < y < \pi$$

BC: $\begin{cases} u(x,0) = 0, \\ u(x, \pi) = \begin{cases} x, & 0 < x < \pi/2 \\ \pi - x, & \pi/2 < x < \pi \end{cases} \\ u(0, y) = \begin{cases} y, & 0 < y < \pi/2 \\ \pi - y, & \pi/2 < y < \pi \end{cases} \\ u(\pi, y) = 0 \end{cases}$

9. Three edges of a square plate of side 2 m are maintained at 100°C and the remaining edge (at $y = 2$) is held at 200°C. Assuming the flat surfaces are insulated,
   (a) Determine the steady-state temperature distribution.
   (b) Use the first term of the series in (a) to estimate the temperature at the center of the plate.
   Hint. Split the problem into two problems, one of which has temperature 100°C on all four sides.

■10. A square plate has its faces and edges $x = 0$ and $x = \pi$ insulated. The edges $y = 0$ and $y = \pi$ are kept at zero temperature and T_0 (constant), respectively. What is the steady-state temperature throughout the plate?

In Problems 11 to 13, solve the given potential problem for a circle.

11. $u_{rr} + \frac{1}{r} u_r + \frac{1}{r^2} u_{\theta\theta} = 0, 0 < r < \rho, -\pi < \theta < \pi$

BC: $\qquad u(\rho, \theta) = \cos^2 \theta$

12. $u_{rr} + \frac{1}{r} u_r + \frac{1}{r^2} u_{\theta\theta} = 0, 0 < r < \rho, -\pi < \theta < \pi$

BC: $\quad u(\rho, \theta) = |\theta|, \qquad -\pi < \theta < \pi$

13. $u_{rr} + \frac{1}{r} u_r + \frac{1}{r^2} u_{\theta\theta} = 0, 0 < r < 1, -\pi < \theta < \pi$

BC: $u_r(1, \theta) = \begin{cases} -1, & -\pi < \theta < 0 \\ 1, & 0 < \theta < \pi \end{cases}$

■14. Solve the potential problem in the quadrant

$$u_{rr} + \frac{1}{r} u_r + \frac{1}{r^2} u_{\theta\theta} = 0,$$

$$0 < r < 1, \qquad 0 < \theta < \pi/2$$

BC: $\begin{cases} u(r, 0) = 0, & u(r, \pi/2) = 0 \\ u(1, \theta) = T_0 & \text{(constant)} \end{cases}$

15. Determine the steady-state temperature distribution in the semicircular disk bounded by $y = \sqrt{1 - x^2}, y = 0$, with 0°C prescribed along the diameter $y = 0$ while along the circumference the temperature is prescribed by $u(1, \theta) = \sin \theta$.

*16. By use of the chain rule, show that the polar transformation $x = r \cos \theta, y = r \sin \theta$, leads to

(a) $u_x = \frac{x}{r} u_r - \frac{y}{r^2} u_\theta$

$u_y = \frac{y}{r} u_r + \frac{x}{r^2} u_\theta$

(b) $u_{xx} = \frac{x^2}{r^2} u_{rr} - \frac{2xy}{r^3} u_{r\theta}$

$+ \frac{y^2}{r^4} u_{\theta\theta} + \frac{y^2}{r^3} u_r + \frac{2xy}{r^4} u_\theta$

(c) Find an expression for u_{yy} similar to that for u_{xx} in (b) and show that

$$u_{xx} + u_{yy} = u_{rr} + \frac{1}{r} u_r + \frac{1}{r^2} u_{\theta\theta}$$

[0] 12.5 Generalized Fourier Series

In this section we briefly examine some applications involving eigenfunction expansions that lead to *generalized Fourier series*.

12.5.1 Convective Heat Transfer at One Endpoint

In solving the heat equation in Section 12.2, the boundary conditions were both limited to specifying either u or u_x at the endpoints. However, when heat is lost from an end of the rod into the surrounding medium through convection, the boundary condition involves a linear combination of u and u_x.

Suppose, for example, we consider a long thin rod of length p that is initially heated to the temperature distribution $f(x)$. At time $t = 0$ the end at $x = 0$ is exposed to ice packs at $0°C$ while heat exchange into the surrounding medium through convection takes place at the end $x = p$. Mathematically, the latter condition is prescribed by

$$hu(p, t) + u_x(p, t) = 0 \qquad (83)$$

where h is the heat transfer coefficient. To find the temperature in the rod at all later times, we must solve the initial boundary value problem

$$u_{xx} = a^{-2}u_t, \qquad 0 < x < p, \qquad t > 0$$

BC: $\qquad u(0, t) = 0, \qquad hu(p, t) + u_x(p, t) = 0 \qquad (84)$

IC: $\qquad\qquad u(x, 0) = f(x), \qquad 0 < x < p$

Other than the boundary condition at $x = p$, this problem is similar to those discussed in Section 12.2. Thus, by separating the variables using $u(x, t) = X(x)T(t)$, we are led to the eigenvalue problem

$$X'' + \lambda X = 0, \qquad X(0) = 0, \qquad hX(p) + X'(p) = 0 \qquad (85)$$

and time-dependent equation

$$T' + \lambda a^2 T = 0 \qquad (86)$$

The eigenvalue problem (85) is similar to that solved in Example 9 in Section 11.3 with $h = p = 1$. Following a similar analysis here, we set $\lambda = k^2 > 0$, which yields the general solution

$$X(x) = C_1 \cos kx + C_2 \sin kx$$

The boundary condition $X(0) = 0$ requires that $C_1 = 0$, whereas the second condition leads to (for $C_2 \neq 0$)

$$h \sin kp + k \cos kp = 0$$

If we denote the nth solution of this last equation by k_n, the eigenvalues and eigenfunctions of (85) are

$$\lambda_n = k_n^2, \qquad \phi_n(x) = \sin k_n x, \qquad n = 1, 2, 3, \dots \qquad (87)$$

The substitution of $\lambda = k_n^2$ into (86) leads to the set of solutions

$$T_n(t) = c_n e^{-a^2 k_n^2 t}, \qquad n = 1, 2, 3, \ldots \tag{88}$$

where the c's are arbitrary constants.

By combining solutions (87) and (88) through use of the superposition principle, we obtain the formal series solution

$$u(x, t) = \sum_{n=1}^{\infty} c_n \sin(k_n x) e^{-a^2 k_n^2 t} \tag{89}$$

The initial condition at $t = 0$ leads to the **generalized Fourier series**

$$u(x, 0) = f(x) = \sum_{n=1}^{\infty} c_n \sin k_n x, \qquad 0 < x < p \tag{90}$$

where, by recalling Theorem 11.6 in Section 11.6, we deduce that

$$c_n = \frac{\int_0^p f(x) \sin k_n x \, dx}{\int_0^p \sin^2 k_n x \, dx}, \qquad n = 1, 2, 3, \ldots \tag{91}$$

EXAMPLE 5

Find the temperature distribution in a rod of unit length for which

$$u_{xx} = a^{-2} u_t, \qquad 0 < x < 1, \qquad t > 0$$

BC: $$u(0, t) = 0, \qquad u(1, t) + u_x(1, t) = 0$$

IC: $$u(x, 0) = T_0, \qquad 0 < x < 1$$

Solution

This is just the problem described by (84) with $p = h = 1$. Thus, the formal series solution is

$$u(x, y) = \sum_{n=1}^{\infty} c_n \sin(k_n x) e^{-a^2 k_n^2 t}$$

where k_n is the nth root of $k = -\tan k$. By imposing the prescribed initial condition, we obtain

$$u(x, 0) = T_0 = \sum_{n=1}^{\infty} c_n \sin k_n x$$

where (recall Example 21 in Section 11.6)

$$c_n = \frac{2T_0(1 - \cos k_n)}{k_n(1 + 2 \cos^2 k_n)}, \qquad n = 1, 2, 3, \ldots$$

Consequently, the solution we seek assumes the form

$$u(x, t) = 2T_0 \sum_{n=1}^{\infty} \frac{1 - \cos k_n}{k_n(1 + 2 \cos^2 k_n)} \sin(k_n x) e^{-a^2 k_n^2 t}$$

12.5.2 Nonhomogeneous Heat Equation

When a heat source is present within the domain of interest, the governing PDE is **nonhomogeneous** [recall Equation (33)]. Thus, the general problem we now wish to discuss can be formulated by

$$u_{xx} = a^{-2}u_t - q(x, t), \qquad 0 < x < p, \qquad t > 0$$

$$\text{BC:} \qquad B_1[u] = 0, \qquad B_2[u] = 0 \qquad\qquad (92)$$

$$\text{IC:} \qquad u(x, 0) = f(x), \qquad 0 < x < p$$

where $q(x, t)$ is proportional to the heat source and where B_1 and B_2 are general boundary operators defined by

$$B_1[u] = h_1 u(0, t) + h_2 u_x(0, t)$$
$$B_2[u] = h_3 u(p, t) + h_4 u_x(p, t) \qquad (93)$$

We begin by assuming the solution of (92) can be written in the product form

$$u(x, t) = \sum_{n=1}^{\infty} E_n(t)\phi_n(x) \qquad (94)$$

where the ϕ's are eigenfunctions belonging to the associated eigenvalue problem

$$X'' + \lambda X = 0, \qquad B_1[X] = 0, \qquad B_2[X] = 0 \qquad (95)$$

and the E's are time-dependent coefficients that must be determined. Equation (95) is the eigenvalue problem that would arise by separating variables for the case $q(x, t) \equiv 0$. Also, note that the assumed form of the solution is similar to that obtained in Section 12.2 except we have now replaced $T_n(t)$ with $E_n(t)$. Assuming that termwise differentiation is permitted, we find

$$u_t = \sum_{n=1}^{\infty} E_n'(t)\phi_n(x) \qquad (96)$$

and

$$u_{xx} = \sum_{n=1}^{\infty} E_n(t)\phi_n''(x) = -\sum_{n=1}^{\infty} \lambda_n E_n(t)\phi_n(x) \qquad (97)$$

where we are recognizing the relation [from (95)]

$$\phi_n''(x) + \lambda_n \phi_n(x) = 0, \qquad n = 1, 2, 3, \ldots$$

Hence, by writing the PDE in (92) in the form

$$a^2 q(x, t) = u_t - a^2 u_{xx} \qquad (98)$$

and substituting (96) and (97) into the right-hand side, we obtain

$$a^2 q(x, t) = \sum_{n=1}^{\infty} [E_n'(t) + a^2 \lambda_n E_n(t)]\phi_n(x) \qquad (99)$$

For a fixed value of t, we can interpret (99) as a **generalized Fourier series** of the function $a^2 q(x, t)$, with Fourier coefficients defined by

$$E'_n(t) + a^2 \lambda_n E_n(t) = a^2 \|\phi_n(x)\|^{-2} \int_0^p q(x, t)\phi_n(x)\, dx \tag{100}$$

for $n = 1, 2, 3, \ldots$, where

$$\|\phi_n(x)\|^2 = \int_0^p [\phi_n(x)]^2\, dx, \qquad n = 1, 2, 3, \ldots \tag{101}$$

Assuming $\lambda_n \neq 0$, then for each n, Equation (100) is a first-order linear ODE with general solution

$$E_n(t) = \left[c_n + a^2 \int_0^t e^{a^2 \lambda_n \tau} Q_n(\tau)\, d\tau \right] e^{-a^2 \lambda_n t}, \qquad n = 1, 2, 3, \ldots \tag{102}$$

where, for notational convenience, we define

$$Q_n(t) = \|\phi_n(x)\|^{-2} \int_0^p q(x, t)\phi_n(x)\, dx, \qquad n = 1, 2, 3, \ldots \tag{103}$$

and where the c's are arbitrary constants. Finally, the substitution of (102) into the solution form (94) provides us with the formal solution

$$u(x, t) = \sum_{n=1}^\infty \left[c_n + a^2 \int_0^t e^{a^2 \lambda_n \tau} Q_n(\tau)\, d\tau \right] \phi_n(x) e^{-a^2 \lambda_n t} \tag{104}$$

To determine the constants c_n ($n = 1, 2, 3, \ldots$), we set $t = 0$ in (104) and use the prescribed initial condition in (92) to get

$$u(x, 0) = f(x) = \sum_{n=1}^\infty c_n \phi_n(x), \qquad 0 < x < p \tag{105}$$

We recognize (105) as a **generalized Fourier series** for which (recall Theorem 11.6)

$$c_n = \|\phi_n(x)\|^{-2} \int_0^p f(x)\phi_n(x)\, dx, \qquad n = 1, 2, 3, \ldots \tag{106}$$

■
EXAMPLE 6

Solve

$$u_{xx} = u_t + (1 - x)\cos t, \qquad 0 < x < 1, \qquad t > 0$$

BC: $u(0, t) = 0, \qquad u(1, t) = 0$

IC: $u(x, 0) = 0, \qquad 0 < x < 1$

Solution

We first identify and solve the related eigenvalue problem (obtained through separation of variables on the homogeneous equation)

$$X'' + \lambda X = 0, \qquad X(0) = 0, \qquad X(1) = 0$$

Hence, we find that

$$\lambda_n = n^2\pi^2, \qquad \phi_n(x) = \sin n\pi x, \qquad n = 1, 2, 3, \ldots$$

To solve the nonhomogeneous PDE, we seek a solution of the form

$$u(x, t) = \sum_{n=1}^{\infty} E_n(t) \sin n\pi x$$

which, when substituted into the PDE written as

$$-(1 - x) \cos t = u_t - u_{xx}$$

yields

$$-(1 - x) \cos t = \sum_{n=1}^{\infty} \left[E'_n(t) + n^2\pi^2 E_n(t) \right] \sin n\pi x$$

Making the observation

$$\|\sin n\pi x\|^2 = \int_0^1 \sin^2 n\pi x \, dx = \tfrac{1}{2}, \qquad n = 1, 2, 3, \ldots$$

we deduce that

$$E'_n(t) + n^2\pi^2 E_n(t) = -2 \int_0^1 (1 - x) \cos t \sin n\pi x \, dx$$

$$= -\frac{2}{n\pi} \cos t, \qquad n = 1, 2, 3, \ldots$$

and solving this ODE for $E_n(t)$ leads to

$$E_n(t) = \left(c_n - \frac{2}{n\pi} \int_0^t e^{n^2\pi^2\tau} \cos \tau \, d\tau \right) e^{-n^2\pi^2 t}$$

$$= c_n e^{-n^2\pi^2 t} + \frac{2}{n\pi(1 + n^4\pi^4)} \left[n^2\pi^2 (e^{-n^2\pi^2 t} - \cos t) - \sin t \right]$$

Finally, by substituting this expression for $E_n(t)$ into the above series for $u(x, t)$ and imposing the initial condition $u(x, 0) = 0$, we see that $c_n = 0$ for all n. Hence, our solution is

$$u(x, t) = \frac{2}{\pi} \sum_{n=1}^{\infty} \frac{\sin n\pi x}{n(1 + n^4\pi^4)} \left[n^2\pi^2 (e^{-n^2\pi^2 t} - \cos t) - \sin t \right] \qquad \blacksquare$$

12.5.3 Applications Involving Bessel Functions

In solving certain heat conduction problems or wave motion problems in circular domains, it often happens that *Bessel functions* arise in the solution process. For illustrative purposes, let us consider the problem of determining the small transverse displacements u of a thin circular membrane (such as a drumhead) of *unit* radius whose edge is rigidly fixed. The governing PDE for this problem is the *two-dimensional* **wave equation**

$$u_{xx} + u_{yy} = c^{-2} u_{tt} \qquad (107)$$

However, the shape of the membrane suggests the use of polar coordinates rather than rectangular. Thus, if $x = r \cos \theta$ and $y = r \sin \theta$, recall from Equation (71) that

$$u_{xx} + u_{yy} = u_{rr} + \frac{1}{r} u_r + \frac{1}{r^2} u_{\theta\theta}$$

Further, if the displacements depend only upon the radial distance r from the center of the membrane and on time t, then $u_{\theta\theta} = 0$ and the governing PDE is the *radial symmetric* form of the wave equation given by

$$u_{rr} + \frac{1}{r} u_r = c^{-2} u_{tt}, \qquad 0 < r < 1, \qquad t > 0 \qquad (108)$$

Let the initial displacement of the membrane be given by $f(r)$ and assume the initial velocity is zero. Thus, since we have taken the case where the membrane is rigidly fixed on the boundary, we have the following boundary and initial conditions

BC: $\qquad\qquad\qquad\qquad u(1, 0) = 0$

IC: $\qquad u(r, 0) = f(r), \qquad u_t(r, 0) = 0, \qquad 0 < r < 1$ $\qquad (109)$

In addition, we need to impose the *implicit* boundary condition

BC: $\qquad\qquad\qquad\qquad\qquad |u(0, t)| < \infty \qquad\qquad\qquad\qquad\qquad (110)$

which requires the solution to be bounded at $r = 0$, that is, requires the solution to be *continuous*.

To solve (108) and (109), we use separation of variables with $u(r, t) = R(r)T(t)$. The substitution of this expression into the radial symmetric wave equation (108) leads to

$$R''T + \frac{1}{r} R'T = c^{-2} RT''$$

or, upon division by RT,

$$\frac{R'' + R'/r}{R} = \frac{T''}{c^2 T} = -\lambda$$

From this last relation, we deduce that

$$rR'' + R' + \lambda rR = 0, \qquad |R(0)| < \infty, \qquad R(1) = 0 \qquad (111)$$

$$T'' + \lambda c^2 T = 0 \qquad\qquad\qquad\qquad (112)$$

where we have also included the boundary conditions.

It can be shown that (111) has only positive eigenvalues. Hence, if we set $\lambda = k^2 > 0$, then Equation (111) is a generalized form of **Bessel's equation** of order zero, that is,

$$rR'' + R' + k^2 rR = 0 \qquad\qquad\qquad\qquad (113)$$

By making the change of variable $s = kr$, this equation reduces to the standard form of Bessel's equation (see Problem 12 in Exercises 12.5) with general solution

$$R(s) = C_1 J_0(s) + C_2 Y_0(s)$$

and therefore the general solution of (113) is

$$R(r) = C_1 J_0(kr) + C_2 Y_0(kr) \tag{114}$$

where $J_0(kr)$ and $Y_0(kr)$ are **Bessel functions of the first and second kinds** of order zero.

The continuity boundary condition at $r = 0$ in (111) requires us to set $C_2 = 0$, since $Y_0(kr)$ becomes unbounded as $r \to 0$. The remaining boundary condition in (111) leads to

$$R(1) = C_1 J_0(k) = 0 \tag{115}$$

It is known that $J_0(k)$ has an infinite number of positive zeros at $k_1, k_2, k_3, \ldots,$ $k_n, \ldots,$ although without some numerical calculations we don't know their exact values (recall Theorem 5.6 in Section 5.6). Regardless, we can say that the eigenvalues and corresponding eigenfunctions of (111) are given by

$$\lambda_n = k_n^2, \qquad \phi_n(r) = J_0(k_n r), \qquad n = 1, 2, 3, \ldots \tag{116}$$

With $\lambda = k_n^2$, the solutions of (112) are

$$T_n(t) = a_n \cos k_n ct + b_n \sin k_n ct, \qquad n = 1, 2, 3, \ldots \tag{117}$$

and by combining results, we obtain the family of solutions

$$u_n(r, t) = (a_n \cos k_n ct + b_n \sin k_n ct) J_0(k_n r), \qquad n = 1, 2, 3, \ldots \tag{118}$$

where the a's and b's are arbitrary constants. These solutions are called **standing waves,** since each can be viewed as having fixed shape $J_0(k_n r)$ with amplitude $T_n(t)$. Thus, the situation is similar to that in Section 12.3, where we discussed standing waves in connection with the vibrating string problem. However, because the zeros of the Bessel function are not regularly spaced (in contrast to the zeros of the sine functions appearing in the vibrating string problem), the sound emitted from a drum, for example, is quite different from that of, say, a violin. In musical tones the zeros are evenly spaced and the frequencies are integral multiples of the fundamental frequency.

Combining the solutions (118) by the superposition principle, we get the formal series solution

$$u(r, t) = \sum_{n=1}^{\infty} (a_n \cos k_n ct + b_n \sin k_n ct) J_0(k_n r) \tag{119}$$

where the constants must satisfy

$$u(r, 0) = f(r) = \sum_{n=1}^{\infty} a_n J_0(k_n r) \tag{120}$$

and

$$u_t(r, 0) = 0 = \sum_{n=1}^{\infty} k_n c b_n J_0(k_n r) \tag{121}$$

These series are **generalized Fourier series,** also known as **Fourier–Bessel series.** Clearly, (121) is satisfied by setting

$$b_n = 0, \qquad n = 1, 2, 3, \ldots \tag{122}$$

In the case of (120), however, the situation is more complicated. First, we recognize that Bessel's equation (111) is in *self-adjoint form* and thus we see that the weighting function is simply r. And although (111) is not a regular Sturm-Liouville problem, it can be shown that the eigenfunctions $\phi_n(r) = J_0(k_n r)$ satisfy the orthogonality property

$$\int_0^1 r J_0(k_m r) J_0(k_n r)\,dr = 0, \qquad m \neq n \tag{123}$$

Thus, by multiplying (120) by $r J_0(k_m r)$ and integrating over the unit interval $[0, 1]$, it can be shown that (see Problem 13 in Exercises 12.5)

$$a_n = \frac{\int_0^1 r f(r) J_0(k_n r)\,dr}{\int_0^1 r [J_0(k_n r)]^2\,dr}, \qquad n = 1, 2, 3, \ldots \tag{124}$$

Moreover, the integral in the denominator of (124), which is the square of the norm of $J_0(k_n r)$, leads to the result (see Problem 14 in Exercises 12.5)

$$\int_0^1 r [J_0(k_n r)]^2\,dr = \tfrac{1}{2}[J_1(k_n)]^2 \tag{125}$$

■——— EXAMPLE 7

A flat circular plate of unit radius is initially heated to constant temperature T_0. If the flat surfaces of the plate are insulated and the circular boundary is held at $0°C$, determine the temperature in the plate at all later times.

Solution

The governing PDE this time is the radial symmetric heat equation

$$u_{rr} + \frac{1}{r} u_r = a^{-2} u_t, \qquad 0 < r < 1, \qquad t > 0$$

The prescribed boundary and initial conditions are given by

BC: $\qquad\qquad |u(0, t)| < \infty, \qquad u(1, t) = 0$

IC: $\qquad\qquad u(r, 0) = T_0, \qquad 0 < r < 1$

where we have incorporated the continuity requirement at $r = 0$.

By setting $u(r, t) = R(r)T(t)$ and separating the variables, we are led to

$$rR'' + R' + \lambda r R = 0, \qquad |R(0)| < \infty, \qquad R(1) = 0$$

and

$$T' + \lambda a^2 T = 0$$

The first equation is recognized as the generalized form of Bessel's equation once again for which the eigenvalues and eigenfunctions are

$$\lambda_n = k_n^2, \qquad \phi_n(r) = J_0(k_n r), \qquad n = 1, 2, 3, \ldots$$

where the k's are solutions of $J_0(k_n) = 0$, $n = 1, 2, 3, \ldots$. With $\lambda = k_n^2$, the second equation has the set of solutions

$$T_n(t) = c_n e^{-a^2 k_n^2 t}, \qquad n = 1, 2, 3, \ldots$$

where the c's are arbitrary constants. Combining results, we obtain the series solution

$$u(r, t) = \sum_{n=1}^{\infty} c_n J_0(k_n r) e^{-a^2 k_n^2 t}$$

By imposing the initial condition, we get

$$u(r, 0) = T_0 = \sum_{n=1}^{\infty} c_n J_0(k_n r)$$

where, based on the above results,

$$c_n = \frac{2T_0}{[J_1(k_n)]^2} \int_0^1 r J_0(k_n r) \, dr, \qquad n = 1, 2, 3, \ldots$$

To evaluate the integral, we need to use the integral formula

$$\int x J_0(x) \, dx = x J_1(x) + C$$

Thus, by making the change of variable $x = k_n r$, we see that

$$\int_0^1 r J_0(k_n r) \, dr = \frac{1}{k_n^2} \int_0^{k_n} x J_0(x) \, dx$$

$$= \frac{1}{k_n^2} [x J_1(x)] \Big|_0^{k_n}$$

$$= \frac{1}{k_n} J_1(k_n)$$

and thus deduce that

$$c_n = \frac{2T_0}{k_n J_1(k_n)}, \qquad n = 1, 2, 3, \ldots$$

Hence, the solution we seek is given by

$$u(r, t) = 2T_0 \sum_{n=1}^{\infty} \frac{1}{k_n J_1(k_n)} J_0(k_n r) e^{-a^2 k_n^2 t} \qquad \blacksquare$$

Exercises 12.5

In Problems 1 to 4, solve the given initial boundary value problem.

1. $u_{xx} = a^{-2} u_t$, $0 < x < 1$, $t > 0$
BC: $u(0, t) - u_x(0, t) = 0$, $u(1, t) = 0$
IC: $u(x, 0) = T_0$

2. $u_{xx} = a^{-2} u_t$, $0 < x < 1$, $t > 0$
BC: $u(0, t) = 0$, $2u(1, t) + u_x(1, t) = 0$
IC: $u(x, 0) = T_0$

3. $u_{xx} = a^{-2} u_t$, $0 < x < 1$, $t > 0$
BC: $u_x(0, t) = 0$, $u(1, t) = T_0$
IC: $u(x, 0) = T_0$
Hint. Assume $u(x, t) = S(x) + v(x, t)$.

*4. $u_{xx} = a^{-2} u_t$, $0 < x < 1$, $t > 0$
BC: $u(0, t) = T_0$, $u_x(1, t) = 0$
IC: $u(x, 0) = x$

In Problems 5 and 6, find a formal series solution of the given problem.

5. $u_{xx} = a^{-2}u_t, \ 0 < x < p, \ t > 0$
BC: $hu(0, t) - u_x(0, t) = 0,$
$\qquad hu(p, t) + u_x(p, t) = 0$
IC: $u(x, 0) = f(x)$

6. $u_{xx} + u_{yy} = 0, \ 0 < x < 1, \ 0 < y < 1$
BC: $u(0, y) = 0, \ u(1, y) + u_x(1, y) = 0$
$\qquad u(x, 0) = 0, \ u(x, 1) = f(x)$

In Problems 7 to 10, solve by the method of Section 12.5.2.

7. $u_{xx} = a^{-2}u_t - A\cos \omega t, \ 0 < x < \pi, \ t > 0$
BC: $u_x(0, t) = 0, \ u_x(\pi, t) = 0$
IC: $u(x, 0) = 0$

■8. $u_{xx} = u_t - e^{-t}, \ 0 < x < 1, \ t > 0$
BC: $u_x(0, t) = 0, \ u(1, t) = 0$
IC: $u(x, 0) = 0$

9. $u_{xx} = u_t + 5\sin t - 2x, \ 0 < x < 1, \ t > 0$
BC: $u(0, t) = 0, \ u(1, t) = 0$
IC: $u(x, 0) = 0$

*10. $u_{xx} = u_t - t, \ 0 < x < \pi, \ t > 0$
BC: $u(0, t) = \cos t, \ u(\pi, t) = \cos t$
IC: $u(x, 0) = 1$
Hint. Assume $u(x, t) = \cos t + v(x, t)$.

*11. A guy wire stretched tightly between fixed supports at $x = 0$ and $x = \pi$ is initially at rest until a gust of wind comes along. Assuming the wind is normal to the wire and can be modeled as a simple sinusoidal function of time with constant amplitude, determine the subsequent motions of the wire. The problem is mathematically characterized by

$$u_{xx} = c^{-2}u_{tt} - P\sin \omega t, \qquad 0 < x < \pi, \qquad t > 0$$

BC: $\qquad u(0, t) = 0, \qquad u(\pi, t) = 0$

IC: $\qquad u(x, 0) = 0, \qquad u_t(x, 0) = 0$

Hint. Use the one-sided Green's function to solve the DE for $E_n(t)$.

12. Show that the change of variable $s = kr$ in

$$rR'' + R' + k^2 rR = 0$$

yields Bessel's equation

$$s\frac{d^2 R}{ds^2} + \frac{dR}{ds} + sR = 0$$

13. Given the series

$$f(r) = \sum_{n=1}^{\infty} a_n J_0(k_n r), \qquad J_0(k_n) = 0$$

show by multiplying the series by $rJ_0(k_m r)$ and integrating over $[0, 1]$, that the Fourier coefficients a_n are those given by (124).
Hint. Use the orthogonality property (123).

*14. Multiply the DE

$$\frac{d}{dx}\left[x\frac{d}{dx}J_0(kx) \right] + k^2 x J_0(kx) = 0$$

by the factor $2x(d/dx)J_0(kx)$ and show that

(a) $\dfrac{d}{dx}\left[x\dfrac{d}{dx}J_0(kx) \right]^2 + k^2 x^2 [J_0(kx)]^2 = 0$

(b) Integrate the expression in (a) over the interval $[0, 1]$ and deduce that

$$\int_0^1 x[J_0(kx)]^2 \, dx = \frac{1}{2k^2}\left\{ [kJ_0'(k)]^2 \right.$$
$$\left. + k[J_0(k)]^2 \right\}$$

(c) Finally, set $k = k_n$, where $J_0(k_n) = 0$, and use the identity $J_0'(x) = -J_1(x)$ to deduce that

$$\int_0^1 x[J_0(k_n x)]^2 \, dx = \tfrac{1}{2}[J_1(k_n)]^2$$

15. Given that $J_0(k_1) = 0$, solve

$$u_{rr} + \frac{1}{r}u_r = c^{-2}u_{tt}, \qquad 0 < r < 1, \qquad t > 0$$

BC: $\qquad\qquad u(1, t) = 0$

IC: $\ u(r, 0) = 0.1 J_0(k_1 r), \qquad u_t(r, 0) = 0$

16. Given

$$u_{rr} + \frac{1}{r} u_r = c^{-2} u_{tt}, \qquad 0 < r < 1, \qquad t > 0$$

BC: $u(1, t) = 0$

IC: $u(r, 0) = 0, \qquad u_t(r, 0) = 1$

show that

$$u(r, t) = \frac{2}{c} \sum_{n=1}^{\infty} \frac{\sin(k_n ct)}{k_n^2 J_1(k_n)} J_0(k_n r)$$

where $J_0(k_n) = 0$, $n = 1, 2, 3, \ldots$

17. Solve the problem described by (108) and (109) for a circle of radius ρ.

■18. The temperature distribution in an insulated circular plate satisfies the initial boundary value problem

$$u_{rr} + \frac{1}{r} u_r = a^{-2} u_t, \qquad 0 < r < 1, \qquad t > 0$$

BC: $u_r(1, t) = 0$

IC: $u(r, 0) = f(r)$

Show that the formal solution is given by

$$u(r, t) = c_0 + \sum_{n=1}^{\infty} c_n J_0(k_n r) e^{-a^2 k_n^2 t}$$

where $J_0'(k_n) = 0$, $n = 1, 2, 3, \ldots$

*19. The temperature distribution $u(r, t)$ in a thin circular plate with heat exchanges from its faces into the surrounding medium at $0°C$ satisfies the initial boundary value problem

$$u_{rr} + \frac{1}{r} u_r - bu = u_t, \qquad 0 < r < 1, \qquad t > 0$$

BC: $u(1, t) = 0$

IC: $u(r, 0) = 1$

where $b > 0$. Show that

$$u(r, t) = 2e^{-bt} \sum_{n=1}^{\infty} \frac{J_0(k_n r)}{k_n J_1(k_n)} e^{-k_n^2 t}$$

where $J_0(k_n) = 0$, $n = 1, 2, 3, \ldots$

*20. By assuming

$$u(r, t) = \sum_{n=1}^{\infty} E_n(t) J_0(k_n r)$$

where $J_0(k_n) = 0$, $n = 1, 2, 3, \ldots$, follow the technique of Section 12.5.2 to find a formal solution of

$$u_{rr} + \frac{1}{r} u_r = u_t - q(r, t), \qquad 0 < r < 1, \qquad t > 0$$

BC: $u(1, t) = 0$

IC: $u(r, 0) = 0$

12.6 Chapter Summary

In this chapter we have given a brief introduction to the solution of PDEs that occur in problems involving heat conduction in homogeneous solids, wave phenomena, and potential theory. Specifically, we have looked at applications involving the **heat equation**

$$u_{xx} = a^{-2} u_t \tag{126}$$

the **wave equation**

$$u_{xx} = c^{-2} u_{tt} \tag{127}$$

and **Laplace's equation**

$$u_{xx} + u_{yy} = 0 \tag{128}$$

In the case of all three equations, the method of solution was **separation of variables** wherein we assume the solution can be represented by a product of functions, each dependent upon only one independent variable. This technique reduces the PDE to a system of two ODEs to solve, one of which is an **eigenvalue problem.** Collecting all solutions of the ODEs by use of the **superposition principle,** we obtain a solution of the original PDE that satisfies all homogeneous boundary conditions. The remaining initial conditions, or nonhomogeneous boundary conditions, are then imposed upon this solution to determine the arbitrary constants. The constants are found to be related to the Fourier coefficients of a **Fourier series representation** of the functions occurring in the initial conditions or nonhomogeneous boundary conditions.

12.7 Historical Comments

In 1727, John Bernoulli treated the vibrating string problem by imagining the string to be a thin thread having a number of equally spaced weights placed along it. However, because his governing equation was not time-dependent, it was not truly a partial differential equation. The French mathematician d'Alembert derived the **one-dimensional wave equation** as we know it today by letting the number of weights in Bernoulli's model become infinite while at the same time allowing the space between them to go to zero. His solution, now called **d'Alembert's solution,** appeared around 1746. Euler, D. Bernoulli, and Lagrange all solved the wave equation in the mid-1700s by what we now call the method of **separation of variables** or, sometimes, the **Bernoulli product method.** The merits of their various solutions were argued in a series of papers extending over more than 25 years, mostly concerning the nature of the kinds of functions that can be represented by trigonometric series.

Laplace's equation, also called the **potential equation,** arose in the study of gravitational attraction problems by Laplace and Legendre, both of whom were professors of mathematics at the Ecole Militaire in France. In fact, it was Legendre's famous 1782 study of the gravitational attraction of spheroids that introduced what are now called **Legendre polynomials.** The work of Legendre on potential theory was continued by Laplace, although a general method of solution was never developed during the eighteenth century.

The first major step toward developing a general method of solution began in the early 1800s when J. Fourier made his famous study of the **heat equation.** Because his work concerning the representation of certain functions by trigonometric series was purely formal, it was originally rejected by the Paris Academy in 1807. Nonetheless, he was eventually recognized for his pioneering effort and his name is the one used today in reference to these series.

Mathematicians of the nineteenth century worked vigorously on problems associated with Laplace's equation, trying to extend the work of Legendre and Laplace. Despite this great effort by mathematicians such as Poisson and Gauss, very little was known about the general properties of the solutions of Laplace's equation until 1828 when G. Green, a self-taught British mathematician whose main interest was in electricity and magnetism, and the Russian mathematician Michel Ostrogradsky (1801– 1861) independently studied properties of a class of solutions known as **harmonic**

functions. In the second half of the nineteenth century, mathematicians began to make progress on problems concerning the existence of solutions to partial differential equations. These investigations involved not only Laplace's equation but the heat and wave equations as well, and were eventually extended to partial differential equations with variable coefficients.

Through the years a host of mathematicians have played a role in the general development of the theory of ordinary and partial differential equations, but their names are too numerous to mention here beyond those already cited. Even today, partial differential equations continues to be an active area of research for mathematicians. In part, this is motivated by the large number of problems in partial differential equations that engineers and scientists are faced with that are seemingly intractable. Many of these equations are nonlinear, and come from such areas of application as fluid and gas dynamics, elasticity, and chemical reactions. Owing to the ever-increasing need in engineering and science to solve more and more complicated problems, it seems quite likely that differential equations will remain an important area of research for many years to come.

References for Additional Reading

Listed below are other standard texts on ordinary differential equations.

W. Boyce and R. Diprima, *Elementary Differential Equations,* 4th ed., Wiley, New York, 1986.

F. Brauer and J. Nohel, *Introduction to Differential Equations with Applications,* Harper & Row, New York, 1986.

W. R. Derrick and S. I. Grossman, *A First Course in Differential Equations with Applications,* 2d ed., Addison-Wesley, Reading, 1981.

C. H. Edwards, Jr., and D. E. Penny, *Elementary Differential Equations with Boundary Value Problems,* 2d ed., Prentice-Hall, Englewood Cliffs, 1989.

S. L. Ross, *Introduction to Ordinary Differential Equations,* 4th ed., Wiley, New York, 1989.

D. Zill, *A First Course in Differential Equations with Applications,* 3d ed., Prindle, Weber, & Schmidt, Boston, 1986.

The following are intermediate to advanced level texts covering more advanced aspects of differential equations.

L. C. Andrews, *Elementary Partial Differential Equations with Boundary Value Problems,* Academic, Orlando, 1986.

L. C. Andrews, *Special Functions for Engineers and Applied Mathematicians,* Macmillan, New York, 1985.

L. C. Andrews and B. K. Shivamoggi, *Integral Transforms for Engineers and Applied Mathematicians,* Macmillan, New York, 1988.

G. Birkhoff and G.-C. Rota, *Ordinary Differential Equations,* 3d ed., Wiley, New York, 1978.

R. V. Churchill, *Operational Mathematics,* 3d ed., McGraw-Hill, New York, 1972.

R. V. Churchill and J. W. Brown, *Fourier Series and Boundary Value Problems,* 3d ed., McGraw-Hill, New York, 1978.

R. Haberman, *Elementary Applied Partial Differential Equations,* 2d ed., Prentice-Hall, Englewood Cliffs, 1987.

E. L. Ince, *Ordinary Differential Equations,* Dover, New York, 1956.

N. N. Lebedev, *Special Functions and Their Applications,* Dover, New York, 1972.

G. F. Simmons, *Differential Equations,* McGraw-Hill, New York, 1972.

Answers to Odd-Numbered Problems

Chapter 1

Exercises 1.2

1. Linear, second-order   **3.** Linear in y, first-order   **5.** Linear in x, first-order
7. Nonlinear in x and y, first-order   **9.** Linear in i, first-order   **11.** Linear, second-order
13. Linear in x, first-order   **15.** Nonlinear, third-order   **17.** Nonlinear, first-order   **19.** Nonlinear, first-order

Exercises 1.3

15. $m = 2$ **17.** $m = 2, -3$ **19.** $m = 0, 1, -4$ **23.** $y' + 2y = 0$ **25.** $xy' - (x + 1)y = 0$
27. $(y') + y^2 = 1$ **29.** $(y')^2 + 1 = 1/y^2$ **31. (b)** $K = \frac{1}{4}$
33. The second equation is nonlinear and so a linear combination of y_1 and y_2 won't necessarily satisfy it.

Exercises 1.4

1. $y = e^{2x}$ **3.** $y = \sinh x$ **5.** $y = \frac{1}{2}x^2$ **7.** $y = x(x - 1)$ **9.** $y = e^x(\cos x - \sin x)$ **11.** $y = x$
13. $y = C_1 \cos x$ (C_1 arbitrary)   **15.** $y = 0$ **17.** $k = \pm n, n = 1, 2, 3, \ldots$

Review Exercises

1. Ordinary, second-order, linear   **3.** Ordinary, first-order, nonlinear   **5.** y_2 **7.** y_1 and y_2
11. Initial value, $C_1 = 2e^3$ **13.** Initial value, $C_1 = 1, C_2 = 2$ **15.** Boundary value, $C_1 = C_2 = 1$

Chapter 2

Exercises 2.2

1. $y^3 = Cx^2$ **3.** $x^2 + y^2 = 10$ **5.** $\dfrac{1 + y}{1 - y} = C\left(\dfrac{1 + x}{1 - x}\right)$ **7.** $2x + \sin 2x - 4 \cos y = C$ **9.** $\dfrac{1}{P} = \dfrac{b}{a} + Ce^{-at}$

11. $\log N = (t - 1)e^{t+2} - t + 1$ **13.** $\sqrt{x^2 + 1} = C(y + 1)^2 e^{y^2/2 - y}$ **15.** $x^x y^y = C$ **17.** $y^2 = x^2 + x + 9$
19. $x + y + \log(y/x^x) = C$ **21.** $v^2 = v_0^2 + 2g(t - t_0)$ **23.** $5x^2 - 2y^2 = 2$ **25.** $y^2 + y = x^2 - 4$

27. $y \cos^2 \frac{1}{2}(x + C) = 1$ **29.** $4y^3 - 9(x + C)^2 = 0$ **31. (a)** $y = \dfrac{1 - Cx^2}{1 + Cx^2}$ **(b)** $y = -1$ (singular solution)

(c) No solution   **33.** Not homogeneous   **35.** Homogeneous; degree zero   **37.** Homogeneous; degree one
39. Homogeneous; degree one   **41.** $y = -x \log|x| + Cx$ **43.** $y = -x \log|1 - \log|x||$

45. $\frac{1}{2} \log(u^2 + v^2) + \tan^{-1} \dfrac{u}{v} = C$ **47.** $y^2(y^2 + 2x^2) = C$ **49.** $y(\log y - \log x - 1) + x \log x + Cx = 0$

Exercises 2.3

1. $x^3 - 3x^2 y - y^2 = C$ **3.** $t(x^2 - 1) = C$ **5.** Not exact   **7.** $3x \cos y + y^3 = C$ **9.** $\sin^2 \theta + r^2(1 - \theta^2) = C$
11. $u^2 - ue^{3v} + \sin 3v = C$ **13.** $\cos x \sin y - \log|\cos x| = C$ **15.** Not exact   **17.** $x^3 y + 4x^2 y^2 + 4y^3 = 4$
19. $2 \cos x \cos y = 1$ **21.** $B = 6$; $x^3 y + 3xy^2 = C$ **23.** $B = 2$; $2x^2 \cos y + 2x^3 y - y^2 = C$
25. $x^2 y(y - 1) = C$ **27.** $(xy + y^2)e^x = C$ **29.** $2x - \log(x^2 + y^2) = C$ **31.** $y = Cx - \frac{1}{2}x^3$

33. $(x^3 + y^3) + \dfrac{3y}{x} = C$ **35.** $x^4 y^3(3 + x^5 y^3) = C$ **39. (a)** $y^4 x = C$ **(b)** $x^2 + y^2 = Cy$ **(c)** $y^4 \sin x + y^3 = C$

Exercises 2.4

3. $y = C_1 e^{-2x}$ **5.** $x = C_1 e^{-y^4}$ **7.** $y = (C_1 + \frac{1}{2}x^2)e^{-x^2}$ **9.** $xy + x \cos x - \sin x = C_1$

11. $x = \dfrac{1}{\sqrt{1 + y^2}}[C_1 + \log(y + \sqrt{1 + y^2})]$ **13.** $y = \frac{1}{5} + C_1 e^{-5x}$ **15.** $y = \dfrac{1}{x}(\sin x + C_1) - \cos x$

17. $y = x^4[(x - 1)e^x + C_1]$ **19.** $z = \frac{1}{4} + C_1 e^{-2x^2}$ **21.** $y = (x + C_1) \csc x$ **23.** $y = 1 + \dfrac{C_1}{\sqrt{x^2 + 9}}$

25. $y = (x + C_1)e^{mx}$ **27.** $y = \dfrac{1}{2x^2}(e^x + C_1 e^{-x})$ **29.** $y = \dfrac{b}{k - a}x^k + C_1 x^a$ **31.** $y = 2e^{-\sin x}$, $-\infty < x < \infty$

33. $i = \dfrac{E}{R}(1 - e^{-Rt/L})$, $-\infty < t < \infty$ **35.** $y = \frac{1}{2}\sqrt{2x + 3} \log(2x + 3)$, $-\frac{3}{2} < x < \infty$
37. $y = 3e^x + 2(x - 1)e^{2x}$, $-\infty < x < \infty$ **39.** $y = xe^x + ex$, $-\infty < x < \infty$

Exercises 2.5

1. $y = C_1 e^{-2x}$ **3.** $y = C_1 \cos x$ **5.** $y = C_1 x$ **7.** $y = C_1 e^{-x^4}$ **9.** $s = \dfrac{C_1}{t}e^{-2t}$ **11.** $x = C_1 e^{-y^4}$

13. $y = \frac{1}{5} + C_1 e^{-5x}$ **15.** $y = \dfrac{1}{x}(\sin x + C_1) - \cos x$ **17.** $y = x^4[(x - 1)e^x + C_1]$ **19.** $z = \frac{1}{4} + C_1 e^{-2x^2}$

21. $y = (x + C_1) \csc x$ **23.** $y = 1 + \dfrac{C_1}{\sqrt{x^2 + 9}}$ **25.** $y = (x + C_1)e^{mx}$ **27.** $y = \dfrac{1}{2x^2}(e^x + C_1 e^{-x})$

29. $y = \dfrac{b}{k - a}x^k + C_1 x^a$ **31.** $y = 2e^{-\sin x}$, $-\infty < x < \infty$ **33.** $i = \dfrac{E}{R}(1 - e^{-Rt/L})$, $-\infty < t < \infty$

35. $y = \frac{1}{2}\sqrt{2x + 3} \log(2x + 3)$, $-\frac{3}{2} < x < \infty$ **37.** $y = 3e^x + 2(x - 1)e^{2x}$, $-\infty < x < \infty$
39. $y = xe^x + ex$, $-\infty < x < \infty$ **41.** $y = \begin{cases} 1 - e^{-x}, & 0 \le x \le 1 \\ (e - 1)e^{-x}, & x > 1 \end{cases}$

43. $y = \begin{cases} \frac{1}{2}x^2/(1 + x^2), & 0 \le x \le 1 \\ (1 - \frac{1}{2}x^2)/(1 + x^2), & x > 1 \end{cases}$

Exercises 2.6

1. $y^2 = \frac{1}{2}x^2 + C_1 x^{-2}$ **3.** $y^{-2} = \frac{2}{5x} + C_1 x^4$ **5.** $x^{-3}y = 2 + C_1\sqrt{y}$ **7.** $y^{-4} = \frac{3}{4}e^{-4x} - x + \frac{1}{4}$

9. $y^{-1} = e^x - \sin x$ **11.** $y_1 = 1 + x$ **13.** $y_1 = 1 + x + x^2/2$
$\qquad\qquad\qquad\qquad\qquad y_2 = 1 + x + x^2/2 \qquad\qquad\qquad y_2 = 1 + x + x^2 + x^3/6$
$\qquad\qquad\qquad\qquad\qquad y_3 = 1 + x + x^2/2 + x^3/6 \qquad\quad y_3 = 1 + x + x^2 + x^3/3 + x^4/24$

15. $y_1 = x$ **17.** $y_1 = 1 + x - x^2$
$\quad\;\; y_2 = x + x^3/3 \qquad\qquad\qquad\qquad\;\; y_2 = 1 + x - x^2 - 2x^3/3 + x^4/2$
$\quad\;\; y_3 = x + x^3/3 + 2x^5/15 + x^7/63 \qquad y_3 = 1 + x - x^2 - 2x^3/3 + x^4/2 + 4x^5/15 - x^6/6$

19. $y_1 = 2x - \cos x$ **25.** $y = x + \dfrac{C_1 e^x}{1 - C_1 e^x}$
$\quad\;\; y_2 = 2x + x^2 - \cos x - \sin x$
$\quad\;\; y_3 = 2x + x^2 + x^3/3 - \sin x$

Review Exercises

1. $4(y + 1)^3 + 3x^4 = C_1$ **3.** $x^3 - 2y^3 = C_1 x$ **5.** $y = 5e^{2\sqrt{x}}$ **7.** $6x^2y + 4x^3 + 3x^4 = C_1$
9. $x^4 y^3 - x^2 y = C_1$ **11.** $x^4 y^3 = C_1 e^y$ **13.** $x^6 y^3 + x^4 y^5 = C_1$

15. $\log|2x^2 + 2xy + y^2| - 4 \tan^{-1}\left(\dfrac{x + y}{x}\right) = C_1$ **17.** $y = C_1 \sec x + \tan x$ **19.** $y^2 + x \log|x| = C_1 x$

Chapter 3

Exercises 3.2

1. $x^2 - y^2 = k$ **3.** $y = ke^{-2x}$ **5.** $xy = k$ **7.** $y^3\left(1 + \dfrac{3x^2}{y^2}\right) = k$ **9.** $y^4 = k(2x^2 - 3y^2)$

11. $y^{5/3} = x^{5/3} + k$ **15.** $r = b(1 + \sin \theta)$ **17.** $r^2 = b \sin \theta$

Exercises 3.3

1. $v(t) = 40(1 - e^{-4t/5})$ **3. (a)** $t = \dfrac{m}{c}\log\left(1 + \dfrac{v_0 c}{mg}\right)$ **(b)** 0.884

5. (a) $v(t) \cong \sqrt{30}\left(\dfrac{1.116 + e^{-11.68t}}{1.116 - e^{-11.68t}}\right)$ **(b)** $v_\infty = \sqrt{30}$ ft/s

7. (a) $v = 320(1 - e^{-t/10}), 0 \le t < 5$ **(b)** $v = 16\left(\dfrac{1 + 0.755e^{-4(t-5)}}{1 - 0.755e^{-4(t-5)}}\right), t > 5$ **(c)** $v_\infty = 320$ ft/s if chute never opens
$\qquad v_\infty = 16$ ft/s if chute opens

9. v_0 **11.** 3.1 mi/s   **13.** $v = 493$ ft/s   **15. (a)** Approximately 5×10^{-4} V   **(b)** 5×10^{-4} V

17. $i(t) = \dfrac{E_0 C\omega}{1 + R^2 C^2 \omega} \cos \omega t + \dfrac{E_0 RC^2\omega^2}{1 + R^2 C^2\omega} \sin \omega t$ **19. (a)** $i(t) = \dfrac{E_0}{R}(1 - e^{-Rt/L}) + i_0 e^{-Rt/L}$

(b) $i(t) = \dfrac{E_0 R}{R^2 + L^2\omega^2} \sin \omega t + \dfrac{E_0 L\omega}{R^2 + L^2\omega^2}(e^{-Rt/L} - \cos \omega t) + i_0 e^{-Rt/L}$ **(c)** $i(t) = \begin{cases} 6, & 0 \le t < 10 \\ 6e^{1-t/10}, & t > 10 \end{cases}$

Exercises 3.4

1. 9.33 g   **3.** 55,809 years   **5.** 4.48×10^6

7. $P(t) = (100b + P_0 - \frac{1000}{3})e^{-3t} + \frac{1000}{3} + 100b(3 \sin t - \cos t)$, steady-state: $\frac{1000}{3} + 100b(3 \sin t - \cos t)$

9. (b) $k = 1$: $M(10) = 1.790\, M_0$ **(c)** $\dfrac{dM}{dt} = 0.06M,\ M(0) = M_0$ **11.** Approximately 9 days

      $k = 4$: $M(10) = 1.814\, M_0$

      $k = 365$: $M(10) = 1.821\, M_0$ $M(t) = M_0 e^{0.06t}$

                             $M(10) = 1.822\, M_0$

Exercises 3.5

1. (a) $u(t) = 72 - 52e^{-(t/2)\log 2}$ **(b)** $65.5°C$ **3.** $85.5°F$ **5. (a)** 12.16 min   **(b)** 28.5 min   **(c)** Forever

7. (a) $S(t) = 50e^{-t/150}$ **(b)** $S(60) = 33.52$ lb   **(c)** $S_\infty = 0$

9. (a) $S(t) = 2.04 \times 10^{-5}(t - 300)^3 - 2(t - 300)$ **(b)** $S(60) = 198.4$ lb   **(c)** $S_\infty = 0$ (after 5 hrs)

11. 7 h, 40 min, 51 s   **13.** Approximately 9 min, 45 s   **15.** Approximately 20 min, 47 s

17. $y = \dfrac{a}{4}\left[\left(\dfrac{x}{a}\right)^2 - 1\right] - \dfrac{a}{2}\log \dfrac{x}{a}$ **19.** $y = \dfrac{a}{2}\left[\dfrac{(x/a)^{k+1} - 1}{k+1} + \dfrac{(a/x)^{k-1} - 1}{k-1}\right]$, $k = \dfrac{v}{w}$

$$D = x\sqrt{1 + \frac{1}{2}\left[\left(\frac{x}{a}\right)^k - \left(\frac{a}{x}\right)^k\right]^2}$$

21. 24 units, $\frac{27}{2}$ units   **23.** $y = \dfrac{a}{2}\left(1 - \dfrac{x^2}{a^2}\right)$; no   **25.** $y^2 = 2Cx + C^2$ (parabolic shape)

Review Exercises

1. $x^2 + y^2 = ky$ **3. (a)** $v = 128(1 - e^{-t/4})$ **(b)** $v_\infty = 128$ ft/s   **5.** $i = \frac{1}{10}(1 - e^{-50t})$ **7.** $8B_0$

9. $t = 13.3$ years

Chapter 4

Exercises 4.2

1. Independent   **3.** Dependent   **5.** Independent   **7.** Independent   **9.** Independent   **13.** -1

15. $(b - a)e^{(a+b)x}$ **17.** 0   **19.** $y = \frac{2}{7}(e^{3x} - e^{-x/2})$, $-\infty < x < \infty$ **21.** $y = 0$, $-\infty < x < \infty$

23. Not a fundamental set   **25.** Ce^{4x} **27.** $C/(1 - x^2)$ **29.** 25

Exercises 4.3

1. $y = C_1 + C_2 e^{-2x}$ **3.** $y = (C_1 + C_2 x)e^{3x}$ **5.** $y = C_1 \sin x + C_2 \cos x$ **7.** $y = e^x(C_1 \cos 2x + C_2 \sin 2x)$

9. $y = C_1 x^3 + C_2 x^{-2}$ **11.** $y = (C_1 \log x + C_2)\sqrt{x}$ **13.** $y = C_1 + C_2 \log \dfrac{1 + x}{1 - x}$

15. $y = \dfrac{1}{\sqrt{x}}(C_1 \sin x + C_2 \cos x)$

Exercises 4.4

1. $y = C_1 + C_2 e^{x/3}$ **3.** $y = C_1 e^x + C_2 e^{-3x}$ **5.** $y = C_1 e^{(5 + 2\sqrt{2})x} + C_2 e^{(5 - 2\sqrt{2})x}$ **7.** $y = (C_1 + C_2 x)e^x$

9. $y = (C_1 + C_2 x)e^{-x/3}$ **11.** $y = (C_1 + C_2 x)e^{2x/3}$ **13.** $y = e^{-x/2}\left(C_1 \cos \dfrac{\sqrt{3}}{2}x + C_2 \sin \dfrac{\sqrt{3}}{2}x\right)$

15. $y = e^{3x}(C_1 \cos 4x + C_2 \sin 4x)$ **17.** $y = e^{x/4}\left(C_1 \cos \dfrac{\sqrt{7}}{4}x + C_2 \sin \dfrac{\sqrt{7}}{4}x\right)$ **23.** $y = C_1 \cosh 2x + C_2 \sinh 2x$

25. $y = e^{5x/4}(C_1 \cosh \tfrac{7}{4}x + C_2 \sinh \tfrac{7}{4}x)$ **27.** $y = e^{5x/3}\left(C_1 \cosh \dfrac{\sqrt{13}}{3}x + C_2 \sinh \dfrac{\sqrt{13}}{3}x\right)$ **29.** $y = (C_1 + C_2 x)e^{4x}$

31. $y = C_1 e^{5x/2} + C_2 e^{x/2}$ **33.** $y = e^{-x} - e^{3x}$ **35.** $y = \dfrac{1}{e^2(e-1)}(e^{2x-1} - e^x)$ **37.** $y = \tfrac{6}{5}e^{-3x} + \tfrac{9}{5}e^{2x}$

39. $y = \dfrac{3x}{4}e^{3x}$

Exercises 4.5

1. $D(3D - 1)$ **3.** $(D - 1)^2$ **5.** $(D - 3 - 4i)(D - 3 + 4i)$ **7.** $D(D + 1)(2D + 1)$

9. $(D + 1)\left(D + \dfrac{1}{2} + \dfrac{\sqrt{3}}{2}i\right)\left(D + \dfrac{1}{2} - \dfrac{\sqrt{3}}{2}i\right)$ **11.** $4D^2 - 17D - 15$ **13.** $D^3 - 1$ **15.** $D^2 + 1 - x^2$

17. $x^2 D^2 + 4xD - 6$ **19.** $x^2 D^2 + (x^3 - 1)D + x^2 - x$ **23.** $y(C_1 + C_2 x)e^{4x}$ **25.** $y = C_1 e^{5x/2} + C_2 e^{x/2}$

Exercises 4.6

1. Dependent   **3.** Independent   **5.** Dependent   **7.** Dependent, all x **9.** Independent, $x \neq 0$

11. $y = C_1 + (C_2 + C_3 x + C_4 x^2)e^x$ **13.** $y = e^{-x/2}\left[(C_1 + C_2 x) \cosh \dfrac{\sqrt{5}}{2}x + (C_3 + C_4 x) \sinh \dfrac{\sqrt{5}}{2}x\right]$

15. $y = C_1 e^x + e^{-x/2}\left(C_1 \cos \dfrac{\sqrt{3}}{2}x + C_2 \sin \dfrac{\sqrt{3}}{2}x\right)$ **17.** $y = (C_1 + C_2 x + C_3 x^2)e^{-x}$

19. $y = C_1 + C_2 x + e^{-3x/2}\left(C_3 \cos \dfrac{\sqrt{5}}{2}x + C_4 \sin \dfrac{\sqrt{5}}{2}x\right)$ **21.** $y = C_1 + (C_2 + C_3 x)e^{-x/2}$

23. $y = C_1 e^x + C_2 e^{2x} + C_3 e^{10x/3}$ **25.** $y = (C_1 + C_2 x)e^{2x} + (C_3 + C_4 x)e^{-3x/2}$

27. $y = C_1 e^{-x} + C_2 e^{2x} + C_3 e^{-x/2} + C_4 e^{3x/2}$ **29.** $y = e^{2x}(C_1 \cos x + C_2 \sin x) + C_3 \cos \sqrt{5}x + C_4 \sin \sqrt{5}x$

31. $y = 2e^{2x} + (3x - 2)e^{-x}$ **33.** $y = e^{-x} - e^{-2x} \cos 3x$ **35.** $y = x - xe^{-3x}$ **37.** $(D + 4)(2D - 1)^2 y = 0$

39. $y = C_1 \sin x + C_2 \cos x + e^{-x}(C_3 \sin 2x + C_4 \cos 2x)$

Exercises 4.7

1. $y_P = Ax^2 + Bx + C$ **3.** $y_P = Ax^2 + Bx + C + x^2(Dx^2 + Ex + F)e^x$

5. $y_P = (Ax^3 + Bx^2 + Cx + D)e^{2x} + Ex^2 + Fx$ **7.** $y_P = x^2(Ax + B)e^{-x} + x(Cx + D)$

9. $y_P = Ax^2 e^x + B \cos x + C \sin x$ **11.** $y = C_1 + C_2 e^{-x} + \tfrac{1}{2}(\cos x - \sin x)$ **13.** $y = (C_1 + C_2 x)e^{3x} + \tfrac{1}{4}e^x$

15. $y = C_1 \cos 2\sqrt{2}x + C_2 \sin 2\sqrt{2}x + \tfrac{2}{9}e^{-x} + \tfrac{5}{8}$ **17.** $y = C_1 \cos 2x + C_2 \sin 2x + \sin x + \tfrac{4}{3}\cos x - 2$

19. $y = C_1 e^{4x} + C_2 e^{-x} + 6xe^{4x}$ **21.** $y = e^x(C_1 \cos 2x + C_2 \sin 2x) + \tfrac{1}{3}e^x \sin x$

23. $y = e^{-x/2}\left(C_1 \cos \dfrac{\sqrt{3}}{2}x + C_2 \sin \dfrac{\sqrt{3}}{2}x\right) + 1 + \dfrac{3}{13}\cos 2x - \dfrac{2}{13}\sin 2x$

25. $y = C_1 e^{2x} + C_2 e^{-2x} + C_3 e^{-x} + \frac{1}{2} + x - xe^{-x}$ **27.** $y = C_1 e^{2x} + C_2 e^{-x} + C_3 e^{x/2} + 1 + \frac{1}{9}xe^{2x} - \frac{1}{20}e^{-2x}$
29. $y = C_1 e^x + e^{-x}(C_2 \cos x + C_3 \sin x) - \frac{1}{2}(1 + x^2) + \frac{10}{9} \cos 2x - \frac{5}{27} \sin 2x$

31. $y = C_1 e^{x/2} + C_2 e^{-x/2} + C_3 \cos \dfrac{x}{2} + C_4 \sin \dfrac{x}{2} + \dfrac{3}{4} xe^{x/2}$ **33.** $y = \frac{3}{2} \sin x - (1 + x/2) \cos x$

35. $y = (2x - \pi)\cos x - \frac{1}{3}(8 \cos 2x + 5 \sin 2x)$ **37.** $y = \frac{7}{6} e^x + \frac{13}{12} \cos 2x + \frac{1}{24} \sin 2x - \frac{1}{4}(x + 1)$

39. $y = \frac{1}{4}(e^x - e^{-x}) + \frac{1}{2} \sin x - x$ **41.** Yes **43.** $y = 1 + x - \dfrac{3 \sin x}{2 \sin 1}$ **45.** $y = (1 - x/2) \cos x - \sin x$

47. $y = 2 \cosh x + \dfrac{3 - 2 \cosh 1}{\sinh 1} \sinh x - x^2 - 2$

Exercises 4.8

1. $y = C_1 e^x + C_2 e^{-x} + \frac{1}{4}(2x - 1)e^x$ **3.** $y = C_1 \cos 3x + C_2 \sin 3x + \frac{1}{36}(\sin 3x - 6x \cos 3x)$
5. $y = (C_1 + \log|\cos x|) \cos x + (C_2 + x) \sin x$
7. $y = C_1 \cos x + C_2 \sin x + \frac{1}{6} \sec^2 x - \frac{1}{2} + \frac{1}{2} \sin x \log|\sec x + \tan x|$
9. $y = C_1 \cos x + (C_2 + \log|\csc x - \cot x|) \sin x$
11. $y = (C_1 - \log|\sec x + \tan x|) \cos x + (C_2 - \log|\csc x - \cot x|) \sin x$ **13.** $y = (C_1 + C_2 x + \frac{1}{2}x^2 + \frac{1}{6}x^3)e^{2x}$
15. $y = (C_1 + C_2 x + \frac{1}{2}x^2 \log|x| - \frac{3}{4}x^2)e^{-x}$ **17.** $y = C_1 e^x + C_2 e^{-x} - \sin(e^{-x}) - e^x \cos(e^{-x})$
19. $y = (C_1 + \frac{1}{2} \cot^2 x) \cos x + (C_2 - \frac{1}{3} \cot^3 x) \sin x$ **21.** $y = \frac{1}{8}(2x^2 - 2x + 9)e^x + \frac{7}{8}e^{-x}$
23. $y = (1 + \pi - 2x) \sin x - (3 + 2 \log|\sin x|) \cos x$ **25.** $y = 3[\text{Si}(\pi/2) - \text{Si}(x)] \cos x - 3[\text{Ci}(\pi/2) - \text{Ci}(x)] \sin x$

27. $y = C_1 x + C_2 e^x + (\frac{1}{2} - x)e^{-x}$ **29.** $y = C_1 x^2 + C_2 e^x + x^2(x - 3)e^x$ **31.** $y = C_1 + C_2 \log \dfrac{1 + x}{1 - x} - x$

33. $y = C_1(1 - 3x^2) + C_2(3x - x^3) - \frac{1}{4}x^2(6 - x^2)(1 - 3x^2) + x(1 - x^2)(3x - x^3)$
35. $y = C_1 + C_2 e^x + C_3 e^{-x} - x^2 - 2$ **37.** $y = C_1 e^x + C_2 e^{-x} + C_3 e^{2x} + \frac{1}{8}e^{3x}$

39. $y = C_1 e^x + (C_2 + C_3 x)e^{-2x} + \dfrac{1}{3}\left(\dfrac{1}{3} - \dfrac{x}{4}\right)e^{2x} + \dfrac{1}{6}\left(\dfrac{x}{3} - \dfrac{1}{2}\right)$ **41.** $y = C_1 x + C_2 x^2 + C_3 x^3 - \dfrac{1}{24x}$

Exercises 4.9

1. $y = C_1 x + C_2 x^5$ **3.** $y = (C_1 + C_2 \log x)x$ **5.** $y = C_1 + C_2 x^{1/3}$
7. $y = x^3[C_1 \cos(4 \log x) + C_2 \sin(4 \log x)]$ **9.** $y = (C_1 + C_2 \log x)x^3 + x^3(\log x)^2$
11. $y = C_1 x^2 + C_2 x^3 + 2x - 1$ **13.** $y = C_1 x^{-1} + C_2 x + \frac{1}{2}x \log x$ **15.** $y = C_1 x + C_2 x^2 + x^3(\log x - \frac{3}{2})$
17. $y = x^3 - 2x^2$ **19.** $y = \frac{1}{10}x^2[9 \cos(3 \log x) - 7 \sin(3 \log x)] + \frac{1}{10}x^3$ **21.** $y = C_1 x^{-1} + C_2 \sqrt{x}$
23. $y = C_1 x^{-5} + C_2 x^{-1}$ **25.** $y = C_1 + (C_2 + C_3 \log x)x^2$ **27.** $y = C_1 x + C_2 \cos(\log x) + C_3 \sin(\log x)$
29. $y = (C_1 + C_2 \log x)x + C_3 x^2 + \frac{1}{4}x^3$ **31.** $y = C_1(x + 5)^{-1} + C_2(x + 5)^3$

Review Exercises

1. Dependent **3.** $y = C_1 e^{-x} + C_2 e^{2x}$ **5.** $y = e^{-4x}(C_1 \cos 3x + C_2 \sin 3x)$ **7.** $y = (C_1 + C_2 x)e^{-x/4}$
9. $y = (C_1 + C_2 x + C_3 x^2)e^{-2x}$ **11.** $y = C_1 + C_2 \cos 2x + C_3 \sin 2x$ **13.** $y = C_1 + C_2 e^{7x} - e^{6x}$
15. $y = C_1 + C_2 x + C_3 \cos 2x + C_4 \sin 2x + 2x^2$ **17.** $y = C_1 e^x + (C_2 - 1)xe^x + xe^x \log|x|$

19. $y = (C_1 + \frac{1}{2})e^x + (C_2 + 1)e^{-x} - 1 + e^{-x} \log(1 + e^{-x})$ **21.** $y = e^{3x/2}\left(\cos \dfrac{\sqrt{7}x}{2} + \sqrt{7} \sin \dfrac{\sqrt{7}x}{2}\right)$

23. $y = x + \cos x - 3 \sin x$ **25.** $y = 4(1 + x)e^{2x} - 7e^x$

Chapter 5

Exercises 5.2

1. $y = 3 \cos 4t + \frac{9}{4} \sin 4t$ **3.** $y = \sqrt{10} \cos \sqrt{\frac{5}{2}}t - \cos \sqrt{\frac{5}{2}}t$ **5. (a)** $2\pi/5$ **(b)** $5/2\pi$ **(c)** $1/2$ **(d)** 0 **7. (a)** $\pi/4$
(b) $4/\pi$ **(c)** $5/12$ **(d)** -0.64 rad   **9.** $t \cong 0.116$ s, 0.509 s   **11.** 8 lb   **13.** $t = \pi/6$ s
15. (a) 29 ticks   **(b)** $T \cong 2.8$ s   **17.** $t = 0.364$ s, $v = 3.2$ rad/s   **19.** $y = -5te^{-4t}$; critically damped

21. $y = e^{-t/2}\left(\cos \frac{\sqrt{3}}{2}t - \frac{1}{\sqrt{3}} \sin \frac{\sqrt{3}}{2}t \right)$; underdamped; $f_0 = \sqrt{3}/4\pi$ Hz, $\tau = 4\pi/\sqrt{3}$ s   **23. (a)** $c > 4\sqrt{6}$

(b) $c = 4\sqrt{6}$ **(c)** $c < 4\sqrt{6}$ **25. (a)** $y = \frac{5 + 2\sqrt{5}}{4\sqrt{5}} \exp[(-6 + 2\sqrt{5})t] - \frac{5 - 2\sqrt{5}}{4\sqrt{5}} \exp[(-6 - 2\sqrt{5})t]$

(b) $y = \frac{1}{2\sqrt{3}} e^{-2t} \sin 2\sqrt{3}t$; $\tau = \pi/\sqrt{3}$ **(c)** $y = (2 + 7t)e^{-4t}$ **27. (a)** $\pi/4\sqrt{3}$ **(b)** $\frac{1}{2}\sqrt{\frac{\pi}{3}}$ **29. (a)** $b = 13$

(b) $y = 8te^{-13t}$ **(c)** $t = 1/13$ s   **31.** $y = \frac{25}{3}(e^{-8t} - e^{-2t})$ **33.** $\tau \cong 0.973$ s

Exercises 5.3

1. (a) $y = \frac{3}{4}(1 - \cos 2t)$ **(b)** $y = \frac{3}{4}(1 - \cos 2t) + \frac{1}{2} \sin 2t$ **(c)** $y = \frac{3}{4}(1 - \cos 2t) + y_0 \cos 2t + \frac{v_0}{2} \sin 2t$

3. (a) $y = \sin 5t - \frac{1}{25} t \cos 2t$ **(b)** $y = (1 - t/5) \cos 5t + \frac{6}{25} \sin 5t$ **(c)** $y = (y_0 - t/5) \cos 5t + \frac{1}{5}(v_0 + 1/5) \sin 5t$

7. $y = (\frac{1}{4} - t) \cos 8t + \frac{1}{8} \sin 8t$ **9.** $y = \begin{cases} 2(1 - \cos 2t), & 0 \le t \le 4 \\ 2 \cos 2(t - 4) - 2 \cos 2t, & t > 4 \end{cases}$

11. (a) 0.0056 Hz **(b)** 2 **(c)** Approximately 89.76 s   **13.** $y = e^{-2t}(C_1 \cos 3t + C_2 \sin 3t) + \frac{P}{4\sqrt{13}} \sin \sqrt{13}t$

(a) $C_1 = y_0$, $C_2 = \frac{1}{3}(2y_0 - P/4)$ **(b)** $C_1 = 0$, $C_2 = \frac{1}{3}(v_0 - P/4)$ **(c)** $C_1 = y_0$, $C_2 = \frac{1}{3}(2y_0 + v_0 - P/4)$
15. (a) 5.56 s **(b)** 227.5 m   **17.** $e^{-t} \sin t$ (transient); $2 \sin t$ (steady-state)   **19.** $P/6$

Exercises 5.4

3. (a) 0 **(b)** $i_P(t) \cong \cos 2t - 0.3 \sin 2t$ **5. (a)** $i(t) = e^{-3t}(3 \sin 4t - 4 \cos 4t) + 4 \cos 5t$ **(b)** $i(t) = 5e^{-2t} \sin t$
7. $q(t) = \frac{3}{2} - \frac{1}{2}e^{-10t}(\cos 10t + \sin 10t)$, $\frac{3}{2}$ coulombs   **9.** $R = 2.19$ Ω

Exercises 5.5

1. $g_1(t, \tau) = t - \tau$ **3.** $g_1(t, \tau) = \frac{1}{\sqrt{5}} \sin \sqrt{5}(t - \tau)$ **5.** $g_1(t, \tau) = 2e^{t-\tau} \sin \frac{1}{2}(t - \tau)$ **7.** $g_1(t, \tau) = \frac{\tau}{8}\left[\left(\frac{t}{\tau}\right)^4 - \left(\frac{\tau}{t}\right)^4 \right]$

9. $g_1(t, \tau) = \frac{1}{2}(1 - \tau)^2 \log \frac{(1 + t)(1 - \tau)}{(1 - t)(1 + \tau)}$ **11.** $y = e^t - 1$ **13.** $y = \frac{1}{25}[(60e^2 + 9 - 5t)e^{-t} + (15e^{-8} + e^{-10})e^{4t}]$

15. $y = e^{-2t} + e^t + 2$ **17.** $y = \frac{1}{6} + \frac{1}{3}e^{-t} + e^{-2t}(\frac{1}{2} \cos \sqrt{2}t - \frac{4}{3}\sqrt{2} \sin \sqrt{2}t)$ **19.** $y = (4 + 6t + \frac{1}{2}t^3)e^{-t}$
21. $y = \frac{1}{24}t^{-5} - \frac{1}{8}t^{-1} + \frac{1}{12}t$ **23.** $y = \frac{1}{18}t^3 + \frac{1}{12}t^{-2} - \frac{1}{6} \log t + \frac{1}{36}$ **25.** $y = t^{-1/2}[\log t + \frac{1}{4}(\log t)^3]$
31. (a) $g_1(t, \tau) = \frac{1}{2}(t - \tau)^2$ **(b)** $g_1(t, \tau) = \frac{1}{2} \sin^2(t - \tau)$ **(c)** $g_1(t, \tau) = \frac{1}{2}(e^{t-\tau} - e^{\tau-t}) - (t - \tau)$

Exercises 5.6

5. According to Theorem 5.2, the zeros of $y_1 = e^t$ and those of $y_2 = e^{-t}$ must alternate on the entire axis $-\infty < t < \infty$. Although true, the assertion is vacuous since neither solution has a zero for any value of t.

9. $u'' + Q(t)u = 0$, where $Q(t) = \dfrac{1}{m}(k - c^2/4m)$. Hence $Q(t) > 0$ only if $c^2 < 4mk$.

11. $u'' + \dfrac{5 - 6t^2}{(1 - t^2)^2}\,u = 0$ **13.** $u'' + \dfrac{8t^2 + 8t - 1}{4(t + 1)^2}\,u = 0$ **19.** $y'' + (1 + t^2)y = 0$

Review Exercises

1. (a) $f = 2\sqrt{6}/\pi$ **(b)** $y = \tfrac{1}{4}\cos 4\sqrt{6}t$ **3.** $y \cong \cos\sqrt{245}t - 0.0319\sin\sqrt{245}t$ **5.** $y = \tfrac{4}{3}e^{-4t}\sin 3t$
7. (a) $y = \tfrac{2}{5}e^{-2t}(\cos 2t - 3\sin 2t) + \tfrac{2}{5}(2\sin 4t - \cos 4t)$ **(b)** $y = \tfrac{2}{5}(2\sin 4t - \cos 4t)$, $A = \sqrt{20}/5$

9. $y = \tfrac{1}{6}\cos 8t + 4t\sin 8t$ **11. (a)** $q = \dfrac{11}{250} - e^{-50t}\left(\dfrac{11}{250}\cos 50\sqrt{19}t + \dfrac{11\sqrt{19}}{4750}\sin 50\sqrt{19}t\right)$

(b) $i = \dfrac{44\sqrt{19}}{19}e^{-50t}\sin 50\sqrt{19}t$ **(c)** $q = \dfrac{11}{250}$

Chapter 6

Exercises 6.2

1. $\dfrac{1}{s^2}$ **3.** $\dfrac{n!}{s^{n+1}}$ **5.** $\dfrac{s}{s^2 + k^2}$ **7.** $\dfrac{b - a}{(s + a)(s + b)}$ **9.** $\dfrac{s}{s^2 - k^2}$ **11.** $\dfrac{s\sin b + a\cos b}{s^2 + a^2}$

13. $\dfrac{s^2 - 2k^2}{s(s^2 - 4k^2)}$ **15.** $\dfrac{s - a}{(s - a)^2 - k^2}$ **17.** $\dfrac{1}{(s - a)^2}$ **19.** $\dfrac{2s(s^2 - 3k^2)}{(s^2 + k^2)^3}$ **21.** $\dfrac{s^2 + 2k^2}{s(s^2 + 4k^2)}$ **23.** $\dfrac{k}{s^2 + 4k^2}$

Exercises 6.3

1. $\dfrac{3}{(s - 2)^2}$ **3.** $\dfrac{2k(3s^2 - k^2)}{(s^2 + k^2)^3}$ **5.** $\dfrac{s^3}{s^4 + 4k^4}$ **7.** $\dfrac{3(s + 4)}{(s + 4)^2 + 16} - \dfrac{24(s + 4)}{[(s + 4)^2 + 16]^2}$ **9.** $\log\dfrac{s + 1}{s - 1}$

11. $\dfrac{(s + 1)^2 + 2}{(s + 1)[(s + 1)^2 + 4]}$ **13. (b)** $\dfrac{s}{s^2 + 16}$

Exercises 6.4

1. $\tfrac{1}{2}t^2$ **3.** $\tfrac{1}{12}t^4 e^{3t}$ **5.** $e^{3t}\sin t$ **7.** $13te^{-4t}$ **9.** $e^{-2t}(7\cos 5t - \tfrac{17}{5}\sin 5t)$

11. $\dfrac{1}{3}e^{-2t/3}\left(5\cos\dfrac{2\sqrt{5}}{3}t - \dfrac{8\sqrt{5}}{5}\sin\dfrac{2\sqrt{5}}{3}t\right)$ **13.** $1 - e^{-t}$ **15.** $e^{-2t}(1 - 4t + 2t^2)$

17. $-\tfrac{1}{100} + \tfrac{5}{324}e^{4t} - e^{-5t}(\tfrac{11}{2025} + \tfrac{4}{45}t)$ **19.** $\tfrac{1}{3}(e^{-t} - e^{t}) + \tfrac{5}{12}(e^{2t} - e^{-2t})$ **21.** $-\tfrac{1}{6} - \tfrac{2}{15}e^{-3t} + \tfrac{3}{10}e^{2t}$

23. $\tfrac{3}{50}e^{3t} - \tfrac{1}{25}e^{-2t} + \tfrac{1}{50}e^{-t}(9\sin 2t - \cos 2t)$ **29. (b)** $\dfrac{\pi}{2}$

Exercises 6.5

1. $y = \tfrac{1}{2}(e^{2t} - 3)$ **3.** $y = \sinh t$ **5.** $y = \sin t$ **7.** $y = e^{t} - \cos t - \sin t$ **9.** $y = e^{-t}\sin t - \tfrac{1}{2}\sin 2t$

11. $y = \dfrac{1}{6} + \dfrac{1}{3}e^{-t} + e^{-2t}\left(\dfrac{1}{2}\cos\sqrt{2}t - \dfrac{8}{3\sqrt{2}}\sin\sqrt{2}t\right)$ **13.** $y = \tfrac{1}{4}(1 + t) + \tfrac{1}{4}(3 - 7t)e^{2t}$

15. $y = 8e^{-t} - 9e^{-2t} + 3e^{-3t}$ **17.** $y = \tfrac{5}{18}e^{t} - \tfrac{8}{9}e^{-t/2} + \tfrac{1}{2}e^{-t} + \tfrac{1}{9}e^{-2t}$ **19.** $y = \tfrac{1}{2}(\sinh t + \sin t) - t$

Exercises 6.6

1. $e^{2t} - e^t$ **3.** $\frac{1}{2}(t^2 - t) + \frac{1}{4}(1 - e^{-2t})$ **5.** $\frac{1}{2}t \sin t$ **7.** $e^{-t} + t - 1$ **9.** $\frac{1}{6}t^3 - t + \sin t$

11. $\frac{1}{25}e^{-t}(2 + 5t) - \frac{1}{50}(4 \cos 2t + 3 \sin 2t)$ **17.** $\frac{1}{a} \sin at$ **19. (a)** $y = t + \frac{3}{2} \sin 2t$ **(b)** $y = \frac{1}{2}$

23. (a) $y = \dfrac{P}{k}(1 - \cos \omega_0 t)$ **(b)** $y = \dfrac{P}{m(\omega_0^2 - \omega^2)}(\cos \omega t - \cos \omega_0 t)$ **(c)** $y = \dfrac{P}{2m\omega_0} t \sin \omega_0 t$

Exercises 6.7

9. $f(t)$

11. $f(t)$

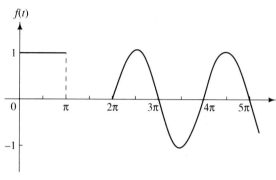

13. $f(t)$

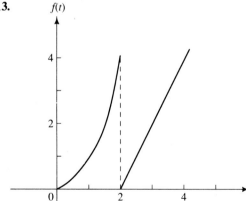

15. $\dfrac{2}{s} + e^{-s}\left(\dfrac{1}{s^2} - \dfrac{1}{s}\right)$ **17.** $\dfrac{1 + e^{-\pi s}}{s^2 + 1}$ **19.** $\dfrac{e^{-3(s+1)}}{s + 1}$

21. $\dfrac{2}{s^3} + \left(\dfrac{2}{s} - \dfrac{2}{s^2} - \dfrac{2}{s^3}\right)e^{-s} - \dfrac{3}{s}e^{-4s}$

23. $5h(t - 3) - h(t - 1)$ **25.** $3(t - 2)h(t - 2) - t$

27. $\frac{1}{2}(t - 3)^2 e^{-2(t-3)}h(t - 3)$ **29.** $\cos 2t[1 + h(t - \pi)]$

31. $y = e^{-t} + [1 - e^{-(t-1)}]h(t - 1)$

33. $y = \cos 2t + t - \frac{1}{2}\sin 2t - [t - 1 - \frac{1}{2}\sin 2(t - 1)]h(t - 1)$

35. $y = \frac{1}{12}(\cos 2t - \cos 4t)[1 - h(t - \pi)] + \frac{1}{2}\sin 2t$

37. $y = \frac{1}{2}\sin^2(t - 3)h(t - 3)$

39. $y = e^{-2t} + e^{-t} + [(t - 3)e^{-(t-2)} - e^{-2(t-2)}]h(t - 2)$

Exercises 6.8

1. $y = \cos 2t[1 + h(t - 2\pi)]$ **3.** $y = e^{-t}(\cos t + \sin t) - e^{-(t-\pi)} \sin t \, h(t - \pi)$

5. $y = 2te^{-t} + [1 - e^{-(t-2\pi)} - (t - 2\pi)e^{-(t-2\pi)}]h(t - 2\pi)$ **11.** $G(s) = \dfrac{1}{2s^2 - 5s - 3}$, $g(t) = \dfrac{1}{7}(e^{3t} - e^{-t/2})$

$$y = \frac{1}{7}\int_0^t (e^{3u} - e^{-u/2})f(t - u)\, du$$

13. $G(s) = \dfrac{1}{9s^2 + 6s + 1}$, $g(t) = \dfrac{1}{9} te^{-t/3}$ **15.** $G(s) = \dfrac{1}{s^2 - 6s + 25}$, $g(t) = \dfrac{1}{4} e^{3t} \sin 4t$

$$y = \frac{1}{9} \int_0^t ue^{-u/3} f(t - u)\, du \qquad\qquad y = \frac{1}{4} \int_0^t e^{3u} \sin 4u f(t - u)\, du$$

17. $g_1(t, u) = t - u$ **19.** $g_1(t, u) = \tfrac{1}{2} e^{-(t-u)} \sin 2(t - u)$ **21.** $g_1(t, u) = \dfrac{(t - u)^{n-1}}{(n - 1)!}$

23. $g_1(t, u) = \tfrac{1}{2} [\sinh(t - u) - \sin(t - u)]$

Review Exercises

1. $F(s) = \dfrac{2a^3}{s^4 - a^4}$ **3.** $F(s) = \dfrac{6ks^2 - 2k^3}{(s^2 + k^2)^3}$ **5.** $F(s) = \dfrac{5}{s} + e^{-3s}\left(\dfrac{1}{s^2} - \dfrac{2}{s}\right)$ **7.** $f(t) = \cos\sqrt{5}\,t + \dfrac{3}{\sqrt{5}} \sin\sqrt{5}\,t$

9. $f(t) = 2e^{2t}(3\cos 4t + \sin 4t)$ **11.** $f(t) = 2(e^{-t} - \cos t + \sin t)$ **13.** $y = 5\cos 3t - \sin 3t$

15. $y = 4e^{2t} + 5e^{-t} - 2t^2 + 2t - 3$ **17.** $y = \tfrac{12}{5} e^{-10t} \sin 10t$ **19.** $y = \tfrac{1}{3} e^{-t}(\sin t + \sin 2t)$

Chapter 7

Exercises 7.2

1. $-1 \le x < 1$ **3.** Diverges, all x **5.** Converges, $2 < x < 8$ **7.** Converges, all x **9.** $\displaystyle\sum_{n=0}^{\infty} \dfrac{(-1)^n}{n!} x^n$

11. $\displaystyle\sum_{n=0}^{\infty} \dfrac{(-1)^n}{(2n)!} (x - \pi/2)^{2n}$ **13.** $\displaystyle\sum_{n=0}^{\infty} \dfrac{n + 5}{(n + 1)!} x^n$ **15.** $\displaystyle\sum_{n=0}^{\infty} \dfrac{(-1)^{n-1}}{(2n - 2)!} x^{2n-2}$

Exercises 7.3

1. -1 **3.** None   **5.** $-1, 2$ **7.** None   **9.** $y = c_0 \displaystyle\sum_{n=0}^{\infty} (-1)^n x^n$ **11.** $y = x - 1 + c_0 \displaystyle\sum_{n=0}^{\infty} \dfrac{(-1)^n}{n!} x^n$

13. $y = c_0 \displaystyle\sum_{n=0}^{\infty} \dfrac{(-1)^n}{(2n)!} (2x)^n + c_1 \displaystyle\sum_{n=1}^{\infty} \dfrac{(-1)^{n-1}}{(2n - 1)!} x^{2n-1}$, $|x| < \infty$

15. $y = c_0(1 + 4x^2) + c_1(x + \tfrac{4}{3} x^3 - \tfrac{16}{15} x^5 + \tfrac{64}{35} x^7 + \cdots)$, $|x| < 2$

17. $y = c_0(1 - 6x^2 + 3x^4 + \tfrac{4}{5} x^6 + \cdots) + c_1(x - \tfrac{5}{3} x^3)$, $|x| < 1$

19. $y = c_0(1 - x^2 + \tfrac{1}{4} x^4 - \tfrac{1}{12} x^6 + \cdots) + c_1(x - \tfrac{1}{2} x^3 + \tfrac{3}{40} x^5 - \tfrac{1}{1680} x^7 + \cdots)$, $|x| < \infty$

21. $y = c_0 + c_1(x + \tfrac{1}{2} x^2 + \tfrac{1}{3} x^3 + \tfrac{1}{4} x^4 + \cdots)$, $|x| < 1$ **23.** $y = 1 + \tfrac{1}{2} x^2 + \tfrac{1}{8} x^4 + \tfrac{1}{48} x^6 + \cdots$

25. $y = 4 + 6x + \tfrac{11}{3} x^3 + \tfrac{1}{2} x^4 + \tfrac{11}{4} x^5 + \cdots$

27. $y = c_0[1 - \tfrac{1}{6}(x - 1)^3 + \tfrac{1}{180}(x - 1)^6 + \cdots] + c_1[x - 1 - \tfrac{1}{12}(x - 1)^4 + \tfrac{1}{504}(x - 1)^7 + \cdots]$

29. $y = c_0[1 - 3(x - 1)^2] + c_1[(x - 1) - \tfrac{1}{3}(x - 1)^3]$

31. $y = c_0(1 + \tfrac{1}{6} x^3 + \tfrac{1}{72} x^6 + \cdots) + c_1(x + \tfrac{1}{6} x^4 + \tfrac{13}{1260} x^7 + \cdots) + c_2(x^2 + \tfrac{7}{60} x^5 + \tfrac{1}{180} x^8 + \cdots)$

33. **(a)** $y = c_0(1 - x^2 - \tfrac{1}{6} x^4 - \tfrac{1}{30} x^6 - \cdots) + c_1 x$ **(b)** $y = c_0(1 - 4x^2 + \tfrac{4}{3} x^4) + c_1(x - x^3 + \tfrac{1}{10} x^5 - \cdots)$

Exercises 7.4

1. RSP at $x = \pm 1$; ISP at $x = 0$ **3.** RSP at $x = \pm i$, 1; ISP at $x = 0$ **5.** RSP at $x = 0$; ISP at $x = 4$

7. RSP at $x = 0, 3$ **9.** $s = 0, 0$; Case II   **11.** $s = 1, 1/2$; Case I   **13.** $s = -1, 2$; Case III

15. $y = A(1 + 2x + \tfrac{1}{3} x^2) + Bx^{1/2}(1 + \tfrac{1}{2} x + \tfrac{1}{40} x^2 + \cdots)$ **17.** $y = A(1 - x + \tfrac{15}{14} x^2 - \cdots) + Bx^{-3/2}(1 - 10x)$

19. $y = Ax^{-1} + Bx^{1/2}$ **21.** $y = Ax^{1/2}(1 - x + \frac{1}{6}x^2 - \cdots) + Bx(1 - \frac{1}{3}x + \frac{1}{30}x^2 - \cdots)$
23. $y = Ax^{-1}(1 + \frac{1}{2}x^2 - \frac{1}{24}x^4 + \cdots) + Bx^{3/2}(1 - \frac{1}{18}x^2 + \frac{1}{936}x^4 - \cdots)$

Exercises 7.5

1. $y'(s) = -y(s)\left(\dfrac{1}{s} + \dfrac{1}{s+1} + \dfrac{1}{s+2} + \cdots + \dfrac{1}{s+n-1} - \dfrac{1}{s+n}\right)$

3. $y'(s) = y(s)\left[\dfrac{3}{s+n} - 2\left(\dfrac{1}{s+1} + \dfrac{1}{s+2} + \cdots + \dfrac{1}{s+n-1}\right)\right]$

5. $y = (A + B\log x)(1 + 4x + 4x^2 + \cdots) - B(8x + 12x^2 + \frac{176}{27}x^3 + \cdots)$
7. $y = (A + B\log x)xe^x - B(x^2 + \frac{3}{4}x^3 + \frac{11}{36}x^4 + \cdots)$ **9.** $y = (A + B\log x)x^{-2}$
11. $y = (A + B\log x)x - B(x^2 - \frac{1}{4}x^3 + \frac{1}{18}x^4 + \cdots)$ **13.** $y = (A + B\log x)x^{1/2}$
15. $y = c_0(x - 2x^2 + 2x^3) + c_3(x^4 - \frac{1}{2}x^5 + \frac{1}{5}x^6 + \cdots)$ **17.** $y = c_0(1 + \frac{2}{3}x + \frac{1}{6}x^2) + c_4(x^4 + \frac{2}{5}x^5 + \frac{1}{10}x^6 + \cdots)$
19. $y = c_0x^{-1} + c_3(x^2 - \frac{3}{4}x^3 + \frac{3}{10}x^4 - \cdots)$ **21.** $y = (A + B\log x)(-x + \frac{1}{2}x^2 - \frac{1}{12}x^3 + \cdots) + B(1 + x - \frac{5}{4}x^2 + \cdots)$
23. $y = (A + B\log x)(-3x^2 - 4x^3 - \frac{5}{2}x^4 - \cdots) + B(x^{-1} - 2 - x + 3x^2 + \frac{13}{6}x^3 - \cdots)$
25. $y = (A + B\log x)(-\frac{1}{8} + \frac{1}{48}x - \frac{3}{192}x^2 + \cdots) + B(x^{-2} + \frac{1}{2}x^{-1} + \frac{3}{16} - \frac{17}{288}x + \cdots)$

Exercises 7.6

1. $P_0(x)$ **3.** $P_3(x)$ **5.** $W(y_1, y_2)(x) = \dfrac{C}{1 - x^2}$

Exercises 7.7

11. $y = C_1J_1(x) + C_2Y_1(x)$ **13.** $y = C_1J_{1/2}(x) + C_2J_{-1/2}(x)$ **15.** $W(y_1, y_2)(x) = \dfrac{C}{x}$
21. $y = x^{-2}[C_1J_4(3x) + C_2Y_4(3x)]$ **23.** $y = x^{1/2}[C_1J_1(2\sqrt{x}) + C_2Y_1(2\sqrt{x})]$

Review Exercises

1. No singular points **3.** RSP at $x = \pm 2$; ISP at $x = 0$
5. $y = c_0\left(1 - \dfrac{x^2}{2!} + \dfrac{x^4}{4!} - \dfrac{x^6}{6!} + \cdots\right) + c_1\left(x - \dfrac{x^3}{3!} + \dfrac{x^5}{5!} - \dfrac{x^7}{7!} + \cdots\right), |x| < \infty$

$\quad c_n = -\dfrac{1}{n(n-1)}c_{n-2}, n = 2, 3, 4, \ldots$

7. $y = c_0(1 - \frac{1}{2}x^2 - \frac{1}{8}x^4 - \frac{1}{16}x^8 - \cdots) + c_1x, |x| < 1$

$\quad c_n = \dfrac{n-3}{n}c_{n-2}, n = 2, 3, 4, \ldots$

9. $y = 2(1 + \frac{1}{2}x^2 + \frac{1}{24}x^4 + \frac{1}{20}x^5 + \cdots) + 3(x + \frac{1}{6}x^3 + \frac{1}{12}x^4 + \frac{1}{120}x^5 + \cdots), |x| < \infty$

$\quad c_n = \dfrac{(n-3)c_{n-3} + c_{n-2}}{n(n-1)}, n = 3, 4, 5, \ldots$

11. $s_1 = \frac{1}{2}, s_2 = -3$; Case I **13.** $s_1 = s_2 = 0$; Case II **15.** $y = Ax + Bx^{1/3}$

17. $y = (A + B\log x)\left(1 - x + \dfrac{x^2}{4} - \cdots\right) + 2B\left[x - \dfrac{x^2}{4}\left(1 + \dfrac{1}{2}\right) + \dfrac{x^3}{36}\left(1 + \dfrac{1}{2} + \dfrac{1}{3}\right) + \cdots\right]$

19. $y = A(1 + \frac{1}{10}x^2 + \frac{1}{280}x^4 + \cdots) + Bx^{-3}(1 - \frac{1}{2}x^2 - \frac{1}{8}x^4 - \cdots)$

Chapter 8

Exercises 8.2

1. $y_1' = y_2$
$\quad y_2' = -k^2 y_1 + P \sin \omega t$

3. $y_1' = y_2$
$\quad y_2' = y_3$
$\quad y_3' = y_4$
$\quad y_4' = k^4 y_1 + f(t)$

5. $u' = w$
$\quad v' = -u + t - 10$
$\quad w' = t - 10 + e^t$

7. $z_1' = z_2$
$\quad z_2' = -\dfrac{1}{m_1}(k_1 + k_2)z_1 + \dfrac{k_2}{m_1} z_3$
$\quad z_3' = z_4$
$\quad z_4' = \dfrac{k_1}{m_2} z_1 - \dfrac{k_2}{m_2} z_3$

9. $x' = -2x + y + 4t$
$\quad y' = 2x + y + t$

11. Dependent   **13.** Independent

Exercises 8.3

1. $x = C_1 e^t + C_2 e^{-t}$
$\quad y = C_1 e^t + 3C_2 e^{-t}$

3. $x = C_1 + C_2 e^{-2t}$
$\quad y = \frac{4}{3}C_1 + 2C_2 e^{-2t}$

5. $x = e^t(C_1 \cos 2t + C_2 \sin 2t)$
$\quad y = -\frac{1}{2}e^t(C_2 \cos 2t - C_1 \sin 2t)$

7. $x = e^{4t}[(2C_1 + C_2)\cos t + (2C_2 - C_1)\sin t]$
$\quad y = e^{4t}(C_1 \cos t + C_2 \sin t)$

9. $x = C_1 + C_2 e^{-4t} + C_3 e^{3t}$
$\quad y = 6C_1 - 2C_2 e^{-4t} + \frac{3}{2}C_3 e^{3t}$
$\quad z = -13C_1 - C_2 e^{-4t} - C_3 e^{3t}$

11. $x = C_1 e^{(a+b)t} + C_2 e^{(a-b)t}$
$\quad y = C_1 e^{(a+b)t} - C_2 e^{(a-b)t}$

13. $x = C_1 e^{-3t} + C_2 e^{2t}$
$\quad y = -3C_1 e^{-3t} - \frac{1}{2}C_2 e^{2t}$

15. $x = \frac{4}{15}e^{-2t} + \frac{9}{10}e^{3t} - \frac{1}{6}e^t$
$\quad y = -\frac{16}{15}e^{-2t} + \frac{9}{10}e^{3t} - \frac{5}{6}e^t$

17. $x = 2C_1 + 6C_2 e^{-t}$
$\quad y = C_1 + C_2 e^{-t} + C_3 e^{2t}$

19. $x = C_1 e^{2t}$
$\quad y = C_2 e^{-2t}$

21. $x = C_1 e^t + C_2 e^{-3t} - \frac{1}{3}t - \frac{11}{36}$
$\quad y = -\frac{1}{2}C_1 e^t + \frac{3}{2}C_2 e^{-3t} + \frac{1}{8}t + \frac{5}{12}$

23. $x = C_1 e^t - \frac{1}{2}\sin t$
$\quad y = -\frac{1}{3}C_1 e^t + \frac{1}{2}\sin t$

25. $x = C_1 e^{-t} + C_2 e^{2t} + 3C_3 e^{4t} + \frac{1}{4}$
$\quad y = 3C_1 e^{-t} - C_3 e^{4t} + \frac{1}{4}$

27. $x = -6C_1 e^{-t} - 3C_2 e^{-2t} + 2C_3 e^{3t}$
$\quad y = C_1 e^{-t} + C_2 e^{-2t} + C_3 e^{3t}$
$\quad z = 5C_1 e^{-t} + C_2 e^{-2t} + C_3 e^{3t}$

29. No solution   **31.** Infinitely many solutions

Exercises 8.4

1. **(a)** $f_1 = 1/2\pi$ Hz, $f_2 = \sqrt{6}/2\pi$ Hz   **(b)** $y_1 = \dfrac{2}{5}\cos t + \dfrac{1}{5}\sin t - \dfrac{2}{5}\cos\sqrt{6}t + \dfrac{2\sqrt{6}}{15}\sin\sqrt{6}t$

$\quad y_2 = \dfrac{4}{5}\cos t + \dfrac{2}{5}\sin t + \dfrac{1}{5}\cos\sqrt{6}t - \dfrac{\sqrt{6}}{15}\sin\sqrt{6}t$

3. $y_1 = \frac{13}{42}\cos t + \frac{145}{42}\sin 2t - \frac{25}{14}\sin 3t$
$\quad y_2 = \frac{13}{21}\sin t - \frac{145}{42}\sin 2t + \frac{10}{7}\sin 3t$

7. $i_1 = 1 + e^{-50t}\left(\dfrac{1}{\sqrt{3}}\sin 50\sqrt{3}t - \cos 50\sqrt{3}t\right)$

$\quad i_2 = 1 - e^{-50t}\left(\cos 50\sqrt{3}t + \dfrac{1}{\sqrt{3}}\sin 50\sqrt{3}t\right)$

9. $i_1 \cong 0.1402(e^{-50t} - \cos 60\pi t) + 0.0043 \sin 60\pi t + 6.2\, te^{-50t}$
$\quad i_2 \cong 0.0324(e^{-50t} - \cos 60\pi t) - 0.057 \sin 60\pi t + 12.4\, te^{-50t}$
$\quad i_3 \cong 0.1078(e^{-50t} - \cos 60\pi t) + 0.0613 \sin 60\pi t - 6.2\, te^{-50t}$

11. $i_1 = \frac{10}{13}e^{-2t} + \frac{625}{1469}e^{-15t} - \frac{135}{113}\cos t + \frac{895}{113}\sin t$
$\quad i_2 = -\frac{20}{13}e^{-2t} + \frac{375}{1469}e^{-15t} + \frac{145}{113}\cos t + \frac{85}{113}\sin t$

13. **(a)** $f_1 = 0.69$ Hz, $f_2 = 1.66$ Hz   **(b)** $f_1 = 0.72$ Hz, $f_2 = 1.38$ Hz   **(c)** $f_1 = 0.67$ Hz, $f_2 = 2.10$ Hz
$\quad$ **(d)** $f_1 = 0.56$ Hz, $f_2 = 1.46$ Hz   **(e)** $f_1 = 0.56$ Hz, $f_2 = 1.46$ Hz

15. $x \cong 200 - 54.09e^{-0.0264t} - 145.9e^{-0.1136t}$
$\quad y \cong 200 - 114.7e^{-0.0264t} + 114.7e^{-0.1136t}$

Exercises 8.5

1. $x = \frac{1}{3}(e^t - e^{-2t})$
$y = \frac{1}{3}(2e^t + e^{-2t})$

3. $x = -\cos 3t - \frac{5}{3}\sin 3t$
$y = 2\cos 3t - \frac{7}{3}\sin 3t$

5. $x = -e^t - \frac{1}{10}e^{-t} + \frac{1}{10}e^{2t}(21\cos t - 13\sin t)$
$y = e^t - \frac{1}{5}e^{-t} + \frac{1}{5}e^{2t}(-4\cos t + 7\sin t)$

7. $x = 5e^{-t} - 4 + 5t - 2t^2 + \frac{1}{3}t^3$
$y = -5e^{-t} + 5 - 5t + 2t^2$

9. $x = -\frac{1}{2}t - \frac{3}{4}\sqrt{2}\sin\sqrt{2}t$
$y = -\frac{1}{2}t + \frac{3}{4}\sqrt{2}\sin\sqrt{2}t$

11. $x = 2 + t + e^{-2t} + \sin t$
$y = 1 - t - 3e^{-2t} - \cos t$

Exercises 8.6

1. (a) $\begin{pmatrix} 2 & 6 \\ 5 & 9 \end{pmatrix}$ **(b)** $\begin{pmatrix} 4 & 2 \\ 5 & 3 \end{pmatrix}$ **(c)** $\begin{pmatrix} -9 & -2 \\ -10 & -3 \end{pmatrix}$

3. (a) $\begin{pmatrix} 1 & -1 \\ 4 & 3 \\ 3 & 1 \end{pmatrix}$ **(b)** $\begin{pmatrix} -5 & 1 \\ 4 & -1 \\ 11 & 5 \end{pmatrix}$ **(c)** $\begin{pmatrix} 13 & -3 \\ -8 & 4 \\ -26 & -12 \end{pmatrix}$

5. $\begin{pmatrix} -3 & 18 \\ -5 & 28 \end{pmatrix}$ **7.** $\begin{pmatrix} 7 & 4 & 9 \\ 3 & 1 & 11 \end{pmatrix}$ **9.** -2 **11.** Singular **13.** $\begin{pmatrix} \frac{5}{4} & 2 \\ -\frac{3}{4} & -1 \end{pmatrix}$ **15.** $\begin{pmatrix} \frac{1}{12} & -\frac{1}{6} & \frac{1}{6} \\ \frac{5}{12} & \frac{1}{6} & -\frac{1}{6} \\ -\frac{1}{4} & \frac{1}{2} & \frac{1}{2} \end{pmatrix}$

17. $\begin{pmatrix} -1 & \frac{1}{3} & \frac{1}{3} \\ 1 & 0 & 0 \\ -2 & -\frac{1}{3} & \frac{2}{3} \end{pmatrix}$ **21.** $\frac{1}{5}\begin{pmatrix} 3e^t & -2e^t \\ e^{-4t} & e^{-4t} \end{pmatrix}$ **23.** $\begin{pmatrix} -1 & 8 \\ 1 & 12 \end{pmatrix}$ **25.** $\begin{pmatrix} 1-e^{-t} & \frac{1}{2}(e^{4t}-1) \\ e^{-t}-1 & \frac{3}{4}(e^{4t}-1) \end{pmatrix}$

Exercises 8.7

1. $Y' = \begin{pmatrix} 2 & -1 \\ 3 & -2 \end{pmatrix}Y$ **3.** $Y' = \begin{pmatrix} 4 & -3 \\ 8 & -6 \end{pmatrix}Y$ **5.** $Y' = \begin{pmatrix} 1 & 0 & 1 \\ 1 & 1 & 0 \\ -2 & 0 & -1 \end{pmatrix}Y$

7. $Y' = \begin{pmatrix} 2 & -5 \\ 1 & -2 \end{pmatrix}Y + \begin{pmatrix} -\sin 2t \\ t \end{pmatrix}, \quad Y(0) = \begin{pmatrix} 0 \\ 1 \end{pmatrix}$ **9.** $Y' = \begin{pmatrix} 1 & -1 & 3 \\ 2 & 3 & -4 \\ 3 & -1 & -1 \end{pmatrix}Y + \begin{pmatrix} e^t \\ e^{-t} \\ -1 \end{pmatrix}, \quad Y(0) = \begin{pmatrix} 1 \\ -5 \\ 2 \end{pmatrix}$

11. $x' = -10x + 6y + 10e^{-3t}, \ x(0) = -1$
$y' = -12x + 7y + 18e^{-3t}, \ y(0) = 2$

13. $x' = x + 2y + z - e^{2t}, \ x(0) = 0$
$y' = 6x - y + 3e^{2t}, \ y(0) = -1$
$z' = -x - 2y - z, \ z(0) = 5$

21. $Y = Y^{(2)}$

23. $Y = \frac{1}{2}Y^{(1)} + \frac{5}{6}Y^{(2)}$ **25.** $Y = -\frac{1}{12}Y^{(1)} - \frac{1}{4}Y^{(2)} + \frac{2}{3}Y^{(3)}$

Exercises 8.8

1. $Y = C_1\begin{pmatrix} 1 \\ 1 \end{pmatrix}e^t + C_2\begin{pmatrix} 1 \\ 3 \end{pmatrix}e^{-t}$ **3.** $Y = C_1\begin{pmatrix} 1 \\ 4 \\ 3 \end{pmatrix} + C_2\begin{pmatrix} 1 \\ 2 \end{pmatrix}e^{-2t}$ **5.** $Y = \begin{pmatrix} 1 \\ 0 \end{pmatrix}e^t\cos 2t + \begin{pmatrix} 0 \\ \frac{1}{2} \end{pmatrix}e^t\sin 2t$

7. $Y = C_1\begin{pmatrix} 2\cos t - \sin t \\ \cos t \end{pmatrix}e^{4t} + C_2\begin{pmatrix} 2\cos t + \sin t \\ \cos t \end{pmatrix}e^{4t}$ **9.** $Y = C_1\begin{pmatrix} 1 \\ \frac{1}{3} \end{pmatrix}e^{-3t} + C_2\begin{pmatrix} \frac{1}{2} + 3t \\ t \end{pmatrix}e^{-3t}$

11. $Y = 2\begin{pmatrix} 3 \\ 1 \end{pmatrix}e^{2t} - 6\begin{pmatrix} 1 \\ -1 \end{pmatrix}e^{-2t}$ **13.** $Y = C_1\begin{pmatrix} \frac{1}{2} \\ 1 \end{pmatrix}e^t + C_2\begin{pmatrix} -1 \\ 1 \end{pmatrix}e^{-2t}$

15. $Y = C_1\begin{pmatrix} \cos t \\ \frac{1}{2}(\sin t - \cos t) \\ -\cos t - \sin t \end{pmatrix} + C_2\begin{pmatrix} \sin t \\ -\frac{1}{2}(\sin t + \cos t) \\ \cos t - \sin t \end{pmatrix} + C_3\begin{pmatrix} 0 \\ 1 \\ 0 \end{pmatrix}e^t$

17. $Y = C_1\begin{pmatrix} 1 \\ 6 \\ -13 \end{pmatrix} + C_2\begin{pmatrix} 1 \\ -2 \\ -1 \end{pmatrix}e^{-4t} + C_3\begin{pmatrix} 1 \\ \frac{3}{2} \\ -1 \end{pmatrix}e^{3t}$ **19.** $Y = C_1\begin{pmatrix} 1 \\ 1 \\ 1 \end{pmatrix}e^{4t} + C_2\begin{pmatrix} 1 \\ 0 \\ -1 \end{pmatrix}e^{-2t} + C_3\begin{pmatrix} 0 \\ 1 \\ -1 \end{pmatrix}e^{-2t}$

21. (a) $\begin{pmatrix} 1.5 & 1 \\ 1 & 1.5 \end{pmatrix}$ **(b)** $\begin{pmatrix} 1.543 & 1.175 \\ 1.175 & 1.543 \end{pmatrix}$

Exercises 8.9

1. $\mathbf{Y} = C_1 \begin{pmatrix} 1 \\ -4 \end{pmatrix} e^{-2t} + C_2 \begin{pmatrix} 1 \\ 1 \end{pmatrix} e^{3t} - \frac{1}{6} \begin{pmatrix} 1 \\ 5 \end{pmatrix} e^{t}$

3. $\mathbf{Y} = C_1 \begin{pmatrix} \cos t \\ \frac{2}{5}\cos t + \frac{1}{5}\sin t \end{pmatrix} + C_2 \begin{pmatrix} \sin t \\ -\frac{1}{5}\cos t + \frac{2}{5}\sin t \end{pmatrix} + \frac{1}{3} \begin{pmatrix} 2\cos 2t + 2\sin 2t \\ \sin 2t \end{pmatrix} - \begin{pmatrix} 5t \\ 2t - 1 \end{pmatrix}$

5. $\mathbf{Y} = C_1 \begin{pmatrix} \sin t \\ \frac{1}{2}\cos t \end{pmatrix} + C_2 \begin{pmatrix} -2\cos t \\ \sin t \end{pmatrix} + \begin{pmatrix} -2\cos t \, \log|\sec t + \tan t| \\ -1 + \sin t \, \log|\sec t + \tan t| \end{pmatrix} e^{t}$

7. $\mathbf{Y} = 17 \begin{pmatrix} 2 \\ 3 \end{pmatrix} e^{-t} - 13 \begin{pmatrix} 3 \\ 4 \end{pmatrix} e^{-2t} + \begin{pmatrix} 4 \\ 3 \end{pmatrix} e^{-3t}$ **9.** $\mathbf{Y} = \frac{1}{49} \begin{pmatrix} 2 \\ 1 \end{pmatrix} e^{5t} + \frac{2}{49} \begin{pmatrix} -1 \\ 3 \end{pmatrix} e^{-2t} + \frac{1}{7} \begin{pmatrix} -2t \\ 6t - 1 \end{pmatrix} e^{-2t}$

11. $\mathbf{Y}_P = \begin{pmatrix} 0 \\ 1 \\ 1 \end{pmatrix} te^{t}$

Exercises 8.10

1 to 19. See answers 1 to 19 in Exercises 8.8.   **23 to 31.** See answers 1 to 9 in Exercises 8.9.

Review Exercises

1. $x = 5e^{-t} + 3e^{4t}$ **3.** $x = \frac{2}{5}t + \frac{2}{25}$ **9.** $\mathbf{Y}' = \begin{pmatrix} 2 & -3 \\ -2 & 1 \end{pmatrix} \mathbf{Y}, \ \mathbf{Y}(0) = \begin{pmatrix} 8 \\ 3 \end{pmatrix}$
  $y = 5e^{-t} - 2e^{4t}$ $y = \frac{16}{5}t + \frac{1}{25}$

11. $\mathbf{Y}' = \begin{pmatrix} 4 & 2 \\ 3 & -1 \end{pmatrix} \mathbf{Y} + \begin{pmatrix} -8t \\ 2t + 3 \end{pmatrix}, \ \mathbf{Y}(0) = \frac{1}{25} \begin{pmatrix} 2 \\ 1 \end{pmatrix}$ **13.** $\lambda_1 = \lambda_2 = -8$ **15.** $\lambda_1 = -4, \lambda_2 = 2$

Chapter 9

Exercises 9.2

1. $y = \beta e^{-2t}$; $y = 0$ is a stable critical point.

3. $y = \dfrac{\beta}{\beta + (1 - \beta)e^{-t}}$; $y = 1$ is a stable critical point, $y = 0$ is an unstable critical point.

5. $y = \dfrac{3(\beta - 1) - (\beta - 3)e^{-2t}}{(\beta - 1) - (\beta - 3)e^{-2t}}$; $y = 3$ is a stable critical point, $y = 1$ is an unstable critical point.

9. (a) a/b **(b)** $a/2b$

Exercises 9.3

1. $y^2 - (x + 1)^2 = C$; critical point is $(-1, 0)$ **3.** $(x - y)^2 = C(x + y)$; critical point is $(0, 0)$
5. $x^3 - 3xy + y^3 = C$; critical points are $(0, 0), (1, 1)$ **7.** $x^2 + y^2 = C$; critical points are $(0, 0)$ and line $x = 1$
9. $3y - 2x^3 = C$; critical points are lines $x = 0$ and $y = -4$

11. The trajectory is the circle $x^2 + y^2 = 9$ (counterclockwise).

13. The trajectory is the left half of the hyperbola $x^2 - y^2 = 1$ (clockwise).

15. **(a)** $x = e^t$, $y = (t + 3)e^t$ **(b)** $x = e^{t-3}$, $y = te^{t-3}$ 17. **(a)** $2 \log x = \log^2 y$ **(b)** $2 \log x = \log^2 y - 2 \log y$

(c) $2 \log x = \log^2 y - 4 \log y$

Exercises 9.4

1. Saddle point (unstable)   3. Center (stable)   5. Node (asymptotically stable)   7. Saddle point (unstable)

9. Focus (unstable)   11. Saddle point (unstable)   13. $a < 0$: asymptotically stable focus

$0 \le a < 1$: asymptotically stable node

$a > 1$: saddle point (unstable)

15. $a \le -2$: asymptotically stable node

$-2 < a < 0$: asymptotically stable focus

$a = 0$: stable center

$0 < a < 2$: unstable focus

$a > 2$: unstable node

17. **(a)** Asymptotically stable node   **(b)** Asymptotically stable node   **(c)** Asymptotically stable focus

Exercises 9.5

1. $(0, 0)$ 3. $(0, 0)$ 5. $y = 2$ and $x = -3$ 7. $(0, 0)$ and $(1, 3)$ 9. $(0, 0)$ and $(-1/2, 1)$

11. Saddle point (unstable)   13. Asymptotically stable node   15. Saddle point (unstable)

17. No conclusion   19. Asymptotically stable node   21. $(0, 0)$ unstable; $(2, 4)$ unstable

23. $(1, 1)$ unstable; $(-1, -1)$ asymptotically stable   25. **(b)** $(6, 12)$ **(c)** $(12, 6)$

Chapter 10

Exercises 10.2

1. 2.9278 (Euler)
   3.4509 (improved Euler)

3. 3.2261 (Euler)
   3.8254 (improved Euler)

5. 1.8371 (Euler)
   2.0488 (improved Euler)

7. 1.2194 (Euler)
   1.3260 (improved Euler)

9. 0.4198 (Euler)
   0.4053 (improved Euler)

11. to **20.** See answers 1 to 9

21. $y(1) = 2.4883$
    $e = 2.71828 \ldots$ (actual value)

23. **(a)** $y(1) = 0.8337$
    (b) $\pi \cong 3.3348$

25. **(a)** 2.2368 (Euler)
        4.4715 (improved Euler)

(b) 2.9060 (Euler)
        8.7988 (improved Euler)

(c) 4.1048 (Euler)
        51.873 (improved Euler)

Exercises 10.3

1. 3.4902   3. 3.9078   5. 2.0670   7. 1.3333   9. 0.4055   11. $y(1) = 2.7183$
$e = 2.71828 \ldots$ (actual value)

13. **(a)** 2.2368   **(b)** 4.4715   **(c)** 22.453

Exercises 10.4

1. $x(0.1) = 1.1$, $y(0.1) = -0.1$
   $x(0.2) = 1.25$, $y(0.2) = -0.22$

3. $x(0.1) = 1.2$, $y(0.1) = 1.1$
   $x(0.2) = 1.451$, $y(0.2) = 1.32$

5. $x(0.1) = 0.8$, $y(0.1) = 1.1$
   $x(0.2) = 0.582$, $y(0.2) = 1.18$

7. $x(0.1) = 1.3110$, $y(0.1) = 1.9197$ **9.** $x(0.1) = 1.1305$, $y(0.1) = 0.3851$ **11.** $x(0.1) = 1.2$, $x(0.2) = 1.39$
13. $x(0.1) = 1.1947$

Chapter 11

Exercises 11.2

1. $y = 0$, unique   **3.** $y = C \cos 3x$, not unique   **5.** $y = C_1 \cos \pi x + C_2 \sin \pi x$, not unique   **7.** $y = 0$

9. No solution   **11.** $y = 2\dfrac{\sin kx}{\sin k}$, unique   **13.** $y = \frac{1}{2}e^{x-1} - e^{-(x+1)}$, unique   **15.** $y = 3xe^{3(x-1)}$, unique

17. $y = Cx^3 \sin(4 \log x)$, not unique   **19.** $y = (1 - \frac{1}{2}\log x)x$, unique   **21.** No solution

23. $y = -2 + \dfrac{4}{\pi}x - \dfrac{3}{\pi^2}\sin \pi x$, unique   **25.** $y = -\frac{6}{5}\cos 2x + \frac{1}{5}e^{-x}$, unique   **27.** $y = e^{-x}(x - 1) + x - 2$, unique

29. $y = \dfrac{1}{6}\left[\left(\dfrac{e^3 - e - 6}{e^3 - e^{-2}}\right)e^{-2x} + \left(\dfrac{e - e^{-2} + 6}{e^3 - e^{-2}}\right)e^{3x} - e^x\right]$, unique   **31.** $y = \frac{4}{5}\sqrt{x}(x - 1)$, unique

33. $y = \dfrac{x}{9}(5 - 2x - 3x^2)$

Exercises 11.3

1. $\lambda_n = n^2\pi^2/4$, $\phi_n(x) = \sin n\pi x/2$, $n = 1, 3, 5, \ldots$ **3.** $\lambda_0 = 0$, $\phi_0(x) = 1$; $\lambda_n = \dfrac{n^2\pi^2}{p^2}$, $\phi_n(x) = \cos \dfrac{n\pi x}{p}$, $n = 1, 2, 3, \ldots$

5. $\lambda_n = -n^2\pi^2$, $\phi_n(x) = e^{-x}\sin n\pi x$, $n = 1, 2, 3, \ldots$ **7.** $\lambda_n = n^2\pi^2 + \frac{1}{4}$, $\phi_n(x) = e^{-x/2}\sin n\pi x$, $n = 1, 2, 3, \ldots$

9. $\lambda_n = \dfrac{n^2\pi^2}{2} + \dfrac{9}{8}$, $\phi_n(x) = e^{3x/2}\sin n\pi x$, $n = 1, 2, 3, \ldots$ **11.** $\lambda_n = n^2\pi^2$, $\phi_n(x) = \sin(n\pi \log x)$, $n = 1, 2, 3, \ldots$

13. $\lambda_n = \dfrac{n^2\pi^2}{(\log 4)^2}$, $\phi_n(x) = \cos\left(\dfrac{n\pi \log x}{2 \log 2}\right)$, $n = 1, 3, 5, \ldots$ **15.** $k/m = n^2\pi^2$, $n = 1, 2, 3, \ldots$

19. (a) $\lambda_n = \dfrac{(2n - 1)^2\pi^2}{4b^2}$, $\phi_n(x) = \cos \dfrac{(2n - 1)\pi x}{2b}$, $n = 1, 2, 3, \ldots$ (b) $\omega_1 = \sqrt{\dfrac{T}{\rho}}\dfrac{\pi}{2b}$

Exercises 11.4

1. $\dfrac{4}{\pi}\displaystyle\sum_{\substack{n=1 \\ (\text{odd})}}^{\infty}\dfrac{\sin n\pi x}{n}$ **3.** $\dfrac{4}{\pi}\displaystyle\sum_{n=1}^{\infty}\dfrac{(-1)^{n-1}}{n}\sin \dfrac{n\pi x}{2}$ **5.** $\dfrac{8}{\pi^3}\displaystyle\sum_{\substack{n=1 \\ (\text{odd})}}^{\infty}\dfrac{\sin n\pi x}{n^3}$ **7.** $\sin x - 4 \sin 3x$

9. $2\pi\displaystyle\sum_{n=1}^{\infty}\dfrac{[1 - (-1)^n e]n}{n^2\pi^2 + 1}\sin n\pi x$ **11.** 1 **13.** $1 - 8\displaystyle\sum_{\substack{n=1 \\ (\text{odd})}}^{\infty}\dfrac{1}{n^2\pi^2}\cos \dfrac{n\pi x}{2}$ **15.** $\dfrac{1}{6} - \dfrac{2}{\pi^2}\displaystyle\sum_{n=1}^{\infty}\dfrac{(-1)^n}{n^2}\cos n\pi x$

17. $-\dfrac{2}{3\pi} + \dfrac{4}{\pi}\displaystyle\sum_{n=1}^{\infty}\left(\dfrac{1}{1 - 4n^2} - \dfrac{12}{9 - 4n^2}\right)\cos 2nx$ **19.** $e - 1 + 2\displaystyle\sum_{n=1}^{\infty}\dfrac{[(-1)^n e - 1]}{1 + n^2\pi^2}\cos n\pi x$

Exercises 11.5

5. $-\dfrac{1}{2} - \dfrac{2}{\pi}\displaystyle\sum_{\substack{n=1 \\ (\text{odd})}}^{\infty}\dfrac{\sin n\pi x}{n}$ **7.** $2\displaystyle\sum_{n=1}^{\infty}\dfrac{(-1)^n}{n}\sin nx$ **9.** $-\dfrac{1}{2} + \displaystyle\sum_{n=1}^{\infty}\left[\dfrac{4}{(2n - 1)^2\pi^2}\cos \dfrac{(2n - 1)\pi x}{2} - \dfrac{(-1)^n 2}{n\pi}\sin \dfrac{n\pi x}{2}\right]$

11. $\dfrac{1}{\pi}(1 - e^{-\pi}) + \dfrac{2}{\pi}\displaystyle\sum_{n=1}^{\infty}\dfrac{1 - e^{-\pi}(-1)^n}{n^2 + 1}\cos nx$ **13.** $-\dfrac{p^2}{3} - \dfrac{4}{\pi}\displaystyle\sum_{n=1}^{\infty}\dfrac{(-1)^n}{n}\left(p\cos\dfrac{n\pi x}{p} + \sin\dfrac{n\pi x}{p}\right)$

15. (a) $\frac{1}{2}$ **(b)** $\frac{5}{2}$ **(c)** 2 **(d)** 2 **(e)** 1   **19.** Odd   **21.** Odd   **23.** Neither

27. $\dfrac{1}{17}(1 - e^{-\pi}) + \dfrac{2}{\pi}\displaystyle\sum_{n=1}^{\infty}\dfrac{1 - e^{-\pi}}{4n^2 + 1}(\cos 2nx + 2n\sin 2nx)$

29. (a) $2\displaystyle\sum_{n=1}^{\infty}\dfrac{\sin nx}{n}$ **(b)** $\dfrac{\pi}{2} + \dfrac{4}{\pi}\displaystyle\sum_{n=1}^{\infty}\dfrac{\cos(2n-1)x}{(2n-1)^2}$ **(c)** $\dfrac{\pi}{2} + \displaystyle\sum_{n=1}^{\infty}\dfrac{\sin 2nx}{n}$

Exercises 11.6

3. $\dfrac{d}{dx}(e^{-x}y') + \lambda e^{-x}y = 0$ **5.** $\dfrac{d}{dx}(xy') + \left(-\dfrac{n^2}{x} + \lambda x\right)y = 0$ **9.** $1/\sqrt{2}$

11. $1/\sqrt{2\pi},\ 1/\sqrt{\pi}$ **13.** $\dfrac{4}{\pi}\displaystyle\sum_{n=1}^{\infty}\dfrac{1}{2n-1}\sin\dfrac{(2n-1)\pi x}{2}$ **15.** $1 - 2\displaystyle\sum_{n=1}^{\infty}\dfrac{(-1)^n}{n}\sin nx$

Chapter 12

Exercises 12.2

1. $u(x, t) = 3\sin(\pi x)e^{-a^2\pi^2 t} - 5\sin(4\pi x)e^{-16a^2\pi^2 t}$ **3.** $u(x, t) = \dfrac{8p^2}{\pi^2}\displaystyle\sum_{\substack{n=1\\(\text{odd})}}^{\infty}\dfrac{1}{n^3}\sin\left(\dfrac{n\pi x}{p}\right)e^{-a^2 n^2\pi^2 t/p^2}$

5. $S(x) = T_1 + (T_2 - T_1)x/(1 + p)$ **7.** $u(x, t) = 1 - x + \dfrac{4}{\pi}\displaystyle\sum_{n=1}^{\infty}\dfrac{(-1)^{n-1}}{n}\sin(n\pi x)e^{-a^2 n^2\pi^2 t}$

9. $u(x, t) = \dfrac{T_0}{2}\left[1 - \cos\left(\dfrac{2\pi x}{p}\right)e^{-4a^2\pi^2 t/p^2}\right]$ **11.** $u(x, t) = T_0 + \dfrac{8}{\pi^3}\displaystyle\sum_{\substack{n=1\\(\text{odd})}}^{\infty}\dfrac{1}{n^3}\sin(n\pi x)e^{-a^2 n^2\pi^2 t}$

13. $u(x, t) = \dfrac{\pi}{2} - \dfrac{4}{\pi}\displaystyle\sum_{\substack{n=1\\(\text{odd})}}^{\infty}\dfrac{1}{n^2}\cos(nx)e^{-a^2 n^2 t}$

15. $u(x, t) = 100 + \displaystyle\sum_{n=1}^{\infty}c_n(\sin k_n x + k_n\cos k_n x)e^{-a^2 k_n^2 t}$, where $\cos k_n\pi - k_n\sin k_n\pi = 0$ and $c_n = 400/\{k_n[(1 + k_n^2)\pi + 1]\}$, $n = 1, 2, 3, \dots$

17. $t = \dfrac{4}{a^2\pi^2}\log(40/3\pi^2)$

Exercises 12.3

1. $u(x, t) = \dfrac{4v_0}{c\pi^2}\displaystyle\sum_{n=1}^{\infty}\dfrac{1}{n^2}\sin(nc\pi t)\sin(n\pi x)$ **3.** $u(x, t) = \dfrac{8}{\pi^3}\displaystyle\sum_{\substack{n=1\\(\text{odd})}}^{\infty}\dfrac{1}{n^2}\cos(nc\pi t)\sin(n\pi x)$ **5.** $\omega = \dfrac{\pi}{p}\sqrt{\dfrac{\tau}{m}}$

7. Maximum displacement is 10 cm at $x = \pi/2$ and $t = \pi/80$ s.

9. $u(x, t) = Ae^{-kt}\left(\cos\mu t + \dfrac{k}{\mu}\sin\mu t\right)\sin\pi x$,   where $\mu = \sqrt{\pi^2 c^2 - k^2}$

11. $u(x, t) = \dfrac{1}{2}\left[\dfrac{1}{1 + 2(x + ct)^2} + \dfrac{1}{1 + 2(x - ct)^2}\right]$

Exercises 12.4

1. $u(x, y) = \dfrac{\sinh y}{\sinh 1} \sin x$ **3.** $u(x, y) = \dfrac{1}{3} + \dfrac{4}{\pi^2} \displaystyle\sum_{n=1}^{\infty} \dfrac{(-1)^n}{n^2 \cosh n\pi} \cosh[n\pi(1 - y)] \cos(n\pi x)$

7. $\alpha_n = \dfrac{2}{a \sinh(n\pi b/a)} \displaystyle\int_0^a f_2(x) \sin \dfrac{n\pi x}{a}\, dx$

$\beta_n = \dfrac{2}{a \sinh(n\pi b/a)} \displaystyle\int_0^a f_1(x) \sin \dfrac{n\pi x}{a}\, dx, n = 1, 2, 3, \ldots$ **9. (a)** $u(x, y) = 100 + \dfrac{400}{\pi} \displaystyle\sum_{\substack{n=1 \\ (\text{odd})}}^{\infty} \dfrac{\sinh(n\pi y/2)}{n \sinh n\pi} \sin \dfrac{n\pi x}{2}$

(b) $u(1, 1) \cong 125.4°C$

11. $u(r, \theta) = \dfrac{1}{2}\left[1 + \left(\dfrac{r}{\rho}\right)^2 \cos 2\theta\right]$ **13.** $u(r, \theta) = \dfrac{1}{2} a_0 + \dfrac{4}{\pi} \displaystyle\sum_{\substack{n=1 \\ (\text{odd})}}^{\infty} \dfrac{r^n}{n^2} \sin n\theta$ **15.** $u(r, \theta) = r \sin \theta$

Exercises 12.5

1. $u(x, t) = 2T_0 \displaystyle\sum_{n=1}^{\infty} \dfrac{1 - \cos k_n}{k_n(1 + 2 \cos^2 k_n)} \sin[k_n(1 - x)]e^{-a^2 k_n^2 t}$, where $k_n = -\tan k_n, n = 1, 2, 3, \ldots$ **3.** $u(x, t) = T_0$

5. $u(x, t) = \displaystyle\sum_{n=1}^{\infty} c_n \phi_n(x)e^{-a^2 k_n^2 t}$, where $2k_n h = (k_n^2 - h^2) \tan k_n p$, $\phi_n(x) = k_n \cos k_n x + h \sin k_n x, n = 1, 2, 3, \ldots$, and

$c_n = \|\phi_n(x)\|^{-2} \displaystyle\int_0^p f(x)\phi_n(x)\, dx, n = 1, 2, 3, \ldots$

7. $u(x, t) = \dfrac{a^2 A}{\omega} \sin \omega t$

9. $u(x, t) = \displaystyle\sum_{n=1}^{\infty} \left\{\dfrac{4}{n^3\pi^3}(-1)^{n-1}(1 - e^{-n^2\pi^2 t}) - \dfrac{10}{n\pi}\dfrac{[1 - (-1)^n]}{1 + n^2\pi^2}(n^2\pi^2 \sin t - \cos t + e^{-n^2\pi^2 t})\right\} \sin n\pi x$

11. $u(x, t) = \dfrac{4Pc}{\pi} \displaystyle\sum_{\substack{n=1 \\ (\text{odd})}}^{\infty} \dfrac{\sin nx}{n^2(\omega^2 - n^2 c^2)} (\omega \sin nct - nc \sin \omega t)$ **15.** $u(r, t) = 0.1J_0(k_1 r) \cos k_1 ct$

17. $u(r, t) = \displaystyle\sum_{n=1}^{\infty} (a_n \cos k_n ct + b_n \sin k_n ct)J_0(k_n r)$, where $J_0(k_n \rho) = 0, n = 1, 2, 3, \ldots,$

$a_n = \dfrac{2}{\rho^2[J_1(k_n \rho)]^2} \displaystyle\int_0^\rho rf(r)J_0(k_n r)\, dr, n = 1, 2, 3, \ldots, k_n cb_n = \dfrac{2}{\rho^2[J_1(k_n \rho)]^2} \displaystyle\int_0^\rho rg(r)J_0(k_n r)\, dr, n = 1, 2, 3, \ldots$

Index